Mathematical Foundations of Quantum Field Theory and Perturbative String Theory

Proceedings of Symposia in
PURE MATHEMATICS

Volume 83

Mathematical Foundations of Quantum Field Theory and Perturbative String Theory

Hisham Sati
Urs Schreiber
Editors

American Mathematical Society
Providence, Rhode Island

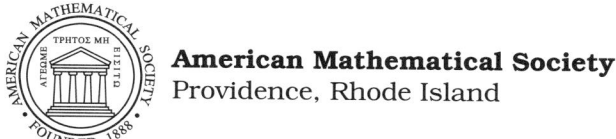

2010 *Mathematics Subject Classification.* Primary 81T45, 81T40, 81T30, 81T05, 81T60, 70S05, 18D05, 18D50, 55N34, 55U40.

Library of Congress Cataloging-in-Publication Data

Mathematical foundations of quantum field theory and perturbative string theory / Hisham Sati, Urs Schreiber, editors.

p. cm. — (Proceedings of symposia in pure mathematics ; v. 83)

Includes bibliographical references.

ISBN 978-0-8218-5195-1 (alk. paper)

1. Quantum field theory—Mathematics—Congresses. 2. Topological fields—Congresses. 3. String models—Mathematics—Congresses. I. Sati, Hisham, 1973– II. Schreiber, Urs, 1974–

QC174.45.A1M375 2011
530.14′3—dc23

2011030793

∞ The paper used in this book is acid-free and falls within the guidelines
established to ensure permanence and durability.
Visit the AMS home page at http://www.ams.org/

10 9 8 7 6 5 4 3 2 1 16 15 14 13 12 11

Contents

Preface

Conceptual progress in fundamental theoretical physics is linked with the search for suitable mathematical structures that model the physics in question. There are a number indications that today we are in a period where the fundamental mathematical nature of quantum field theory (QFT) and of the worldvolume aspects of string theory is being identified. It is not unlikely that future generations will think of the turn of the millennium and the beginning of the 21st century as the time when it was fully established that QFT in general and worldvolume theories in particular are precisely the representations of higher categories of cobordisms with structure or, dually, encoded by copresheaves of local algebras of observables, vertex operator algebras, factorization algebras and their siblings.

While significant insights on these matters have been gained in the last several years, their full impact has possibly not yet received due attention, notably not among most of the theoretical but pure physicists for whom it should be of utmost relevance. At the same time, those who do appreciate the mathematical structures involved may wonder how it all fits into the big physical picture of quantum field and string theory.

This volume is aimed at trying to improve on this situation by collecting original presentations as well as reviews and surveys of recent and substantial progress in the unravelling of mathematical structures underlying the very nature of quantum field and worldvolume string theory. All contributions have been carefully refereed.

It is reassuring that some of the conferences on fundamental and mathematical physics these days begin to witness a new, more substantial interaction between theoretical physicists and mathematicians, where the latter no longer just extract the isolated remarkable conjectures that the black box *string theory* has been producing over the decades, but finally hold in their hands a workable axiom system that allows one to genuinely consider core aspects of QFT in a formal manner. This book has grown out of the experience of such meetings.

The editors express their thanks to the authors who kindly made their work available for this volume. We also acknowledge the hard work of the referees. We thank Sergei Gelfand, Christine Thivierge, and the dedicated staff at the American Mathematical Society for their effort in publishing this volume. We also thank Arthur Greenspoon for carefully proofreading the papers and for his input on the volume as a whole.

Hisham Sati
Urs Schreiber

Proceedings of Symposia in Pure Mathematics
Volume **83**, 2011

Introduction

Hisham Sati and Urs Schreiber

Abstract. The contributions in this volume are intended to indicate core aspects of a firm and workable mathematical foundation for quantum field theory and perturbative string theory. Here we provide some motivational background, as well as the overall picture in which the various articles fit.

The history of theoretical fundamental physics is the story of a search for the suitable mathematical notions and structural concepts that naturally model the physical phenomena in question. It may be worthwhile to recall a few examples:

(1) the identification of symplectic geometry as the underlying structure of classical Hamiltonian mechanics;
(2) the identification of (semi-)Riemannian differential geometry as the underlying structure of gravity;
(3) the identification of group and representation theory as the underlying structure of the zoo of fundamental particles;
(4) the identification of Chern-Weil theory and differential cohomology as the underlying structure of gauge theories.

All these examples exhibit the identification of the precise mathematical language that naturally captures the physics under investigation. While each of these languages upon its introduction into theoretical physics originally met with some skepticism or even hostility, we do know in retrospect that the modern insights and results in the respective areas of theoretical physics would have been literally unthinkable without usage of these languages. A famous historical example is the Wigner-Weyl approach and its hostile dismissal from mainstream physicists of the time ("*Gruppenpest*"); we now know that group theory and representation theory have become indispensible tools for every theoretical and mathematical physicist.

Much time has passed since the last major such formalization success in theoretical physics. The rise of quantum field theory (QFT) in the middle of the last century and its stunning successes, despite its notorious lack of formal structural underpinnings, made theoretical physicists confident enough to attempt an attack on the next open structural question – that of the quantum theory of gauge forces

2010 *Mathematics Subject Classification.* Primary 81T40; secondary 81T45, 81T30, 81T60, 81T05, , 57R56, 70S05, 18D05, 55U40, 18D50, 55N34, 19L50, 53C08.
Keywords and phrases. Topological field theory, conformal field theory, supersymmetric field theory, axiomatic quantum field theory, perturbative string theory, conformal nets, monoidal categories, higher categories, generalized cohomology, differential cohomology, quantization, operads.

including gravity – without much more of a structural guidance than the folklore of the path integral, however useful that had otherwise proven to be.

While everyone involved readily admitted that nobody knew the full answer to

What is string theory?

perhaps it was gradually forgotten that nobody even knew the full answer to

What is quantum field theory?

While a huge discussion ensued on the "*landscape*" moduli space of backgrounds for string theory, it was perhaps forgotten that nobody even had anything close to a full answer to

What is a string theory background?

or even to what should be a simpler question:

What is a classical string theory background?

which in turn is essentially the question:

What is a full 2-dimensional σ-model conformal field theory?

Most of the literature on 2-dimensional conformal field theory (2d CFT) describes just what is called *chiral conformal field theory*, formalized in terms of vertex operator algebras or local conformal nets. But this only captures the holomorphic and low-genus aspect of conformal field theory and is just one half of the data required for a full CFT, the remaining piece being the full solution of the sewing constraints that makes the theory well defined for all genera.

With these questions – fundamental as they are for perturbative string theory – seemingly too hard to answer, a plethora of related model and toy model quantum field theoretic systems found attention instead. A range of topological (quantum) field theories (T(Q)FTs) either approximates the physically relevant CFTs as in the topological A-model and B-model, or encodes these holographically in their boundary theory as for Chern-Simons theory and *its* toy model, the Dijkgraaf-Witten theory.

In this way a wealth of worldvolume QFTs appears that in some way or another is thought to encode information about string theory. Furthermore, in each case what really matters is the *full* worldvolume QFT: the rule that assigns correlators to all possible worldvolume cobordisms, because this is what is needed even to write down the corresponding second quantized perturbation series. However, despite this urgent necessity for understanding QFT on arbitrary cobordisms, the tools to study or even formulate this precisely were for a long time largely unavailable. Nevertheless, proposals for how to make these questions accessible to the development of suitable mathematical machinery already existed.

Early on it was suggested, based on topological examples, that the path integral and the state-propagation operators that it is supposed to yield are nothing but a representation of a category of cobordisms [**At88**]. It was further noticed that this prescription is not restricted to TQFTs, and in fact CFTs were proposed to be axiomatized as representations of categories of conformal cobordisms [**Se04**]. In parallel to this development, another school developed a dual picture, now known as *local* or *algebraic quantum field theory* (AQFT) [**Ha92**], where it is not the state-propagation – the *Schrödinger picture* – of QFT that is axiomatized and made accessible to high-powered machinery, but rather the assignment of algebras of observables – the *Heisenberg picture* of QFT.

While these axiomatizations were known and thought of highly by a few select researchers who worked on them, they were mostly happily ignored by the quantum field theory and string theory community at large, and to a good degree rightly so: nobody should trust an axiom system that has not yet proven its worth by providing useful theorems and describing nontrivial examples of interest. But neither the study of cobordism representations nor that of systems of algebras of observables could for a long time – apart from a few isolated exceptions – claim to add much to the world-view of those who value formal structures in physics, but not a priori formal structures in mathematics. It is precisely this that is changing now.

Major structural results have been proven about the axioms of functorial quantum field theory (FQFT) in the form of cobordism representations and dually those of local nets of algebras (AQFT) and factorization algebras. Furthermore, classes of physically interesting examples have been constructed, filling these axiom systems with life. We now provide a list of such results, which, while necessarily incomplete, may serve to give an impression of the status of the field, and serve to put the contributions of this book into perspective.

I. Cobordism representations

(i) Topological case. The most foundational result in TQFT is arguably the formulation and proof [**Lur09b**] of the cobordism hypothesis [**BaDo95**] which classifies *extended* (meaning: "fully local") n-dimensional TQFT by the "fully dualizability"-structure on the "space" of states (an object in a symmetric monoidal (∞, n)-category) that it assigns to the point. (In this volume the contribution by Bergner surveys the formulation and proof of the cobordism hypothesis). This hugely facilitates the construction of interesting examples of extended n-dimensional TQFTs. For instance

- recently it was understood that the state-sum constructions of 3d TQFTs from fusion categories (e.g. [**BaKi00**]) are subsumed by the cobordism hypothesis-theorem and the fact [**DSS11**] that fusion categories are the fully dualizable objects in the $(\infty, 3)$-category of monoidal categories with bimodule categories as morphisms;
- the Calabi-Yau A_∞-categories that Kontsevich conjectured [**Ko95**] encode the 2d TQFTs that participate in homological mirror symmetry have been understood to be the "almost fully dualizable" objects (*Calabi-Yau objects*) that classify extended open/closed 2-dimensional TQFTs on cobordisms with non-empty outgoing boundary with values in the $(\infty, 1)$-category of chain complexes ("TCFTs" [**Cos07a**], [**Lur09b**]);

  In this context crucial aspects of Witten's observation in [**Wi92**] have been made precise [**Cos07b**], relating Chern-Simons theory to the effective target space theory of the A- and B-model topological string, thus providing a rigorous handle on an example of the effective background theory induced by a string perturbation series over all genera.

(ii) Conformal case. A complete classification of *rational* full 2d CFTs on cobordisms of all genera has been obtained in terms of Frobenius algebra objects in modular tensor categories [**FRS06**]. While the rational case is still "too simple" for the most interesting applications in string theory, its full solution shows that already here considerably more interesting structure is to be found than suggested by the naive considerations in much of the physics literature. (The contributions by Kapustin-Saulina and by Kong in this volume discuss aspects of this.)

(iii) Supergeometric case. There is now a full proof available, starting from the axioms, that the partition function of a (2|1)-dimensional supersymmetric 2d-QFT indeed is a modular form, as suggested by Witten's work [**Wi86**] on the partition function of the heterotic string and the index of the Dirac operator on loop space. (A formalization and proof of this fact in terms of supergeometric cobordism representations is described in the contribution by Stolz-Teichner to this volume.) This suggests a deep relationship between superstrings and the generalized cohomology theory called *tmf* (for *topological modular forms*) – in a sense, the universal elliptic cohomology theory – which lifts the more familiar relation between superparticles (spinors) and K-theory to higher categorical dimension. (This is the content of the contribution by Douglas-Henriques in this volume.)

(iv) Boundary conditions and defects/domain walls. One simple kind of extra structure on cobordisms that is of profound importance is boundary labels and decompositions of cobordisms into domains, meeting at *domain walls* ("defects"). (The definition of QFT with defects is part of the content of the contribution by Davydov-Runkel-Kong to this volume). That cobordism representations with boundaries for the string encode D-branes on target space was originally amplified by Moore and Segal [**MoSe06**]. Typically open-closed QFTs are entirely determined by their open sectors and boundary conditions, a fact that via [**Cos07a**] led to Lurie's proof of the cobordism hypothesis. (A survey of a list of results on presentation of 2d CFT by algebras of boundary data is in the contribution by Kong to this volume.)

(v) Holographic principle. A striking aspect of the classification of rational CFT mentioned above is that it proceeds – rigorously – by a version of the *holographic principle*. This states that under some conditions the partition function and correlators of an n-dimensional QFT are encoded in the *states* of an $(n+1)$-dimensional TQFT in codimension 1. The first example of this had been the holographic relation between 3-dimensional Chern-Simons theory and the 2-dimensional WZW CFT in the seminal work [**Wi89**], which marked the beginning of the investigation of TQFT in the first place. A grand example of the principle is the AdS/CFT conjecture, which states that type II string theory itself is holographically related to super Yang-Mills theory. While mathematical formalizations of AdS/CFT are not available to date, lower dimensional examples are finding precise formulations. (The contribution by Kapustin-Saulina in this volume discusses how the construction of rational 2d CFT by [**FRS06**] is naturally induced from applying the holographic principle to Chern-Simons theory with defects).

One of the editors once suggested that, in the formalization by cobordism representations holography corresponds to the fact that *transformations* between $(n+1)$-functors are in components themselves essentially given by n-functors. A formalization of this observation for extended 2d QFT has been given in [**SP10**]. (The contribution by Stolz-Teichner to this volume crucially uses transformations between higher dimensional QFTs to *twist* lower dimensional QFTs.)

II. Systems of algebras of observables

(i) Nets of algebras. In the form of the *Haag-Kastler axioms*, the description of QFT through its local algebras of observables had been given a clean mathematical formulation [**HaMü06**] a long time ago [**HaKa64**]. This approach had long produced fundamental structural results about QFT, such as the PCT theorem and the spin-statistics theorem (cf. [**StWi00**]). Only recently has it finally been shown in detail [**BDF09**] how examples of AQFT nets can indeed be constructed along the lines of perturbation theory and Wilsonian effective field theory, thus connecting the major tools of practicing particle physicists with one of the major formal axiom systems. Using an operadic variant of Haag-Kastler nets in the case of Euclidean ("Wick rotated") QFT – called *factorization algebras* – a similar discussion is sketched in [**CoGw**]. At the same time, the original axioms have been found to naturally generalize from Minkowski spacetime to general (globally hyperbolic) curved and topologically nontrivial spacetimes [**BFV01**].

(ii) Boundaries and defects. The Haag-Kastler axioms had been most fruitful in the description of 2 dimensional and conformal field theory ("conformal nets"), where they serve to classify chiral 2d CFTs [**KaLo03**][**Ka03**], construct integral 2d QFTs [**Le06**] and obtain insights into boundary field theories (open strings) [**LoRe04**]. Remarkably, the latter has recently allowed a rigorous re-examination [**LoWi10**] of old arguments about the background-independence of string field theory. (The contribution by Douglas-Henriques in this volume presents a modern version of the Haag-Kastler axioms for conformal nets and extends the discussion from boundary field theory to field theory with defects.)

(iii) Higher chiral algebras. The geometric reformulation of vertex operator algebras in terms of *chiral algebras* [**BeDr04**] has proven to be fruitful, in particular in its higher categorical generalizations [**Lur11**] by factorizable cosheaves of ∞-algebras. While the classical AQFT school restricted attention to QFT over trivial topologies, it turns out that also topological QFTs can be described and constructed by local assignments of algebras "of observables". In [**Lur09b**] n-dimensional extended TQFTs are constructed from E_n*-algebras* – algebras over the little n-cubes operad – by a construction called *topological chiral homology*, which is a grand generalization of Hochschild homology over arbitrary topologies. (The contribution by Weiss in this volume discusses the theory of homotopy algebras over operads involved in these constructions.)

This last work is currently perhaps the most formalized and direct bridge between the two axiom systems, the functorial and the algebraic one. This indicates the closure of a grand circle of ideas and makes the outline of a comprehensive fundamental formalization of full higher-genus QFT visible.

III. Quantization of classical field theories

While a realistic axiomatization is the basis for all mathematical progress in QFT, perhaps even more important in the long run for physics is that with the supposed *outcome* of the (path integral) quantization process thus identified precisely by axioms for QFT, it becomes possible to consider the nature of the *quantization* process itself. This is particularly relevant in applications of QFT as worldvolume theories in string theory, where one wishes to explicitly consider QFTs that arise as the quantization of sigma-models with specified gauge background fields. A good understanding of this quantization step is one of the links between the worldvolume theory and the target space theory and hence between the abstract algebraic description of the worldvolume QFT and the phenomenological interpretation of its correlators in its target space, ultimately connecting theory to experiment. We now indicate some of the progress in mathematically understanding the process of quantization in general and of sigma-models in particular.

(i) Path integral quantization. It has been suggested (e.g. [**Fre06**]) that the path integral is to be understood abstractly as a pull-push operation – an *integral transform* – acting on states in the form of certain cocycles, by first pulling them up to the space of worldvolume configurations along the map induced by the incoming boundary, and then pushing forward along the map induced by the outgoing boundary. This is fairly well understood for Dijkgraaf-Witten theory [**FrQu93**]. In [**FHLT10**] it is claimed that at least for all the higher analogs of Dijkgraaf-Witten theory (such as the Yetter model [**MaPo07**]) a formal pull-push path integral quantization procedure exists in terms of colimits of n-categorical algebras, yielding fully extended TQFTs.

A more geometric example for which pull-push quantization is well understood is Gromov-Witten theory [**Ka06**]. More recently also Chas-Sullivan's string topology operations have been understood this way, for strings on a single brane in [**Go07**] and recently for arbitrary branes in [**Ku11**]. In [**BZFNa11**] it is shown that such integral transforms exist on stable ∞-categories of quasicoherent sheaves for all target spaces that are perfect derived algebraic stacks, each of them thus yielding a 2-dimensional TQFT from background geometry data.

(ii) Higher background gauge fields. Before even entering (path integral) quantization, there is a fair bit of mathematical subtleties involved in the very definition of the string's action functional in the term that describes the coupling to the higher background gauge fields, such as the Neveu-Schwarz (NS) B-field and the Ramond-Ramond (RR) fields. All of these are recently being understood systematically in terms of *generalized differential cohomology* [**HS05**].

Early on it had been observed that the string's coupling to the B-field is globally occurring via the higher dimensional analog of the line holonomy of a circle bundle: the surface holonomy [**GaRe02**][**FNSW09**] of a *circle 2-bundle with connection* [**Sch11**]: a *bundle gerbe* with connection, classified by degree-3 ordinary differential cohomology. More generally, on orientifold target space backgrounds it is the nonabelian $(\mathbb{Z}_2//U(1))$-surface holomomy [**ScWa08**][**Ni11**] over unoriented surfaces [**SSW05**].

After the idea had materialized that the RR fields have to be regarded in K-theory [**MoWi00**] [**FrHo00**], it eventually became clear [**Fre01**] that all the higher abelian background fields appearing in the effective supergravity theories of string theory are properly to be regarded as cocycles in *generalized differential cohomology* [**HS05**] – the RR-field being described by *differential K-theory* [**BuSch11**] – and even more generally in *twisted* such theories: the presence of the B-field makes the RR-fields live in *twisted K-theory* (cf. [**BMRZ08**]).

A perfectly clear picture of twisted generalized cohomology theory in terms of associated E_∞-module spectrum ∞-bundle has been given in [**ABG10**]. This article in particular identifies the twists of tmf-theory, which are expected [**Sa10**] [**AnSa11**] to play a role in M-theory in the higher analogy of twisted K-theory in string theory.

(iii) Quantum anomaly cancellation. The cancellation of the quantum anomaly of fermions on the superstring's worldvolume – the (differential) class of their Pfaffian line bundles on the bosonic configuration space – imposes subtle conditions on the background gauge fields on spacetime to which the string couples.

By means of the machinery of generalized differential cohomology, recently [**Bu09**] makes fully precise the old argument of Killingback about the worldsheet version of the celebrated *Green-Schwarz anomaly cancellation* (the effect that initiated the "First superstring revolution"), using a model for *twisted differential string structures* [**SSS10**] [**FSS11**] in terms of bundle gerbes, due to [**Wa09**]. These differential string structures – controlled by the higher Lie and Chern-Weil theory of the *smooth string 2-group* [**Hen08**][**BCSS07**] – are the higher superstring analogs in higher smooth geometry [**Sch11**] of the spin-bundles with connection that control the dynamics of spinning/superparticles.

(In our volume the contribution by Distler-Freed-Moore presents what is to date the most accurate description of the conditions on the differential cohomology classes of the superstring's background gauge fields for general orbifold and orientifold target spaces.)

Taken together, all these developments should go a long way towards understanding the fundamental nature of QFT on arbitrary cobordisms and of the string perturbation series defined by such 2d QFTs. However, even in the light of all these developments, the reader accustomed to the prevailing physics literature may still complain that none of this progress in QFT on cobordisms of all genera yields a definition of what string theory really is. Of course this is true if by "string theory" one understands its non-perturbative definition. But this supposed non-perturbative definition of string theory is beyond reach at the moment. Marvelling – with a certain admiration of their audacity – at how ill-defined this is has made the community forget that something much more mundane, the perturbation series over CFT correlators that defines *perturbative string theory*, has been ill-defined all along: only the machinery of full CFT in terms of cobordism representations gives a precise meaning to what exactly it is that the string pertubation series is a series over. Perhaps it causes feelings of disappointment to be thrown back from the realm of speculations about non-perturbative string theory to just the perturbation series. But at least this time one lands on solid ground, which is the only ground that serves as a good jumping-off point for further speculation.

In string theory it has been the tradition to speak of major conceptual insights into the theory as *revolutions* of the theory. The community speaks of a first and a second superstring revolution and a certain longing for the third one to arrive can be sensed. With a large part of the community busy attacking grand structures with arguably insufficient tools, it does not seem farfetched that when the third one does arrive, it will have come out of mathematics departments. [1]

[1] See in this context for instance the opening and closing talks at the *Strings 2011* conference.

Summary

We now outline the contents of the volume, highlighting how the various articles are related and emphasizing how they fit into the big picture that we have drawn above.

I. Foundations of Quantum Field Theory

1. Models for (∞, n)-Categories and the Cobordism Hypothesis – by *Julia Bergner.*

The Schrödinger picture of extended topological quantum field theory of dimension n is formalized as being an (∞, n)-functor on the (∞, n)-category of cobordisms of dimension n. This article reviews the definition and construction of the ingredients of this statement, due to [**Lur09b**].

This picture is the basis for the formulation of QFTs on cobordisms with structure. Contributions below discuss cobordisms with defect structure, with conformal structure and with flat Riemannian structure.

2. From operads to dendroidal sets – by *Ittay Weiss.*

The higher algebra that appears in the algebraic description of QFT – by local nets of observables, factorization algebra or chiral algebras – is in general operadic. For instance the vertex operator algebras appearing in the description of CFT (see Liang Kong's contribution below) are algebras over an operad of holomorphic punctured spheres.

This article reviews the theory of operads and then discusses a powerful presentation in terms of dendroidal sets – the operadic analog of what simplicial sets are for $(\infty, 1)$-categories. This provides the homotopy theory for $(\infty, 1)$-operads, closely related to the traditional model by topological operads.

3. Field theories with defects and the centre functor – by *Alexei Davydov, Liang Kong and Ingo Runkel.*

This article gives a detailed discussion of cobordism categories for cobordisms with defects/domain walls. An explicit construction of a lattice model of two-dimensional TQFT with defects is spelled out. The authors isolate a crucial aspect of the algebraic structure induced by defect TQFTs on their spaces of states: as opposed to the algebra of ordinary bulk states, that of defect states is in general non-commutative, but certain worldsheet topologies serve to naturally produce the centre of these algebras.

Below in *Surface operators in 3d TQFT* topological field theories with defects are shown to induce, by a holographic principle, algebraic models for 2-dimensional CFT. In *Topological modular forms and conformal nets* conformal field theories with defects are considered.

II. Quantization of Field Theories

1. Homotopical Poisson reduction of gauge theories – by *Frédéric Paugam.*

The basic idea of quantization of a Lagrangian field theory is simple: one forms the covariant phase space given as the critical locus of the action functional, then forms the quotient by gauge transformations and constructs the canonical symplectic form. Finally, one applies deformation quantization or geometric quantization to the resulting symplectic manifold.

However, to make this naive picture work, care has to be taken to form both the intersection (critical locus) and the quotient (by symmetries) not naively but up to homotopy in derived geometry [**Lur09a**]. The resulting derived covariant phase space is known in physics in terms of its Batalin-Vilkovisky–Becchi-Rouet-Stora-Tyutin (BV-BRST) complex. This article reviews the powerful description of variational calculus and the construction of the BV-BRST complex in terms of D-geometry [**BeDr04**] – the geometry over de Rham spaces – and uses this to analyze subtle finiteness conditions on the BV-construction.

2. Orientifold précis – by *Jacques Distler, Daniel Freed, and Gregory Moore.*

The consistent quantization of the sigma model for the (super-)string famously requires the target space geometry to satisfy the Euler-Lagrange equations of an effective supergravity theory on target space. In addition there are subtle cohomological conditions for the cancellation of fermionic worldsheet anomalies.

This article discusses the intricate conditions on the differential cohomology of the background fields – namely the Neveu-Schwarz B-field in ordinary differential cohomology (or a slight variant, which the authors discuss) and the RR-field in differential K-theory twisted by the B-field – in particular if target space is allowed to be not just a smooth manifold but more generally an orbifold and even more generally an orientifold. Among other things, the result shows that the "landscape of string theory vacua" – roughly the moduli space of consistent perturbative string backgrounds (cf. [**Do10**]) – is more subtle an object than often assumed in the literature.

III. Two-dimensional Quantum Field Theories

1. Surface operators in 3d TFT and 2d Rational CFT – by *Anton Kapustin and Natalia Saulina.*

Ever since Witten's work on 3-dimensional Chern-Simons theory it was known that by a holographic principle this theory induces a 2d CFT on 2-dimensional boundary surfaces. This article amplifies that if one thinks of the 3d Chern-Simons TQFT as a topological QFT with defects, then the structures formed by codimension-0 defects bounded by codimension-1 defects naturally reproduce, holographically, the description of 2d CFT by Frobenius algebra objects in modular tensor categories [**FRS06**].

2. Conformal field theory and a new geometry – by *Liang Kong.*

While the previous article has shown that the concept of TQFT together with the holographic principle naturally imply that 2-dimensional CFT is encoded by monoid objects in modular tensor categories, this article reviews a series of strong results about the details of this encoding. In view of these results and since every 2d CFT also induces an effective target space geometry – as described in more detail in the following contribution – the author amplifies the fact that stringy geometry is thus presented by a categorified version of the familiar duality between spaces and algebras: now for algebra objects internal to suitable monoidal categories.

3. Collapsing Conformal Field Theories, spaces with non-negative Ricci curvature and non-commutative geometry – by *Yan Soibelman.*

The premise of perturbative string theory is that every suitable 2d (super-)CFT describes the quantum sigma model for a string propagating in *some* target space geometry, if only we understand this statement in a sufficiently general context of geometry, such as spectral noncommutative geometry. In this article the author analyzes the geometries induces from quantum strings in the point-particle limit ("collapse limit") where only the lowest string excitations are relevant. In the limit the algebraic data of the SCFT produces a spectral triple, which had been shown by Alain Connes to encode generalized Riemannian geometry in terms of the spectrum of Hamiltonian operators. The author uses this to demonstrate compactness results about the resulting moduli space of "quantum Riemann spaces".

4. Supersymmetric field theories and generalized cohomology – by *Stephan Stolz and Peter Teichner.*

Ever since Witten's derivation of what is now called the *Witten genus* as the partition function of the heterotic superstring, there have been indications that superstring physics should be governed by the generalized cohomology theory called topological modular forms (tmf) in analogy to how super/spinning point particles are related to K-theory. In this article the authors discuss the latest status of their seminal program of understanding these cohomological phenomena from a systematic description of functorial 2d QFT with metric structure on the cobordisms.

After noticing that key cohomological properties of the superstring depend only on supersymmetry and not actually on conformal invariance, the authors simplify to cobordisms with flat super-Riemannian structure, but equipped with maps into some auxiliary target space X. A classification of such QFTs by generalized cohomology theories on X is described: a relation between $(1|1)$-dimensional flat Riemannian field theories and K-theory and between $(2|1)$-dimensional flat Riemannian field theories and tmf.

5. Topological modular forms and conformal nets – by *Christopher Douglas and André Henriques.*

Following in spirit the previous contribution, but working with the AQFT-description instead, the authors of this article describe a refinement of conformal nets, hence of 2d CFT, incorporating defects. Using this they obtain a tricategory of fermionic conformal nets ("spinning strings") which constitutes a higher analog of the bicategory of Clifford algebras. Evidence is provided which shows that these *categorified* spinors are related to tmf in close analogy to how ordinary Clifford algebra is related to K-theory, providing a concrete incarnation of the principle by which string physics is a form of categorified particle physics.

Acknowledgements. The authors would like to thank Arthur Greenspoon for his very useful editorial input.

References

[ABG10] M. Ando, A. Blumberg, and D. Gepner, *Twists of K-theory and TMF*, Superstrings, geometry, topology, and C^*-algebras, 27–63, Proc. Sympos. Pure Math., 81, Amer. Math. Soc., Providence, RI (2010), [arXiv:1002.3004].

[AnSa11] M. Ando, H. Sati, *M-brane charges and twisted tmf*, in preparation.

[At88] M. Atiyah, *Topological quantum field theories*, Inst. Hautes Études Sci. Publ. Math. **68** (1988), 175-186.

[BCSS07] J. Baez, A. Crans, U. Schreiber, and D. Stevenson, *From loop groups to 2-groups*, Homology, Homotopy Appl. **9** (2007), 101–135, [arXiv:math/0504123].

[BaDo95] J. Baez and J. Dolan, *Higher-dimensional algebra and topological quantum field theory*, J. Math. Phys **36** (1995), 6073–6105, [arXiv:q-alg/9503002].

[BaKi00] B. Bakalov and A. Kirillov, *Lectures on tensor categories and modular functors*, University Lecture Series, Amer. Math. Soc., Providence, RI (2000).

[BeDr04] A. Beilinson, V. Drinfeld, *Chiral algebras*, Amer. Math. Soc., Providence, RI (2004).

[BZFNa11] D. Ben-Zvi, J. Francis, and D. Nadler, *Integral transforms and Drinfeld centers in derived algebraic geometry*, J. Amer. Math. Soc. **23** (2010), no. 4, 909–966, [arXiv:0805.0157].

[BMRZ08] J. Brodzki, V. Mathai, J. Rosenberg, and R. J. Szabo, *D-branes, RR-fields and duality on noncommutative manifolds*, Commun. Math. Phys. **277** (2008), 643–706, [arXiv:hep-th/0607020].

[BDF09] R. Brunetti, M. Dütsch, and K. Fredenhagen, *Perturbative algebraic quantum field theory and the renormalization groups*, Adv. Theor. Math. Physics **13** (2009), 1541–1599, [arXiv:0901.2038].

[BFV01] R. Brunetti, K. Fredenhagen, and R. Verch, *The generally covariant locality principle – a new paradigm for local quantum field theory*, Commun. Math. Phys. **237** (2001), 31–68, [arXiv:math-ph/0112041].

[Bu09] U. Bunke, *String structures and trivialisations of a Pfaffian line bundle*, preprint, [arXiv:0909.0846].

[BuSch11] U. Bunke and T. Schick, *Differential K-theory: A survey*, preprint, [arXiv:1011.6663].

[CoKr00] A. Connes and D. Kreimer, *Renormalization in quantum field theory and the Riemann-Hilbert problem I: the Hopf algebra structure of graphs and the main theorem*, Commun. Math. Phys. **210** (2000), 249–273, [arXiv:hep-th/9912092].

[Cos07a] K. Costello, *Topological conformal field theories and Calabi-Yau categories*, Adv. Math. **210** (2007), 165–214, [arXiv:math/0412149].

[Cos07b] K. Costello, *Topological conformal field theories and gauge theories*, Geom. Top. **11** (2007), 1539–1579, [arXiv:math/0605647].

[CoGw] K. Costello and O. Gwilliam, *Factorization algebras in perturbative quantum field theory*, preprint, [`http://math.northwestern.edu/~costello/factorization_public.html`].

[DSS11] C. Douglas, C. Schommer-Pries, and N. Snyder, *The Structure of Fusion Categories via 3D TQFTs*, preprint (2011), [`http://ncatlab.org/nlab/files/DSSFusionSlides.pdf`].

[Do10] M.R. Douglas, *Spaces of quantum field theories*, Open Access Journal of Physics: Conference Series, Institute of Physics Publishing, [arXiv:1005.2779].

[FSS11] D. Fiorenza, U. Schreiber, and J. Stasheff, *Čech cocycles for differential characteristic classes– – An ∞-Lie theoretic construction*, preprint, [arXiv:1011.4735].

[Fre01] D. Freed, *Dirac charge quantization and generalized differential cohomology*, in Surv. Diff. Geom. **VII**, 129–194, Int. Press, Somerville, MA (2000), [arXiv:hep-th/0011220].

[Fre06] D. Freed, *Twisted K-theory and the Verlinde ring*, Andrejewski Lecture, Leipzig (2006), [`http://www.ma.utexas.edu/users/dafr/Andrejewski%20Lectures.html`].

[FrHo00] D. Freed and M. Hopkins, *On Ramond-Ramond fields and K-theory*, J. High Energy Phys. 5 (2000), 44, [arXiv:hep-th/0002027].

[FHLT10] D. Freed, M. Hopkins, J. Lurie, and C. Teleman, *Topological quantum field theories from compact Lie groups*, A celebration of the mathematical legacy of Raoul Bott, 367–403, Amer. Math. Soc., Providence, RI (2010), [arXiv:0905.0731].

[FrQu93] D. Freed and F. Quinn, *Chern-Simons theory with finite gauge group*, Commun. Math. Phys. **156** (1993), 435–472, [arXiv:hep-th/9111004].

[FrRe11] K. Fredenhagen and K. Rejzner, *Batalin-Vilkovisky formalism in the functional approach to classical field theory*, preprint, [arXiv:1101.5112].

[FNSW09] J. Fuchs, T. Nikolaus, C. Schweigert, and K. Waldorf, *Bundle gerbes and surface holonomy*, Proceedings of the Fifth European Congress of Mathematics, Amsterdam (2008), [arXiv:0901.2085].

[FRS06] J. Fuchs, I. Runkel, and C. Schweigert, *Categorification and correlation functions in conformal field theory*, Proceedings of the ICM 2006, European Mathematical Society Publishing House (2007), [arXiv:math.CT/0602079].

[GaRe02] C. Gawędzki and N. Reis, *WZW branes and gerbes*, Rev. Math. Phys. **14** (2002) 1281–1334, [arXiv:hep-th/0205233].

[Go07] V. Godin, *Higher string topology operations*, (2007), preprint, [arXiv:0711.4859].

[Ha92] R. Haag, *Local quantum physics – Fields, particles, algebras*, Springer, Berlin, 1992.

[HaKa64] R. Haag, D. Kastler, *An algebraic approach to quantum field theory*, J. Math. Phys. **5** (1964), 848–861.

[HaMü06] H. Halvorson (with an appendix by M. Müger), *Algebraic quantum field theory*, in Philosophy of Physics, North Holland (2006), [arXiv:math-ph/0602036].

[Hen08] A. Henriques, *Integrating L_∞-algebras*, Compos. Math. **144** (2008), 1017–1045, [arXiv:math/0603563].

[HS05] M. Hopkins and I. Singer, *Quadratic functions in geometry, topology, and M-theory*, , J. Differential Geom. **70** (2005), 329–452, [arXiv:math/0211216].

[Ka06] S. Katz, *Enumerative geometry and string theory*, Amer. Math. Soc., Providence, RI (2006).

[Ka03] Y. Kawahigashi, *Classification of operator algebraic conformal field theories in dimensions one and two*, Proceedings of XIV International Congress on Mathematical Physics, J.-C. Zambrini (ed.), World Scientific, Singapore (2006), [arXiv:math-ph/0308029].

[KaLo03] Y. Kawahigashi and R. Longo, *Classification of Local Conformal Nets. Case $c < 1$*, Ann. Math. **160** (2004), 493–522, [arXiv:math-ph/0201015].

[Ko95] M. Kontsevich, *Homological algebra of mirror symmetry*, in Proceedings of the International Congress of Mathematicians, (Zürich, 1994), pages 120–139, Basel, Birkhäuser (1995), [arXiv:alg-geom/9411018].

[Ku11] A. P. M. Kupers, *String topology operations*, MSc thesis, Utrecht University, The Netherlands (2011), [`http://igitur-archive.library.uu.nl/student-theses/2011-0706-200829//UUindex.html`].

[Le06] G. Lechner, *Construction of quantum field theories with factorizing S-matrices*, Commun. Math. Phys. **277** (2008), 821–860, [arXiv:math-ph/0601022].

[LoRe04] R. Longo,K.-H. Rehren, *Local fields in boundary conformal QFT*, Rev. Math. Phys. **16** (2004), 909–960, [arXiv:math-ph/0405067].

[LoWi10] R. Longo and E. Witten, *An Algebraic construction of boundary quantum field theory*, Commun. Math. Phys. **303** (2011), 213–232, [arXiv:1004.0616].

[Lur09a] J. Lurie, *Structured Spaces*, preprint, [arXiv:0905.0459].

[Lur09b] J. Lurie *On the classification of topological field theories*, Current developments in mathematics, 2008, 129–280, Int. Press, Somerville, MA (2009), [arXiv:0905.0465].

[Lur11] J. Lurie, *Higher Algebra*, preprint (2011), [`http://www.math.harvard.edu/~lurie/papers/higheralgebra.pdf`].

[MaPo07] J. Martins and T. Porter, *On Yetter invariants and an extension of the Dijkgraaf-Witten invariant to categorical groups*, Theory Appl. Categ. **18** (2007), 118–150, [arXiv:math/0608484].

[MoWi00] G. Moore and E. Witten, *Self-duality, Ramond-Ramond fields, and K-theory*, J. High Energy Phys. **0005** (2000) 032, [arXiv:hep-th/9912279].

[MoSe06] G. Moore and G. Segal, *D-branes and K-theory in 2D topological field theory*, preprint (2006), [arXiv:hep-th/0609042].

[Ni11] T. Nikolaus, *Higher categorical structures in QFT – General theory and applications to QFT*, PhD thesis, Hamburg University (2011), [`http://ediss.sub.uni-hamburg.de/volltexte/2011/5200/`].

[Sa10] H. Sati, *Geometric and topological structures related to M-branes*, Proc. Symp. Pure Math. **81** (2010), 181–236, [arXiv:1001.5020].

[SSS10] H. Sati, U. Schreiber, and J. Stasheff, *Twisted differential String- and Fivebrane structures*, preprint (2009), [arXiv:0910.4001].

[Se04] G. Segal, *The definition of conformal field theory*, in Topology, Geometry and Quantum Field Theory, 421–577, U. Tillmann (ed.), Cambridge University Press, Cambridge (2004).

[SP10] C. Schommer-Pries, *Topological defects and classifying local topological field theories in low dimensions*, preprint (2010), [http://ncatlab.org/nlab/files/SchommerPriesDefects.pdf].

[Sch11] U. Schreiber, *Differential cohomology in a cohesive topos*, preprint 2011, http://ncatlab.org/schreiber/show/differential+cohomology+in+a+cohesive+topos .

[SSW05] U. Schreiber, C. Schweigert, and K. Waldorf, *Unoriented WZW models and holonomy of bundle gerbes*, Commun. Math. Phys. **274** (2007), 31–64, [arXiv:hep-th/0512283].

[ScWa08] U. Schreiber and K. Waldorf, *Connections on non-abelian gerbes and their holonomy*, preprint (2008), [arXiv:0808.1923].

[StWi00] R. F. Streater and A. S. Wightman, *PCT, Spin Statistics, and All That*, Princeton University Press, Princeton, NJ (2000).

[Wa09] K. Waldorf, *String connections and Chern-Simons theory*, preprint (2009), [arXiv:0906.0117].

[Wi86] E. Witten, *The index of the Dirac operator In loop space*, Elliptic curves and modular forms in algebraic topology (Princeton, NJ, 1986), 161–181, Lecture Notes in Math. **1326**, Springer, Berlin (1988).

[Wi89] E. Witten, *Quantum field theory and the Jones polynomial*, Commun. Math. Phys. **121** (1989), 351–399.

[Wi92] E. Witten, *Chern-Simons gauge theory as a string theory*, The Floer memorial volume, 637–678, Prog. Math. **133**, Birkhäuser, Basel (1995), [arXiv:hep-th/9207094].

Current address: Department of Mathematics, University of Pittsburgh, 139 University Place, Pittsburgh, PA 15260

E-mail address: hsati@pitt.edu

Current address: Department of Mathematics, Utrecht University, Budapestlaan 6, 3584 CD Utrecht, The Netherlands

E-mail address: urs.schreiber@gmail.com

Foundations for Quantum Field Theory

Proceedings of Symposia in Pure Mathematics
Volume **83**, 2011

Models for (∞, n)-categories and the cobordism hypothesis

Julia E. Bergner

ABSTRACT. In this paper we introduce the models for (∞, n)-categories which have been developed to date, as well as the comparisons between them that are known and conjectured. We review the role of (∞, n)-categories in the proof of the Cobordism Hypothesis.

1. Introduction

The role of higher categories is not new in the study of topological quantum field theories. However, recent work of Lurie has introduced a homotopical approach to higher categories, that of (∞, n)-categories, to the subject with his recent paper on the Cobordism Hypothesis. The aim of this paper is to describe some of the known models for (∞, n)-categories and the comparisons between them and to give a brief exposition of how they are used in Lurie's work.

For any positive integer n, an n-category consists of objects, 1-morphisms between objects, 2-morphisms between 1-morphisms, and so forth, up to n-morphisms between $(n-1)$-morphisms. One can even have such higher morphisms for all n, leading to the idea of an ∞-category. If associativity and identity properties are required to hold on the nose, then we have a strict n-category or strict ∞-category, and there is no problem with this definition. However, in practice the examples that we find throughout mathematics are rarely this rigid. More often we have associativity holding only up to isomorphism and satisfying some kinds of coherence laws, leading to the idea of a weak n-category or a weak ∞-category. Many definitions have been proposed for such higher categories, but showing that they are equivalent to one another has proven to be an enormously difficult task.

Interestingly enough, the case of weak ∞-groupoids, where all morphisms at all levels are (weakly) invertible, can be handled more easily. Given any topological space, one can think of it as an ∞-groupoid by regarding its points as objects, paths between the points as 1-morphisms, homotopies between the paths as 2-morphisms, homotopies between the homotopies as 3-morphisms, and continuing thus for all k-morphisms. In fact, it is often taken as a definition that an ∞-groupoid *is* a topological space.

2010 *Mathematics Subject Classification.* Primary 55U35; Secondary 18D20, 18G30, 18G55, 57R56.

Key words and phrases. (∞, n)-categories, topological quantum field theory, cobordism hypothesis.

The author was partially supported by NSF grant DMS-0805951.

If we take categories enriched in topological spaces, then we obtain a model for $(\infty,1)$-categories, in which we have k-morphisms for all $k \geq 1$, but now they are only invertible for $k > 1$; the points of the spaces now play the role of 1-morphisms. Using the homotopy-theoretic equivalence between topological spaces and simplicial sets, it is common instead to consider categories enriched over simplicial sets, often called simplicial categories. Topological or simplicial categories are good for many applications, but for others they are still too rigid, since composition of the mapping spaces is still required to have strict associativity and inverses.

There are different ways to weaken the definition of simplicial category so that composition is no longer strictly defined. Segal categories, complete Segal spaces, and quasi-categories were all developed as alternatives to simplicial categories. In sharp contrast to definitions of weak n-categories, these models are known to be equivalent to one another in a precise way. More specifically, there is a model category corresponding to each model, and these model categories are Quillen equivalent to one another. Further details on these models and the equivalences between them are given in the next section.

One might ask if we could move from $(\infty,1)$-categories to more general (∞,n)-categories using similar methods. Not surprisingly, there are many more possible definitions of models for more general n. In this paper, we describe several of these approaches as well as some of the known comparisons between them. Many more of these relationships are still conjectural, although they are expected to be established in the near future. Indeed, the treatment of different models in this paper is regrettably unbalanced, due to the fact that much of the work in this area is still being done.

In the world of topological quantum field theory, these kinds of higher-categorical structures have become important due to their important role in Lurie's recent proof of the Baez-Dolan Cobordism Hypothesis [**23**]. In the last section of this paper, we give a brief description of how to obtain a definition of a cobordism (∞,n)-category. We also give an introduction to the Cobordism Hypothesis as originally posed by Baez and Dolan and as proved by Lurie.

Throughout this paper we freely use the language of model categories and simplicial sets; readers unfamiliar with these methods are encouraged to look at [**15**] and [**16**] for further details.

ACKNOWLEDGMENTS. Many thanks are due to the people who shared their knowledge of their work on this subject with me, including Clark Barwick, Jacob Lurie, Chris Schommer-Pries, and Claire Tomesch. Helpful comments on the paper from the anonymous referee as well as editing suggestions from Arthur Greenspoon are also gratefully acknowledged.

2. Comparison of models for $(\infty,1)$-categories

In this section we review the various models for $(\infty,1)$-categories, their model structures, and the Quillen equivalences between them. A more extensive survey is given in [**6**].

In some sense, the most basic place to start is with $(\infty,0)$-categories, or ∞-groupoids. Topological spaces or, equivalently, simplicial sets, are generally taken to be the definition of ∞-groupoids. Certainly one can think of a topological space as an ∞-groupoid by regarding the points of the space as objects, the paths between points as 1-morphisms, homotopies between paths to be 2-morphisms, homotopies between homotopies as 3-morphisms, and so forth for all natural numbers. While it can be argued that this explanation is one-directional, there is great difficulty in pinning down an actual precise definition of an ∞-groupoid, so it is common to take the above as a definition.

When moving one level higher to $(\infty, 1)$-categories, the most natural definition is to take topological categories, or categories enriched over topological spaces. Now points of mapping spaces are 1-morphisms, paths are 2-morphisms, and so forth, so that now there is no reason for 1-morphisms to be invertible, but all higher morphisms are. More commonly, we use *simplicial categories*, or categories enriched over simplicial sets.

Given two objects x and y of a simplicial category $\mathcal{C}$, we denote the mapping space between them by $\mathrm{Map}(x, y)$. For a simplicial category $\mathcal{C}$, its *category of components* $\pi_0\mathcal{C}$ is the ordinary category with objects the same as those of $\mathcal{C}$ and with objects given by

$$\mathrm{Hom}_{\pi_0\mathcal{C}}(x, y) = \pi_0\mathrm{Map}_{\mathcal{C}}(x, y).$$

A simplicial functor $f\colon \mathcal{C} \to \mathcal{D}$ is a *Dwyer-Kan equivalence* if

- for any objects x, y of $\mathcal{C}$, the map
$$\mathrm{Map}_{\mathcal{C}}(x, y) \to \mathrm{Map}_{\mathcal{D}}(fx, fy)$$
is a weak equivalence of simplicial sets, and
- the induced functor on component categories
$$\pi_0 f\colon \pi_0\mathcal{C} \to \pi_0\mathcal{D}$$
is an equivalence of categories.

A simplicial functor $f : \mathcal{C} \to \mathcal{D}$ is a *fibration* if

- for any objects x and y in $\mathcal{C}$, the map
$$\mathrm{Hom}_{\mathcal{C}}(x, y) \to \mathrm{Hom}_{\mathcal{D}}(fx, fy)$$
is a fibration of simplicial sets, and
- for any object x_1 in $\mathcal{C}$, y in $\mathcal{D}$, and homotopy equivalence $e : fx_1 \to y$ in $\mathcal{D}$, there is an object x_2 in $\mathcal{C}$ and homotopy equivalence $d : x_1 \to x_2$ in $\mathcal{C}$ such that $fd = e$.

THEOREM 2.1. [**5**] *There is a model structure $\mathcal{SC}$ on the category of small simplicial categories with weak equivalences the Dwyer-Kan equivalences and the fibrations as above.*

However, there are other possible models for $(\infty, 1)$-categories. One complaint that one might have about simplicial categories is that the requirement that the composition of mapping spaces be associative is too strong. We are thus led to the definition of Segal categories

First, recall that a simplicial category $\mathcal{C}$ as we have defined it is a special case of a more general simplicial object in the category of small categories, where we assume that all face and degeneracy maps are the identity on the objects so that the objects form a discrete simplicial set. We can take the simplicial nerve to obtain a simplicial space with 0-space discrete. For any simplicial space X recall that we can define Segal maps

$$X_k \to \underbrace{X_1 \times_{X_0} \cdots \times_{X_0} X_1}_{k}.$$

In the case of a nerve of a simplicial category, these maps are isomorphisms of simplicial sets. We obtain the notion of Segal category when we weaken this restriction on simplicial spaces.

DEFINITION 2.2. [**17**] A simplicial space X is a *Segal precategory* if X_0 is discrete. It is a *Segal category* if, in addition, the Segal maps are weak equivalences of simplicial sets.

In a Segal category X, we can consider the discrete space X_0 as the set of "objects" and, given a pair of objects (x, y), the "mapping space" $\operatorname{map}_X(x, y)$ defined as the fiber over (x, y) of the map $(d_1, d_0)\colon X_1 \to X_0 \times X_0$. In this way, we can use much of the language of simplicial categories in the Segal category setting. Furthermore, a Segal category X has a corresponding homotopy category $\operatorname{Ho}(X)$ with the same objects as X but with the sets of components of mapping spaces as the morphisms.

There is a functorial way of "localizing" a Segal precategory to obtain a Segal category in such a way that the set at level zero is unchanged. We denote this functor by L. We then define a map $f\colon X \to Y$ of Segal precategories to be a *Dwyer-Kan equivalence* if

(1) for any x and y in X_0, the map $\operatorname{map}_{LX}(x, y) \to \operatorname{map}_{LY}(fx, fy)$ is a weak equivalence of spaces, and
(2) the map $\operatorname{Ho}(LX) \to \operatorname{Ho}(LY)$ is an equivalence of categories

THEOREM 2.3. [**7**], [**24**] *There are two model structures, $\mathcal{S}e\mathcal{C}at_c$ and $\mathcal{S}e\mathcal{C}at_f$, on the category of Segal precategories such that the fibrant objects are Segal categories and such that the weak equivalences are the Dwyer-Kan equivalences.*

The need for two different model structures with the same weak equivalences arises in the comparison with other models; they are Quillen equivalent to one another via the identity functor. In the model structure $\mathcal{S}e\mathcal{C}at_c$, the cofibrations are the monomorphisms and hence every object is cofibrant. In $\mathcal{S}e\mathcal{C}at_f$, while it is not true that the fibrations are levelwise, the cofibrations are what one would expect them to be if they were; the discrepancy arises from technicalities in working with Segal precategories rather than with all simplicial spaces.

The following theorem can be regarded as a rigidification result, showing that weakening the condition on the Segal maps did not make much of a difference from the homotopy-theoretic point of view.

THEOREM 2.4. [**7**] *The model categories $\mathcal{SC}$ and $\mathcal{S}e\mathcal{C}at_f$ are Quillen equivalent.*

However, for many purposes the condition that the space at level zero be discrete is an awkward one. Thus we come to our third model, that of complete Segal spaces. These objects will again be simplicial spaces, and for technical reasons we require them to be fibrant on the Reedy model structure on simplicial spaces.

DEFINITION 2.5. [**26**] A Reedy fibrant simplicial space W is a *Segal space* if the Segal maps are weak equivalences of simplicial sets.

Like simplicial categories, Segal spaces have objects (this time the set $W_{0,0}$) and mapping spaces between them. Using the weak composition between mapping spaces, we can define homotopy equivalences and consider the subspace W_h of such sitting inside of W_1. It is not hard to see that the degeneracy map $W_0 \to W_1$

has image in W_h, since the image consists of "identity maps" which are certainly homotopy equivalences.

DEFINITION 2.6. [**26**] A Segal space is *complete* if the map $s_0 \colon W_0 \to W_h$ is a weak equivalence of simplicial sets.

THEOREM 2.7. [**26**] *There is a model structure $\mathcal{CSS}$ on the category of simplicial spaces in which the fibrant objects are the complete Segal spaces and the weak equivalences between fibrant objects are levelwise weak equivalences of simplicial sets.*

THEOREM 2.8. [**7**] *The model categories $\mathcal{S}e\mathcal{C}at_c$ and $\mathcal{CSS}$ are Quillen equivalent.*

We approach the fourth model, that of quasi-categories, a bit differently. If we begin with an ordinary category $\mathcal{C}$, its nerve is a simplicial set nerve($\mathcal{C}$). Again we can think of a Segal condition, here where the maps are isomorphisms of sets. However, this description doesn't lend itself well to weakening, since we are dealing with sets rather than simplicial sets. Alternatively, we can describe the "composites" in the nerve of a category via what is commonly called a horn-filling condition. Consider the inclusions $V[m,k] \to \Delta[m]$ for any $m \geq 1$ and $0 \leq k \leq m$. A simplicial set K is the nerve of a category if and only if any map $V[m,k] \to K$ extends uniquely to a map $\Delta[m] \to X$ for any $0 < k < m$. Because we don't include the cases where $k = 0$ and $k = m$, this property is called the *unique inner horn filling condition.*

In the special case where $\mathcal{C}$ is a groupoid, then nerve($\mathcal{C}$) has the property that such a unique extension exists for $0 \leq k \leq m$, i.e., has the *unique horn filling condition.* However, in homotopy theory, it has long been common to consider simplicial sets that are a bit weaker than the nerves of groupoids. If we have a simplicial set K such that any map $V[m,k] \to K$ extends to a map $\Delta[m] \to K$, but this extension is no longer required to be unique, it is called a *Kan complex.* Such simplicial sets are significant in that they are the fibrant objects in the standard model structure on simplicial sets. Therefore, they can be regarded as particular models for ∞-groupoids.

Here we return to the inner horn filling condition and call a simplicial set K an *inner Kan complex* or *quasi-category* if it has the above non-unique extension property for $0 < k < m$, a notion that was first defined by Boardman and Vogt [**10**]. Then, just as a Kan complex is a homotopy version of a groupoid, a quasi-category is a homotopy version of a category, in fact a model for an $(\infty, 1)$-category.

THEOREM 2.9. [**12**], [**18**], [**22**] *There is a model structure $\mathcal{QC}at$ on the category of simplicial sets such that the fibrant objects are the quasi-categories.*

One can actually define mapping spaces, for example, in a quasi-category ([**12**] goes into particular detail on this point), and the weak equivalences between quasi-categories again have the same flavor as the Dwyer-Kan equivalences of simplicial sets. Making this relationship more precise, there is a coherent nerve functor $\mathcal{SC} \to \mathcal{QC}at$, first defined by Cordier and Porter [**11**]. Proofs for using this functor to obtain a Quillen equivalence between $\mathcal{SC}$ and $\mathcal{QC}at$ are given in Lurie's book and in unpublished work of Joyal; Dugger and Spivak have given a substantially shorter proof using different methods.

THEOREM 2.10. [**12**], [**19**], [**22**] *The model categories $\mathcal{SC}$ and $\mathcal{QC}at$ are Quillen equivalent.*

While the previous theorem was sufficient to prove that all four models are equivalent, Joyal and Tierney have established multiple direct Quillen equivalences between $\mathcal{QC}at$ and the other models.

THEOREM 2.11. [**20**] *There are two different Quillen equivalences between $\mathcal{CSS}$ and $\mathcal{QC}at$, and there are two analogous Quillen equivalences between $\mathcal{S}e\mathcal{C}at_c$ and $\mathcal{QC}at$.*

A fifth approach has long been of interest from the perspective of homotopy theory, namely that of viewing a simplicial category as a model for a homotopy theory, where the essential data of a "homotopy theory" is a category with some specified class of weak equivalences. This idea was made precise by Dwyer and Kan via their methods of simplicial localization [**13**], [**14**]. Recent work of Barwick and Kan includes a model structure $\mathcal{CWE}$ on the category of small categories with weak equivalence, together with a Quillen equivalence between $\mathcal{CWE}$ and $\mathcal{CSS}$ [**4**].

We should remark that these models are by no means the only ones which have been proposed; for example, A_∞ categories are conjectured to be equivalent as well.

3. Multisimplicial models: Segal n-categories and n-fold complete Segal spaces

In this section we give definitions of the earliest defined models for (∞, n)-categories. The idea behind them is to iterate the simplicial structure, so an (∞, n)-category is given by a functor $(\mathbf{\Delta}^{op})^n \to \mathcal{SS}ets$, satisfying some properties.

DEFINITION 3.1. An n-fold simplicial space is a functor $X\colon (\mathbf{\Delta}^{op})^n \to \mathcal{SS}ets$.

Notice that there are different ways to regard such an object as a functor. One useful alternative is to think of an n-fold simplicial space as a functor

$$X\colon \mathbf{\Delta}^{op} \to \mathcal{SS}ets^{(\mathbf{\Delta}^{op})^{n-1}}$$

where $\mathcal{SS}ets^{(\mathbf{\Delta}^{op})^{n-1}}$ denotes the category of functors $(\mathbf{\Delta}^{op})^{n-1} \to \mathcal{SS}ets$. This perspective is useful in that it makes use of the idea that an (∞, n)-category should somehow resemble a category enriched in $(\infty, n-1)$-categories. In particular from this viewpoint we can consider the Reedy model structure.

The first definition for (∞, n)-categories was that of Segal n-categories, first given by Hirschowitz and Simpson [**17**]. It is given inductively, building from the definition of Segal category given in the previous section. We denote by

$$\mathcal{SS}ets^{\mathbf{\Delta}^{op}}_{disc}$$

the category of Segal precategories, or functors $Y\colon \mathbf{\Delta}^{op} \to \mathcal{SS}ets$ such that Y_0 is discrete. Then, define inductively the category $\mathcal{SS}ets^{(\mathbf{\Delta}^{op})^n}_{disc}$ of functors $X\colon \mathbf{\Delta}^{op} \to \mathcal{SS}ets^{(\mathbf{\Delta}^{op})^{n-1}}_{disc}$ such that X_0 is discrete. In particular, notice that discreteness conditions are built in at several levels.

DEFINITION 3.2. An n-fold simplicial space $X\colon \mathbf{\Delta}^{op} \to \mathcal{SS}ets^{(\mathbf{\Delta}^{op})^{n-1}}_{disc}$ is a *Segal n-precategory* if X_0 is discrete. It is a *Segal n-category* if, in addition, the Segal maps

$$X_k \to \underbrace{X_1 \times_{X_0} \cdots \times_{X_0} X_1}_{k}$$

are weak equivalences of Segal $(n-1)$-categories for $n \geq 2$.

THEOREM 3.3. [**24**] *There is a model structure* $nSeCat$ *on the category of Segal* n*-precategories in which the fibrant objects are Segal* n*-categories.*

To obtain a higher-order version of complete Segal spaces, we can work inductively, beginning with the definition of complete Segal spaces as given in the previous section. Hence, in the following definitions we can assume that $n \geq 2$. The definitions we give here are as stated by Lurie in [**23**]; he gives a more general treatment of them in [**21**].

DEFINITION 3.4. A Reedy fibrant n-fold simplicial space is an *n-fold Segal space* if

- each Segal map
$$X_k \to \underbrace{X_1 \times_{X_0} \cdots \times_{X_0} X_1}_{k}$$
is a weak equivalence of $(n-1)$-fold Segal spaces for each $k \geq 2$,
- X_k is an $(n-1)$-fold Segal space for each $k \geq 0$, and
- the $(n-1)$-fold Segal space X_0 is essentially constant.

Recall that an $(n-1)$-fold simplicial space X is *essentially constant* if there exists a weak equivalence $Y \to X$ where Y is given by a constant diagram.

DEFINITION 3.5. An n-fold Segal space X is *complete* if

- each X_k is an $(n-1)$-fold complete Segal space, and
- the simplicial space $X_{k,0,\ldots,0}$ is a complete Segal space for all $k \geq 0$.

It is expected that there is a model category, which we denote $n\mathcal{CSS}$, on the category of n-fold simplicial spaces in which the fibrant objects are the n-fold complete Segal spaces. Such a model structure seems to have been developed by Barwick but is not currently in the literature; it will be given precisely in [**9**].

4. Models given by Θ_n-diagrams

As an alternative to n-fold complete Segal spaces, Rezk proposed a new model, that of Θ_n-spaces [**25**]. The idea is to use a new diagram Θ_n rather than iterating simplicial diagrams. We begin by defining the diagrams Θ_n inductively, using a more general construction on categories.

Given a category $\mathcal{C}$, define a category $\Theta\mathcal{C}$ with objects $[m](c_1, \ldots, c_m)$, where $[m]$ is an object of $\mathbf{\Delta}$ and $c_1, \ldots, c_m$ objects of $\mathcal{C}$. A morphism
$$[m](c_1, \ldots, c_m) \to [p](d_1, \ldots, d_p)$$
is given by (δ, f_{ij}), where $\delta \colon [m] \to [p]$ is a morphism in $\mathbf{\Delta}$ and $f_{ij} \colon c_i \to d_j$ is defined for every $1 \leq i \leq m$ and $1 \leq j \leq q$ where $\delta(i-1) < j \leq \delta(i)$ [**25**, 3.2].

Let Θ_0 be the terminal category with one object and only the identity morphism. Inductively define $\Theta_n = \Theta\Theta_{n-1}$. Notice that Θ_1 is just $\mathbf{\Delta}$.

One perspective on the objects of Θ_n is that they are "basic" (strict) n-categories in the same way that objects of $\mathbf{\Delta}$ are "basic" categories, in the sense that they encode the basic kinds of composites that can take place. Therefore, if we take functors $\Theta_n^{op} \to \mathcal{SS}ets$ and require conditions guaranteeing composition up to homotopy and some kind of completeness, we get models for (∞, n)-categories. In the case $n = 1$, we obtain complete Segal spaces; for higher values of n, describing these conditions becomes more difficult but can be done in an inductive manner.

The underlying category for this model is $\mathcal{SS}ets^{\Theta_n^{op}}$, the category of functors $\Theta_n^{op} \to \mathcal{SS}ets$. The model structure we want is obtained as a localization of the injective model structure on this category.

First, given an object $[m](c_1, \ldots, c_m)$ in Θ_n, we obtain a corresponding "simplex" $\Theta[m](c_1, \ldots, c_m)$ in the category $\mathcal{S}ets^{\Theta_n^{op}}$; making it constant in the additional simplicial direction gives an object of $\mathcal{SS}ets^{\Theta_n^{op}}$. This object should be regarded as the analogue of the m-simplex $\Delta[m]$ arising from the object $[m]$ of $\mathbf{\Delta}$.

Given $m \geq 2$ and $c_1, \ldots, c_m$ objects of Θ_{n-1}, define the object

$$G[m](c_1, \ldots, c_m) = \operatorname{colim}(\Theta[1](c_1) \leftarrow \Theta[0] \rightarrow \cdots \leftarrow \Theta[0] \rightarrow \Theta[1](c_m)).$$

There is an inclusion map

$$se^{(c_1, \ldots, c_m)} \colon G[m](c_1, \ldots, c_n) \rightarrow \Theta[n](c_1, \ldots, c_m).$$

We define the set

$$Se_{\Theta_n} = \{se^{(c_1, \ldots, c_m)} \mid m \geq 2, c_1, \ldots c_m \in \operatorname{ob}(\Theta_{n-1})\}.$$

Localizing with respect to this set of maps gives composition up to homotopy, but only on the level of n-morphisms. We need to localize additionally in such a way that lower-level morphisms also have this property, and we can do so inductively.

In [**25**, 4.4], Rezk defines an intertwining functor

$$V \colon \Theta(\mathcal{SS}ets_c^{\Theta_{n-1}^{op}}) \rightarrow \mathcal{SS}ets_c^{\Theta_n^{op}}$$

by

$$V[m](A_1, \ldots, A_m)([q](c_1, \ldots, c_q)) = \coprod_{\delta \in \operatorname{Hom}_{\mathbf{\Delta}}([q],[m])} \prod_{i=1}^{q} \prod_{j=\delta(i-1)+1}^{\delta(i)} A_j(c_i)$$

where the A_j are objects of $\mathcal{SS}ets^{\Theta_{n-1}^{op}}$ and the c_i are objects of Θ_n. This functor can be used to "upgrade" sets of maps in $\mathcal{SS}ets^{\Theta_{n-1}^{op}}$ to sets of maps in $\mathcal{SS}ets^{\Theta_n^{op}}$. Given a map $f \colon A \rightarrow B$ in $\mathcal{SS}ets^{\Theta_{n-1}^{op}}$, we obtain a map $V[1](f) \colon V[1](A) \rightarrow V[1](B)$.

Let $\mathcal{S}_1 = Se_{\mathbf{\Delta}}$, and for $n \geq 2$, inductively define $\mathcal{S}_n = Se_{\Theta_n} \cup V[1](\mathcal{S}_{n-1})$. Localizing the model structure $\mathcal{SS}ets_c^{\Theta_n^{op}}$ with respect to $\mathcal{S}_n$ results in a cartesian model category whose fibrant objects are higher-order analogues of Segal spaces.

However, we need to incorporate higher-order completeness conditions as well. To define the maps with respect to which we need to localize, we make use of an adjoint relationship with simplicial spaces as described by Rezk in [**25**, 4.1]. First, define the functor $T \colon \mathbf{\Delta} \rightarrow \mathcal{SS}ets^{\Theta_n^{op}}$ by

$$T[q]([m](c_1, \ldots, c_m)) = \operatorname{Hom}_{\mathbf{\Delta}}([m], [q]).$$

We use this functor T to define the functor $T^* \colon \mathcal{SS}ets^{\Theta_n^{op}} \rightarrow \mathcal{SS}ets^{\mathbf{\Delta}^{op}}$ defined by

$$T^*(X)[m] = \operatorname{Map}_{\mathcal{SS}ets^{\Theta_n^{op}}}(T[m], X),$$

which has a left adjoint $T_\#$. This adjoint pair is in fact a Quillen pair with respect to the injective model structures.

Now, define $Cpt_{\mathbf{\Delta}} = \{E \rightarrow \Delta[0]\}$ and, for $n \geq 2$,

$$Cpt_{\Theta_n} = \{T_\# E \rightarrow T_\# \Delta[0]\}.$$

Let $\mathcal{T}_1 = Se_{\Theta_1} \cup Cpt_{\Theta_1}$ and, for $n \geq 2$,

$$\mathcal{T}_n = Se_{\Theta_n} \cup Cpt_{\Theta_n} \cup V[1](\mathcal{T}_{n-1}).$$

THEOREM 4.1. [**25**, 8.5] *Localizing $\mathcal{SS}ets_c^{\Theta_n^{op}}$ with respect to the set $\mathcal{T}_n$ gives a cartesian model category, which we denote by $\Theta_n Sp$. Its fibrant objects are higher-order analogues of complete Segal spaces, and therefore it is a model for (∞, n)-categories.*

In his work on the categories Θ_n, Joyal suggested that there should be models for (∞, n)-categories given by functors $\Theta_n \to \mathcal{S}ets$ satisfying higher-order inner horn-filling conditions, but he was unable to find the right way to describe these conditions. More recently, Barwick has been able to use the relationship between complete Segal spaces and quasi-categories to formulate a higher-dimensional version.

DEFINITION 4.2. [**3**] A *quasi-n-category* is a functor $\Theta_n^{op} \to \mathcal{S}ets$ satisfying appropriate inner horn-filling conditions.

THEOREM 4.3. [**3**] *There is a cartesian model structure $n\mathcal{Q}\mathcal{C}at$ on the category $\mathcal{S}ets^{\Theta_n^{op}}$ in which the fibrant objects are quasi-n-categories.*

5. (∞, n)-categories as enriched categories: strict and weak versions

Intuitively, one would like to think of (∞, n)-categories as categories enriched over $(\infty, n-1)$-categories. In practice, this approach can be problematic. In particular, if we want our models for (∞, n)-categories to be objects in a model category, then at the very least we need our model structure on $(\infty, n-1)$-categories to be cartesian. Since the model structure $\mathcal{SC}$ for simplicial categories is not cartesian, in that the product is not compatible with the model structure, we cannot continue the induction using that model. However, we can enrich over other models.

The model category $\Theta_n Sp$ is cartesian, and therefore we can consider categories enriched in $\Theta_{n-1}Sp$ as another model for (∞, n)-categories. In doing so, we have a way to realize the intuitive idea that (∞, n)-categories are categories enriched in $(\infty, n-1)$-categories.

Let $\mathrm{Map}_{\mathcal{C}}(x, y)$ denote the mapping object in $\Theta_{n-1}Sp$ between objects x and y of a category $\mathcal{C}$ enriched in $\Theta_{n-1}Sp$.

DEFINITION 5.1. [**8**] Let $\mathcal{C}$ and $\mathcal{D}$ be categories enriched in $\Theta_{n-1}Sp$. An enriched functor $f \colon \mathcal{C} \to \mathcal{D}$ is a *weak equivalence* if

(1) $\mathrm{Map}_{\mathcal{C}}(x, y) \to \mathrm{Map}_{\mathcal{D}}(fx, fy)$ is a weak equivalence in $\Theta_{n-1}Sp$ for any objects x, y, and
(2) $\pi_0\mathcal{C} \to \pi_0\mathcal{D}$ is an equivalence of categories, where $\pi_0\mathcal{C}$ has the same objects as $\mathcal{C}$ and $\mathrm{Hom}_{\pi_0\mathcal{C}}(x, y) = \mathrm{Hom}_{\mathrm{Ho}(\Theta_{n-1}Sp)}(1, \mathrm{Map}_{\mathcal{C}}(x, y))$.

THEOREM 5.2. [**8**] *There is a model structure $\Theta_{n-1}Sp - \mathcal{C}at$ on the category of small categories enriched in $\Theta_{n-1}Sp$ with weak equivalences defined as above.*

Just as we had a simplicial nerve functor taking a simplicial category to a simplicial diagram of simplicial sets, we have a nerve functor taking a category enriched in $\Theta_{n-1}Sp$ to a simplicial diagram of objects in $\Theta_{n-1}Sp$. If we call the resulting simplicial object X, we can observe that the strict Segal condition holds in this setting, in that the maps

$$X_k \to \underbrace{X_1 \times_{X_0} \cdots \times_{X_0} X_1}_{k}$$

are isomorphisms of objects in $\Theta_{n-1}Sp$. Weakening this condition as before leads to the following definition.

DEFINITION 5.3. [**8**] A *Segal precategory object in* $\Theta_{n-1}Sp$ is a functor $X\colon \mathbf{\Delta}^{op} \to \Theta_{n-1}Sp$ such that the functor $X_0\colon \Theta_{n-1}^{op} \to \mathcal{S}Sets$ is discrete. It is a *Segal category object* if the Segal maps are weak equivalences in $\Theta_{n-1}Sp$.

We can again define a functor L taking a Segal precategory object to a Segal category object, and define mapping objects (now Θ_{n-1}-spaces rather than simplicial sets) and a homotopy category for a Segal category object such as we did for a Segal category. A functor of Segal precategory objects $f\colon X \to Y$ is a *Dwyer-Kan equivalence* if

(1) for any x and y in X_0, the map $\operatorname{map}_{LX}(x,y) \to \operatorname{map}_{LY}(fx, fy)$ is a weak equivalence in $\Theta_{n-1}Sp$, and
(2) the map $\operatorname{Ho}(LX) \to \operatorname{Ho}(LY)$ is an equivalence of categories.

THEOREM 5.4. [**8**] *There are two model structures on the category of Segal precategory objects,* $\mathcal{S}e(\Theta_{n-1}Sp)^{\mathbf{\Delta}^{op}}_{disc,f}$ *and* $\mathcal{S}e(\Theta_{n-1}Sp)^{\mathbf{\Delta}^{op}}_{disc,c}$, *in which the weak equivalences are the Dwyer-Kan equivalences.*

The notion of more general Segal category objects has also been developed by Simpson [**29**].

6. Comparisons between different models

With so many models for (∞, n)-categories, we need to establish that they are all equivalent to one another. At this time, we have partial results in this area and many more conjectures.

We begin with known results, which can be summarized by the following diagram:

$$(\Theta_{n-1}Sp) - Cat \leftrightarrows \mathcal{S}e(\Theta_{n-1}Sp)^{\mathbf{\Delta}^{op}}_{disc,f} \rightleftarrows \mathcal{S}e(\Theta_{n-1}Sp)^{\mathbf{\Delta}^{op}}_{disc,c}.$$

The leftmost Quillen equivalence is given by the following theorem, which is a generalization of the Quillen equivalence between $\mathcal{SC}$ and $\mathcal{S}e\mathcal{C}at_f$.

THEOREM 6.1. [**8**] *The enriched nerve functor is the right adjoint for a Quillen equivalence between* $(\Theta_{n-1}Sp) - Cat$ *and* $\mathcal{S}e(\Theta_{n-1}Sp)^{\mathbf{\Delta}^{op}}_{disc,f}$.

Moving to the right, the next Quillen equivalence is the easiest.

PROPOSITION 6.2. [**8**] *The identity functor gives a Quillen equivalence between* $\mathcal{S}e(\Theta_{n-1}Sp)^{\mathbf{\Delta}^{op}}_{disc,f}$ *and* $\mathcal{S}e(\Theta_{n-1}Sp)^{\mathbf{\Delta}^{op}}_{disc,c}$.

To continue the comparison, we make use of a definition of *complete Segal space objects in* $(\Theta_{n-1}Sp)^{\mathbf{\Delta}^{op}}$. There is a model structure $L_{CS}(\Theta_{n-1}Sp)^{\mathbf{\Delta}^{op}}_c$ on the category of functors $\mathbf{\Delta}^{op} \to \Theta_{n-1}Sp$ in which the fibrant objects satisfy Segal and completeness conditions. The Segal condition is as given for Segal category objects above, but the completeness condition is subtle, and the comparison with Segal category objects is still work in progress [**9**].

CONJECTURE 6.3. *There are Quillen equivalences*

$$\mathcal{S}e(\Theta_{n-1}Sp)^{\mathbf{\Delta}^{op}}_{disc,c} \rightleftarrows L_{CS}(\Theta_{n-1}Sp)^{\mathbf{\Delta}^{op}} \rightleftarrows \Theta_n Sp.$$

The second equivalence in this chain should in fact be the first in a chain of Quillen equivalences between $\Theta_n Sp$ and $n\mathcal{CSS}$, induced by the chain of functors

$$\mathbf{\Delta}^n = \mathbf{\Delta}^{n-1} \times \mathbf{\Delta} \to \mathbf{\Delta}^{n-1} \times \Theta_2 \to \cdots \to \mathbf{\Delta} \times \Theta_{n-1} \to \Theta_n.$$

CONJECTURE 6.4. *There is a chain of Quillen equivalences*

$$n\mathcal{CSS} \rightleftarrows L_{CS}(\Theta_2 Sp)^{(\mathbf{\Delta}^{op})^{n-2}} \rightleftarrows \cdots \rightleftarrows L_{CS}(\Theta_{n-1}Sp)^{\mathbf{\Delta}^{op}} \rightleftarrows \Theta_n Sp.$$

There are other models, and proposed comparisons between them, that are currently being developed. Since the model structure $n\mathcal{Q}\mathcal{C}at$ is also cartesian, we can define categories enriched in $(n-1)\mathcal{Q}\mathcal{C}at$. With an appropriate model structure on the category of such, we conjecture that it is Quillen equivalent to $(\Theta_{n-1}Sp) - Cat$, using the Quillen equivalence between $\Theta_n Sp$ and $n\mathcal{Q}\mathcal{C}at$.

It is expected that there is yet another method of connecting $\mathcal{S}e(\Theta_{n-1}Sp)^{\mathbf{\Delta}^{op}}_{disc,c}$ to $\Theta_n Sp$ with Quillen equivalences, where the intermediate model category has objects functors $\Theta_n^{op} \to \mathcal{SS}ets$ satisfying discreteness conditions. In other words, such a model structure would be a Segal category version of $\Theta_n Sp$. This model structure should be connected to $n\mathcal{S}e\mathcal{C}at$ via a chain of Quillen equivalences analogous to those connecting $\Theta_n Sp$ and $n\mathcal{CSS}$. The comparison between $n\mathcal{S}e\mathcal{C}at$ and other models is also being investigated by Tomesch [**30**].

7. (∞, n)-categories and the cobordism hypothesis

In this section, we give a brief account of the Cobordism Hypothesis, which was originally posed in the context of weak n-categories by Baez and Dolan in [**2**] and proved in the setting of (∞, n)-categories by Lurie [**23**]. This result not only gives a purely algebraic description of higher categories defined in terms of cobordisms of manifolds, but is usually interpreted as a statement about topological quantum field theories.

We begin with Atiyah's definition of topological quantum field theory [**1**].

DEFINITION 7.1. For $n \geq 1$, denote by $Cob(n)$ the category with objects closed framed $(n-1)$-dimensional manifolds and morphisms the diffeomorphism classes of framed cobordisms between them.

Notice that composition is defined by gluing together cobordisms; since we are taking diffeomorphism classes on the level of morphisms there is no difficulty defining composition. This category is also equipped with a symmetric monoidal structure given by disjoint union of $(n-1)$-dimensional manifolds.

DEFINITION 7.2. For a field $\Bbbk$, a *topological quantum field theory* of dimension n is a symmetric monoidal functor $Z \colon Cob(n) \to Vect(\Bbbk)$, where $Vect(\Bbbk)$ is the category of vector spaces equipped with the usual tensor product.

An important observation about topological quantum field theories is that the structure of $Cob(n)$ affects which vector spaces can be in the image of one. Since a framed manifold has a dual, the vector space to which it is assigned must also have a well-behaved dual; in particular it must be finite-dimensional.

In low dimensions, topological quantum field theories are well-understood, since the manifolds appearing in the category $Cob(n)$ are easily described. In particular, one can cut them into simpler pieces in a nice way and understand the entire functor just by knowing what happens to these pieces. However, when $n \geq 3$, these pieces may not just be manifolds with boundary but instead manifolds with corners. We need a higher-categorical structure to encode this larger range of dimensions of manifolds.

DEFINITION 7.3. For $n \geq 1$, define a weak n-category $Cob_n(n)$ to have objects 0-dimensional closed framed 0-dimensional manifolds, 1-morphisms framed cobordisms between them, 2-morphisms framed cobordisms between the cobordisms, up to n-morphisms which are diffeomorphism classes of framed cobordisms of dimension n.

Notice that, unlike for $Cob(n)$, we cannot take diffeomorphism classes except at the top dimension, since cobordisms must be defined between actual manifolds, not diffeomorphism classes of them. Therefore, composition of morphisms is no longer strictly defined, and hence we have a weak, rather than strict, n-category.

To define a generalized topological field theories using $Cob_n(n)$ rather than $Cob(n)$, we need a higher-categorical version of $Vect(\Bbbk)$. While there are several proposed definitions for such a weak 2-category, it is unknown how to continue to still higher dimensions. Fortunately, we do not really need a particular structure and therefore can replace $Vect(\Bbbk)$ with an arbitrary symmetric monoidal weak n-category.

DEFINITION 7.4. Let $\mathcal{C}$ be a symmetric monoidal weak n-category. An *extended $\mathcal{C}$-valued topological field theory of dimension n* is a symmetric monoidal functor $Z\colon Cob_n(n) \to \mathcal{C}$.

We are almost ready to state Baez and Dolan's original conjecture, but we first need to comment on duality in higher categories. As mentioned above, the image of a topological quantum field theory had to consist of finite-dimensional vector spaces, since they have well-behaved duals. In the extended world, duality becomes more complicated, in that we need to consider objects as well as morphisms of all levels. Objects having the appropriate properties are called *fully dualizable objects*. We do not give a precise definition of fully dualizable here, but refer the reader to Lurie's paper [**23**, 2.3].

THEOREM 7.5. *(Baez-Dolan Cobordism Hypothesis) Let $\mathcal{C}$ be a symmetric monoidal weak n-category and Z a topological quantum field theory. Then the evaluation functor $Z \mapsto Z(*)$ determines a bijection between isomorphism classes of framed extended $\mathcal{C}$-valued topological quantum field theories and isomorphism classes of fully dualizable objects of $\mathcal{C}$.*

This version of the Cobordism Hypothesis has been proved in the case where $n = 2$ by Schommer-Pries [**28**], but it seems difficult to prove for higher values of n. Not least is the difficulty of knowing how to handle weak n-categories, and how strict or weak the symmetric monoidal structure should be. However, Lurie's main insight, allowing a sidestep to this problem, was to prove an alternative version using (∞, n)-categories. The original version follows via truncation methods.

Here, we focus on the definition of an (∞, n)-category $Bord_n^{fr}$ which generalizes $Cob_n(n)$. Following Lurie [**23**], we use the n-fold complete Segal space model. An alternative definition in the setting of Θ_n-spaces is being developed by Rozenblyum and Schommer-Pries [**27**].

DEFINITION 7.6. [**23**, 2.2] Let V be a finite-dimensional vector space. Define an n-fold simplicial space $PBord_n^{fr,V}$ by

$$(PBord_n^{fr,V})_{k_1,\ldots,k_n} = \{(M, \{t_0^1 \leq \cdots, \leq t_{k_1}^1\}, \ldots, \{t_0^1 \leq \cdots, t_{k_n}^n\})\}$$

where:

- M is a closed n-dimensional framed submanifold of $V \times \mathbb{R}^n$ such that the projection $M \to \mathbb{R}^n$ is proper,
- for every $S \subseteq \{1, \ldots, n\}$ and every $\{0 \leq j_i \leq k_i\}_{i \in S}$, the composite map
$$M \to \mathbb{R}^n \to \mathbb{R}^S$$
does not have $(t_{j_i})_{i \in S}$ as a critical value, and
- the map $M \to \mathbb{R}^{\{i+1, \ldots, n\}}$ is submersive at every $x \in M$ whose image in $\mathbb{R}^{\{i\}}$ belongs to the set $\{t_{i_0}, \ldots, t_{i_k}\}$.

As V ranges over all finite-dimensional subspaces of $\mathbb{R}^\infty$, let
$$PBord_n^{fr} = \mathrm{colim}_V PBord_n^{fr,V}.$$

PROPOSITION 7.7. *The n-fold simplicial space $PBord_n^{fr}$ is an n-fold Segal space which is not necessarily complete.*

To define $Bord_n^{fr}$, we use the fact that any n-fold simplicial space has a completion, or weakly equivalent n-fold complete Segal space.

DEFINITION 7.8. Define $Bord_n^{fr}$ to be the completion of $PBord_n^{fr}$.

With this definition in place, we can now state Lurie's version of the Cobordism Hypothesis.

THEOREM 7.9. [**23**, 1.4.9] *Let $\mathcal{C}$ be a symmetric monoidal (∞, n)-category. The evaluation functor $Z \mapsto Z(*)$ determines a bijection between isomorphism classes of symmetric monoidal functors $Bord_n^{fr} \to \mathcal{C}$ and isomorphism classes of fully dualizable objects of $\mathcal{C}$.*

We conclude with some comments about Lurie's method of proof, in particular his use of (∞, n)-categories. Using intuitive definitions, it would seem that (∞, n)-categories would be more complicated than weak n-categories and therefore more difficult to use in practice. However, the homotopical nature of (∞, n)-categories makes the opposite true. While in each case composition is defined weakly, for (∞, n)-categories the resulting coherence data is nicely packaged into the definition. This way of dealing with the inherent complications of weak n-categories proved to be an effective technique in establishing this result.

References

[1] Michael Atiyah, Topological quantum field theories. *Inst. Hautes Études Sci. Publ. Math.* No. 68 (1988), 175-186 (1989).

[2] John C. Baez and James Dolan, Higher-dimensional algebra and topological quantum field theory. *J. Math. Phys.* 36 (1995), no. 11, 6073-6105.

[3] Clark Barwick, Homotopy coherent algebra II: Iterated wreath products of O and (∞, n)-categories, in preparation.

[4] C. Barwick and D.M. Kan, Relative categories as another model for the homotopy theory of homotopy theories I: The model structure, preprint available at http://math.mit.edu/~clarkbar/papers/barwick-kan.pdf.

[5] Julia E. Bergner, A model category structure on the category of simplicial categories, *Trans. Amer. Math. Soc.* 359 (2007), 2043-2058.

[6] Julia E. Bergner, A survey of $(\infty, 1)$-categories, in J. Baez and J. P. May, *Towards Higher Categories*, *IMA Volumes in Mathematics and Its Applications*, Springer, 2010, 69-83.

[7] Julia E. Bergner, Three models for the homotopy theory of homotopy theories, *Topology* 46 (2007), 397-436.

[8] Julia E. Bergner and Charles Rezk, Comparison of models (∞, n)-categories, I, in preparation.

[9] Julia E. Bergner and Charles Rezk, Comparison of models (∞, n)-categories, II, work in progress.

[10] J.M. Boardman and R.M. Vogt, *Homotopy invariant algebraic structures on topological spaces. Lecture Notes in Mathematics, Vol. 347.* Springer-Verlag, 1973.

[11] J.M. Cordier and T. Porter, Vogt's theorem on categories of homotopy coherent diagrams, *Math. Proc. Camb. Phil. Soc.* (1986), 100, 65-90.

[12] Daniel Dugger and David I. Spivak, Mapping spaces in quasicategories, preprint available at math.AT/0911.0469.

[13] W.G. Dwyer and D.M. Kan, Function complexes in homotopical algebra, *Topology* 19 (1980), 427-440.

[14] W.G. Dwyer and D.M. Kan, Simplicial localizations of categories, *J. Pure Appl. Algebra* 17 (1980), no. 3, 267–284.

[15] W.G. Dwyer and J. Spalinski, Homotopy theories and model categories, in *Handbook of Algebraic Topology*, Elsevier, 1995.

[16] P.G. Goerss and J.F. Jardine, *Simplicial Homotopy Theory, Progress in Math*, vol. 174, Birkhauser, 1999.

[17] A. Hirschowitz and C. Simpson, Descente pour les n-champs, preprint available at math.AG/9807049.

[18] A. Joyal, Simplicial categories vs quasi-categories, in preparation.

[19] A. Joyal, The theory of quasi-categories I, in preparation.

[20] André Joyal and Myles Tierney, Quasi-categories vs Segal spaces, *Contemp. Math.* 431 (2007) 277-326.

[21] Jacob Lurie, $(\infty, 2)$-categories and Goodwillie calculus, preprint available at math.CT/09050462.

[22] Jacob Lurie, *Higher topos theory. Annals of Mathematics Studies*, 170. Princeton University Press, Princeton, NJ, 2009.

[23] Jacob Lurie, On the classification of topological field theories. *Current developments in mathematics*, 2008, 129-280, Int. Press, Somerville, MA, 2009.

[24] Regis Pelissier, Catégories enrichies faibles, preprint available at math.AT/0308246.

[25] Charles Rezk, A cartesian presentation of weak n-categories, *Geom. Topol.* 14 (2010) 521571.

[26] Charles Rezk, A model for the homotopy theory of homotopy theory, *Trans. Amer. Math. Soc.* , 353(3), 973-1007.

[27] Nick Rozenblyum and Chris Schommer-Pries, work in progress.

[28] Chris Schommer-Pries, Classification of extended 2D topological field theories, PhD thesis, UC Berkeley, 2009.

[29] Carlos Simpson, Homotopy theory of higher categories, preprint available at math.CT/1001.4071.

[30] Claire Tomesch, work in progress.

DEPARTMENT OF MATHEMATICS, UNIVERSITY OF CALIFORNIA, RIVERSIDE, CA 92521
E-mail address: bergnerj@member.ams.org

Proceedings of Symposia in Pure Mathematics
Volume **83**, 2011

From Operads to Dendroidal Sets

Ittay Weiss

ABSTRACT. Dendroidal sets offer a formalism for the study of ∞-operads akin to the formalism of ∞-categories by means of simplicial sets. We present here an account of the current state of the theory while placing it in the context of the ideas that led to the conception of dendroidal sets. We briefly illustrate how the added flexibility embodied in ∞-operads can be used in the study of A_∞-spaces and weak n-categories in a way that cannot be realized using strict operads.

1. Introduction

This work aims to be a conceptually self-contained introduction to the theory and applications of dendroidal sets, surveying the current state of the theory and weaving together ideas and results in topology to form a guided tour that starts with Stasheff's work [**37**] on H-spaces, goes on to Boardman and Vogt's work [**5**] on homotopy invariant algebraic structures followed by the generalization [**1, 2, 3**] of their work by Berger and Moerdijk, arrives at the birth of dendroidal sets [**34, 35**] and ends with the establishment, by Cisinski and Moerdijk in [**8, 9, 10**], of dendroidal sets as models for homotopy operads. With this aim in mind we adopt the convention of at most pointing out core arguments of proofs rather than detailed proofs that can be found elsewhere.

We assume basic familiarity with the language of category theory and mostly follow [**30**]. Regarding enriched category theory we assume little more than familiarity with the definition of a category enriched in a symmetric monoidal category as can be found in [**21**]. The elementary results on presheaf categories that we use can be found in [**31**]. Some comfort of working with simplicial sets is needed for which the first chapter of [**14**] suffices. Some elementary understanding of Quillen model categories is desirable with standard references being [**17, 18**].

Operads arose in algebraic topology and have since found applications across a wide range of fields including Algebra, Theoretical Physics, and Computer Science. The reason for the success of operads is that they offer a computationally effective formalism for treating algebraic structures of enormous complexity, usually involving some notion of (abstract) homotopy. As such, the first operads to be

2010 *Mathematics Subject Classification.* Primary 55P48, 55U10, 55U35; Secondary 18D50, 18D10, 18G20.

Key words and phrases. Operads, Homotopy Theory, Weak Algebras.

introduced were topological operads and most of the other variants are similarly enriched in other categories. However, the presentation we give here of operads treats them as a rather straightforward generalization of the notion of category. It is that viewpoint that quite naturally leads to defining dendroidal sets to serve as the codomain category for a nerve construction for operads, extending the usual nerve of categories.

The path we follow is the following one. We first examine the expressive power of non-enriched symmetric operads. We find that by considering operad maps it is possible to classify a wide range of strict algebraic structures such as associative and commutative monoids and to show that operads carry a closed monoidal structure that, via the internal hom, internalizes algebraic structures. We show that the rather trivial fact that algebraic structures can be transferred along isomorphisms, which we call the isomorphism invariance property, is a result of symmetric operads supporting a Quillen model structure compatible with the monoidal structure. We then turn to the much more challenging homotopy invariance property for algebraic structures in the presence of homotopy notions. We show how the theory of operads is used to adequately handle this more subtle situation, however at a cost. The internalization of these so-called weak algebras, via an internal hom construction, is lost. The sequel can be seen as a presentation of the successful attempt to develop a formalism for weak algebraic structures in which the internalization of algebras is restored. This formalism is given by dendroidal sets and a suitable Quillen model structure which can be used to give a proof of the homotopy invariance property (which is completely analogous to the case of non-enriched operads). The consequences and applicability of the added flexibility of dendroidal sets is portrayed by considering n-fold A_∞-spaces and weak n-categories.

Section 2 introduces in the first half non-enriched symmetric operads and presents their basic theory. The second half is concerned with enriched operads and the Berger-Moerdijk generalization of the Boardman-Vogt W-construction. Section 3 is a parallel development of the ideas in Section 2. The first half introduces dendroidal sets and presents their basic theory while the second half is concerned with the homotopy coherent nerve construction with applications to A_∞-spaces and weak n-categories. Section 4 is devoted to the Cisinski-Moerdijk model structure on dendroidal sets and the way it is used to prove the homotopy invariance property. Section 5 closes this work with a brief presentation of a planar dendroidal Dold-Kan correspondence and discusses the yet unsolved problem of obtaining a satisfactory geometric realization for dendroidal sets.

Remark. Below we work in a convenient category of topological spaces Top. In some places it is important that this category be closed monoidal, in which case the category of compactly generated Hausdorff spaces would suffice. We will not remark about such issues further.

Acknowledgements. The author wishes to thank the referee for the diligent reading of an earlier version of this work and for the numerous helpful remarks and comments that helped shape the article.

2. Operads and algebraic structures

Remark. The reader already familiar with operads who reads this section just to familiarize herself with the notation is strongly advised to look at Remark 2.19

and Fact 2.30 below. For her convenience the opening paragraph below recounts the contents of the entire section.

Our journey starts with non-enriched symmetric operads, also known as symmetric multicategories (originating in Lambek's study of deductive systems in logic [**23**]) or symmetric coloured operads (e.g., [**3, 24**]). In the literature on operads these structures are underrepresented probably due to the fact that the first operads, introduced by May in [**33**], were enriched in topological spaces and many of the most important uses of operads require enrichment. The point of view of operads we adopt is that operads generalize categories. Consequently, just as a study of categories starts with non-enriched categories, with enrichment usually treated at some later stage, we first present non-enriched symmetric operads. The operadically versed reader will immediately recognize that our definition of algebra differs slightly from the standard one. We define the Boardman-Vogt tensor product of symmetric operads and the notion of natural transformations for symmetric operads that endows the category of symmetric operads with the structure of a symmetric closed monoidal category. We then address the isomorphism invariance property and treat it in the context of a suitable Quillen model structure on symmetric operads. We then turn to the much more subtle and interesting case of the homotopy invariance property and give an expository treatment of the theory developed by Berger and Moerdijk relevant for the rest of the presentation.

2.1. Trees. Symmetric (also called 'non-planar') rooted trees are useful in the study of symmetric operads. There is no standard definition of 'tree' that is commonly used (Ginzburg and Kapranov in [**12**] use a topological definition while Leinster in [**24**] uses a combinatorial one) but all approaches are essentially the same. More recently, Joachim Kock in [**22**] established a close connection between trees and polynomial functors, offering yet another formalism of trees while shedding a different light on the symbiosis between operads and trees.

We present here the formalism of trees we use and introduce terminology for commonly occurring trees as well as grafting of trees. We end the section by presenting a generalization of posets that shows trees to be analogues of finite linear orders.

2.1.1. *Symmetric rooted trees.*

DEFINITION 2.1. A *tree* (short for symmetric rooted tree) is a finite poset $(T, \leq)$ which has a smallest element and such that for each $e \in T$ the set $\{y \in T \mid y \leq e\}$ is linearly ordered. The elements of T are called *edges* and the unique smallest edge is called the *root*. Part of the information of a tree is a subset $L = L(T)$ of maximal elements, which are called *leaves*. An edge is *outer* if it is either the root or it belongs to L, otherwise it is called *inner*.

Given edges $e, e' \in T$ we write e/e' if $e' < e$ and if for any $x \in T$ for which $e' \leq x \leq e$ holds that either $x = e'$ or $x = e$. For a non-leaf edge e the set $in(e) = \{t \in T \mid t/e\}$ is called the set of *incoming edges* into e. For such an edge e the set $v = \{e\} \cup in(e)$ is called the *vertex* above e and we define $in(v) = in(e)$ and $out(v) = e$ which are called, respectively, the set of *incoming edges* and the *outgoing edge* associated to v. The *valence* of v is equal to $|in(v)|$ and could be 0. Note that there is no vertex associated to a leaf. We will draw trees by the graph dual of their Hesse diagrams with the root at the bottom and will use a $\bullet$

for vertices. For example, in the tree

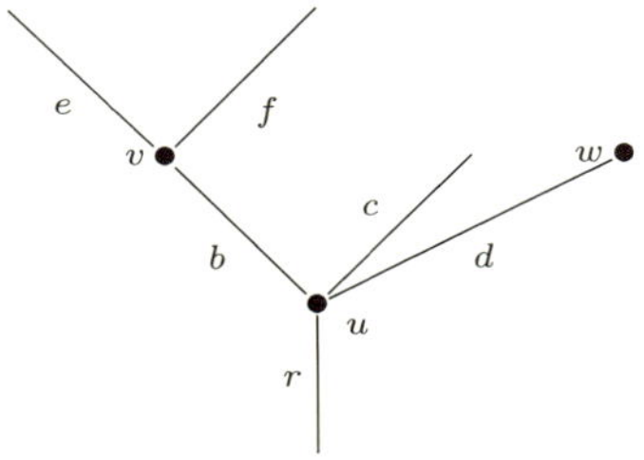

there are three vertices of valence 2,3, and 0 and three leaves $L = \{e, f, c\}$. The outer edges are e, f, c, and r, where r is the root. The inner edges are then b and d.

2.1.2. *Some common trees.* The following types of trees appear often enough in the theory of dendroidal sets to merit their own notation.

DEFINITION 2.2. For each $n \geq 0$, a tree L_n of the form

with one leaf and n vertices, all unary (i.e., each vertex has valence equal to 1), will be called a *linear tree of order n*. The special case of the tree L_0

consisting of just one edge and no vertices is called the *unit* tree. We denote this tree by η, or η_e if we wish to explicitly name its unique edge. In this tree, the only edge is both the root and a leaf.

DEFINITION 2.3. For each $n \geq 0$, a tree C_n of the form

that has just one vertex and n leaves will be called an *n-corolla*. Note that the case $n = 0$ results in a tree different than η.

2.1.3. *Grafting.*

DEFINITION 2.4. Let T and S be two trees whose only common edge is the root r of S which is also one of the leaves of T. The *grafting*, $T \circ S$, of S on T along r is the poset $T \cup S$ with the obvious poset structure and set of leaves equal to $(L(S) \cup L(T)) - \{r\}$.

Pictorially, the grafted tree $T \circ S$ is obtained by putting the tree S on top of the tree T by identifying the output edge of S with the input edge r of T. By repeatedly grafting, one can define a full grafting operation $T \circ (S_1, \cdots, S_n)$ in the obvious way.

We now state a useful decomposition of trees that allows for inductive proofs on trees. The proof is trivial.

PROPOSITION 2.5. *Let T be a tree. Suppose T has root r and $\{r, e_1, \cdots, e_n\}$ is the vertex above r. Let T_{e_i} be the tree that contains the edge e_i as root and everything above it in T. Then*

$$T = T_{root} \circ (T_{e_1}, \cdots, T_{e_n})$$

where T_{root} is the n-corolla consisting of r as root and $\{e_1, \cdots, e_n\}$ as the set of leaves.

2.1.4. *Trees and dendroidally ordered sets.* The trees we defined above are going to be the objects of the category Ω whose presheaf category Set_Ω is the category of dendroidal sets. Recall that the simplicial category Δ (whose presheaf category is the category of simplicial sets) can be defined as (a skeleton of) the category of totally ordered finite sets with order preserving maps. In this section we present an extension of the notion of totally ordered finite sets closely related to trees. The content of this section is not used anywhere in the sequel and is presented for the sake of completeness. Consequently we give no proofs and refer the reader to [**38**] for more details.

First we extend the notion of a relation and that of a poset to what we call broad relation and broad poset. For a set A we denote by $A^+ = (A^+, +, 0)$ the free commutative monoid on A. A *broad relation* is a pair (A, R) where A is a set and R is a subset of $A \times A^+$. As is common with ordinary relations, we use the notation $aR(a_1 + \cdots + a_n)$ instead of $(a, (a_1 + \cdots + a_n)) \in R$.

DEFINITION 2.6. A *broad poset* is a broad relation (A, R) satisfying:

(1) Reflexivity: aRa holds for any $a \in A$.
(2) Transitivity: For all $a_0, \cdots, a_n \in A$ and $b_1, \cdots, b_n \in A^+$ such that $a_i R b_i$ for $1 \leq i \leq n$, holds that if $a_0 R(a_1 + \cdots + a_n)$ then $aR(b_1 + \cdots + b_n)$.
(3) Anti-symmetry: For all $a_1, a_2 \in A$ and $b_1, b_2 \in A^+$ if $a_1 R(a_2 + b_2)$ and $a_2 R(a_1 + b_1)$ then $a_1 = a_2$.

When (A, R) is a broad poset we denote R by $\leq$. The meaning of $<$ is then defined in the usual way.

A *map* of broad posets $f : A \to B$ is a set function preserving the broad poset structure, that is if $a \leq (a_1 + \cdots + a_n)$ then $f(a) \leq (f(a_1) + \cdots + f(a_n))$.

DEFINITION 2.7. We denote by *BrdPoset* the category of all broad posets and their maps.

Let $\star$ be a singleton set $\{*\}$ with the broad poset structure given by $* \le *$. Note that $\star$ is not a terminal object in *BrdPoset*.

LEMMA 2.8. *(Slicing lemma for broad posets) There is an isomorphism of categories between $BrdPoset/\star$ and the category Poset of posets and order preserving maps. Moreover, along this isomorphism one obtains a functor $k_! : Poset \to BrdPoset$ which has a right adjoint $k^* : BrdPoset \to Poset$ which itself has a right adjoint $k_* : Poset \to BrdPoset$.*

As motivation for the following definition recall that a finite ordinary poset A is linearly ordered if, and only if, it has a smallest element and for every $a \in A$ the set $a_\uparrow = \{x \in A \mid a < x\}$ is either empty or has a smallest element.

DEFINITION 2.9. A finite broad poset A is called *dendroidally ordered* if

(1) There is an element $r \in A$ such that for every $a \in A$ there is $b \in A^+$ such that $r \le a + b$.
(2) For every $a \in A$ the set $a_\uparrow = \{b \in A^+ \mid a < b\}$ is either empty or it contains an element $s(a) = a_1 + \cdots + a_n$ such that every $b \in a_\uparrow$ can be written as $b = b_1 + \cdots + b_n$ with $a_i \le b_i$ for all $1 \le i \le n$.
(3) For every $a_0, \cdots, a_n \in A$, if $a_0 \le a_1 + \cdots + a_n$ then for $i \ne j$ there holds $a_i \ne a_j$.

Trees are related to finite dendroidally ordered sets as follows. Given a tree T define a broad relation on the set $E(T)$ of edges by declaring $e \le e_1 + \cdots + e_n$ precisely when there is a vertex v such that $in(v) = \{e_1, \cdots, e_n\}$ (without repetitions) and $out(v) = e$. The transitive closure of this broad relation is then a dendroidally ordered set. This constructions can be used to give an equivalence of categories between the full subcategory *DenOrd* of *BrdPoset* spanned by the dendroidally ordered sets and the dendroidal category Ω defined below. It is easily seen that *DenOrd*, upon slicing over $\star$, is isomorphic to the category of all finite linearly ordered sets and order preserving maps.

2.2. Operads and algebras. We now present symmetric operads viewed as a generalization of categories where arrows are allowed to have domains of arity n for any $n \in \mathbb{N}$. We then define the notion of $\mathcal{P}$-algebras for a symmetric operad $\mathcal{P}$ which are often referred to as the raison d'être of operads. We deviate here from the common definition of algebras noting that our definition encompasses the standard one. We define an algebra to simply be a morphism between symmetric operads, the difference being purely syntactic. The assertion that symmetric operads exist in order to define algebras thus agrees with the idea that in any category the objects' raison d'être is to serve as domains and codomains of arrows.

DEFINITION 2.10. A *planar operad* $\mathcal{P}$ consists of a class $\mathcal{P}_0$ whose elements are called the *objects* of $\mathcal{P}$ and to each sequence $P_0, \cdots, P_n \in \mathcal{P}_0$ a set $\mathcal{P}(P_1, \cdots, P_n; P_0)$ whose elements are called *arrows* depicted by $\psi : (P_1, \cdots, P_n) \to P_0$. With this notation $(P_1, \cdots, P_n)$ is the *domain* of ψ, P_0 its *codomain*, and n its *arity* (which is allowed to be 0). The domain and codomain are assumed to be uniquely determined by ψ. There is for each object $P \in \mathcal{P}_0$ a chosen arrow $id_P : P \to P$ called the *identity* at P. There is a specified composition rule: Given $\psi_i : (P_1^i, \cdots, P_{m_i}^i) \to P_i$, $1 \le i \le n$, and an arrow $\psi : (P_1, \cdots, P_n) \to P_0$ their composition is denoted by $\psi \circ (\psi_1, \cdots, \psi_n)$ and has domain $(P_1^1, \cdots, P_{m_1}^1, P_1^2, \cdots, P_{m_2}^2, \cdots, P_1^n, \cdots, P_{m_n}^n)$ and codomain P_0. The composition is to obey the following unit and associativity laws:

- Left unit axiom: $id_P \circ \psi = \psi$
- Right unit axiom: $\psi \circ (id_{P_1}, \cdots, id_{P_n}) = \psi$
- Associativity axiom: the composition

$$\psi \circ (\psi_1 \circ (\psi_1^1, \cdots, \psi_{m_1}^1), \cdots, \psi_n \circ (\psi_1^n, \cdots, \psi_{m_n}^n))$$

is equal to

$$(\psi \circ (\psi_1, \cdots, \psi_n)) \circ (\psi_1^1, \cdots, \psi_{m_1}^1, \cdots, \psi_1^n, \cdots, \psi_{m_n}^n).$$

The morphisms of planar operads are the obvious structure preserving maps. A map of operads will also be referred to as a *functor*.

DEFINITION 2.11. A *symmetric operad* is a planar operad $\mathcal{P}$ together with actions of the symmetric groups in the following sense: for each $n \in \mathbb{N}$, objects $P_0, \cdots, P_n \in \mathcal{P}_0$, and a permutation $\sigma \in \Sigma_n$ a function $\sigma^* : \mathcal{P}(P_1, \cdots, P_n; P_0) \to \mathcal{P}(P_{\sigma(1)}, \cdots, P_{\sigma(n)}; P_0)$. We write $\sigma^*(\psi)$ for the value of the action of σ on $\psi : (P_1, \cdots, P_n) \to P_0$ and demand that for any two permutations $\sigma, \tau \in \Sigma_n$ there holds $(\sigma\tau)^*(\psi) = \tau^*\sigma^*(\psi)$. Moreover, these actions of the permutation groups are to be compatible with compositions in the obvious sense (see [**24, 33**] for more details). *Functors* of symmetric operads $\mathcal{P} \to \mathcal{Q}$ are functors of the underlying planar operads that respect the actions of the symmetric groups.

When dealing with operads we make a distinction between small and large ones according to whether the class of objects is, respectively, a set or a proper class. If more care is needed and size issues become important we implicitly assume working in the formalism of Grothendieck universes ([**6**]) similarly to the way such issues are avoided in category theory. We now obtain the category Ope_π of small planar operads and their functors as well as the category Ope of small symmetric operads and their functors. There is clearly a forgetful functor $Ope \to Ope_\pi$ which has an easily constructed left adjoint $S : Ope_\pi \to Ope$ called the *symmetrization* functor.

REMARK 2.12. We note that our symmetric operads are also called symmetric multicategories (see e.g., [**24**]) as well as symmetric coloured operads. The composition as given above is sometimes called full $\circ$ composition. Using the identities in an operad we can then define what is known as the $\circ_i$ composition as follows. Given an arrow ψ of arity n and $1 \leq i \leq n$ one can compose an arrow φ onto the i-th place of the domain of ψ, provided the object at the i-th place is equal to the codomain of φ, by means of $\psi \circ_i \varphi = \psi \circ (id, \cdots, id, \varphi, id, \cdots id)$ with φ appearing in the i-th place. Some authors consider operads defined in terms of the $\circ_i$ operations rather than the full $\circ$ composition. In the presence of identities there is no essential difference but if identities are not assumed than one obtains a slightly weaker structure called a *pseudo-operad* (see [**32**]). The operads we consider always have identities so that the full $\circ$ and partial $\circ_i$ compositions differ only cosmetically and will be used interchangeably as convenient.

Operads are closely related to categories. Indeed, one trivially sees that a category is an operad where each arrow has arity equal to 1.

A slightly less trivial and more useful fact is the following. Call a symmetric operad *reduced* if it has no 0-ary operations. We denote by $\star$ an operad with one object and only the identity arrow on it.

LEMMA 2.13. *(Slicing lemma for symmetric operads) There is an isomorphism between the category Cat of small categories and the slice category Ope/$\star$. Moreover, there are functors $j_!$: Cat $\to$ Ope and j^* : Ope $\to$ Cat such that j^* is right adjoint to $j_!$. The functor j^* does not preserve pushouts and thus does not have a right adjoint. However, the restriction of j_* to the subcategory of reduced operads does have a right adjoint. Under the isomorphism Cat $\cong$ Ope/$\star$ the functor $j_!$ is the forgetful functor Cat = Ope/$\star$ $\to$ Ope.*

PROOF. We explicitly describe the functors, omitting any details. Given a category $\mathcal{C}$ the operad $j_!(\mathcal{C})$ has $j_!(\mathcal{C})_0 = \mathcal{C}_0$ (here $\mathcal{C}_0$ stands for the class of objects of the category $\mathcal{C}$) and the arrows in $j_!(\mathcal{C})$ are given for $P_0, \cdots, P_n \in j_!(\mathcal{C})_0$ as follows:

$$j_!(\mathcal{C})(P_1, \cdots, P_n; P_0) = \begin{cases} \mathcal{C}(P_1, P_0) & \text{if } n = 1 \\ \emptyset & \text{if } n \neq 1 \end{cases}$$

The composition is the same as in $\mathcal{C}$. The right adjoint j^* is given for an operad $\mathcal{P}$ as follows. $j^*(\mathcal{P})_0 = \mathcal{P}_0$ and the arrows in $j^*(\mathcal{P})$ are given for $C, D \in j^*(\mathcal{P})_0$ by:

$$j^*(\mathcal{P})(C, D) = \mathcal{P}(C; D).$$

The composition is the same as in $\mathcal{P}$. Finally, the functor j_*, right adjoint to the restriction of j^* to reduced operads, is defined for a category $\mathcal{C}$ as follows. $j_*(\mathcal{C})_0 = \mathcal{C}_0$ and the arrows in $j_*(\mathcal{C})$ are given for $P_0, \cdots, P_n \in j_*(\mathcal{C})$ as follows:

$$j_*(\mathcal{C})(P_1, \cdots, P_n; P_0) = \begin{cases} \mathcal{C}(P_1, P_0) & \text{if } n = 1 \\ \{(P_1, \cdots, P_n; P_0)\} & \text{if } n \neq 1 \end{cases}$$

Composition of unary arrows is given as in $\mathcal{C}$. Composition of two arrows at least one of which is not unary is uniquely determined since the hom set of where that arrow is to be found consists of just one object. It is therefore automatic that the composition so defined is associative. □

REMARK 2.14. The construction of the three functors above follows from general abstract nonsense and is related to locally cartesian closed categories. Indeed, if $\mathcal{C}$ is a category with a terminal object $*$, then for any object $A \in \mathcal{C}_0$ the unique arrow $A \to *$ induces a functor between the slice categories $F_! : \mathcal{C}/A \to \mathcal{C}/*$. It is then a general result that $F_!$ has a right adjoint F^* if, and only if, $\mathcal{C}$ admits products with A. Moreover, F^* has a right adjoint F_* if, and only if, A is exponentiable in $\mathcal{C}$. The case we had at hand is when $\mathcal{C}$ is the category of symmetric operads, or its subcategory of reduced symmetric operads, and $A = \star$.

Due to this intimate connection between symmetric operads and categories we will employ category theoretic terminology in the context of symmetric operads. For example, we will refer to morphisms of operads as functors, and feel free to use category theoretic terminology within the 'category part' $j^*(\mathcal{P})$ of an operad $\mathcal{P}$. So the notion of a unary arrow f in $\mathcal{P}$ being, for instance, an isomorphism, a monomorphism, or a split idempotent simply means that f has the same property in the category $j^*(\mathcal{P})$. In this spirit we give the following definition of equivalence of operads.

DEFINITION 2.15. Let $\mathcal{P}$ and $\mathcal{Q}$ be symmetric operads and $F : \mathcal{P} \to \mathcal{Q}$ a functor. We say that F is an *equivalence of operads* if F is fully faithful (which means that it is bijective on each hom-set) and essentially surjective (which means that $j^*(F)$ is an essentially surjective functor of categories).

REMARK 2.16. We make a few remarks to emphasize differences and similarities between the categories *Ope* and *Cat*:

- *Ope* is small complete and small cocomplete.
- There is a unique initial operad which is, of course, equal to $j_!(\emptyset)$.
- For the operad $\star$ above and a terminal category $*$ there holds that $\star \cong j_!(*)$ and $* \cong j^*(\star)$.
- $\star$ is not terminal but is exponentiable in the category of reduced symmetric operads.
- The terminal object in *Ope* is the operad $Comm = j_*(*)$ consisting of one object and one n-ary operation for every $n \in \mathbb{N}$.
- The subobjects of the terminal operad $Comm$ are all of the following form. An operad with one object and for every $n \geq 0$ at most one arrow of arity n such that if an arrow of arity m and an arrow of arity k exist then there is also an arrow of arity $m + k - 1$.

A typical example of category is obtained by fixing some mathematical object and considering the totality of those objects and their naturally occurring morphisms. In many cases these objects also have a notion of 'morphism of several variables' in which case the totality of objects and their multivariable arrows will actually form an operad. One case in which this is guaranteed is the following.

LEMMA 2.17. *Let $(\mathcal{E}, \otimes, I)$ be a symmetric monoidal category and consider for every $x_0, \cdots, x_n \in \mathcal{E}_0$ the set $\hat{\mathcal{E}}(x_1, \cdots, x_n; x_0) = \mathcal{E}(x_1 \otimes \cdots \otimes x_n, x_0)$. With the obvious definitions of composition and identities this construction defines a symmetric operad $\hat{\mathcal{E}}$ with $(\hat{\mathcal{E}})_0 = \mathcal{E}_0$.*

PROOF. The associativity of the composition in $\hat{\mathcal{E}}$ is a result of the coherence in $\mathcal{E}$. □

REMARK 2.18. Certainly not every symmetric operad is obtained in that way from a symmetric monoidal category (e.g., any of the proper subobjects of $Comm$ or any operad of the form $j_!(\mathcal{C})$ for a category $\mathcal{C}$). It is possible to internally characterize those symmetric operads that do arise in that way from symmetric monoidal categories, as is explained in detail in [**16**] and indicated in [**24**].

Another type of category that arises naturally is one that encodes some properties of arrows abstractly. For example, the free-living isomorphism $0 \leftrightarrows 1$ is a category with two distinct objects and, except for the two identities, two other arrows between the objects, each of which is the inverse of the other. A functor from the free-living isomorphism to any category $\mathcal{C}$ corresponds exactly to a choice of an isomorphism in $\mathcal{C}$ and can be seen as the abstract free-living isomorphism becoming concrete in the category $\mathcal{C}$. A similar phenomenon is true in operads, where one readily sees the much greater expressive power of operads compared to categories. Consider for example the terminal operad $Comm$, for which it is straightforward to prove that any functor of operads $Comm \to \hat{\mathcal{E}}$ is the same as a commutative monoid in $\mathcal{E}$, for any symmetric monoidal category $\mathcal{E}$. There is no category $\mathcal{C}$ with the property that functors $\mathcal{C} \to \mathcal{E}$ correspond to commutative monoids in $\mathcal{E}$.

REMARK 2.19. To distinguish between symmetric operads such as $Comm$ thought of as encoding properties of arrows and symmetric operads such as $\hat{\mathcal{E}}$ thought of as environments where operads $\mathcal{P}$ are interpreted concretely we will use letters near $\mathcal{P}$ for abstract symmetric operads and letters near $\mathcal{E}$ for symmetric operads as environments (whether they come from a symmetric monoidal category or not). We will also call symmetric operads $\mathcal{E}$ *environment operads.* The distinction is purely syntactic.

The utility of operads is in their ability to codify quite a wide range of algebraic structures in the way described above. The usual terminology one uses is that of an *algebra* of an operad. The following definition of algebra is more general than the usual one (e.g., [**32, 33**]).

DEFINITION 2.20. Let $\mathcal{P}$ and $\mathcal{E}$ be symmetric operads and consider a functor $F : \mathcal{P} \to \mathcal{E}$. If $F_0 : \mathcal{P}_0 \to \mathcal{E}_0$ is the object part of the functor F we say that F is a *$\mathcal{P}$-algebra* structure on the collection of objects $\{F_0(P)\}_{P \in \mathcal{P}_0}$ in the environment operad $\mathcal{E}$.

Many basic properties of $\mathcal{P}$-algebras are captured efficiently by the introduction of a closed monoidal structure on *Ope*. The appropriate tensor product of symmetric operads is the Boardman-Vogt tensor product which was first introduced in [**5**] for (certain structures that are essentially equivalent to) symmetric operads enriched in topological spaces. The construction is general enough that it can be performed for operads enriched in other monoidal categories and certainly also in the non-enriched case, which is the version we give now.

DEFINITION 2.21. Let $\mathcal{P}$ and $\mathcal{Q}$ be two symmetric operads. Their *Boardman-Vogt tensor product* is the symmetric operad $\mathcal{P} \otimes \mathcal{Q}$ with $(\mathcal{P} \otimes \mathcal{Q})_0 = \mathcal{P}_0 \times \mathcal{Q}_0$ given in terms of generators and relations as follows. For each $Q \in \mathcal{Q}_0$ and each operation $\psi \in \mathcal{P}(P_1, \cdots, P_n; P)$ there is a generator $\psi \otimes Q$ with domain $(P_1, Q), \cdots, (P_n, Q)$ and codomain (P, Q). For each $P \in \mathcal{P}_0$ and an operation $\varphi \in \mathcal{Q}(Q_1, \cdots, Q_m; Q)$ there is a generator $P \otimes \varphi$ with domain $(P, Q_1), \cdots, (P, Q_m)$ and codomain (P, Q). There are five types of relations among the arrows (σ and τ below are permutations whose roles are explained below):

1) $(\psi \otimes Q) \circ ((\psi_1 \otimes Q), \cdots, (\psi_n \otimes Q)) = (\psi \circ (\psi_1, \cdots, \psi_n)) \otimes Q$
2) $\sigma^*(\psi \otimes Q) = (\sigma^* \psi) \otimes Q$
3) $(P \otimes \varphi) \circ ((P \otimes \varphi_1), \cdots, (P \otimes \varphi_m)) = P \otimes (\varphi \circ (\varphi_1, \cdots, \varphi_m))$
4) $\sigma^*(P \otimes \varphi) = P \otimes (\sigma^* \varphi)$
5) $(\psi \otimes Q) \circ ((P_1 \otimes \varphi), \cdots, (P_n \otimes \varphi)) = \tau^*((P \otimes \varphi) \circ ((\psi, Q_1), \cdots, (\psi, Q_m)))$

By the relations above we mean every possible choice of arrows $\psi, \varphi, \psi_i, \varphi_j$ for which the compositions are defined. The relations of type 1 and 2 ensure that for any $Q \in \mathcal{P}_0$, the map $P \mapsto (P, Q)$ naturally extends to a functor $\mathcal{P} \to \mathcal{P} \otimes \mathcal{Q}$. Similarly, the relations of type 3 and 4 guarantee that for each $P \in \mathcal{P}_0$, the map $Q \mapsto (P, Q)$ naturally extends to a functor $\mathcal{Q} \to \mathcal{P} \otimes \mathcal{Q}$. The relation of type 5 can be visualized as follows. The left hand side can be drawn as

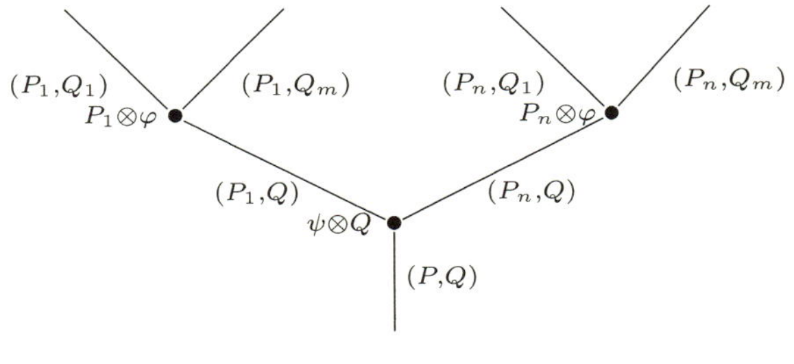

while the right hand side can be drawn as

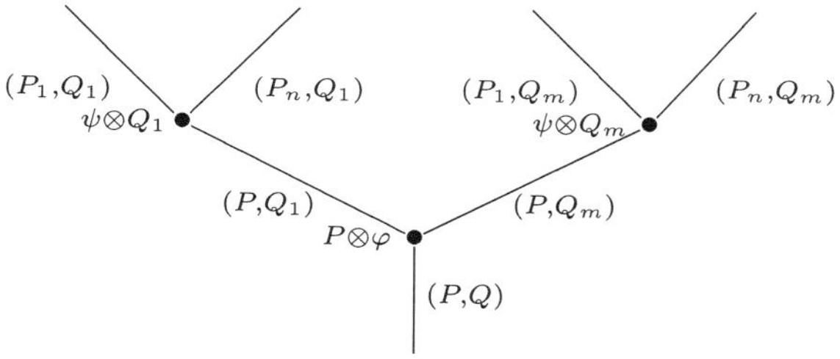

As given, the operations cannot be equated since their domains do not agree. There is however an evident permutation τ that equates the domains and it is that permutation τ that is used in the equation of type 5 above.

THEOREM 2.22. *The category* $(Ope, \otimes, \star)$ *is a symmetric closed monoidal category.*

PROOF. The internal hom operad $[\mathcal{P}, \mathcal{Q}]$ has as objects all morphisms of operads $F : \mathcal{P} \to \mathcal{Q}$ and the arrows with domain $F_1, \cdots, F_n$ and codomain F_0 are analogues of natural transformations as follows. A *natural transformation* α from $(F_1, \cdots, F_n)$ to F_0 is a family $\{\alpha_P\}_{P \in \mathcal{P}_0}$, with $\alpha_P \in \mathcal{Q}(F_1(P), \cdots, F_n(P); F_0(P))$, satisfying the following property. Given any operation $\psi \in \mathcal{P}(P_1, \cdots, P_m; P)$ consider the following diagrams in $\mathcal{Q}$:

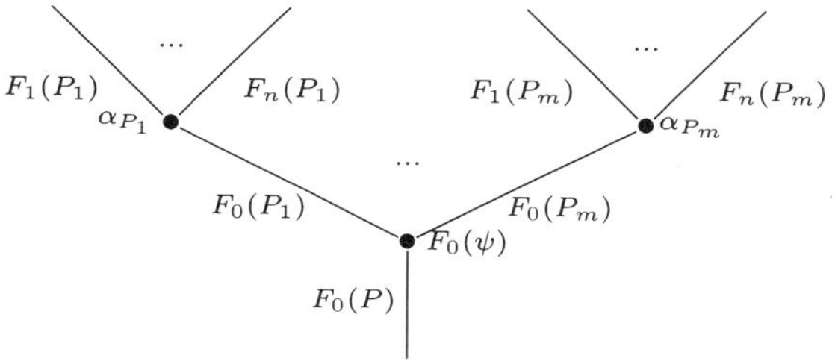

and

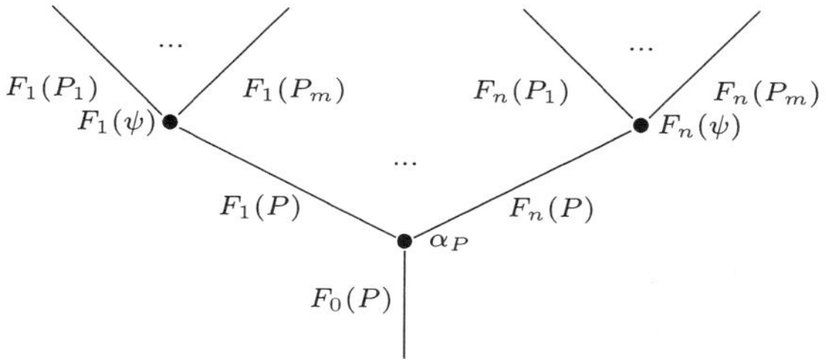

and let φ_1 and φ_2 be their respective compositions. Then $\varphi_2 = \sigma^*(\varphi_1)$, where σ is the evident permutation equating the domain of φ_1 with that of φ_2. The interested reader is referred to [**38**] for more details on horizontal and vertical compositions of natural transformations leading to the construction of the strict 2-category of small operads in which the strict 2-category of small categories embeds. □

We now return to our general notion of $\mathcal{P}$-algebras in $\mathcal{E}$ and notice the very simple result:

LEMMA 2.23. *Let $\mathcal{P}$ and $\mathcal{E}$ by symmetric operads. The internal hom $[\mathcal{P}, \mathcal{E}]$ is rightfully to be called the operad of $\mathcal{P}$-algebras in $\mathcal{E}$ in the sense that the objects of $[\mathcal{P}, \mathcal{E}]$ are the $\mathcal{P}$-algebras in $\mathcal{E}$, the unary arrows are the morphisms of such algebras, and the n-ary arrows are 'multivariable' morphisms of algebras (with 0-ary morphisms thought of as constants).*

It is trivial to verify for example that $[Comm, Set]$ is isomorphic to the operad obtained from the symmetric monoidal category $CommMon(Set)$ of commutative monoids in Set by means of the construction given in Lemma 2.17. Here Set can be replaced by any symmetric monoidal category. This motivates the following definition.

DEFINITION 2.24. Let $\mathcal{E}$ be a symmetric operad and S some notion of an algebraic structure on objects of $\mathcal{E}$ together with a notion of (perhaps multivariable) morphisms between such structures. We call a symmetric operad $\mathcal{P}$ a *classifying* operad for S (in $\mathcal{E}$) if the operad $[\mathcal{P}, \mathcal{E}]$ satisfies that $[\mathcal{P}, \mathcal{E}]_0$ is precisely the set of S-structures in $\mathcal{E}$ and the arrows in $[\mathcal{P}, \mathcal{E}]$ correspond precisely to the notion of morphisms between such structures.

EXAMPLE 2.25. The symmetric operad $Comm$ is a classifying operad for commutative monoids in a symmetric operad $\mathcal{E}$. There is a symmetric operad As that classifies monoids in $\mathcal{E}$ (i.e., an object with an associative binary operation with a unit) which the reader is invited to find. A *magma* is a set together with a binary operation, not necessarily associative, and there is a symmetric operad that classifies magmas. There is also a symmetric operad that classifies non-unital monoids as well as one that classifies non-unital commutative monoids. It is a rather unfortunate fact that there is no symmetric operad that classifies all small categories. However, given a fixed set A consider the category Cat_A of *categories over A*, in which the objects are categories having A as set of objects and where the arrows are functors between such categories whose object part is the identity. Then there is a symmetric operad C_A that classifies categories over A. Similarly, with the obvious definition, there is a symmetric operad O_A that classifies symmetric operads over A.

REMARK 2.26. In general, there can be two non-equivalent operads $\mathcal{P}$ and $\mathcal{Q}$ that classify the same algebraic structure. We will not get into the question of detecting when two symmetric operads have equivalent operads of algebras.

A well-known phenomenon in category theory is the interchangeability of repeated structures. Thus, for example, a category object in Grp is the same as a group object in Cat. With the formalism of symmetric operads we have thus far we can easily prove a whole class of such cases (but in fact not the case just mentioned, since group objects cannot be classified by symmetric operads).

LEMMA 2.27. *Let $\mathcal{P}_1$ and $\mathcal{P}_2$ be two symmetric operads and let $\mathcal{E}$ be an environment operad. Then $\mathcal{P}_1$-algebras in $\mathcal{P}_2$-algebras in $\mathcal{E}$ are the same as $\mathcal{P}_2$-algebras in $\mathcal{P}_1$-algebras in $\mathcal{E}$.*

PROOF. The precise formulation of the lemma is that there is an isomorphism of symmetric operads $[\mathcal{P}_1, [\mathcal{P}_2, \mathcal{E}]] \cong [\mathcal{P}_2, [\mathcal{P}_1, \mathcal{E}]]$. The proof is trivial from the symmetry of the Boardman-Vogt tensor product. □

Consider the operad As that classifies monoids: it has just one object and its arrows of arity n is the set Σ_n of permutations on n-symbols. It is not hard to show that $As \otimes As \cong Comm$ which essentially is Eckman-Hilton duality proving that associative monoids in associative monoids are commutative monoids, except that it is done at the level of classifying operads rather than algebras.

We conclude our review of the basics of operad theory by noting that in the same way that categories can be enriched in a symmetric monoidal category $\mathcal{E}$ (see [**21**]) so can operads be so enriched. With the evident definitions one then obtains the category $Ope(\mathcal{E})$ of all small operads enriched in $\mathcal{E}$.

REMARK 2.28. In the presence of coproducts in $\mathcal{E}$ any non-enriched symmetric operad $\mathcal{P}$ gives rise to an operad $Dis(\mathcal{P})$ enriched in $\mathcal{E}$ in which each hom-object is a coproduct, indexed by the corresponding hom-set in $\mathcal{P}$, of the unit I of $\mathcal{E}$. We will usually refer to $Dis(\mathcal{P})$ as the corresponding discrete operad in $\mathcal{E}$ and call it again $\mathcal{P}$.

Our main interest in symmetric operads is in their use in the theory of homotopy invariant algebraic structures where enrichment plays a vital role. However, before we embark on the subtleties of homotopy invariance we briefly treat the isomorphism invariance property for non-enriched symmetric operads.

2.3. The isomorphism invariance property. It is a triviality that an algebraic structure can be transferred, uniquely, along an isomorphism. To be more precise and to formulate this in the language of operads, let $\mathcal{P}$ and $\mathcal{E}$ be symmetric operads and $F : \mathcal{P} \to \mathcal{E}$ an algebra structure on $\{F_0(P)\}_{P\in\mathcal{P}_0}$. Assume that we are given a family $\{f_P : F_0(P) \to G_0(P)\}_{P\in\mathcal{P}_0}$ of isomorphisms in $\mathcal{E}$. Then there exists a unique $\mathcal{P}$-algebra structure $G : \mathcal{P} \to \mathcal{E}$ on $\{G_0(P)\}_{P\in\mathcal{P}_0}$ for which the family $\{f_P\}_{P\in\mathcal{P}_0}$ forms a natural isomorphism from F to G and thus an isomorphism between the algebras. We call this the *isomorphism invariance property* of algebras.

We can reformulate this property diagrammatically as follows. Let 0 be a one-object symmetric operad with the identity arrow only, and $0 \to (0 \leftrightarrows 1)$ the inclusion $0 \mapsto 0$ into the free-living isomorphism. Then a choice of functor $F : \mathcal{P} \to \mathcal{E}$ is the same as a functor $0 \to [\mathcal{P}, \mathcal{E}]$ while a functor $(0 \leftrightarrows 1) \to [\mathcal{P}, \mathcal{E}]$ can be identified with two functors $\mathcal{P} \rightrightarrows \mathcal{E}$ and a natural isomorphism between them. The set of objects $\mathcal{P}_0$ seen as a category with only identity arrows can be seen as a symmetric operad. One then has the evident inclusion functor $\mathcal{P}_0 \to \mathcal{P}$ which induces a functor $[\mathcal{P}, \mathcal{E}] \to [\mathcal{P}_0, \mathcal{E}]$. The isomorphism invariance property for $\mathcal{P}$-algebras in $\mathcal{E}$ is then the statement that in the following diagram:

$$\begin{array}{ccc} 0 & \xrightarrow{\forall F} & [\mathcal{P},\mathcal{E}] \\ \downarrow & \overset{\exists\alpha}{\nearrow} & \downarrow \\ 0 \leftrightarrows 1 & \xrightarrow[\forall\{F_0(P)\to G_0(P)\}_{P\in\mathcal{P}_0}]{} & [\mathcal{P}_0,\mathcal{E}] \end{array}$$

the diagonal filler exists (and is unique) for any functor $F : \mathcal{P} \to \mathcal{E}$ and any family of isomorphisms $\{F_0(P) \to G_0(P)\}_{P\in\mathcal{P}_0}$.

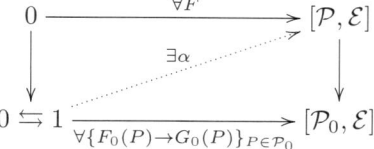

In the formalism of Quillen model structures there is a conceptual way to see why a lift in the diagram above exists. To present it we recall that a functor $F : \mathcal{C} \to \mathcal{D}$ of categories is an *isofibration* if it has the right lifting property with respect to the inclusion $0 \to (0 \leftrightarrows 1)$. Similarly, a functor $F : \mathcal{P} \to \mathcal{Q}$ of symmetric operads is an *isofibration* of symmetric operads if it has the right lifting property with respect to the same inclusion $0 \to (0 \leftrightarrows 1)$ with each category seen as an operad. Equivalently, $F : \mathcal{P} \to \mathcal{Q}$ is an isofibration (of operads) if, and only if, $j^*(F)$ is an isofibration of categories. We now recall the operadic Quillen model structure on symmetric operads.

THEOREM 2.29. *The category Ope of symmetric operads with the Boardman-Vogt tensor product admits a cofibrantly generated closed monoidal model structure in which the weak equivalences are the operadic equivalences, the cofibrations are those functors $F : \mathcal{P} \to \mathcal{Q}$ such that the object part of F is injective, and the fibrations are the isofibrations. All operads are fibrant and cofibrant. The Quillen model structure induced on $Cat \cong Ope/\star$ is the categorical one (also known as the 'folk' or 'natural' model structure).*

PROOF. A direct verification of the axioms of a model category is not difficult and not too tedious. Further details can be found in [**38**]. □

Now, in the diagram above the left vertical arrow is a trivial cofibration and the right vertical arrow is, by the monoidal model structure axiom, a fibration and hence the lift exists. We summarize the above discussion:

FACT 2.30. *The notion of algebras of operads is internalized to the category Ope by it being closed monoidal with respect to the Boardman-Vogt tensor product. The isomorphism invariance property of algebras is captured by the operadic Quillen model structure and its compatibility with the Boardman-Vogt tensor product.*

2.4. The homotopy invariance property. In the presence of homotopy in $\mathcal{E}$ one can ask if a stronger property than the isomorphism invariance property holds. Namely, if one merely asks for the arrows $F_0(P) \to G_0(P)$ to be weak equivalences instead of isomorphisms is it still possible to transfer the algebra structure? A simple example is when one considers a topological monoid X and a topological space Y together with continuous mappings $f : X \to Y$ and $g : Y \to X$ such that $f \circ g$ and $g \circ f$ are homotopic to the respective identities. It is evident that if f and g are not actual inverses of each other then the monoid structure on X will not, in general, induce a monoid structure on Y. The question as to what kind of structure is induced goes back to Stasheff's study of H-spaces and his famous associahedra that are used to describe the kind of structure that arises [**37**]. The more general problem for algebraic structures on topological spaces can be addressed by using enriched symmetric operads as is done by Boardman and Vogt in [**5**]. Their techniques and results were generalized by Berger and Moerdijk in a series of three papers [**1, 2, 3**] and below we present an expository account of the constructions we will need.

First we give a slightly vague definition of the homotopy invariance property. In the context of dendroidal sets below we will give a precise definition that is completely analogous to the definition of the isomorphism invariance property.

DEFINITION 2.31. Let $\mathcal{E}$ be a symmetric monoidal model category and $\mathcal{Q}$ a symmetric operad enriched in $\mathcal{E}$. We say that $\mathcal{Q}$-algebras have the *homotopy invariance property* if given an algebra $F : \mathcal{Q} \to \hat{\mathcal{E}}$ on $\{F_0(Q)\}_{Q \in \mathcal{Q}_0}$ and a family $\{f_Q : F_0(Q) \to G_0(Q)\}_{Q \in \mathcal{Q}_0}$ of weak equivalences in $\mathcal{E}$ (with perhaps some extra conditions) there exists an essentially unique $\mathcal{Q}$-algebra structure $G : \mathcal{Q} \to \hat{\mathcal{E}}$ on $\{G_0(Q)\}_{Q \in \mathcal{Q}_0}$.

It is evident that an arbitrary symmetric operad $\mathcal{P}$ need not have the homotopy invariance property and the problem of sensibly replacing $\mathcal{P}$ by another operad $\mathcal{Q}$ that does have this property is referred to as the problem of finding the up-to-homotopy version of the algebraic structure classified by $\mathcal{P}$. To make this notion precise we recall that in [**1, 2, 3**] Berger and Moerdijk establish the following result.

THEOREM 2.32. *Let $\mathcal{E}$ be a cofibrantly generated symmetric monoidal model category. Under mild conditions the category $Ope(\mathcal{E})_A$ of symmetric operads enriched in $\mathcal{E}$ with fixed set of objects equal to A and whose functors are the identity on all objects admits a Quillen model structure in which the weak equivalences are hom-wise weak equivalences and the fibrations are hom-wise fibrations.*

We refer to this model structure as the Berger-Moerdijk model structure on $Ope(\mathcal{E})_A$.

REMARK 2.33. The Berger-Moerdijk model structure on symmetric operads over a singleton $A = \{*\}$ (given in [**1**]) settles one of the open problems listed by Hovey in [**18**].

Among the consequences of the model structure Berger and Moerdijk prove the following.

THEOREM 2.34. *If $\mathcal{Q}$ is cofibrant in the Berger-Moerdijk model structure on $Ope(\mathcal{E})_A$ then $\mathcal{Q}$-algebras in $\mathcal{E}$ have, under mild conditions, the homotopy invariance property.*

PROOF. See Theorem 3.5 in [**1**] for more details. □

Thus, the problem of finding the up-to-homotopy version of the algebraic structure classified by a symmetric operad $\mathcal{P}$ enriched in $\mathcal{E}$ reduces to finding a cofibrant replacement $\mathcal{Q}$ of $\mathcal{P}$ in the Berger-Moerdijk model structure on $Ope(\mathcal{E})_{\mathcal{P}_0}$. Of course, a cofibrant replacement always exists just by the presence of the Quillen model structure. However, in order to actually compute with it one needs an efficient construction of it, and this is the aim of the W-construction.

2.4.1. *The original Boardman-Vogt W-construction for topological operads.* The W-construction is a functor $W : Ope(Top) \to Ope(Top)$ equipped with a natural transformation (an augmentation) $W \to id$. A detailed account (albeit in a slightly different language than that of operads) can be found in [**5**] where it first appeared. We give here an expository presentation aiming at explaining the ideas important to us.

For simplicity let us describe the planar version of the W-construction, that is, we describe a functor taking a planar operad enriched in Top to another such planar operad. We now fix a topological planar operad $\mathcal{P}$ and describe the operad $W\mathcal{P}$. The objects of $W\mathcal{P}$ are the same as those of $\mathcal{P}$. To describe the arrow spaces we consider standard planar trees (a tree is planar when it comes with an orientation of the edges at each vertex and *standard* means that a choice was made of a single planar tree of each isomorphism class of planar isomorphisms of planar trees) whose edges are labelled by objects of $\mathcal{P}$ and whose vertices are labelled by arrows of $\mathcal{P}$ according to the rule that the objects labelling the input edges of a vertex are equal (in their natural order) to the input of the operation labelling that vertex. Similarly the object labelling the output of the vertex is the output object of the operation at the vertex. Moreover, each inner edge in such a tree is given a length $0 \leq t \leq 1$. For objects $P_0, \cdots, P_n \in (W\mathcal{P})_0$ let $A(P_1, \cdots, P_n; P_0)$ be the topological space whose underlying set is the set of all such planar labelled trees $\bar{T}$ for which the leaves of $\bar{T}$ are labelled by $P_1, \cdots, P_n$ (in that order) and the root of $\bar{T}$ is labelled by P_0. The topology on $A(P_1, \cdots, P_n; P_0)$ is the evident one induced by the topology of the arrow spaces in $\mathcal{P}$ and the standard topology on the unit interval $[0, 1]$.

The space $W\mathcal{P}(P_1, \cdots, P_n; P_0)$ is the quotient of $A(P_1, \cdots, P_n; P_0)$ obtained by the following identifications. If $\bar{T} \in A(P_1, \cdots, P_n; P_0)$ has an inner edge e whose length is 0 then we identify it with the tree $\bar{T}/e$ obtained from $\bar{T}$ by contracting the edge e and labelling the newly formed vertex by the corresponding $\circ_i$-composition of the operations labelling the vertices at the two sides of e (the other labels are as in $\bar{T}$). Thus pictorially we have that locally in the tree a configuration

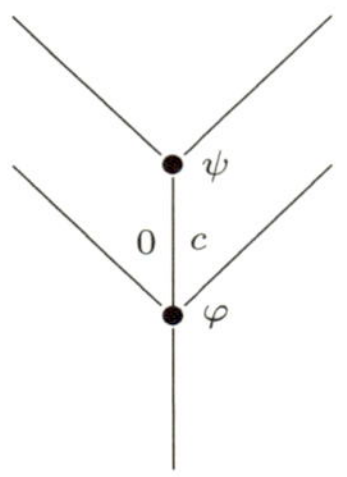

is identified with the configuration

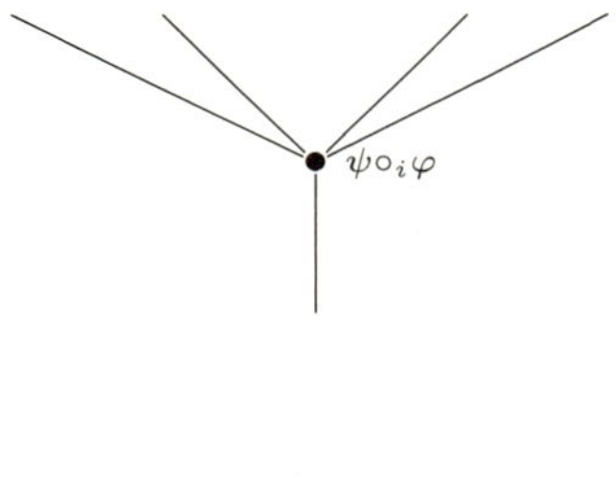

Another identification is in the case of a tree $\bar{S}$ with a unary vertex v labelled by an identity. We identify such a tree with the tree $\bar{R}$ obtained by removing the vertex v and identifying its input edge with its output edge. The length assigned to the new edge is determined as follows. If it is an outer edge then it has no length. If it is an inner edge then it is assigned the maximum of the lengths of s and t (where if either s or t does not have a length, i.e., it is an outer edge, then its length is considered to be 0). The labelling is as in $\bar{S}$ (notice that the label of the newly formed edge is unique since v was labelled by an identity which means that its input and output were labelled by the same object). Pictorially, this identification

identifies the labelled tree

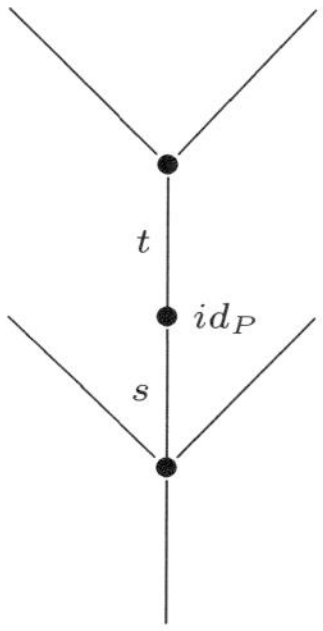

with the tree

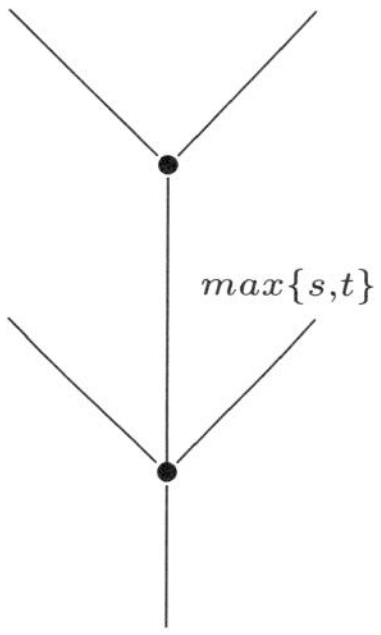

The composition in $W\mathcal{P}$ is given by grafting such labelled trees, giving the newly formed inner edge length 1. The augmentation $W\mathcal{P} \to \mathcal{P}$ is the identity on objects and sends an arrow represented by such a labeled tree to the operation obtained by contracting all lengths of internal edges to 0 and composing in $\mathcal{P}$. We leave the necessary adaptations needed for obtaining the symmetric version of the W-construction to the reader.

EXAMPLE 2.35. Let $\mathcal{P}$ be the planar operad with a single object and a single n-ary operation in each arity $n \geq 1$ and no arrows of arity 0. We consider $\mathcal{P}$ to be a discrete operad in Top. It is easily seen that a functor $\mathcal{P} \to Top$ corresponds to a non-unital topological monoid (we treat this case for simplicity). Let us now calculate the first few arrow spaces in $W\mathcal{P}$. Firstly, $W\mathcal{P}$ too has just one object. We thus use the notation of classical operads, namely $W\mathcal{P}(n)$ for the space of operations of arity n. Clearly $W\mathcal{P}(0)$ is just the empty space. The space $W\mathcal{P}(1)$ consists of labelled trees with one input. Since in such a tree the only possible label at a vertex is the identify, the identification regarding identities implies that $W\mathcal{P}(1)$ is again just a one-point space. In general, since every unary vertex in a labelled tree in $W\mathcal{P}(n)$ can only be labelled by the identity, and those are then identified with trees not containing unary vertices, it suffices to only consider reduced trees, namely trees with no unary vertices. To calculate $W\mathcal{P}(2)$ we need to consider all reduced trees with two inputs, but there is just one such tree, the 2-corolla, and it has no inner edges, thus $W\mathcal{P}(2)$ is also a one-point space. Things become more interesting when we calculate $W\mathcal{P}(3)$. We need to consider reduced trees with three inputs. There are three such trees, namely

The middle tree contributes a point to the space $W\mathcal{P}(3)$. Each of the other trees has one inner edge and thus contributes the interval $[0,1]$ to the space. The only identification to be made is when the length of one of those inner edges is 0, in which case it is identified with the point corresponding to the middle tree. The space $W\mathcal{P}(3)$ is thus the gluing of two copies of the interval $[0,1]$ where we identify both ends named 0 to a single point. The result is then just a closed interval, $[-1,1]$. However, it is convenient to keep in mind the trees corresponding to each point of this interval. Namely, the tree corresponding to the middle point, 0, is the middle tree. With a point $0 < t \leq 1$ corresponds the tree on the right where the length of the inner edge is t, and with a point $-1 \leq -t < 0$ corresponds the tree on the left where its inner edge is given the length t. In this way one can calculate the entire operad $W\mathcal{P}$. It can then be shown that the spaces $\{W\mathcal{P}(n)\}_{n=0}^{\infty}$, reproduce, up to homeomorphism, the Stasheff associahedra. An A_∞-space is then an algebra over $W\mathcal{P}$ and $W\mathcal{P}$ classifies A_∞-spaces and their *strong* morphisms.

2.4.2. *The Berger-Moerdijk generalization of the W-construction to operads enriched in a homotopy environment.* Observe that in the W-construction given above one can construct the space $W\mathcal{P}(P_1, \cdots, P_n; P_0)$ as follows. For each labelled planar tree $\bar{T}$ as above let $H^{\bar{T}}$ be H^k where k is the number of inner edges in $\bar{T}$ and $H = [0,1]$, the unit interval. Further, for each vertex v of $\bar{T}$ let $\mathcal{P}(v) = \mathcal{P}(x_1, \cdots, x_n; x_0)$ where $x_1, \cdots, x_n$ are (in that order) the inputs of v and x_0 its output. Finally, let $\mathcal{P}(\bar{T})$ be the product of $\mathcal{P}(v)$ where v ranges over the vertices of $\bar{T}$. Now, The space $A(P_1, \cdots, P_n; P_0)$ constructed above is homeomorphic to $\coprod_{\bar{T}}(H^{\bar{T}} \times \mathcal{P}(\bar{T}))$ where $\bar{T}$ varies over all labelled standard planar trees $\bar{T}$ whose leaves are labelled by $P_1, \cdots, P_n$ and whose root is labelled by P_0. The identifications that are then made to construct the space $W\mathcal{P}(P_1, \cdots, P_n; P_0)$ are completely determined by the combinatorics of the various trees $\bar{T}$. This observation is the key to generalizing the W-construction to symmetric operads in monoidal model categories $\mathcal{E}$ other than *Top* and is carried out in [**2, 3**]. What is needed is a suitable replacement for the unit interval $[0,1]$ used above to assign lengths to the inner edges of the trees. Such a replacement is the notion of an interval object in a monoidal model category $\mathcal{E}$ given in [**2**].

DEFINITION 2.36. Let $\mathcal{E}$ be a symmetric monoidal model category $\mathcal{E}$ with unit I. An *interval* object in $\mathcal{E}$ (see Definition 4.1 in [**2**]) is a factorization of the codiagonal $I \coprod I \to I$ into a cofibration $I \coprod I \to H$ followed by a weak equivalence $\epsilon : H \to I$ together with an associative operation $\vee : H \otimes H \to H$ which has a neutral element, an absorbing element, and for which ϵ is a counit. For convenience, when an interval element is chosen we will refer to $(\mathcal{E}, H)$ as a *homotopy environment.*

Relevant examples to our presentation are the ordinary unit interval in Top with the standard model structure (with $x \vee y = max\{x, y\}$) and the free-living isomorphism $0 \leftrightarrows 1$ in Cat with the categorical model structure.

In such a setting the topological W-construction can be mimicked by gluing together objects $H^{\otimes_k}$ instead of cubes $[0,1]^k$. This is done in detail in [**2**], to which the interested reader is referred. We thus obtain a functor $W_H : Ope(\mathcal{E}) \to Ope(\mathcal{E})$ for any homotopy environment $\mathcal{E}$. Usually we will just write W instead of W_H, which is quite a harmless convention since Proposition 6.5 in [**2**] guarantees that under mild conditions a different choice of interval object yields essentially equivalent W-constructions.

EXAMPLE 2.37. Consider the category Cat with the categorical model structure. In this monoidal model category we can choose the category H to be the free-living isomorphism $0 \leftrightarrows 1$ as interval object, with the obvious structure maps. Let us again consider the planar operad $\mathcal{P}$ classifying non-unital associative monoids, this time as a discrete operad in Cat. To calculate $W\mathcal{P}(n)$ we should again consider labelled standard planar trees with lengths. The same argument as above implies that we should only consider reduced trees, and a similar calculation shows that $W\mathcal{P}(n)$ is a one-point category for $n = 1, 2$. Now, to calculate $W\mathcal{P}(3)$ we again consider the three trees as given above. This time the middle tree contributes the category $H^0 = I$. Each of the other trees contributes the category H. The identifications identify the object named 0 in each copy of H with the unique object of I. The result is a contractible category with three objects. In general, the category $W\mathcal{P}(n)$ is a contractible category with $tr(n)$ objects, where $tr(n)$ denotes the number of reduced standard planar trees with n leaves. The composition in $W\mathcal{P}$ is given by grafting of such trees. The operad $W\mathcal{P}$ classifies unbiased monoidal categories and strict monoidal functors (an unbiased monoidal category is a category with an n-ary multiplication functor for each $n \geq 0$ together with some coherence conditions. See [**24**] for more details as well as a discussion about the equivalence of such categories and ordinary weak monoidal categories).

The generalized Boardman-Vogt W-construction thus provides a computationally tractable way to classify weak algebras for a wide variety of structures in a homotopy environment. However, $W\mathcal{P}$ tends to classify weak $\mathcal{P}$-algebra with their strong morphisms and not with their weak morphisms. Indeed, for some fixed homotopy environment $\mathcal{E}$ assume that $Ope(\mathcal{E})$ is closed monoidal with respect to a Boardman-Vogt type tensor product. If we now consider for a symmetric operad $\mathcal{P} \in Ope(\mathcal{E})_0$ the internal hom $[W\mathcal{P}, \mathcal{E}]$ then the elements of $[W\mathcal{P}, \mathcal{E}]_0$ are precisely the weak $\mathcal{P}$-algebras in $\mathcal{E}$. However, a unary arrow in $[W\mathcal{P}, \mathcal{E}]$ corresponds to a map of symmetric operads $(W\mathcal{P}) \otimes [1] \to \mathcal{E}$ (where $[1]$ is the operad $0 \to 1$ considered as a discrete operad in $\mathcal{E}$). This already shows that the notion one gets is of strong (because W does not act on $[1]$) morphisms between weak (because W does act on $\mathcal{P}$) $\mathcal{P}$-algebras.

2.5. Weak maps between weak algebras. Luckily, to arrive at the right notion of weak morphisms between weak algebras no extra work is needed. Following on the observation above we make the following definition.

DEFINITION 2.38. Let $\mathcal{P}$ be an operad in Set and $\mathcal{E}$ a homotopy environment. A *weak $\mathcal{P}$-algebra* in $\mathcal{E}$ is a functor of symmetric $\mathcal{E}$-enriched operads $W(\mathcal{P}) \to \hat{\mathcal{E}}$. A *weak map* between up-to-homotopy $\mathcal{P}$-algebras in $\mathcal{E}$ is a functor of symmetric $\mathcal{E}$-enriched operads $W(\mathcal{P} \otimes [1]) \to \hat{\mathcal{E}}$.

An obvious question now is whether the collection of all weak $\mathcal{P}$-algebras and their weak maps forms a category. The answer is that they usually do not. A simple example is provided by A_∞-spaces where it is known that weak A_∞-maps do not compose associatively. The theory so far already suggests a solution to that problem. We denote by $[n]$ the operad $0 \to 1 \to \cdots \to n$ seen as a discrete operad in $\mathcal{E}$. For a symmetric operad $\mathcal{P}$ in Set consider the symmetric operad $\mathcal{P} \otimes [n]$. An algebra for such a symmetric operad is easily seen to be a sequence $X_0, \cdots, X_n$ of $\mathcal{P}$-algebras together with weak $\mathcal{P}$-algebra maps:

$$X_0 \to X_1 \to \cdots \to X_n$$

and all their possible compositions.

PROPOSITION 2.39. *Let $\mathcal{P}$ be a symmetric operad in Set and $\mathcal{E}$ a homotopy environment. For each $n \geq 0$ let X_n be the set of maps*

$$W(\mathcal{P} \otimes [n]) \to \hat{\mathcal{E}}$$

of symmetric operads enriched in $\mathcal{E}$. Then the collection $X = \{X_n\}_{n=0}^{\infty}$ can be canonically made into a simplicial set.

PROOF. The proof follows easily by noting that the sequence $\{\mathcal{P} \otimes [n]\}_{n=0}^{\infty}$ is a cosimplicial object in Ope. □

DEFINITION 2.40. We refer to the simplicial set constructed above as the *simplicial set of weak $\mathcal{P}$-algebras* in $\mathcal{E}$ and denote it by $wAlg[\mathcal{P}, \mathcal{E}]$.

Recall that for strict algebras one could easily iterate structures simply by considering $[\mathcal{P}, [\mathcal{P}, \mathcal{E}]]$ which are classified by $\mathcal{P} \otimes \mathcal{P}$. Our journey into weak algebras in a homotopy environment $\mathcal{E}$ led us to the formation of the simplicial set $wAlg[\mathcal{P}, \mathcal{E}]$ with the immediate drawback that we cannot, at least not in any straightforward manner, iterate. This problem disappears in the dendroidal setting, as we will see below, and is one of the technical advantages of dendroidal sets over enriched operads in the study of weak algebraic structures.

3. Dendroidal sets - a formalism for weak algebras

We now return to non-enriched symmetric operads and introduce the category of dendroidal sets, which is the natural category in which to define nerves of symmetric operads. The category of dendroidal sets is a presheaf category on the dendroidal category Ω and as such one might expect it to be adequate only for the study of non-enriched symmetric operads. However, we will see that it is in fact versatile enough to treat enriched operads quite efficiently by means of the homotopy coherent nerve construction. We do mention that for weak algebraic structures in certain homotopy environments (such as differentially graded vector spaces) dendroidal sets are inappropriate. One might then consider dendroidal objects instead of dendroidal sets as is explained in [**34**], which also contains all of the results below.

3.1. The dendroidal category Ω. To define the dendroidal category Ω recall the definition of symmetric rooted trees given above. It is evident that any such tree T can be thought of as a picture of a symmetric operad $\Omega(T)$: The objects of $\Omega(T)$ are the edges of T and the arrows are freely generated by the vertices of T. In more detail, consider the tree T given by

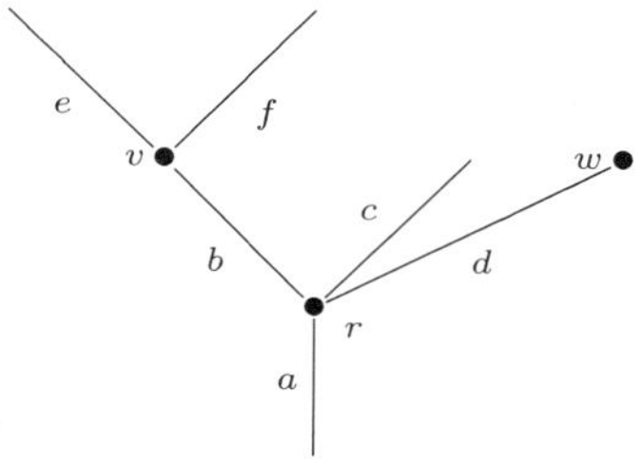

then $\Omega(T)$ has six objects, $a, b, \cdots, f$ and the following generating operations:

$$r \in \Omega(T)(b, c, d; a),$$

$$w \in \Omega(T)(-; d)$$

and

$$v \in \Omega(T)(e, f; b).$$

The other operations are units (such as $1_b \in \Omega(T)(b; b)$), arrows obtained freely by the Σ_n actions, and formal compositions of such arrows.

DEFINITION 3.1. Fix a countable set X. The *dendroidal category* Ω has as objects all symmetric rooted trees T whose edges $E(T)$ satisfy $E(T) \subseteq X$. The arrows $S \to T$ in Ω are arrows $\Omega(S) \to \Omega(T)$ of symmetric operads.

REMARK 3.2. The role of the set X above should be thought of as the role variables play in predicate calculus. The edges are only there to be carriers of symbols and countably many such carriers will always be enough. Of course, another choice of X would result in an isomorphic category. Note that the dendroidal category Ω is thus small (in fact is itself countable).

Recall the linear trees L_n and that the simplicial category Δ is a skeleton of the category of finite linearly ordered sets and order preserving maps. The subcategory of Ω spanned by all trees of the form L_n with $n \geq 0$ is easily seen to be equivalent to the simplicial category Δ. In fact, Δ can be recovered from Ω in a more useful way.

PROPOSITION 3.3. *(Slicing lemma for the dendroidal category) The simplicial category Δ is obtained (up to equivalence) from the dendroidal category Ω by slicing over the linear tree with one edge and no vertices: $\Delta \cong \Omega/L_0$.*

We now describe several types of arrows that generate all of the arrows in Ω. Let T be a tree and v a vertex of valence 1 with $in(v) = e$ and $out(v) = e'$. Consider the tree T/v, obtained from T by deleting the vertex v and the edge e', pictured

locally as

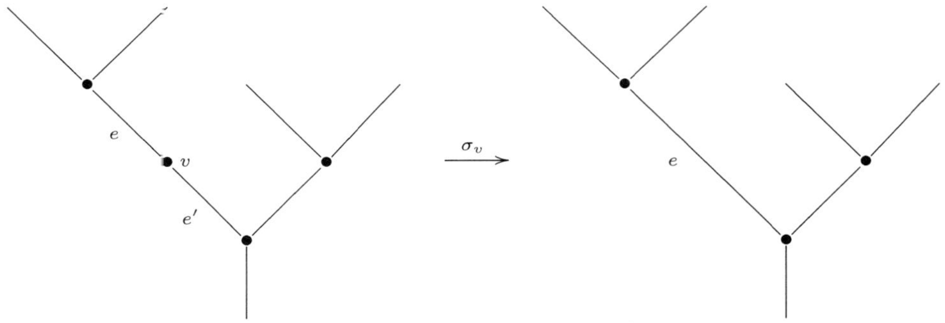

There is then a map in Ω, denoted by $\sigma_v : T \to T/v$, which sends e and e' in T to e in T/v. An arrow in Ω of this kind is called a *degeneracy.*

Consider now a tree T and a vertex v in T with exactly one inner edge attached to it. One can obtain a new tree T/v by deleting v and all the outer edges attached to it to obtain, by inclusion of edges, the arrow $\partial_v : T/v \to T$ in Ω called an *outer face.* For example,

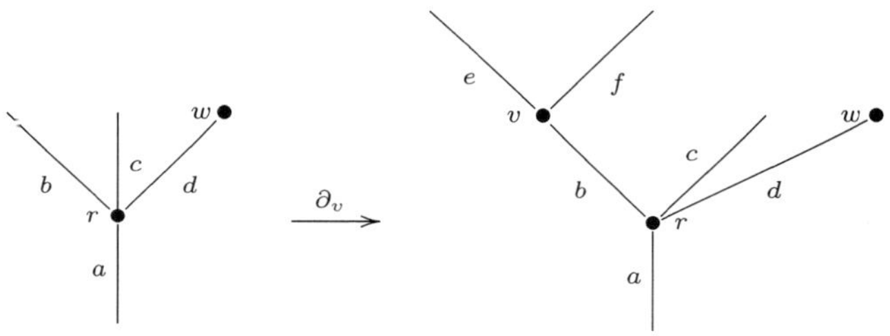

and (to emphasize that it is sometimes possible to remove the root of the tree T)

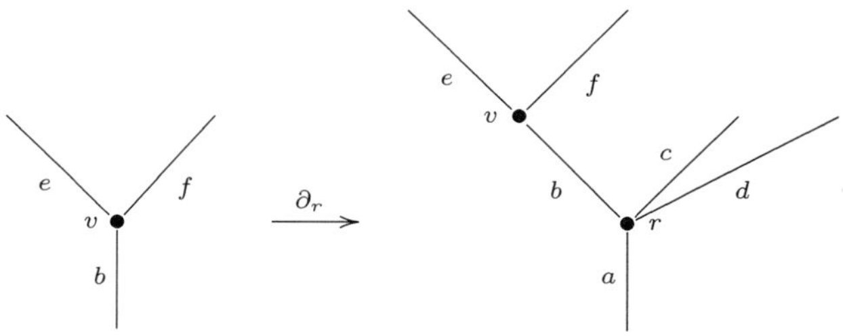

are both outer faces.

Given a tree T and an inner edge e in T, one can obtain a new tree T/e by contracting the edge e. One then obtains, by inclusion of edges, the map $\partial_e : \Omega(T/e) \to \Omega(T)$ in Ω called an *inner face.* For example,

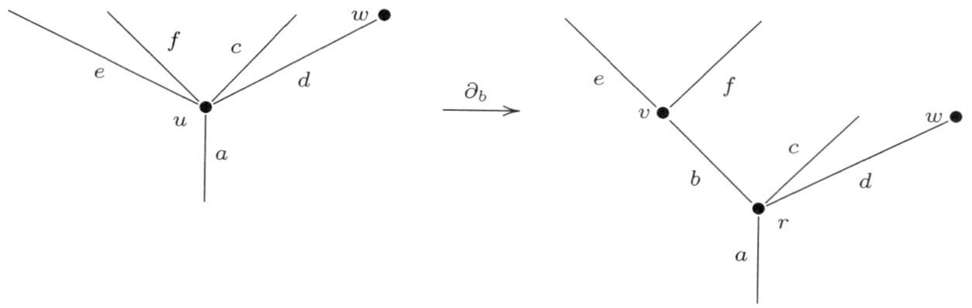

THEOREM 3.4. *Any map* $T \xrightarrow{f} T'$ *in* Ω *factors uniquely as* $f = \varphi\pi\delta$, *where* δ *is a composition of degeneracy maps,* π *is an isomorphism, and* φ *is a composition of (inner and outer) face maps.*

This result generalizes the familiar simplicial relations in the definition of a simplicial set.

3.1.1. *The category of dendroidal sets.*

DEFINITION 3.5. The category of *dendroidal sets* is the presheaf category $dSet = Set_{\Omega}$. Thus a dendroidal set X consists of a collection of sets $\{X_T\}_{T \in \Omega_0}$ together with various maps between them. An element $x \in X_T$ is called a *dendrex of shape* T, or a T*-dendrex.*

For each tree $T \in \Omega_0$ there is associated the *representable dendroidal set* $\Omega[T] = \Omega(-,T)$ which, by the Yoneda Lemma, serves to classify T dendrices in X via the natural bijection $X_T \cong dSet(\Omega[T], X)$. The functor $\Omega \to Ope$ which sends T to $\Omega(T)$ induces an adjunction $dSet \underset{N_d}{\overset{\tau_d}{\rightleftarrows}} Ope$, of which N_d, called the *dendroidal nerve functor*, is given explicitly, for a symmetric operad $\mathcal{P}$, by

$$N_d(\mathcal{P})_T = Ope(\Omega(T), \mathcal{P}).$$

For linear trees L_n we write somewhat ambiguously X_n instead of X_{L_n}. This is a harmless convention since for any two linear trees L_n and $L_n^{'}$ there is a *unique* isomorphism $L_n \to L_n^{'}$ in Ω. Consider the dendroidal set $\star = \Omega[L_0]$.

LEMMA 3.6. *(Slicing lemma for dendroidal sets) There is an equivalence of categories* $dSet/\star \cong sSet$. *If we identify* $sSet$ *as a subcategory of* $dSet$ *then the forgetful functor* $i_! : sSet \to dSet$ *has a right adjoint* i^* *which itself has a right adjoint* i_*.

PROOF. We omit the details and just remark that the adjunctions mentioned can be obtained (equivalently) in one of two ways. The first is to consider Δ as a subcategory of Ω via an embedding functor $i : \Delta \to \Omega$. This functor i then induces a functor $i^* : dSet \to sSet$ which, from the general theory of presheaf categories (see e.g., [**31**]), has a left adjoint $i_!$ and a right adjoint i_*. The second way to obtain the adjunctions is to use Remark 2.14, with $\mathcal{C} = dSet$ and $A = \Omega[L_0]$. □

PROPOSITION 3.7. *Slicing the adjunction* $dSet \underset{N_d}{\overset{\tau_d}{\rightleftarrows}} Ope$ *over* $\star$ *gives the usual adjunction* $sSet \underset{N}{\overset{\tau}{\rightleftarrows}} Cat$ *with* N *the nerve functor and* τ *the fundamental category functor.*

Proof. The precise meaning of the statement is that denoting a one-object operad with just the identity arrow again by $\star$ and for the dendroidal set $\star = \Omega[L_0]$ one has, by slight abuse of notation, that $N_d(\star) = \star$ and $\tau_d(\star) = \star$; thus the functors N_d and τ_d restrict to the respective slices $dSet/\star$ and $Ope/\star$. Then under the identifications $sSet \cong dSet/\star$ and $Cat \cong Ope/\star$ these restrictions give the nerve functor $N : Cat \to sSet$ and its left adjoint τ. □

A general rule of thumb is that any definition or theorem of dendroidal sets will yield, by slicing over $\star = \Omega[L_0]$, a corresponding definition or theorem of simplicial sets. A similar principle is true for operads and categories. We will loosely refer to this process as 'slicing' and say, in the example above for instance, that the usual nerve functor of categories is obtained by slicing the dendroidal nerve functor.

Definition 3.8. Let X and Y be two dendroidal sets. Their tensor product is given by the colimit

$$X \otimes Y = \lim_{\Omega[T]\to X, \Omega[S]\to Y} N_d(\Omega(T) \otimes \Omega(S)),$$

Here we use the canonical expression of a presheaf as a colimit of representables.

As an example of our convention about slicing we mention that slicing the tensor product of dendroidal sets yields the cartesian product of simplicial sets. Note however, that the tensor product in $dSet$ is not the cartesian product.

Theorem 3.9. *The category $dSet$ with the tensor product defined above is a closed monoidal category.*

Proof. This follows by general abstract nonsense. The internal hom is given for two dendroidal sets X and Y by

$$[X, Y]_T = dSet(X \otimes \Omega[T], Y).$$

□

Slicing this theorem proves that $sSet$ is cartesian closed with the usual formula for the internal hom.

Theorem 3.10. *In the diagram*

$$\begin{array}{ccc} Cat & \underset{j^*}{\overset{j_!}{\rightleftarrows}} & Ope \\ {\scriptstyle\tau}\uparrow\downarrow{\scriptstyle N} & & {\scriptstyle\tau_d}\uparrow\downarrow{\scriptstyle N_d} \\ sSet & \underset{i^*}{\overset{i_!}{\rightleftarrows}} & dSet \end{array}$$

all pairs of functors are adjunctions with the left adjoint on top or to the left. Furthermore, the following canonical commutativity relations hold:

$$\tau N \cong id$$
$$\tau_d N_d \cong id$$
$$i^* i_! \cong id$$
$$j^* j_! \cong id$$
$$j_! \tau \cong \tau_d i_!$$
$$N j^* \cong i^* N_d$$
$$i_! N \cong N_d j_!.$$

If we consider the cartesian structures on Cat and sSet, the Boardman-Vogt tensor product on Ope, and the tensor product of dendroidal sets then the four categories are symmetric closed monoidal categories and the functors $i_!, N, \tau, j_!$ and τ_d are strong monoidal.

REMARK 3.11. The dendroidal nerve functor N_d is not monoidal, a fact that plays a vital role in the applicability of dendroidal sets to iterated weak algebraic structures, as we will see below.

We do have the following property.

PROPOSITION 3.12. *For symmetric operads $\mathcal{P}$ and $\mathcal{Q}$ there is a natural isomorphism*

$$\tau_d(N_d(\mathcal{P}) \otimes N_d(\mathcal{Q})) \cong \mathcal{P} \otimes \mathcal{Q}.$$

LEMMA 3.13. *The dendroidal nerve functor commutes with internal Homs in the sense that for any two operads $\mathcal{P}$ and $\mathcal{Q}$ we have*

$$N_d([\mathcal{P}, \mathcal{Q}]) \cong [N_d(\mathcal{P}), N_d(\mathcal{Q})].$$

Moreover, for simplicial sets X and Y we have

$$[i_!(X), i_!(Y)] \cong i_!([X, Y]).$$

The proofs of these results are not hard.

3.2. Algebras in the category of dendroidal sets. We again introduce a syntactic difference between dendroidal sets thought of as encoding structure and dendroidal sets as environments to interpret structures in.

DEFINITION 3.14. Let E and X be dendroidal sets. The dendroidal set $[X, E]$ is called the dendroidal set of *X-algebras* in E. An element in $[X, E]_{L_0}$ is called an *X-algebra* in E. An element of $[X, E]_{L_1}$ is called a *map of X-algebras in E*.

Let us first note that this definition extends the notion of $\mathcal{P}$-algebras in $\mathcal{E}$ for symmetric operads in the sense that for symmetric operads $\mathcal{P}$ and $\mathcal{E}$ there is a natural isomorphism

$$[N_d(\mathcal{P}), N_d(\mathcal{E})] \cong N_d([\mathcal{P}, \mathcal{E}]).$$

Indeed, this is just the statement that N_d commutes with internal Homs.

We thus see that the dendroidal nerve functor embeds *Ope* in *dSet* in such a way that the notion of algebras is retained and in both cases is internalized in the form of an internal Hom with respect to a suitable tensor product. We now wish to study homotopy invariance of algebra structures in a dendroidal set E. The first step is to specify those arrows along which such algebras are to be invariant.

Recall that for symmetric operads the diagram

$$\begin{array}{ccc} 0 \to 1 & \xrightarrow{f} & \mathcal{P} \\ \downarrow & \nearrow_{(f,g)} & \\ 0 \leftrightarrows 1 & & \end{array}$$

(where the vertical arrow is the inclusion of the free-living arrow into the free-living isomorphism) admits a lift precisely when the arrow f admits an inverse g. Taking the dendroidal nerve of this diagram and replacing in it $N_d(\mathcal{P})$ by an arbitrary dendroidal set X we arrive at the following definition.

DEFINITION 3.15. Let X be a dendroidal set. An *equivalence* is a dendrex $x : \Omega[L_1] \to X$ such that in the diagram

$$\begin{array}{ccc} \Omega[L_1] & \xrightarrow{x} & X \\ \downarrow & \nearrow_{\hat{x}} & \\ N_d(0 \leftrightarrows 1) & & \end{array}$$

a lift $\hat{x}$ exists.

Note, that since $\Omega[L_1] \cong i_!(\Delta[1])$ and $N_d(0 \leftrightarrows 1) = i_!(N(0 \leftrightarrows 1))$, by adjunction the dendrex $x : \Omega[L_1] \to X$ is an equivalence if, and only if, in the corresponding diagram of simplicial sets

$$\begin{array}{ccc} \Delta[1] & \xrightarrow{x} & i^*(X) \\ \downarrow & \nearrow & \\ N(0 \leftrightarrows 1) & & \end{array}$$

a lift exists. Thus, being a weak equivalence in the dendroidal set X is actually a property of the simplicial set $i^*(X)$.

REMARK 3.16. Note that $N(0 \leftrightarrows 1)$ is the simplicial infinite dimensional sphere S^∞ and thus a lift $S^\infty \to i^*(X)$ is a rather complicated object. Intuitively, it is a coherent choice of a homotopy inverse of the simplex $x : \Delta[1] \to i^*(X)$, together with coherent choices of homotopies, homotopies between homotopies, etc. See [**19**] for more details. Note moreover, that an equivalence in X is in some sense as weak as X would allow it to be. If $X = N_d(\mathcal{P})$ then a dendrex $\Omega[L_1] \to X$ is an equivalence if, and only if, the corresponding unary arrow in $\mathcal{P}$ is an isomorphism. We will see below a more refined nerve construction in which equivalences correspond to a notion weaker than isomorphism.

We can now formulate the homotopy invariance property in the language of dendroidal sets. Let X and E be dendroidal sets. We identify, somewhat ambiguously, the set X_η with the dendroidal set $\coprod_{x \in X_\eta} \Omega[\eta]$. Then there is a map of dendroidal sets $X_\eta \to X$ which induces a mapping $[X, E] \to [X_\eta, E]$. Consider now a family $\{f_x\}_{x \in X_\eta}$ where each f_x is an equivalence in X. Then this family can be extended (usually in many different ways) to give a map $\hat{f} : N_d(0 \leftrightarrows 1) \to [X_\eta, E]$.

DEFINITION 3.17. Let X and E be a dendroidal sets. We say that X-algebras in E have the *homotopy invariance property* if for every X-algebra in E, given by $F : X \to E$, and any family $\{f_x\}_{x \in X_\eta}$ and any extension of it to $\hat{f}$ as above that fit into the commutative diagram

$$\begin{array}{ccc} \Omega[\eta] & \xrightarrow{\forall F} & [X, E] \\ \downarrow & \nearrow_{\exists \alpha} & \downarrow \\ N_d(0 \leftrightarrows 1) & \xrightarrow{\hat{f}} & [X_\eta, E] \end{array}$$

a lift α exists.

Intuitively, the lift α consists of two X-algebras in E, the first being F and the second one being obtained by transferring the X-algebra structure given by F along the equivalences $\{f_x\}_{x \in X_\eta}$.

3.3. The homotopy coherent nerve and weak algebras. We now show how dendroidal sets enter the picture in the context of operads enriched in a symmetric closed monoidal model category $\mathcal{E}$ with a chosen interval object (which we call a homotopy environment). Recall then that the Berger-Moerdijk generalization of the Boardman-Vogt W-construction sends a symmetric operad $\mathcal{P}$ enriched in $\mathcal{E}$ to a cofibrant replacement $W\mathcal{P}$. Recall as well that any non-enriched symmetric operad can be seen as a discrete symmetric operad enriched in $\mathcal{E}$.

DEFINITION 3.18. Fix a homotopy environment $\mathcal{E}$. Given a symmetric operad $\mathcal{P}$ enriched in $\mathcal{E}$ its *homotopy coherent dendroidal nerve* is the dendroidal set whose set of T-dendrices is

$$hcN_d(\mathcal{P})_T = Ope(\mathcal{E})(W(\Omega(T)), \mathcal{P})$$

of $\mathcal{E}$-enriched functors between $\mathcal{E}$-enriched operads, where $\Omega(T)$ is seen as a discrete operad enriched in $\mathcal{E}$.

The homotopy coherent dendroidal nerve construction, together with the closed monoidal structure on $dSet$ given above, allows for the internalization of the notion of weak algebras. We illustrate this:

DEFINITION 3.19. Let $\mathcal{P}$ be a non-enriched symmetric operad and $\mathcal{E}$ a homotopy environment. The dendroidal set $[N_d(\mathcal{P}), hcN_d(\hat{\mathcal{E}})]$ is called the dendroidal set of *weak $\mathcal{P}$-algebras in $\mathcal{E}$*. Here we view $\mathcal{E}$ as an operad enriched in itself (since $\mathcal{E}$ is assumed closed) and thus $\hat{\mathcal{E}}$ as a symmetric operad enriched in $\mathcal{E}$ is well-defined.

It can be shown that the L_0 dendrices in $[N_d(\mathcal{P}), hcN_d(\hat{\mathcal{E}})]$ correspond to symmetric operad maps $W(\mathcal{P}) \to \hat{\mathcal{E}}$ and thus are weak $\mathcal{P}$-algebras. Moreover, the L_1 dendrices can be seen to correspond to symmetric operad maps $W(\mathcal{P} \otimes [1]) \to \hat{\mathcal{E}}$ and thus are weak maps of weak $\mathcal{P}$-algebras. We have thus recovered an internalization of weak algebras and their weak maps and can now consider iterated weak algebraic structures completely analogously to the way this can be done in the context of non-enriched symmetric operads. We illustrate how this works in two examples below.

3.4. Application to the study of A_∞-spaces and weak n-categories. Recall that an A_∞-space is an algebra for the topologically enriched operad $W(As)$, where As is the non-enriched symmetric operad that classifies monoids. Let $A = N_d(As)$; then, by definition, $[A, hcN_d(Top)]$ is the dendroidal set of A_∞-spaces and their weak (multivariable) mappings. In the classical definition of A_∞-spaces it is not at all clear how to define n-fold A_∞-spaces. However, we now have a perfectly natural such definition.

DEFINITION 3.20. The dendroidal set nA_∞ of *n-fold A_∞-spaces* is defined recursively as follows. For $n = 1$ we set $1A_\infty = [A, hcN_d(Top)]$ and for $n \geq 1$: $(n+1)A_\infty = [A, nA_\infty]$.

Thus, we obtain at once notions of weak multivariable mappings of n-fold A_∞-spaces. And, since $dSet$ is closed monoidal, we can immediately classify n-fold A_∞-spaces.

PROPOSITION 3.21. *For any $n \geq 1$ the dendroidal set $A^{\otimes n}$ classifies n-fold A_∞-spaces.*

It is at this point not known exactly how n-fold A_∞-spaces relate to n-fold loop spaces. However, the recent work **[11]** of Fiedorowicz and Vogt on interchanging A_∞ and E_n structures is a first step towards a full comparison of the dendroidal and classical approaches.

REMARK 3.22. Note that were the dendroidal nerve functor monoidal our definition of n-fold A_∞-spaces would stabilize at $n = 2$. Indeed, we would then have $A^{\otimes n} = N_d(As)^{\otimes n} = N_d(As^{\otimes n}) = N_d(Comm)$.

A similar application, but technically slightly more complicated, is to obtain an iterative definition of weak n-categories. First notice that the fact that categories, as well as symmetric operads, can be enriched in a symmetric monoidal category $\mathcal{E}$ is a consequence of the ability to in fact enrich in an arbitrary symmetric operad. We leave the details of defining what a category (or operad) enriched in a symmetric operad $\mathcal{E}$ is to the reader and only mention that this is related to the idea of enriching in an fc-multicategory (see **[26]**). We now show how in fact categories (and operads) can be enriched in a dendroidal set. Recall from Example 2.25 that for any set A there is a symmetric operad C_A that classifies categories over A.

Once more, the ability to easily iterate within the category of dendroidal sets naturally leads to a definition of weak n-categories enriched in a dendroidal set X as follows.

DEFINITION 3.23. Let X be a dendroidal set. The dendroidal set $[N_d(\mathcal{C}_A), X]$ is called the dendroidal set of *categories over A enriched in X* and is denoted by $Cat(X)_A$.

It can easily be verified that enriching in the dendroidal nerve of $\hat{\mathcal{E}}$ for $\mathcal{E}$ a symmetric monoidal category agrees with the notion of enrichment in the usual sense.

At this point we would like to collate the various dendroidal sets $Cat(X)_A$ into a single dendroidal sets. There is here a technical difficulty and so as not to interrupt the flow of the presentation we refer the reader to Section 4.1 of **[34]** for the details of the construction. One then obtains the dendroidal set $Cat(X)$ of categories enriched in X. Similarly, using the operad O_A classifying symmetric operads over A, we can obtain the dendroidal set $Ope(X)$ of symmetric operads enriched in X.

DEFINITION 3.24. Let X be a dendroidal set. Let ${}_0Cat(X) = X$ and define recursively ${}_{n+1}Cat(X) = Cat({}_nCat(X))$ for each $n \geq 1$. We call ${}_nCat(X)$ the dendroidal set of *n-categories enriched in X*.

In particular, considering the category Cat with its categorical model structure and taking $X = hcN_d(Cat)$ we obtain for each $n \geq 0$ the dendroidal set ${}_nCat = {}_nCat(X)$, which we call the dendroidal set of weak n-categories. In **[27]** the dendrices in ${}_2Cat(X)_\eta$ and ${}_3Cat(X)_\eta$ are compared with other definitions of weak 2-categories and weak 3-categories to show that the notions are in fact equivalent. The complexity of such comparisons increases rapidly with n and is currently, as is the case with many definitions of weak n-categories (see **[25]** for a survey of such), not settled. We mention that we can also consider categories weakly enriched in other dendroidal sets such as $hcN_d(Top)$ or $hcN_d(sSet)$, where again a full comparison with existing structures is yet to be completed. Of course, we can also consider weak n-operads of various sorts as well.

We conclude this section by considering the Baez-Breen-Dolan stabilization hypothesis for weak n-categories as defined above. With every reasonable definition of weak n-categories there is usually associated a notion of k-monoidal n-categories for every $k \geq 0$. These are weak $(n+k)$-categories having trivial information in all dimensions up to and including k. The stabilization hypothesis is that for fixed n the complexity of these structures stabilizes at $k = n+2$. Given a concrete definition of weak n-categories this hypothesis can be made exact and it becomes a conjecture. In our case we proceed as follows.

DEFINITION 3.25. Let $n \geq 0$ be fixed. For $k \geq 0$ we define recursively the dendroidal set $wCat_k^n$ of *weak k-monoidal n-categories* as follows. For $k = 0$ we set $wCat_0^n = {}_nCat$ and for $k > 0$ we define $wCat_k^n = [A, wCat_{k-1}^n]$, where $A = N_d(As)$. A dendrex of shape η in $wCat_k^n$ is called a *k-monoidal n-category.*

CONJECTURE 3.26. *(The Baez-Breen-Dolan stabilization hypothesis for our notion of n-categories) For a fixed $n \geq 0$, there is an isomorphism of dendroidal sets between $wCat_k^n$ and $wCat_{n+2}^n$ for any $k \geq n+2$.*

4. Dendroidal sets - models for ∞-operads

In this section we show how dendroidal sets are used to model ∞-operads. As very brief motivation for the concepts to follow we first discuss ∞-categories, then we present that part of the theory of dendroidal sets needed to define the Cisinski-Moerdijk model structure on dendroidal sets which establishes dendroidal sets as models for homotopy operads and illustrate some of its consequences. The proofs of the results below can be found in [**8, 35**].

4.1. ∞-categories briefly. We have seen above that in general, weak $\mathcal{P}$-algebras in a homotopy environment $\mathcal{E}$ and their weak maps fail to form a category and that in fact one is immediately led to define the simplicial set of weak algebras $wAlg[\mathcal{P}, \mathcal{E}]$. The failure of this simplicial set to be the nerve of a category is a reflection of composition not being associative. However, the composition of weak maps is associative up to coherent homotopies, a fact which induces some extra structure on the simplicial set $wAlg[\mathcal{P}, \mathcal{E}]$. Boardman and Vogt in [**5**] formulated this extra structure by means of a condition called the restricted Kan condition. To define it recall that a horn in a simplicial set X is a mapping $\Lambda^k[n] \to X$, where $\Lambda^k[n]$ is the union in $\Delta[n]$ of all faces except the one opposite the vertex k. A horn is called *inner* when $0 < k < n$. Boardman and Vogt in [**5**], page 102, define

> A simplicial set X is said to satisfy the restricted Kan condition
> if every inner horn $\Lambda^k[n] \to X$ has a filler.

and so ∞-categories were born. They consequently prove that in the context of topological operads the simplicial set of weak algebras satisfies the restricted Kan condition. Simplicial sets satisfying the restricted Kan condition are extensively studied by Joyal (in [**19, 20**] under the name 'quasicategories') and by Lurie (in e.g., [**28**] under the name '$(\infty, 1)$-categories' or more simply '∞-categories').

There are several ways to model ∞-categories, of which the above restricted Kan condition is one. Three other models are complete Segal spaces, Segal categories, and simplicial categories. For each of these models there is an appropriate

Quillen model structure rendering the four different models Quillen equivalent (see [4] for a detailed survey). By considering dendroidal sets instead of simplicial sets Cisinski and Moerdijk in [9, 10] introduce the analogous dendroidal notions: complete dendroidal Segal spaces, Segal operads, and simplicial operads. Moreover, they establish Quillen model structures for each of these notions, proving they are all Quillen equivalent to a Quillen model structure on dendroidal sets they establish in [8]. All of these model structures and equivalences, upon slicing over a suitable object, reduce to the equivalence of the simplicial based structures mentioned above.

There is yet another approach to ∞-operads, taken by Lurie [29], which defines an ∞-operad to be a simplicial set with extra structure. In Lurie's approach the highly developed theory of simplicial sets and quasicategories is readily available to provide a rich theory of ∞-operads. However, the extra structure that makes a simplicial set into an ∞-operad is quite complicated, rendering working with explicit examples of ∞-operads difficult. The approach via dendroidal sets replaces the relative simplicity of the combinatorics of linear trees by the complexity of the combinatorics of trees which renders existing simplicial theory unusable but offers very many explicit examples of ∞-operads. We believe that this trade-off in complexity will result in these two approaches mutually enriching each other as a future comparisons unfold.

Below, following [8] we give a short presentation of the approach to ∞-operads embodied in the Cisinski-Moerdijk model structure on *dSet* that slices to the Joyal model structure on *sSet* and use this model structure to prove a homotopy invariance property for algebras in *dSet*. From this point on ∞-category means a quasicategory.

4.2. Horns in *dSet*. We first introduce some concepts needed for the definition referring the reader to [34, 35] for more details.

DEFINITION 4.1. Let T be a tree and $\alpha : S \to T$ a face map in Ω. The *α-face* of $\Omega[T]$, denoted by $\partial_\alpha \Omega[T]$, is the dendroidal subset of $\Omega[T]$ which is the image of the map $\Omega[\alpha] : \Omega[S] \to \Omega[T]$. Thus we have that

$$\partial_\alpha \Omega[T]_R = \{\, R \longrightarrow S \xrightarrow{\ \alpha\ } T \mid R \to S \in \Omega[S]_R\}.$$

When α is obtained by contracting an inner edge e in T we denote ∂_α by ∂_e.

Let T be a tree. The *boundary* of $\Omega[T]$ is the dendroidal subset $\partial\Omega[T]$ of $\Omega[T]$ obtained as the union of all the faces of $\Omega[T]$:

$$\partial\Omega[T] = \bigcup_{\alpha \in \Phi_1(T)} \partial_\alpha \Omega[T].$$

where $\Phi_1(T)$, is the set of all faces of T.

DEFINITION 4.2. Let T be a tree and $\alpha \in \Phi_1(T)$ a face of T. The *α-horn* in $\Omega[T]$ is the dendroidal subset $\Lambda^\alpha[T]$ of $\Omega[T]$ which is the union of all the faces of T except $\partial_\alpha \Omega[T]$:

$$\Lambda^\alpha[T] = \bigcup_{\beta \neq \alpha \in \Phi_1(T)} \partial_\beta \Omega[T].$$

The horn is called an *inner horn* if α is an inner face, otherwise it is called an *outer horn*. We will denote an inner horn $\Lambda^\alpha[T]$ by $\Lambda^e[T]$, where e is the contracted inner edge in T that defines the inner face $\alpha = \partial_e : \Omega[T/e] \to \Omega[T]$. A horn in a dendroidal set X is a map of dendroidal sets $\Lambda^\alpha[T] \to X$. It is inner (respectively outer) if the horn $\Lambda^\alpha[T]$ is inner (respectively outer).

REMARK 4.3. It is trivial to verify that these notions for dendroidal sets extend the common ones for simplicial sets in the sense, for example, that for the simplicial horn $\Lambda^k[n] \subseteq \Delta[n]$, the dendroidal set

$$i_!(\Lambda^k[n]) \subseteq i_!(\Delta[n]) = \Omega[L_n]$$

is a horn in the dendroidal sense. Furthermore, the horn $\Lambda^k[n]$ is inner (i.e., $0 < k < n$) if, and only if, the horn $i_!(\Lambda^k[n])$ is inner.

Both the boundary $\partial\Omega[T]$ and the horns $\Lambda^\alpha[T]$ in $\Omega[T]$ can be described as colimits as follows.

DEFINITION 4.4. Let $T_1 \to T_2 \to \cdots \to T_n$ be a sequence of n face maps in Ω. We call the composition of these maps a *subface* of T_n of *codimension* n.

PROPOSITION 4.5. *Let $S \to T$ be a subface of T of codimension 2. The map $S \to T$ decomposes in precisely two different ways as a composition of faces.*

Let $\Phi_i(T)$ be the set of all subfaces of T of codimension i. The proposition implies that for each $\beta : S \to T \in \Phi_2(T)$ there are precisely two face maps $\beta_1 : S \to T_1$ and $\beta_2 : S \to T_2$ that factor β as a composition of face maps. Using these maps we can form two maps γ_1 and γ_2

$$\coprod_{S\to T\in\Phi_2(T)} \Omega[S] \rightrightarrows \coprod_{R\to T\in\Phi_1(T)} \Omega[R]$$

where γ_i $(i = 1, 2)$ has component $\Omega[S] \xrightarrow{\Omega[\beta_i]} \Omega[T_i] \longrightarrow \coprod \Omega[R]$ for each $\beta : S \to T \in \Phi_2(T)$.

LEMMA 4.6. *Let T be a tree in Ω. With notation as above we have that the boundary $\partial\Omega[T]$ is a coequalizer*

$$\coprod_{S\to T\in\Phi_2(T)} \Omega[S] \rightrightarrows \coprod_{R\to T\in\Phi_1(T)} \Omega[R] \to \partial\Omega[T]$$

of the two maps γ_1, γ_2 constructed above.

COROLLARY 4.7. *A map of dendroidal sets $\partial\Omega[T] \to X$ corresponds exactly to a sequence $\{x_R\}_{R\to T\in\Phi_1(T)}$ of dendrices whose faces match, in the sense that for each subface $\beta : S \to T$ of codimension 2 we have $\beta_1^*(x_{T_1}) = \beta_2^*(x_{T_2})$.*

A similar presentation for horns holds as well. For a fixed face $\alpha : S \to T \in \Phi_1(T)$ consider the parallel arrows defined by making the following diagram commute

$$\begin{array}{ccc}
\Omega[S] & \xrightarrow{\beta_1} & \Omega[T_1] \\
\downarrow & & \downarrow \\
\coprod_{\beta:S\to T\in\Phi_2(T)} \Omega[S] & \rightrightarrows & \coprod_{R\to T\neq\alpha\in\Phi_1(T)} \Omega[R] \\
\uparrow & & \uparrow \\
\Omega[S] & \xrightarrow{\beta_2} & \Omega[T_2]
\end{array}$$

where the vertical arrows are the canonical injections into the coproduct and where we use the same notation as above.

LEMMA 4.8. *Let T be a tree in Ω and α a face of T. In the diagram*

$$\coprod_{S\to T\in\Phi_2(T)} \Omega[S] \rightrightarrows \coprod_{R\to T\neq\alpha\in\Phi_1(T)} \Omega[R] \to \Lambda^\alpha[T]$$

the dendroidal set $\Lambda^\alpha[T]$ is the coequalizers of the two maps constructed above.

COROLLARY 4.9. *A horn $\Lambda^\alpha[T] \to X$ in X corresponds exactly to a sequence $\{x_R\}_{R\to T\neq\alpha\in\Phi_1(T)}$ of dendrices that agree on common faces in the sense that if $\beta : S \to T$ is a subface of codimension 2 which factors as*

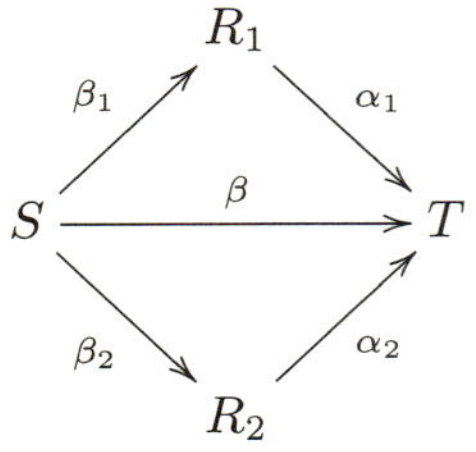

then $\beta_1^*(x_{R_1}) = \beta_2^*(x_{R_2})$.

REMARK 4.10. In the special case where the tree T is linear we obtain the equivalent results for simplicial sets. Namely, the presentation of the boundary $\partial\Delta[n]$ and of the horn $\Lambda^k[n]$ as colimits of standard simplices, and the description of a horn $\Lambda^k[n] \to X$ in a simplicial set X (see [**14**]).

We are now able to define the dendroidal sets that model ∞-operads.

DEFINITION 4.11. A dendroidal set X is an ∞-*operad* if every inner horn $h : \Lambda^e[T] \to X$ has a filler $\hat{h} : \Omega[T] \to X$ making the diagram

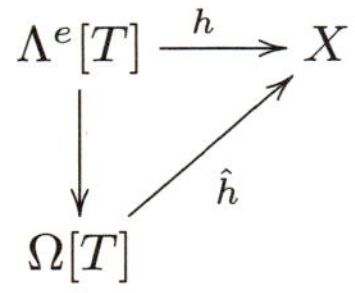

commute.

The following relation between ∞-categories and ∞-operads is trivial to prove:

PROPOSITION 4.12. *If X is an ∞-category then $i_!(X)$ is an ∞-operad. If Y is an ∞-operad then $i^*(Y)$ is an ∞-category.*

It is not hard to see that given any symmetric operad $\mathcal{P}$ its dendroidal nerve $N_d(\mathcal{P})$ is an ∞-operad. In fact we can characterize those dendroidal sets occurring as nerves of operads as follows.

DEFINITION 4.13. An ∞-operad X is called *strict* if any inner horn in X as above has a unique filler.

LEMMA 4.14. *A dendroidal set X is a strict ∞-operad if, and only if, there is an operad $\mathcal{P}$ such that $N_d(\mathcal{P}) \cong X$.*

A family of examples of paramount importance of ∞-operads are given by the following. Recall that when $\mathcal{E}$ is a symmetric monoidal model category a symmetric operad $\mathcal{P}$ enriched in $\mathcal{E}$ is called *locally fibrant* if each hom-object in $\mathcal{P}$ is fibrant in $\mathcal{E}$.

THEOREM 4.15. *Let $\mathcal{P}$ be a locally fibrant symmetric operad in $\mathcal{E}$, where $\mathcal{E}$ is a homotopy environment. The homotopy coherent nerve $hcN_d(\mathcal{P})$ is an ∞-operad.*

4.3. The Cisinski-Moerdijk model category structure on ***dSet*****.** The objective of this section is to present the Cisinski-Moerdijk model structure on *dSet*. All of the material in this section is taken from [**8**], to which the reader is referred to for more information and the proofs. In this model structure ∞-operads are the fibrant objects and it is closely related to the operadic model structure on *Ope* and to the Joyal model structure on *sSet*.

We note immediately that the Cisinski-Moerdijk model structure is not a Cisinski model structure (i.e., a model structure on a presheaf category such that the cofibrations are precisely the monomorphisms) due to a technical complication that prevents the direct application of the techniques developed in [**7**]. Indeed, the cofibrations in the model structure are the so-called *normal monomorphisms.*

DEFINITION 4.16. A monomorphism of dendroidal sets $f : X \to Y$ is *normal* if for every dendrex $t \in Y_T$ that does not factor through f the only isomorphism of T that fixes t is the identity.

An important property of dendroidal sets, proved in [**35**], is the following.

THEOREM 4.17. *Let X be a normal dendroidal set (i.e., $\emptyset \to X$ is normal) and Y an ∞-operad. The dendroidal set $[X, Y]$ is again an ∞-operad.*

PROOF. The proof uses the technique of anodyne extensions, as commonly used in the theory of simplicial sets (e.g., [**13, 14**]), suitably adapted to dendroidal sets. Technically though, the dendroidal case is much more difficult. For simplicial sets there is a rather simple description of the non-degenerate simplices of $\Delta[n] \times \Delta[k]$. But for trees S and T a similar description of the non-degenerate dendrices of $\Omega[S] \otimes \Omega[T]$ is given by the so called poset of percolation trees associated with S and T. Complete details can be found in [**35**] and we just briefly illustrate the construction for the trees S and T:

$S =$ 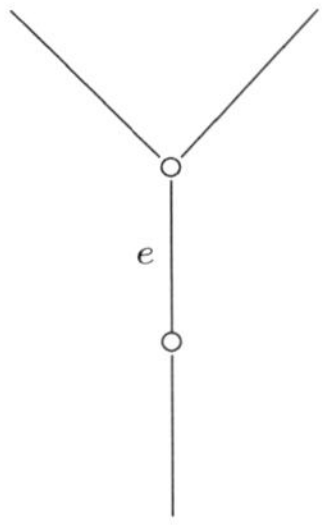

$T =$ 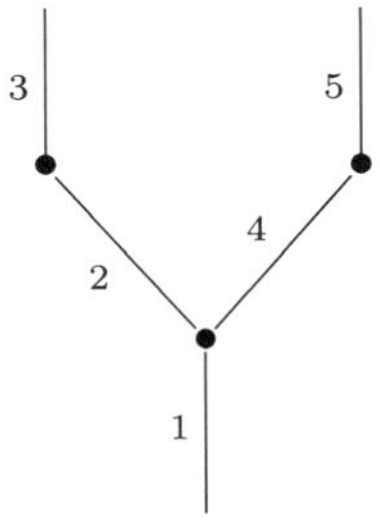

on the following page.

The presentation of $\Omega[T] \otimes \Omega[S]$ is given by the 14 trees $T_1, \cdots, T_{14}$:

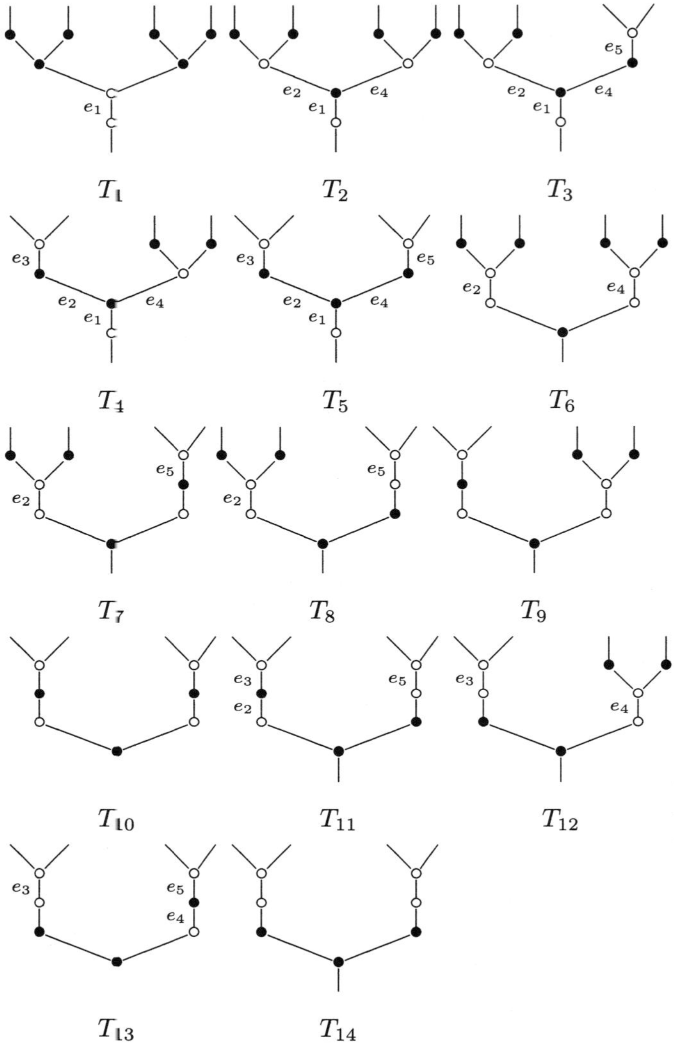

The poset structure on these trees is

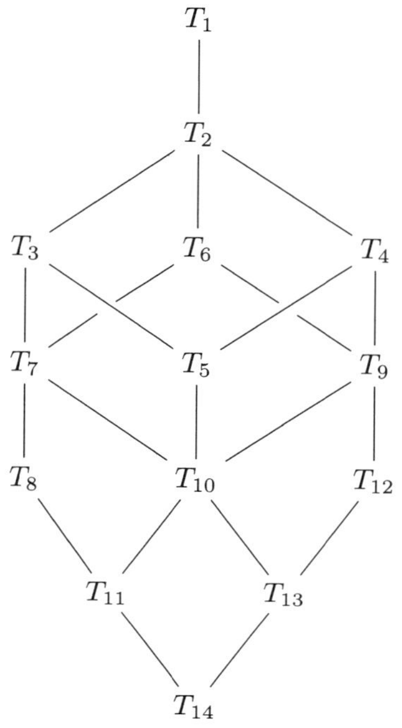

□

As a special case of this result we may now recover Boardman and Vogt's result that the simplicial set $wAlg[\mathcal{P}, \mathcal{E}]$ of weak $\mathcal{P}$-algebras in $\mathcal{E}$ is an ∞-category, as follows.

THEOREM 4.18. *Let $\mathcal{P}$ be a symmetric operad and $\mathcal{E}$ a homotopy environment. If the dendroidal set $N_d(\mathcal{P})$ is normal and $hcN_d(\hat{\mathcal{E}})$ is an ∞-operad then $wAlg[\mathcal{P}, \mathcal{E}]$ is an ∞-category.*

PROOF. The proof follows by noticing that $i^*([N_d(\mathcal{P}), hcN_d(\hat{\mathcal{E}})]) \cong wAlg[\mathcal{P}, \mathcal{E}]$. □

As we have seen above, local fibrancy of $\hat{\mathcal{E}}$ assures that $hcN_d(\hat{\mathcal{E}})$ is an ∞-operad. See below for a condition on $\mathcal{P}$ sufficient to assure $N_d(\mathcal{P})$ is normal. We now turn to the Cisinski-Moerdijk model structure.

THEOREM 4.19. *The category dSet of dendroidal sets admits a Quillen model structure where the cofibrations are the normal monomorphisms, the fibrant objects are the ∞-operads, and the fibrations between ∞-operads are the inner Kan fibrations whose image under τ_d is an operadic fibration. The class $\mathcal{W}$ of weak equivalences can be characterized as the smallest class of arrows which contains all inner anodyne extensions, all trivial fibrations between ∞-operads and satisfies the 2 out of 3 property. Furthermore, with the tensor product of dendroidal sets, this model structure is a monoidal model category. Slicing this model structure recovers* *the Joyal model structure on sSet and in the diagram*

$$\begin{array}{ccc} Cat & \underset{j^*}{\overset{j_!}{\rightleftarrows}} & Ope \\ \tau \big\uparrow \big\downarrow N & & \tau_d \big\uparrow \big\downarrow N_d \\ sSet & \underset{i^*}{\overset{i_!}{\rightleftarrows}} & dSet \end{array}$$

where the categories are endowed (respectively starting from the top-left going clockwise) with the categorical, operadic, Cisinski-Moerdijk, and Joyal model structures all adjunctions are Quillen adjunction (and none is a Quillen equivalence).

PROOF. The proof of the model structure is quite intricate and is established in [**8**]. □

4.4. Homotopy invariance property for algebras in *dSet*. As a consequence of the Cisinski-Moerdijk model structure on dendroidal sets we obtain the following.

THEOREM 4.20. *Let X be a normal dendroidal set and E an ∞-operad. Then X-algebras in E have the homotopy invariance property.*

PROOF. In the diagram defining the homotopy invariance property in Definition 3.17 above the left vertical arrow is a trivial cofibration and the right vertical arrow is a fibration in the Cisinski-Moerdijk model structure and thus the required lift exists. □

We now recall Fact 2.30 regarding the internalization of strict algebras by the closed monoidal structure on *Ope* given by the Boardman-Vogt tensor product and the isomorphism invariance property for such algebras as captured by the operadic monoidal model structure on *Ope*. We recall that a similar such correspondence for weak algebras and their homotopy invariance property does not seem possible within the confines of enriched operads. We are now in a position to summarize the results recounted above in a form completely analogous to the situation of strict algebras.

FACT 4.21. *The notion of algebras of dendroidal sets is internalized to the category dSet by it being closed monoidal with respect to the tensor product of dendroidal sets. The homotopy invariance property of X-algebras in an ∞-operad, for a normal X, holds and is captured by the fact that dSet supports the Cisinski-Moerdijk model structure which is compatible with the tensor product. The notion of a weak $\mathcal{P}$-algebra in a homotopy environment $\mathcal{E}$, where $\mathcal{P}$ is discrete, is subsumed by the notion of algebras in dSet by means of the dendroidal set $[N_d(\mathcal{P}), hcN_d(\mathcal{E})]$ of weak $\mathcal{P}$-algebras in $\mathcal{E}$.*

4.5. Revisiting applications. Recall the iterative construction of the dendroidal set nA_∞ of n-fold A_∞-spaces and ${}_nCat$ of weak n-categories as well as $wCat^n_k$ of k-monoidal n-categories. It follows from the general theory of dendroidal sets that these are all ∞-operads. To see that, one uses Theorem 4.17 together with the fact that for a Σ-free symmetric operad $\mathcal{P}$ (for instance if $\mathcal{P}$ is obtained from a planar operad $\mathcal{Q}$ by the symmetrization functor) then $N_d(\mathcal{P})$ is normal. Thus, weak maps of n-fold A_∞-spaces can be coherently composed and similarly so can weak functors between weak n-categories. As for k-monoidal n-categories we note that $wCat_k^n$ is, for similar reasons as above, an ∞-operad as well.

We may now reduce the Baez-Breen-Dolan stabilization conjecture as follows.

PROPOSITION 4.22. *If for any $n \geq 0$ the dendroidal set $wCat_n^n$ is a strict ∞-operad then the Baez-Breen-Dolan stabilization conjecture is true.*

PROOF. Recall that $wCat_n^n = [A^{\otimes n}, wCat^n]$ and assume it is a strict ∞-operad. We wish to prove that $[A^{\otimes n+j}, wCat^n] \cong [A^{\otimes n+2}, wCat^n]$ for any fixed $j > 2$. By Lemma 4.14 there is an operad $\mathcal{P}$ such that $[A^{\otimes n}, wCat^n] = N_d(\mathcal{P})$. We now have

$$[A^{\otimes n+j}, wCat^n] = [A^{\otimes j}, [A^{\otimes n}, wCat^n]] = [A^{\otimes j}, N_d(\mathcal{P})]$$

which by adjunction is isomorphic to $N_d([\tau_d(A^{\otimes j}), \mathcal{P}])$. However, A is actually the dendroidal nerve of the symmetric operad As classifying associative monoids. By Proposition 3.12 we have

$$\tau_d(A^{\otimes j}) = \tau_d(N_d(As)^{\otimes j}) \cong As^{\otimes j} \cong Comm.$$

and the result follows. □

5. Dendroidal sets - combinatorial models of unknown spaces

All of the theory of dendroidal sets that directly or indirectly is concerned with algebras (we include the Cisinski-Moerdijk model structure here as well) is very operadic in nature and is closely related to the theory of ∞-categories modeled by quasicategories. Simplicial sets are, however, also models for topological homotopy theory. Indeed, simplicial sets were introduced in the context of algebraic topology as combinatorial models of topological spaces. The appropriate equivalence is established in [**36**] and was the reason to introduce Quillen model categories. To recall the main result recall the singular functor $Sing : Top \to sSet$ and its left adjoint $|-| : sSet \to Top$ given by geometric realization.

THEOREM 5.1. *The category Top supports a Quillen model structure in which the weak equivalences are the weak homotopy equivalences and the fibrations are the Serre fibrations. The category sSet supports a Quillen model structure in which the weak equivalences are those maps $f : X \to Y$ for which the geometric realization $|f|$ is a homotopy weak equivalence and the fibrations are the Kan fibrations. With these model structures the adjunction above is a Quillen equivalence.*

Simplicial sets thus support a topologically flavoured Quillen model structure as well as the Joyal model structure which is categorically flavoured and so simplicial sets play two rather different roles. The close connection between dendroidal sets and simplicial sets raises the question as to the existence of a topologically flavoured interpretation of dendroidal sets as well. This problem is open for debate and interpretation and is certainly far from settled.

We remark first that there is some indication that suggests dendroidal sets do carry topological meaning. Recall the Dold-Kan correspondence that establishes an equivalence of categories between the category sAb of simplicial abelian groups and the category Ch of non-negatively graded chain complexes. This correspondence is useful in the calculation of homotopy groups of simplicial sets and in the definition of Eilenberg-Mac Lane spaces $K(G, n)$ for $n > 1$. In [**15**] it is shown that there is a planar dendroidal version (where one considers a planar version of Ω whose objects are planar trees) of the Dold-Kan correspondence. The equivalence is between the category dAb of planar dendroidal abelian groups and the category dCh of planar dendroidal chain complexes. The definition of the latter requires that for each face map ∂_α between planar trees there is associated a sign $sgn(\partial_\alpha) \in \{\pm 1\}$ such that the following holds. In the planar version of Ω it is still true that a face $S \to T$ of codimension 2 decomposes in precisely two ways as the composition of two faces (see Proposition 4.5 above). Thus we can write $S \to T$ as $\partial_\alpha \circ \partial_\beta$ as well as $\partial_\gamma \circ \partial_\delta$ and we require that $sgn(\partial_\alpha) \cdot sgn(\partial_\beta) = -sgn(\partial_\gamma) \cdot sgn(\partial_\delta)$. One may now wonder whether these dendroidal chain complexes give rise to some sort of generalized Eilenberg-Mac Lane spaces. A first step towards answering this question should be a clearer specification of goals in a broad context, which is the aim of the rest of this section.

As inspiration we consider the Quillen equivalence between topological spaces and simplicial sets mentioned above. The geometric realization plays there a prominent role and thus a significant aspect of understanding the homotopy behind dendroidal sets is to find a category $dTop$ together with functors $Sing_d : dTop \to dSet$ and $| - |_d : dSet \to dTop$. The category $dTop$ of course has to be chosen with care so that it will rightfully be considered to be related to topology. We thus expect that there is a fully faithful functor $h_! : Top \to dTop$ with a right adjoint $h^* : dTop \to Top$ that should be defined 'purely topologically'. We thus expect $dTop$ to be a category of some generalized topological spaces in which ordinary topological spaces embed via $h_!$. To allow sufficient flexibility for working with these objects we expect that $dTop$ be small complete and small cocomplete. Moreover, the functor $| - |_d : dSet \to dTop$ should send a dendroidal set X to some generalized space $|X|_d$ in such a way that the combinatorial information in X is not lost. We thus expect of any such functor $| - |_d$ that if for some $f : X \to Y$ the map $|f|_d$ is an isomorphism then f was already an isomorphism. In other words we expect $| - |_d$ to be conservative.

The term 'purely topologically' above is of course vague and open to discussion. In an attempt to formalize it recall the various slicing lemmas we have seen above: Slicing symmetric operads over $\star$ gives categories, slicing dendroidal sets over $\Omega[\eta] = N_d(\star)$ gives simplicial sets, and slicing Ω over η gives Δ. We thus expect that there is an object $\star \in dTop$ such that slicing $dTop$ over $\star$ gives a category equivalent to Top and that in fact the embedding $h_! : Top \to dTop$ is essentially the forgetful functor $dTop/\star \to dTop$. Moreover, noting that the 'correct' tensor product of dendroidal sets is not the cartesian one we expect $dTop$ to posses a monoidal structure different from the cartesian product. And, just as the tensor product of dendroidal sets slices to the cartesian product of simplicial sets we expect the monoidal structure on $dTop$ to slice to the cartesian product of topological spaces. Lastly, an important property of the ordinary geometric realization functor is that it commutes with finite products. We expect of the dendroidal geometric realization functor $dSet \to dTop$ to be monoidal with respect to the non-cartesian monoidal structure on each category.

We summarize our expectations in the following formulation.

PROBLEM 5.2. Find a category $dTop$ together with a functor $Sing_d : dTop \to dSet$, a left adjoint $| - |_d : dSet \to dTop$, and an object $\star \in dTop_0$ such that:

(1) $dTop$ is small complete and small cocomplete.
(2) (Slicing lemma) $dTop/\star$ is equivalent to Top.
(3) The forgetful functor $h_! : Top \to dTop$ is an embedding.

(4) Slicing $Sing_d$ gives $Sing$ and slicing $|-|_d$ gives $|-|$.
(5) $|-|_d$ is conservative.
(6) $dTop$ admits a non-cartesian monoidal structure that slices over $\star$ to the cartesian product in Top (along $h_!$).
(7) The functor $|-|_d$ is to be a monoidal functor with respect to the tensor structures on $dSet$ and $dTop$.

We would thus obtain the diagram

$$\begin{array}{ccc} sSet & \underset{Sing}{\overset{|-|}{\rightleftarrows}} & Top \\ {\scriptstyle i_!}\downarrow\uparrow{\scriptstyle i^*} & & {\scriptstyle h_!}\downarrow\uparrow{\scriptstyle h^*} \\ dSet & \underset{Sing_d}{\overset{|-|_d}{\rightleftarrows}} & dTop \end{array}$$

where both squares commute.

The quest will be complete with the establishment of Quillen model structures on $dSet$ and $dTop$ that slice respectively to the standard (topological) ones on $sSet$ and Top and such that in the square above all adjunctions are Quillen adjunctions with both horizontal ones Quillen equivalences.

References

[1] C. Berger, I. Moerdijk, *Axiomatic homotopy theory for operads*, Comment. Math. Helv. 78 (2003), no. 4, 805-831.

[2] C. Berger, I. Moerdijk, *The Boardman-Vogt resolution of operads in monoidal model categories*, Topology 45 (2006), no. 5, 807-849.

[3] C. Berger, I. Moerdijk, *Resolution of coloured operads and rectification of homotopy algebras*, Categories in algebra, geometry and mathematical physics, *31-58,* Contemp. Math., 431, Amer. Math. Soc., Providence, RI, (2007).

[4] J. Bergner, *A survey of* $(\infty, 1)$*-categories*, Towards higher categories, 69-83, IMA Vol. Math. Appl., 152, Springer, New York, (2010).

[5] J. M. Boardman, R. M. Vogt, *Homotopy invariant algebraic structures on topological spaces,* Lecture Notes in Mathematics, Vol. 347, Springer-Verlag, (1973).

[6] N. Bourbaki, *Univers.* In M. Artin, A. Grothendieck, J.-L. Verdier, eds. Séminaire de Géométrie Algébrique du Bois Marie - 1963-64 - Théorie des topos et cohomologie étale des schémas - (SGA 4) - vol. 1 (Lecture Notes in Mathematics 269, Berlin, Springer-Verlag, pp. 185-217.

[7] D.-C. Cisinski, *Les préfaisceaux comme modèles des types d'homotopie,* Astérisque 308, (2006).

[8] D.-C. Cisinski, I. Moerdijk, *Dendroidal sets as models for* ∞*-operads*, preprint arXiv:0902.1954; to appear in Journal of Topology.

[9] D.-C. Cisinski, I. Moerdijk, *Dendroidal Segal spaces and* ∞*-operads*, preprint arXiv:1010.4956.

[10] D.-C. Cisinski, I. Moerdijk, *Dendroidal sets and simplicial operads*, in preparation.

[11] Z. Fiedorowicz, R.M. Vogt, *Interchanging* A_∞ *and* E_n *structures*, preprint arXiv:1102.1311.

[12] V. Ginzburg, M.M. Kapranov, *Koszul duality for operads*, Duke Math. J. 76 (1994), no. 1, 203-272.

[13] P. Gabriel, M. Zisman, *Calculus of Fractions and Homotopy Theory*, Ergebnisse der Mathematik und ihrer Grenzgebiete, Band 35 Springer-Verlag New York, Inc., New York, (1967).

[14] P.G. Goerss, J.F. Jardine, *Simplicial Homotopy Theory*, Progress in Mathematics, 174. Birkhäuser Verlag, Basel, (1999).

[15] J. J. Gutiérrez, A. Lukács, I. Weiss, *Dold-Kan correspondence for dendroidal abelian groups*, J. Pure Appl. Algebra 215 (2011), pp. 1669-1687 (2009).

[16] C. Hermida, *Representable multicategories*, Adv. Math. 151 (2000), no. 2, 164-225.

[17] P.S. Hirschhorn, *Model Categories and Their Localizations*, Mathematical Surveys and Monographs, 99. American Mathematical Society, Providence, (2003).
[18] M. Hovey, *Model Categories*, Mathematical Survery and Monographs, 63. American Mathematical Society, Providence, (1999).
[19] A. Joyal, *Quasi-categories and Kan complexes*, J. Pure Appl. Algebra 175 (2002), no. 1-3, 207-222.
[20] A. Joyal, *Theory of quasi-categories*, In preparation.
[21] G.M. Kelly, *Basic Concepts of Enriched Category Theory*, London Mathematical Society Lecture Notes Series, 64. Cambridge University Press, Cambridge-New York, (1982).
[22] J. Kock, *Polynomial functors and trees*, Int. Math. Res. Not. (2011), 609-673.
[23] J. Lambek, *Deductive systems and categories II: standard constructions and closed categories.* 1969 Category Theory, Homology Theory and their Applications, I (Battelle Institute Conference, Seattle, Wash., (1968), Vol. One) pp. 76-122 Springer, Berlin.
[24] T. Leinster, *Higher Operads, Higher Categories*, London Mathematical Society Lecture Note Series 298, Cambridge University Press, Cambridge, (2004).
[25] T. Leinster, *A survey of definitions of n-category*, Theory Appl. Categ. 10, (2002), 1-70.
[26] T. Leinster, *Generalized enrichment of categories*, J. Pure Appl. Algebra 168, (2002), no. 2-3, 391-406.
[27] A. Lukács, *Cyclic Operads, Dendroidal Structures, Higher Categories*, PhD. Thesis, Universiteit Utrecht, (2010).
[28] J. Lurie, *Higher Topos Theory*, Annals of Mathematics Studies, 170. Princeton University Press, Princeton, NJ, (2009).
[29] J. Lurie, *Derived algebraic geometry III*, preprint arXiv:math/0703204.
[30] S. Mac Lane, *Categories for the Working Mathematician.* Second edition. Graduate Texts in Mathematics, 5, Springer-Verlag, New York, (1998).
[31] S. Mac Lane, I. Moerdijk, *Sheaves in Geometry and Logic*, Universitext, Springer-Verlag, New york, (1994).
[32] M. Markl, S. Shnider, J. Stasheff, *Operads in Algebra, Topology, and Physics*, Mathematical Surveys and Monographs 96, American Mathematical Society, Providence, (2002).
[33] J.P. May, *The geometry of iterated loop spaces*, Lectures Notes in Mathematics, Vol. 271, Springer-Verlag, Berlin-New York, (1972).
[34] I. Moerdijk, I. Weiss, *Dendroidal Sets.* Algebr. Geom. Topol. 7, (2007), 1441-1470.
[35] I. Moerdijk, I. Weiss, *On inner Kan complexes in the category of dendroidal sets.* Adv. Math. 221 (2009), no. 2, 343-389.
[36] D. Quillen, *Homotopical Algebra*, Lecture Notes in Mathematics, No. 43, Springer-Verlag, Berlin-New York, (1967).
[37] J. Stasheff, *Homotopy associativity of H-spaces. I, II.* Trans. Amer. Math. Soc. 108 (1963), 275-292; ibid., 108:293-312, 1963.
[38] I. Weiss, *Dendroidal Sets*, PhD thesis, Universiteit Utrecht, (2007).

Mathematical Institute, Utrecht University, Budapestlaan 6, 3584 CD, Utrecht, The Netherlands
E-mail address: I.Weiss@uu.nl

Proceedings of Symposia in Pure Mathematics
Volume **83**, 2011

Field theories with defects and the centre functor

Alexei Davydov, Liang Kong, and Ingo Runkel

Abstract. This paper is intended as an introduction to the functorial formulation of quantum field theories with defects. After some remarks about models in general dimension, we restrict ourselves to two dimensions – the lowest dimension in which interesting field theories with defects exist.

We study in some detail the simplest example of such a model, namely a topological field theory with defects which we describe via lattice TFT. Finally, we give an application in algebra, where the defect TFT provides us with a functorial definition of the centre of an algebra. This involves changing the target category of commutative algebras into a bicategory.

Throughout this paper, we emphasise the role of higher categories – in our case bicategories – in the description of field theories with defects.

Contents

1. Introduction

One way to think about quantum field theory – motivated by conformal field theory and string theory [**FS, Va**] – is as functors from bordisms to vector spaces [**Se, At**]; here, each of the terms 'functor', 'bordism', 'vector space' has to be supplemented with the appropriate qualifiers for the application in mind. In its most basic form, the bordisms for an n-dimensional quantum field theory form a symmetric monoidal category whose objects are $(n-1)$-dimensional manifolds equipped

---

2010 *Mathematics Subject Classification.* Primary 57R56, 18D05; Secondary 16U70, 81T40.

LK is supported by the Basic Research Young Scholars Program of Tsinghua University, Tsinghua University independent research Grant No. 20101081762 and by NSFC Grant No. 20101301479.

IR is supported in part by the SFB 676 'Particles, Strings and the Early Universe' of the DFG.

with 'collars' and whose morphisms are equivalence classes of n-dimensional manifolds with parametrised boundary.

To study quantum field theories beyond this basic functorial definition, it is often appropriate to employ higher categories. There are two natural ways in which such higher categories enter.

(1) The $(n-1)$-manifolds which form the objects in the above bordism category could in turn be obtained by gluing $(n-1)$-manifolds along $(n-2)$-manifolds, and so on, down to 0-manifolds, i.e. points. This process is called 'extending the field theory down to points' [**Fre, Law, BD, Lu1**]. One obtains a higher category whose objects are now points (with extra structure) and which has 1-morphisms, 2-morphisms, ..., up to n-morphisms. The resulting field theories are most studied in the case of topological field theories [**Lu1**].

(2) One can let the bordisms remain a (1-)category but equip them with extra structure, namely with 'defects'. These are submanifolds embedded in the $(n-1)$- and n-dimensional bordisms, decorated with labels which describe different possible 'defect conditions'. A field theory on bordisms with defects equips the set of defect conditions with the structure of a higher category.

Here we want to elaborate on the second point. Some other works which also stress the appearance of higher categories in field theories with defects, and which the reader could consult for further references, are [**SFR, Lu1, BDH, Ka, KK**]. In the present paper, we will concentrate on the simplest interesting class of models, namely two-dimensional field theories with defects. In section 2 we will see how a field theory with a particular type of defects – so called topological defects – gives rise to a 2-category defined in terms of the set of defect conditions. In section 3 we use lattice topological field theory to construct a very simple but still non-trivial example of a field theory with defects. This example will motivate – in section 4 – a nice mathematical construction, namely a method to make the assignment which maps an algebra to its centre functorial. Section 5 contains an outlook on further developments.

These three constructions – the 2-category of topological defects (section 2.4), lattice topological field theory with defects (Theorem 3.8), and the centre functor (Theorem 4.12 and Remark 4.19) – are the main points of this paper. We hope that they provide some intuition on how to work with field theories containing defects and illustrate their usefulness.

2. Field theory with defects

2.1. Bordisms with defects. It is beyond the scope of this article (and the present abilities of the authors) to develop an all-purpose formalism for field theories with defects. In this subsection we briefly sketch the basic features of the functorial formulation of field theory with defects.[1]

[1] In doing so will omit most details. For those who are familiar with the functorial formulation, some of these details are: we should equip our object-$(n-1)$-manifolds with collars to ensure a well-defined gluing operation; objects and morphisms could carry extra geometric data such as a metric, a spin structure, etc.; we should work with families to have a natural notion of continuous or smooth dependence of the functor on the bordism; the functor from bordisms to vector spaces

As usual, a field theory will be a functor from a bordism category to a category of vector spaces. In the presence of defects, the target category of the functor remains unchanged. However, we do modify the source category. The category of *n-dimensional bordisms with defects* contains the following ingredients.

- *Sets of defect conditions:* the bordism category will depend on a choice of $n+1$ (possibly empty) sets, D_k, $k = 0, \ldots, n$. The elements of D_k serve as defect conditions for k-dimensional defects.
- *Objects:* the objects are $(n-1)$-dimensional compact oriented manifolds U with empty boundary, together with a disjoint decomposition into submanifolds. That is, $U = \bigcup_{i=0}^{n-1} U_i$, where each U_i is an i-dimensional oriented submanifold of U and $U_k \cap U_l = \emptyset$ for $k \neq l$.[2] The orientation of U_{n-1} is induced by that of U.

  For example, $U_{n-1} = U$ and $U_k = \emptyset$ for $k < n-1$ would be a possibility, or, if U_i $(i < n-1)$ is a closed submanifold of U, then we can take $U_{n-1} = U \setminus U_i$ and $U_k = \emptyset$ for $k \neq i, n-1$.

  Finally, each connected component of U_k is decorated with a defect condition from D_{k+1}, i.e. we have a collection of maps $d_{k+1} : \pi_0(U_k) \to D_{k+1}$; the reason for the shift in k is that the U_k will appear as boundaries of $(k+1)$-dimensional submanifolds in the n-dimensional manifold making up a morphism.
- *Morphisms:* a morphism $M : U \to V$ has a structure analogous to objects, except in one dimension higher. In more detail, M is an n-dimensional compact oriented manifold, together with a decomposition $M = \bigcup_{i=0}^{n} M_i$, where each M_i is an i-dimensional oriented submanifold, possibly with non-empty boundary ∂M_k, and $M_k \cap M_l = \emptyset$ for $k \neq l$ (and footnote 2 applies analogously). The orientation of M_n is induced by that of M. Each connected component of M_k is labelled by a defect condition, but this time from D_k, that is, we have maps $\hat{d}_k : \pi_0(M_k) \to D_k$.

  The boundary ∂M is identified via an orientation preserving diffeomorphism (which is part of the data of a morphism) with the disjoint union $-U \sqcup V$; we require that $\partial M_k \subset \partial M$, and that the resulting decomposition and labelling of ∂M agrees with the one induced by $-U \sqcup V$.

For example, in $n = 3$ dimensions, a generic morphism would look like a foam, where the interior of each bubble is 'coloured' by an element of D_3, the walls between two bubbles by elements of D_2, lines along which the walls between bubbles meet by elements of D_1, and points where these lines meet by elements of D_0 (somewhat problematic in the foam analogy, but nonetheless allowed, are 1- and 0-dimensional submanifolds not attached to any walls).

---

is symmetric monoidal; the target category of the functor consists of topological vector spaces with an appropriate tensor product. These issues are treated carefully in [**ST**].

[2] We also demand that the partial union $\bigcup_{i=0}^{k} U_i$ is a closed subset of U for $k = 0, \ldots, n-1$; this ensures that $\bar{U}_k \setminus U_k$ (the difference of U_k and its closure) is contained in the union $\bigcup_{i=0}^{k-1} U_i$ of lower dimensional pieces. Let us give a non-example in $U = S^3$, which we present as the one-point compactification of $\mathbb{R}^3$. Take $U_0 = \emptyset$, $U_1 = (-1, 1) \times \{(0, 0)\}$, $U_2 = S^2 \subset \mathbb{R}^3$, $U_3 = U \setminus (U_1 \cup U_2)$. In this case, all U_i are submanifolds, but $U_0 \cup U_1$ is not closed, which is not allowed. (But $U_0 \cup U_1 \cup U_2$ is closed). To turn this into an allowed decomposition, take instead $U_0' = \{(\pm 1, 0, 0)\}$ and $U_2' = S^2 \setminus \{(\pm 1, 0, 0)\}$. Then U_0', U_1, U_2', U_3 is an allowed decomposition of U.

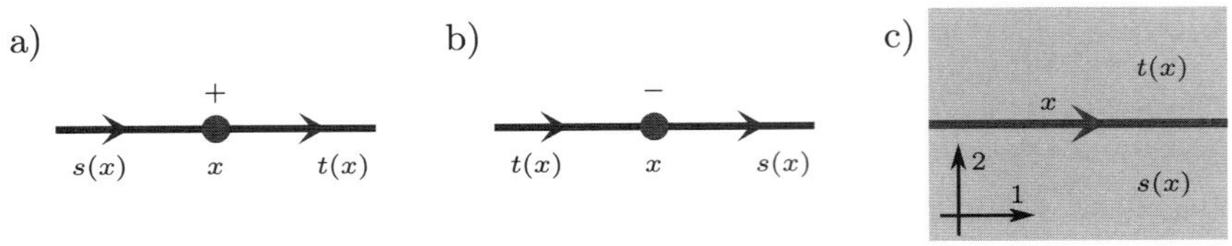

FIGURE 1. Figures a)–c) show open subsets of a bordism in dimension $n = 1$ and $n = 2$. They give our orientation convention in the compatibility condition for the assignment of defect conditions in the case $n = 1$ (figs. a, b) and $n = 2$ (fig. c). The arrows represent positively oriented ordered bases.

While we have given the overall name 'defect conditions' to elements of the sets D_k, more descriptive names in the various dimensions $0 \leq k \leq n$ would be that they are conditions for

- D_n: domains (or phases of the field theory)
- D_{n-1}: domain walls (or phase boundaries)
- $D_{n-2}, \ldots, D_0$: junctions

The sets D_k are equipped with additional structure describing in which geometric configurations the domains can occur. This is complicated in general, but it is easy to state for domain walls. Since the n-dimensional manifold underlying the morphism and the $(n-1)$-dimensional submanifold underlying the domain wall are oriented, we can speak of the 'left and right side' of the domain wall. Accordingly, there are two maps

$$s, t \; : \; D_{n-1} \longrightarrow D_n \tag{2.1}$$

(for 'source' and 'target'), and a domain wall of type $x \in D_{n-1}$ must have a domain labelled by $s(x)$ on its left and $t(x)$ on its right. This gives a restriction on the allowed maps d_n, d_{n-1} in objects and $\hat{d}_n, \hat{d}_{n-1}$ in morphisms. In this work, we will only discuss the cases $n = 1$ and $n = 2$ and our orientation conventions are shown in figure 1.

2.2. One-dimensional topological field theory with domain walls. Before passing to the more interesting two-dimensional situation, let us briefly discuss the simplest one-dimensional field theory with domain walls, namely the case where the field theory is topological.

We fix two sets D_1 and D_0, together with two maps $s, t : D_0 \to D_1$. The objects in the bordism category are finite sets of oriented points U, together with a map $d_1 : U \to D_1$. The morphisms $M : U \to V$ are (diffeomorphism classes of) 1-dimensional manifolds M with a finite set W of marked points in the interior of M. Each connected component of $M \setminus W$ is labelled by an element of D_1, and each element of W by an element of D_0. On the boundary ∂M the D_1-labels have to agree with those of U, resp. V.

A symmetric monoidal functor from this bordism category to (necessarily finite dimensional) k-vector spaces for some field k is then determined

- *on objects:* by a collection of vector spaces $(V_i)_{i \in D_1}$. The value of the functor on a point with orientation '+' and label i is given by V_i, while a point with orientation '−' gets mapped to V_i^*. On 0-dimensional manifolds with more than one point the functor is fixed by the monoidal structure as usual.

- *on morphisms:* by two collections of linear maps $(L_x^+)_{x\in D_0}$ and $(L_x^-)_{x\in D_0}$, where $L_x^+ : V_{s(x)} \to V_{t(x)}$ and $L_x^- : V_{t(x)} \to V_{s(x)}$. Let $\varepsilon \in \{\pm 1\}$. The map L_x^ε is the value of the functor on the interval $[-1,1]$ with standard orientation, together with the 0-dimensional submanifold $\{0\}$ with orientation ε and label $x \in D_0$. If $\varepsilon = +$, the sub-interval $[-1,0)$ is labelled by $s(x) \in D_1$ and $(0,1]$ by $t(x) \in D_1$, while for $\varepsilon = -$, the label of $[-1,0)$ and $(0,1]$ is $t(x)$ and $s(x)$, respectively, as in figure 1 a, b). An arbitrary morphism can be obtained by composing and tensoring the above maps, as well as the cup and cap bordisms, which the functor maps to evaluation and co-evaluation.

We can collect this data in a category $\mathcal{D}$, together with a distinguished subset of arrows, as follows. Take D_1 as objects of $\mathcal{D}$. As space of morphisms $i \to j$, for $i,j \in D_1$, take $\mathcal{D}(i,j) := \mathrm{Hom}_k(V_i, V_j)$, the linear maps from V_i to V_j. Finally, fix a map $D_0 \times \{\pm\} \to \mathrm{Mor}(\mathcal{D})$, which assigns to $(x,\pm)$ the arrow $L_x^\pm$, with source and target as described above.

2.3. Two-dimensional metric bordisms with defects. Let us look in more detail at an instance of a bordism category with defects in two dimensions; the exposition essentially follows [**RS**, Sec. 3]. A note on convention: by manifold we mean smooth manifold, and by a map between manifolds we mean a smooth map; a finite or countable disjoint union has an ordering of its factors, so that for two sets A, B the disjoint unions $A \sqcup B$ and $B \sqcup A$ are isomorphic but not equal.

Sets of defect conditions. We start with the three sets D_2, D_1, and D_0, which are the sets of world sheet phases, domain wall conditions, and junction conditions, respectively. As above we have two maps $s,t : D_1 \to D_2$ giving the phase to the left and right of a domain wall; our orientation conventions are shown in figure 1 c). For a junction in D_0 we need to specify which domain walls can meet with which orientations at a junction point.

The combinatorial description thereof is a bit lengthy: Let $D_1^{(n)}$ be the set of tuples of n *cyclically composable domain walls.* By this we mean the subset of n-fold cartesian product $(D_1 \times \{\pm\})^{\times n}$ selected by the following condition: For $((x_1,+),\dots,(x_n,+))$ we require $t(x_{i+1}) = s(x_i)$ and $t(x_1) = s(x_n)$. If some of the '+' are changed for '−', the role of s and t is exchanged as in figure 2. The group C_n of cyclic permutations acts on the n-tuples in $D_1^{(n)}$. The set D_0 is equipped with a map

$$j : D_0 \longrightarrow \bigsqcup_{n=0}^{\infty} \left(D_1^{(n)}/C_n \right) . \tag{2.2}$$

In words, for each element u of D_0, the map j determines how many domain walls can end at a junction labelled by u and what their orientations and domain wall conditions are, up to cyclic reordering.

The map j is similar in spirit to the relation between D_1 and D_2. There, we can combine the 'source' and 'target' maps into a single map $(s,t) : D_1 \to D_2 \times D_2$, which determines the world sheet phases that must lie on the two sides of a domain wall labelled by a given element of D_1. There is no need to divide by the symmetric

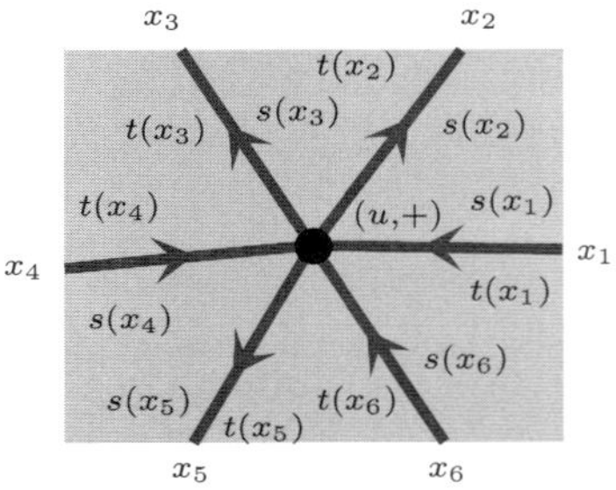

FIGURE 2. Illustration of the condition of cyclic composability of domain walls. Given the n-tuple $((x_1,\varepsilon_1),\dots,(x_n,\varepsilon_n))$, the i th domain wall (counted anti-clockwise) is labelled by x_i and is pointing towards the junction point if $\varepsilon_i = +$ and away from the junction point if $\varepsilon_i = -$. In the present example the 6-tuple is $((x_1,+),(x_2,-),(x_3,-),(x_4,+),(x_5,-),(x_6,+))$. The images under the maps s and t to D_2 have to agree as shown, e.g. $s(x_1) = s(x_2)$ and $t(x_2) = s(x_3)$. The junction point has orientation '+' and is labelled by $u \in D_0$.

group in two elements, because the orientations allow one to distinguish the 'left' and 'right' side of a domain wall.

Objects. In short, an object is a disjoint union of a finite number of unit circles S^1 with marked points, together with a germ of a collar.[3]

In more detail, for a single S^1 the structure is as follows. Take $U = S^1$ to be the unit circle in $\mathbb{C}$, decorated as in section 2.1: a 0-dimensional submanifold $U_0 \subset S^1$ (i.e. a set of points decorated by signs $\pm$), a map $d_1 : \pi_0(U_0) = U_0 \to D_1$, and a map $d_2 : \pi_0(U_1) \to D_2$, where $U_1 = S^1 \setminus U_0$. The maps d_1, d_2 have to be compatible with s,t as in section 2.1 (cf. figure 1 a, b)).

A *collar* is, in short, an extension of the above structure to an open neighbourhood of S^1 in $\mathbb{C}$, see figure 3. Let A be an open neighbourhood of S^1, and let A_1 be a one-dimensional submanifold, closed in A, which intersects S^1 transversally (the tangents to A_1 and S^1 are linearly independent at intersection points). Set $A_2 = A \setminus A_1$ There are maps $\hat{d}_i : \pi_0(A_i) \to D_i$, $i = 1,2$, compatible with s,t as in figure 1 c). The restriction of A_2 and A_1 to S^1 has to reproduce U_1 and U_0 with labelling and orientation, with conventions as in figure 3. Finally, A carries a metric in conformal gauge, i.e. $g(z)_{ij} = e^{\sigma(z)}\delta_{ij}$ for a real-valued function σ on A.

Two collars are equivalent if they agree in some open neighbourhood of S^1; an equivalence class is called a *germ of collars*.

For a disjoint union U of such S^1 with collars, write U_{in} for the subset obtained by taking only points $|z| \geq 1$ in the collar of each S^1, and U_{out} when taking only points with $|z| \leq 1$.

Morphisms. Morphisms $M : U \to V$ are equivalence classes of surfaces with extra structure as in section 2.1, together with a metric. Thus we have a decomposition $M = M_2 \cup M_1 \cup M_0$, maps $\hat{d}_i : \pi_0(M_i) \to D_i$, $i = 0,1,2$. The map $\hat{d}_1$ is

[3]This is more restrictive than allowing general one-dimensional manifolds as in section 2.1 but does not lose any generality and has the advantage that objects form a set, and that the connected components of an object are already ordered by our convention on disjoint unions.

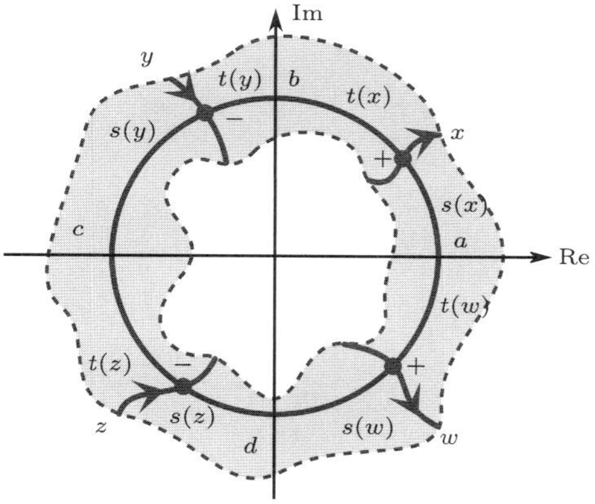

FIGURE 3. Illustration of a collar which forms part of the data for an object in the bordism category. In the notation of the text, the solid (blue) circle is a unit circle $U = S^1$, the shaded area is an open neighbourhood A, the solid (red) short lines form the oriented submanifold A_1 of A which intersects S^1 in U_0. Our convention for the orientation of A_1 induced by that of U_0 (the signs '$\pm$') is as shown. The elements $a, b, c, d \in D_2$ label connected components of U_1 and their extension A_2; these labels have to agree with the source and target maps of the domain wall labels $w, x, y, z \in D_1$ as shown. E.g. $t(w) = a = s(x)$.

compatible with s, t as in figure 1 c). Going beyond the level of detail in section 2.1, we also require the following:

■ *compatibility condition for* $\hat{d}_0$: For a point $p \in M_0$ labelled by $u \in D_0$ (i.e. $\hat{d}_0(p) = u$), let $((x_1, \varepsilon_1), \ldots, (x_n, \varepsilon_n))$ denote the domain wall conditions and orientations in anti-clockwise order (with arbitrary starting point). We require that $j(u)$ is the cyclic permutation equivalence class of $((x_1, \varepsilon_1), \ldots, (x_n, \varepsilon_n))$ if the junction point has orientation '+', cf. figure 2, and that it is in the class of $((x_n, -\varepsilon_n), \ldots, (x_1, -\varepsilon_1))$ if the junction orientation is '−'. Junctions with opposite orientation are dual in the following sense (figure 4): if the bordism is a 2-sphere with two antipodal junction points both labelled by u but with orientations '+' and '−', the domain walls starting at the two junctions can be joined up (intersection-free) by half-circles around the S^2.

■ *boundary parametrisation*: A choice U', V' of collars representing the germs U, V, together with injective maps $f_{\text{in}} : U_{\text{in}} \to M$ and $f_{\text{out}} : V_{\text{out}} \to M$ which preserve the orientation, metric, boundary, 1-dimensional submanifold (with orientation) and labelling. The images of the factors S^1 in U' and V' are disjoint and cover the boundary of M.

Two surfaces are equivalent if they are isometric and the isometry preserves the decomposition $M = M_2 \cup M_1 \cup M_0$ together with orientations and labelling, and commutes with the boundary parametrisation in some open neighbourhood of ∂M.

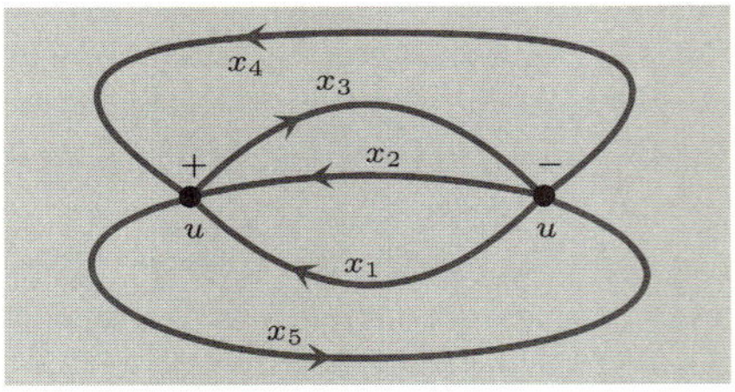

FIGURE 4. Some domain walls and two junctions placed on S^2; we only display a fragment after projection to the plane. The two junctions are labelled by the same junction condition $u \in D_0$ but with opposite orientation '$\pm$'. Here, $j(u)$ is the cyclic permutation equivalence class of $((x_1,+),(x_2,+),(x_3,-),(x_4,+),(x_5,-))$. Thus, the junction labelled by u with orientation '$+$' must have domain walls $(x_1,+),(x_2,+),(x_3,-),(x_4,+),(x_5,-)$ attached in anti-clockwise order, where for $(x_i,+)$ the domain wall is oriented towards the junction and for $(x_i,-)$ it is oriented away from the junction. The junction labelled by u with orientation '$-$' must have domain walls $(x_5,+),(x_4,-),(x_3,+),(x_2,-),(x_1,-)$ attached in anti-clockwise order.

Composition of morphisms is defined by choosing representatives and gluing via the boundary parametrisation; the collars ensure that this does not introduce 'corners' and results again in a surface as described above. The equivalence class of the glued surface is independent of the choice of representatives.

Identities and symmetric structure. So far there are no identity morphisms. We will add these by hand by extending the morphisms to include permutations of the S^1 factors in the disjoint union of a given object U. If we denote the permutation by σ and the permuted disjoint union by $\sigma(U)$, we add morphisms $\sigma : U \to \sigma(U)$. Each morphism $U \to V$ is either a permutation (only possible if $V = \sigma(U)$) or a bordism; composing $\sigma^{-1}(U) \xrightarrow{\sigma} U \xrightarrow{M} V \xrightarrow{\tau} \tau(V)$ produces a bordism $\sigma^{-1}(U) \xrightarrow{M'} \tau(V)$, where M' differs from M only in the boundary parametrisation maps.

This endows the bordism category with a symmetric structure (the tensor product is disjoint union as usual).

Topological domain walls and junctions. Denote the symmetric monoidal category described above by

$$\mathrm{Bord}^{\mathrm{def}}_{2,1}(D_2, D_1, D_0)\ , \tag{2.3}$$

or $\mathrm{Bord}^{\mathrm{def}}_{2,1}$ for short. A two-dimensional quantum field theory with defects can now be defined as a symmetric monoidal functor Q from $\mathrm{Bord}^{\mathrm{def}}_{2,1}$ to topological vector spaces, which depends continuously on the moduli (namely, the metric on a morphism M, the decomposition $M = M_2 \cup M_1 \cup M_0$, and the boundary parametrisation).

REMARK 2.1. It is also easy to say when such a functor Q describes a *conformal field theory with defects.* Namely, the vector space $Q(U)$ assigned to an object U has to be independent of the conformal factor $e^{\sigma(z)}$ giving the metric $g(z)_{ij} = e^{\sigma(z)}\delta_{ij}$ on U, and the linear map $Q(M)$ assigned to a morphism M changes by at most a scalar factor if the metric on M is changed by a conformal factor $g \rightsquigarrow e^f g$ for some $f : M \to \mathbb{R}$. Thus Q would in general only give a projective functor if one passes to conformal equivalence classes of manifolds (it would be a true functor if the so-called central charge vanishes).

With respect to a chosen Q, we can define an interesting subset of domain walls and junctions:

- *topological domain walls* are elements x of D_1 such that
  (1) for all objects U, the vector space $Q(U)$ is unchanged under isotopies moving components of U_0 labelled by x (and their extension into the collars with them) such that no point of U_0 crosses the point $-1 \in S^1$. This condition renders the space of such isotopies contractible (on germs of collars). In particular, a full 2π-rotation is excluded, as it would in general induce a non-trivial endomorphism of $Q(U)$. The metric on U stays fixed.
  (2) for all morphisms M, $Q(M)$ is invariant under isotopies moving components of M_1 labelled by x while leaving M_0 fixed and restricting on ∂M to isotopies respecting the condition in 1.
- *topological junctions* are elements u of D_0 such that $j(u)$ only contains elements of D_1 labelling topological domain walls, and such that $Q(M)$ is invariant under isotopies moving components of M_1 labelled by topological domain wall conditions and points in M_0 labelled by u.

From now on we will concentrate on topological domain walls and junctions. We will denote the corresponding subsets by $D_i^{\mathrm{top}}(Q)$, $i = 0, 1$, or just D_i^{top}.

2.4. 2-categories of defect conditions. Let us fix a two-dimensional field theory with defects as above, i.e. a continuous symmetric monoidal functor Q from $\mathrm{Bord}_{2,1}^{\mathrm{def}}(D_2, D_1, D_0)$ to an appropriate category of topological vector spaces. The construction below will only make use of the sets D_2 and D_1, but not of D_0. To emphasise this, we take $D_0 = \emptyset$ (i.e. no junctions are allowed).

Consider the topological domain walls $s, t : D_1^{\mathrm{top}}(Q) \to D_2$. This is a pre-category (which is nothing but a graph, see e.g. [**ML**, Ch. II.7]), and the aim of this section is to show that Q turns the free category (with conjugates) generated by this pre-category into a 2-category. This 2-category can be thought of as capturing some of the genus-0 information of the field theory Q. Our conventions for bicategories are collected in appendix A.

Recall (e.g. from [**ML**, Ch. II.7]) that the free category is generated by tuples of composable arrows. By the free category with conjugates we mean the category whose objects are D_2 and whose morphisms $a \to b$ (for $a, b \in D_2$) are tuples

$$\underline{x} \equiv ((x_1, \varepsilon_1), \ldots, (x_n, \varepsilon_n)) \tag{2.4}$$

where $x_i \in D_1^{\mathrm{top}}$, $\varepsilon_i \in \{\pm\}$. As for cyclically composable domain walls, if all signs $\varepsilon_i = +$, then we require $s(x_i) = t(x_{i+1})$ and $s(x_n) = a$, $t(x_1) = b$. If some $\varepsilon_i = -$, the roles of s and t change as in figure 2. Composition of $\underline{x} : a \to b$ and $\underline{y} : b \to c$ is by concatenation,

$$\underline{y} \circ \underline{x} = ((y_1, \nu_1), \ldots, (y_m, \nu_n), (x_1, \varepsilon_1), \ldots, (x_n, \varepsilon_n)) \ . \tag{2.5}$$

Let us denote this category by $\mathbf{D} \equiv \mathbf{D}[D_2, D_1^{\mathrm{top}}]$. Morphism spaces $a \to b$ are written as $\mathbf{D}(a, b)$. The conjugation is the involution $* : \mathbf{D}(a, b) \to \mathbf{D}(b, a)$ given

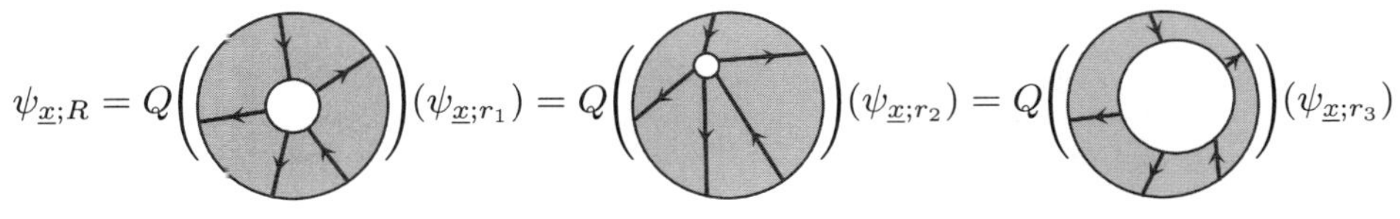

FIGURE 5. Illustration of the condition for scale and translation invariant family of states: Let $\psi_{\underline{x}}$ be such a family. The figure shows Q applied to three annuli, understood as bordisms $O(\underline{x}, r_i) \to O(\underline{x}, R)$, for $i = 1, 2, 3$, where r_i denotes the radius of the inner disc of the ith annulus shown above. All three annuli have the same outer radius R. Applying Q to the bordism and evaluating the resulting linear map on $\psi_{\underline{x},r_i} \in Q(O(\underline{x}, r_i))$, $i = 1, 2, 3$, always results in the vector $\psi_{\underline{x};R} \in Q(O(\underline{x}, R))$ of the same family.

by

$$((x_1, \varepsilon_1), \ldots, (x_n, \varepsilon_n))^* = ((x_n, -\varepsilon_n), \ldots, (x_1, -\varepsilon_1)) \ . \tag{2.6}$$

Note that endomorphisms $\underline{x} : a \to a$ in $\mathbf{D}$ are precisely the tuples of cyclically composable domain walls. In particular, for any $\underline{x}, \underline{y} : a \to b$, the morphism $\underline{y} \circ \underline{x}^*$ is cyclically composable.

To define the 2-category structure on domain walls, we need the notion of translation and scale invariant families of states. Their definition will take us a few paragraphs.

We will only need to know Q on a subset of bordisms, each of which consists of a single disc in $\mathbb{R}^2$ from which a number of smaller discs have been removed (if there were no domain walls, these bordisms would form the little discs operad, see e.g. [**Ma**]). The metric on the bordism is the one induced by $\mathbb{R}^2$ and the boundaries are parametrised by linear maps that are a combination of a translation and a scale transformation $x \mapsto rx + v$, where $x, v \in \mathbb{R}^2$ and $r \in \mathbb{R}_{>0}$.

The objects which serve as source and target of these bordisms are described as follows. Let $\underline{x}$ be cyclically composable and denote by $O(\underline{x}; r)$ an object in the bordism category consisting of a single S^1 with 0-dimensional submanifold given by n points not containing $-1 \in S^1$. These are clockwise cyclically labelled $x_1, \ldots, x_n$, such that x_1 labels the first point in clockwise direction from $-1 \in S^1$. The collar around S^1 is obtained by taking concentric copies to fill a small neighbourhood. The conformal factor defining the metric on the collar is $e^\sigma = r^2$, so that the parametrising map $x \mapsto rx + v$, which takes the S^1 (with radius 1) to a circle of radius r is an isometry. As all x_i are in D_1^{top}, by definition the vector space $Q(O(\underline{x}; r))$ does not depend on the precise position of the n marked points on S^1, as long as they are in the prescribed ordering. However, the vector space $Q(O(\underline{x}; r))$ may still depend on r.[4]

[4]In many examples (but not always), the spaces $Q(O(\underline{x}; r))$ for different values of r are isomorphic, with a preferred isomorphism given by evaluating Q on an annulus with the two radii. But even in this case we do not demand that one passes to a formulation of the theory where these state spaces are actually *equal*.

Let $D(R;r,v) : O(\underline{x};r) \to O(\underline{x};R)$ be the bordism given by a disc of radius R in $\mathbb{R}^2$ centred at the origin, from which a smaller disc of radius r and centre v has been removed. The domain walls are straight lines and the boundary parametrisation is given by scaling and translation as above (see figure 5). A *scale and translation invariant family* $\psi_{\underline{x}}$ is a family of vectors $\{\psi_{\underline{x};r}\}_{r\in\mathbb{R}_{>0}}$ with $\psi_{\underline{x};r} \in Q(O(\underline{x};r))$ such that

$$(2.7) \qquad \psi_{\underline{x};R} = Q\big(D(R;r,v)\big)(\psi_{\underline{x};r}) \quad \text{for all } r,R>0,\ v\in\mathbb{R}^2 \text{ with } r+|v|<R\ .$$

This condition is illustrated in figure 5. The *space of scale and translation invariant families*,

$$(2.8) \qquad H^{\text{inv}}(\underline{x})\ ,$$

is defined to be the vector space of all scale and translation invariant families $\psi_{\underline{x}} \equiv \{\psi_{\underline{x};r}\}_{r\in\mathbb{R}_{>0}}$ for fixed $\underline{x}$. The space $H^{\text{inv}}(\underline{x})$ may be zero-dimensional.

Scale and translation invariant families have the following important property: all amplitudes $Q(M)$ – with M a disc in $\mathbb{R}^2$ with smaller discs removed – are independent of the position and size of an ingoing boundary circle $O(\underline{x};r)$ for which $\psi_{\underline{x};r}$ is inserted as the corresponding argument. This can be seen by using functoriality of Q to cut out a disc $D(R;r,v)$ from M containing such an ingoing boundary, then moving the ingoing boundary circle using the defining property (2.7), and finally gluing the resulting disc $D(R;r',v')$ back.

Given $\underline{x},\underline{y} : a \to b$, we define the space of 2-morphisms from $\underline{x}$ to $\underline{y}$ to be

$$(2.9) \qquad \mathbf{D}_2(\underline{x},\underline{y}) := H^{\text{inv}}(\underline{y}\circ\underline{x}^*)\ .$$

The identity 2-morphisms, and the horizontal and vertical composition are defined by the bordisms shown in figure 6. The identity 1-morphism $\mathbf{1}_a : a \to a$, for $a \in D_2$, is the empty tuple $\mathbf{1}_a = (\,)$.

REMARK 2.2. (i) In order to obtain *families* of states from the bordisms shown in figure 6, one uses that there is an $\mathbb{R}_{>0}$-action on metric bordisms given by rescaling the metric. For the disc shaped bordisms in $\mathbb{R}^2$ relevant here, this amounts to a rescaling by some $R>0$. Consider the bordism in figure 6 b) as an example. Call this bordism M and assume that the radius of its outer disc is 1. Thus

$$(2.10) \qquad M\ :\ O(\underline{z}\circ\underline{y}^*;r_1) \sqcup O(\underline{y}\circ\underline{x}^*;r_2) \longrightarrow O(\underline{z}\circ\underline{x}^*;1)\ .$$

For each $R>0$ this produces a bordism $RM : O(\underline{z}\circ\underline{y}^*;Rr_1) \sqcup O(\underline{y}\circ\underline{x}^*;Rr_2) \to O(\underline{z}\circ\underline{x}^*;R)$. Given two families $\phi_1 \in H^{\text{inv}}(\underline{z}\circ\underline{y}^*)$ and $\phi_2 \in H^{\text{inv}}(\underline{y}\circ\underline{x}^*)$, we obtain a family of vectors

$$(2.11) \qquad \psi_R := Q(RM)\big(\phi_{1;Rr_1}\,,\ \phi_{2;Rr_2}\big) \ \in\ Q\big(O(\underline{z}\circ\underline{x}^*;R)\big)\ .$$

The family $\{\psi_R\}_{R\in\mathbb{R}_{>0}}$ is again scale and translation invariant. This follows by substituting into the defining property (2.7) and using that ϕ_1 and ϕ_2 are scale and translation invariant families.

(ii) For all $r>0$, $v\in\mathbb{R}^2$ with $r+|v|<1$, the bordism $D(1;r,v) : O(\underline{x};r) \to O(\underline{x};1)$ induces the identity map on $H^{\text{inv}}(\underline{x})$. This is just a reformulation of the defining property (2.7) using the prescription in (i). In other words, cylinders give the identity map on H^{inv}, not just idempotents. This is important when verifying that the bordism in figure 6 a) is indeed the unit for the vertical composition in figure 6 b).

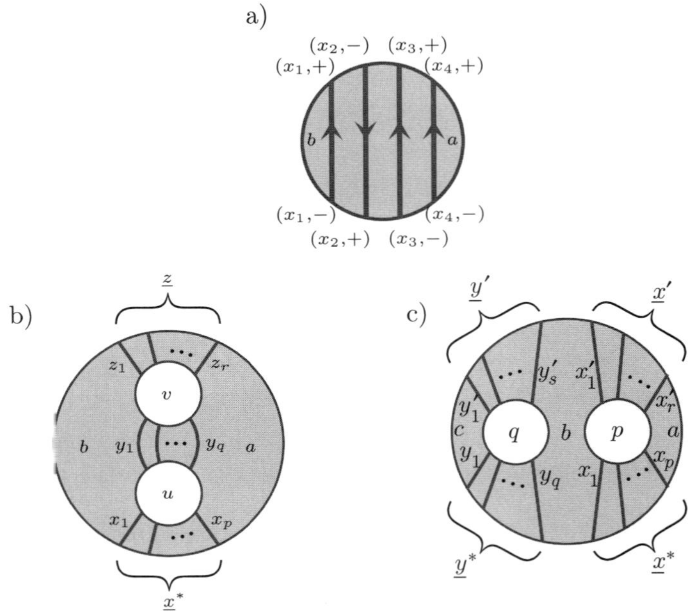

FIGURE 6. Bordisms defining the structure maps for the 2-category. a) the identity on a 1-morphism $\underline{x} : a \to b$; b) vertical composition of 2-morphisms $u \in \mathbf{D}_2(\underline{x}, \underline{y})$ and $v \in \mathbf{D}_2(\underline{y}, \underline{z})$, where $\underline{x}, \underline{y}, \underline{z} \in \mathbf{D}(a, b)$ (drawing the vector inside the cut-out disc means that this vector is to be used as the corresponding argument after applying Q); c) horizontal composition of 2-morphisms $p \in \mathbf{D}_2(\underline{x}, \underline{x}')$, $q \in \mathbf{D}_2(\underline{y}, \underline{y}')$, where $\underline{x}, \underline{x}' : a \to b$ and $\underline{y}, \underline{y}' : b \to c$.

The fact that we are working with topological domain walls and with scale and translation invariant families of states ensures that the properties of a 2-category are satisfied (as composition of 1-morphisms is strictly associative, and the unit 1-morphisms are strict, we indeed have a 2-category and not only a bicategory). Let us denote this 2-category as

$$\mathbf{D}[Q] \equiv \mathbf{D}[D_2, D_1^{\mathrm{top}}; Q] \ . \tag{2.12}$$

As was to be expected, moving one dimension up from the example in section 2.2 also increased the categorial level: in this construction the pre-category $D_1^{\mathrm{top}} \rightrightarrows D_2$ gets extended to a 2-category.

REMARK 2.3. (i) Actually, $\mathbf{D}[Q]$ carries more structure. For example, each 1-morphism $\underline{x} : a \to b$ has a left and a right adjoint, namely $\underline{x}^* : b \to a$, together with adjunction maps as shown in figure 7 (see [**Gr**, Sec. I.6] for more on adjunctions in bicategories). Such rigid and related structures on the category of defects were discussed already in [**Fröl, MN, CR**]. Rigid and pivotal structures on the 2-category $\mathbf{D}[Q]$ were studied in the context of planar algebras[5] in [**Go, DGG**].

[5]A planar algebra [**Jo**] can be understood as a two-dimensional theory with exactly two world sheet phases $D_2 = \{a, b\}$, exactly one topological domain wall type $D_1 = \{a \xrightarrow{x} b\}$, and no

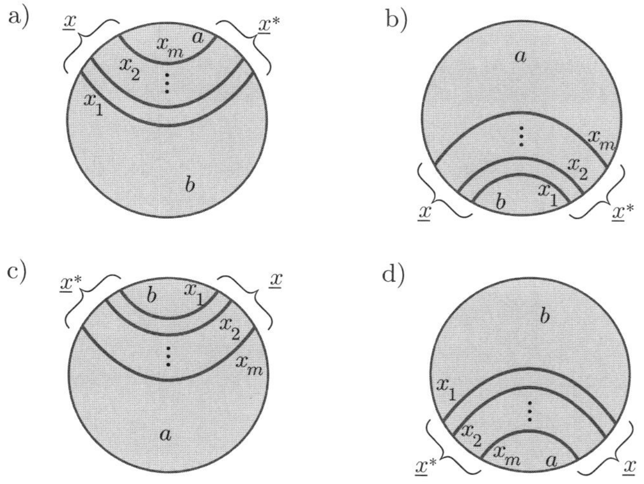

FIGURE 7. Bordisms defining the adjunction maps in $\mathbf{D}[Q]$: The left and right adjoint of $\underline{x} : a \to b$ is $\underline{x}^* : b \to a$ and applying Q to the bordisms shown gives the adjunction maps a) $b_{\underline{x}} : \mathbf{1}_b \to \underline{x} \circ \underline{x}^*$; b) $d_{\underline{x}} : \underline{x}^* \circ \underline{x} \to \mathbf{1}_a$; c) $\tilde{b}_{\underline{x}} : \mathbf{1}_a \to \underline{x}^* \circ \underline{x}$; d) $\tilde{d}_{\underline{x}} : \underline{x} \circ \underline{x}^* \to \mathbf{1}_b$.

(ii) The 2-category $\mathbf{D}[Q]$ is an invariant attached to a quantum field theory with defects. Interestingly, even though only 'topological data' enters its definition (topological domain walls and scale and translation invariant families of states), general quantum field theories do produce more general 2-categories $\mathbf{D}[Q]$ than topological field theories. In a nutshell, the reason is that the rigid structure mentioned in (i) will tend to produce integer quantum dimensions for topological field theories, while for example in rational conformal field theories non-integer quantum dimensions occur.[6]

In slightly more detail, fix a topological field theory with defects and take $k = \mathbb{C}$. Consider the bordism $M : \emptyset \to \emptyset$ given by a torus (say $[0,1] \times [0,1]$ with opposite edges identified) with a single defect line labelled $x \in D_1$ wrapping a non-contractible cycle (say $[0,1] \times \{\frac{1}{2}\}$). We label the unique connected domain of the bordism by $a \in D_2$, so that $x : a \to a$. Then $Q(M) : \mathbb{C} \to \mathbb{C}$ is just a number. This number can be computed in two ways. Let $M_| : O(x) \to O(x)$ be the annulus obtained by cutting M along $\{0\} \times [0,1]$, i.e. by not identifying the vertical edges of M. Then $M_|$ is just the cylinder over $O(x)$. We will learn later in Remark 3.12, that for a topological field theory the space of scale and translation invariant

junctions, $D_0 = \emptyset$. Furthermore, the theory is only defined on genus zero surfaces with exactly one out-going boundary circle, that is, on discs with smaller discs removed, see figures 5–7.

[6]A notion of 2d TFT with domain walls was also studied in [**KPS**] in relation to subfactor planar algebras. Apart from there being exactly two world sheet phases and one type of domain wall – as is usual in the planar algebra setting – there is one important difference: in [**KPS**] a bordism is in addition equipped with a decomposition into 'genus 0 components', and bordisms with different such decompositions are considered distinct, unless they have a common refinement (see [**KPS**, Def. 2.7]). This excludes for example the decomposition of a torus along different cycles to be used below (the 'Cardy condition'). Thus the functors constructed in [**KPS**] are in general not defect TFTs in our sense (cf. eqn. (3.2) below).

states $H^{\rm inv}(x)$ can be identified with the image of $Q(M_|)$ (which is an idempotent in topological field theory) in $Q(O(x))$. Thus, using also functoriality of Q,

$$Q(M) = \mathrm{tr}_{Q(O(x))}\, Q(M_|) = \mathrm{tr}_{H^{\rm inv}(x)}\, \mathrm{id} = \dim(H^{\rm inv}(x)) \in \mathbb{Z}_{\geq 0}\ . \tag{2.13}$$

On the other hand, we can consider $M_- : O(a) \to O(a)$, which is obtained by cutting M along $[0,1] \times \{0\}$, i.e. by not identifying the horizontal edges. The endomorphism $Q(M_-) : Q(O(a)) \to Q(O(a))$ is called the *defect operator* for the defect $x : a \to a$; we will return to this briefly in section 3.5 below. By the same reasoning as above, $Q(M) = \mathrm{tr}_{Q(O(a))}\, Q(M_-)$. Let $C_{O(a)}$ be the cylinder over $O(a)$. By functoriality (and again only for topological field theory) we have $Q(M_-) = Q(C_{O(a)}) \circ Q(M_-) \circ Q(C_{O(a)})$, so that the image of $Q(M_-)$ lies in the image of the idempotent $Q(C_{O(a)})$ and $Q(M_-)$ acts trivially on the kernel of $Q(O(a))$. By restriction we obtain an endomorphism $Q(M_-) : H^{\rm inv}(\mathbf{1}_a) \to H^{\rm inv}(\mathbf{1}_a)$ and the trace can be computed in this restriction,

$$Q(M) = \mathrm{tr}_{Q(O(a))}\, Q(M_-) = \mathrm{tr}_{H^{\rm inv}(\mathbf{1}_a)}\, Q(M_-)\ . \tag{2.14}$$

The fact that the traces (2.13) and (2.14) agree is known as the Cardy condition (because of the paper [**Ca**]) and was first investigated for topological defects in [**PZ**] in the context of rational conformal field theory.

In summary, for topological field theories, the trace of a defect operator over the space of scale and translation invariant states is equal to the dimension of a vector space, and is thus a non-negative integer. In rational conformal field theories with non-degenerate vacuum (this means that the space $H^{\rm inv}(\mathbf{1}_a)$ is one-dimensional), the defect operator acts by multiplying with a number – the (left or right) quantum dimension of x – and the trace $\mathrm{tr}_{H^{\rm inv}(\mathbf{1}_a)}\, Q(M_-)$ is then equal to this number. In many examples, this quantum dimension is not an integer (and not even rational, though still algebraic). This is the case for the examples studied in [**PZ**] and [**FRS1, Frö1**].

3. A simple example: 2d lattice top. field theory

It is difficult to find functors from $\mathrm{Bord}_{2,1}^{\rm def}(D_2, D_1, D_0)$ to topological vector spaces which depend non-trivially on the metric. On the other hand, it is easy to construct examples where the functors are independent of the metric and boundary parametrisation. In this section we describe such a construction.

For the remainder of this section we fix a field k.

3.1. Category of smooth bordisms with defects. Instead of posing restrictions on the functor one can modify the bordism category accordingly. This leads us to define a symmetric monoidal category of *smooth bordisms with defects*, which we denote as

$$\mathrm{Bord}_{2,1}^{\rm def,top}(D_2, D_1, D_0)\ . \tag{3.1}$$

The modifications relative to the definition in section 2.3 are as follows.

- *Objects:* The collar around an S^1 no longer carries a metric.
- *Morphisms:* The manifold does not carry a metric and the parametrising maps are only required to be smooth (rather than isometric). Two bordisms are equivalent if there is a diffeomorphism between them which preserves orientation, decomposition, labelling, as well as the image of the point $-1 \in S^1$ in each connected component of the boundary (rather than

commuting with the parametrising maps in some neighbourhood of the boundary).

The symmetric monoidal structure is as in section 2.3.

Note that this definition is different form the standard 2-bordism category for topological field theories even in the case without domain walls ($D_2 = \{*\}$ and $D_1 = D_0 = \emptyset$) because we have still added the identities (and S_n-action) by hand to the space of morphisms; in particular, the cylinder over a given object U is *not* the identity morphism (but it is still an idempotent).

Denote by $\mathcal{V}ect_f(k)$ the symmetric monoidal category of finite-dimensional k-vector spaces. Fix furthermore sets D_i, $i = 0, 1, 2$, of defect labels with maps as required. The aim of this section is to construct examples of symmetric monoidal functors

$$T : \mathrm{Bord}_{2,1}^{\mathrm{def,top}}(D_2, D_1, D_0) \longrightarrow \mathcal{V}ect_f(k) \ . \tag{3.2}$$

We will do this via a lattice TFT construction which is a straightforward generalisation of the original lattice TFT without domain walls [**BP, FHK**] and of the lattice construction of homotopy TFTs in [**Tu**, Sec. 7]. A construction of field theory correlators on arbitrary world sheets in the presence of domain walls first appeared in [**FRS1, FRS2, Fröl**], where it was carried out in the context of two-dimensional rational conformal field theory. The construction for TFTs given in this section can be extracted from this in the special case that the modular category underlying the CFT is that of vector spaces.

Note, however, that this will not give the most general such functor T (as it does not even do this in the case without domain walls); a classification of functors (3.2) akin to the one in the situation without domain walls in [**Di, Ab**] is at present not known. A classification of 2d TFTs with defects that can be extended down to points has been reported in [**SP**]; related results are given in [**Lu1**, Ex. 4.3.23]. Two dimensional homotopy TFTs over $X = K(G, 1)$, which in the present language correspond to defect TFTs with invertible defects labelled by a group G,[7] have been classified in [**Tu**, Thm. 4.1]. The classification in the case of simply connected X is given in [**BT**, Thm. 4.1].

The construction works in two steps. In the first step, we introduce a larger category, $\mathrm{Bord}_{2,1}^{\mathrm{def,top,cw}}$, where objects and morphisms are in addition endowed with a cell decomposition (section 3.2). It comes with a forgetful functor F : $\mathrm{Bord}_{2,1}^{\mathrm{def,top,cw}} \to \mathrm{Bord}_{2,1}^{\mathrm{def,top}}$, which is surjective (not just essentially surjective) and full (but not faithful). We then construct a symmetric monoidal functor T^{cw} : $\mathrm{Bord}_{2,1}^{\mathrm{def,top,cw}} \to \mathcal{V}ect_f(k)$ (section 3.5). In the second step, we show that T^{cw} is independent of the cell-decomposition in the sense that there exists a symmetric

[7]By a defect TFT with only invertible defects we mean the case that $D_2 = \{*\}$, $D_0 = \emptyset$ and that $D_1 = G$ is a group (more generally we can take $s, t : D_1 \to D_2$ to be a groupoid). We demand that the linear map assigned by T in (3.2) to a bordism does not change if we replace two parallel defect lines with opposite orientation "$\rightleftarrows$" labelled by the same element of D_1 by the 'reconnected' defect lines "$\supset\subset$". Given a bordism M with defect lines labeled by group elements in G, we can construct a principal G-bundle on the bordism by taking the trivial G-bundle over each component of M_2 and choosing the transition functions across M_1 to be multiplication with the group element labelling that component of M_1. This gives a functor from $\mathrm{Bord}_{2,1}^{\mathrm{def,top}}(\{*\}, G, \emptyset)$ to principal G-bundles with two-dimensional base.

monoidal functor T which makes the diagram

$$
\begin{array}{ccc}
\mathrm{Bord}_{2,1}^{\mathrm{def,top,cw}} & \xrightarrow{T^{\mathrm{cw}}} & \mathcal{V}ect_f(k) \\
{\scriptstyle F}\downarrow & \nearrow_{\exists!\,T} & \\
\mathrm{Bord}_{2,1}^{\mathrm{def,top}} & &
\end{array}
\tag{3.3}
$$

commute on the nose (section 3.6). Since F is surjective and full, this diagram defines T uniquely.

3.2. Category of bordisms with cell decomposition. We will use a class of cell decompositions introduced in [**Ki**, Def. 5.1], called PLCW decompositions there. These are less general than CW-complexes, which for example allow one to decompose S^2 into a 0-cell (a point on S^2) and 2-cell (S^2 minus that point). However, they are more general than regular CW-complexes, where one cannot identify different faces of one given cell, or even triangulations, where in addition each cell is a simplex. It is shown in [**Ki**] that any two PLCW decompositions are related by a simple collection of local moves.

Given a compact n-dimensional manifold M, possibly with non-empty boundary, we will consider decompositions of M into a finite number of mutually disjoint open k-cells, $k = 0, \dots, n$, with the following properties. Let B^k be the closed unit ball in $\mathbb{R}^k$ and let $\mathring{B}^k$ be its interior. For each k-cell C there has to exist a continuous[8] map $\varphi : B^k \to M$ such that

- C is the homeomorphic image of $\mathring{B}^k$,
- there exists a decomposition of the boundary S^{k-1} of B^k which gets mapped by φ to the decomposition in M,
- φ is a homeomorphism when restricted to the interior of each cell on S^{k-1}.

For more details we refer to [**Ki**]. We just note that this last condition excludes the decomposition of S^2 into a single 0-cell and a single 2-cell mentioned above.

By abuse of terminology, we will refer to a decomposition of M as just described simply as a *cell decomposition* $C(M)$. The set of i-dimensional cells (i-cells for short) will be called $C_i(M)$.

The category $\mathrm{Bord}_{2,1}^{\mathrm{def,top,cw}}(D_2, D_1, D_0)$ is the same as $\mathrm{Bord}_{2,1}^{\mathrm{def,top}}(D_2, D_1, D_0)$, except that objects and morphisms are equipped with the following extra structure:

- *Objects:* The 1-dimensional part of an object U (the disjoint union of circles, not their collars) is equipped with a cell decomposition $C(U)$ such that each point of the set U_0 of marked points lies in a 1-cell (recall that a 1-cell is homeomorphic to an open interval), and such that each 1-cell contains at most one point of U_0.
- *Morphisms:* By a *bordism with cell decomposition* we mean a bordism $M = M_2 \cup M_1 \cup M_0$ in $\mathrm{Bord}_{2,1}^{\mathrm{def,top}}$, which is equipped with a cell decomposition $C(M)$ such that
  - the 1-dimensional submanifold M_1 only intersects 1-cells and 2-cells, but not 0-cells. Each 1-cell intersects M_1 in at most one point.

[8]To match [**Ki**] we should work in the PL setting. We thus impose the additional condition that the manifold M with the decomposition as defined is homeomorphic to a PLCW complex. This is automatically satisfied if the manifold and the cell maps are already PL.

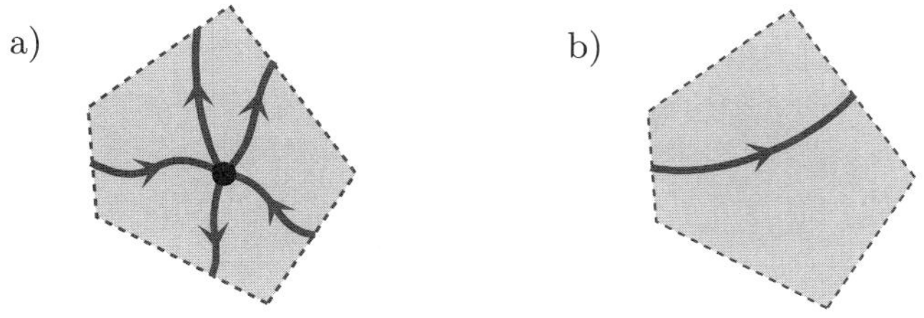

FIGURE 8. Allowed configurations of 2-cells and defects: (a) 2-cell containing a point from M_0; here only a star-shaped pattern of domain walls is allowed and each edge has to be crossed by precisely one domain wall. (b) 2-cell containing no point from M_0 but part of M_1; here only one segment of M_1 may lie in the 2-cell, and it must enter and leave the 2-cell via distinct edges.

- each point of M_0 lies in a 2-cell, and each 2-cell contains at most one point of M_0. A 2-cell containing a point of M_0 must be homeomorphic to one of the type shown in figure 8 a), i.e. it may only contain a 'star-shaped' configuration of domain walls, such that each edge of this 2-cell is traversed by exactly one domain wall.
- a 2-cell containing no point from M_0 but a part of M_1 must be homeomorphic to one of the type shown in figure 8 b), i.e. its intersection with M_1 is a single open interval.

If M has non-empty boundary ∂M, we demand that ∂M is a union of 0-cells and 1-cells (and hence does not intersect 2-cells). Furthermore, the decomposition of ∂M has to be the image of the cell decomposition of the source and target object under the parametrising maps.

Morphisms are equivalence classes of bordisms with cell decomposition. Two bordisms with cell decomposition are equivalent if their underlying bordisms are equivalent in $\mathrm{Bord}_{2,1}^{\mathrm{def,top}}$ and if the two cell decompositions are related by an isotopy (which at each instance has to give a bordism with cell decomposition).

All cells in $C(U)$ and $C(M)$ are a priori unoriented. However, the 2-cells of $C(M)$ have an orientation induced by that of M.

3.3. Algebraic preliminaries. By an *algebra* we shall always mean a unital, associative algebra over k. The *centre* $Z(A)$ of an algebra A is the (commutative, unital) subalgebra

$$Z(A) = \{\, z \in A \,|\, za = az \text{ for all } a \in A \,\} \,. \tag{3.4}$$

Given a right A-module M and a left A-module N, the tensor product $M \otimes_A N$ is defined as the cokernel

$$M \otimes A \otimes N \xrightarrow{l-r} M \otimes N \xrightarrow{\pi_\otimes} M \otimes_A N \,, \tag{3.5}$$

where $l(m \otimes a \otimes n) = m \otimes (a.n)$ and $r(m \otimes a \otimes n) = (m.a) \otimes n$ denote the left and right action. Given an A-A-bimodule X, we shall also require the 'cyclic tensor product' which identifies the left and right action of A on X. We denote this tensor product by $\circlearrowleft_A X$; it is defined as the cokernel

$$A \otimes X \xrightarrow{l-r} X \xrightarrow{\pi_\otimes} \circlearrowleft_A X \ , \tag{3.6}$$

where $l(a \otimes x) = a.x$ and $r(a \otimes x) = x.a$. Note that $\circlearrowleft_A X$ is in general no longer an A-A-bimodule; it does, however, carry a coinciding left and right action of $Z(A)$. Denote by $[A, A]$ the linear subspace of A (not the ideal) generated by all elements of the form $ab - ba$. Then by definition $\circlearrowleft_A A = A/[A, A]$.[9]

A *Frobenius algebra* is a finite-dimensional algebra A together with a linear map $\varepsilon : A \to k$, called the *counit*, such that the bilinear pairing

$$\langle a, b\rangle := \varepsilon(ab) \tag{3.7}$$

on $A \times A$ is non-degenerate. We denote by $\beta : k \to A \otimes A$ the unique linear map such that $((\varepsilon \circ m) \otimes \mathrm{id}_A) \circ (\mathrm{id} \otimes \beta) = \mathrm{id}_A$. In other words, if a_i is a basis of A and a'_i the dual basis in the sense that $\langle a_i, a'_j\rangle = \delta_{i,j}$, then $\beta(1) = \sum_i a'_i \otimes a_i$. The defining property can be written as

$$\sum_i \langle x, a'_i\rangle\, a_i = x \quad \text{for all } x \in A\ . \tag{3.8}$$

By associativity of A, the pairing (3.7) is always *invariant*, i.e. $\langle a, bc\rangle = \langle ab, c\rangle$.

For $a \in A$ let $L_a : A \to A$ be the left multiplication by a, $L_a(b) = ab$. We say that A is a *Frobenius algebra with trace pairing* if it is a Frobenius algebra whose counit is given by[10] $\varepsilon(a) = \mathrm{tr}_A(L_a)$. Note that 'Frobenius' is extra structure on an algebra, while 'Frobenius with trace pairing' is a property of an algebra.

In the following we list some of the special properties of Frobenius algebras with trace pairing. Firstly, it is automatically *symmetric*: $\langle a, b\rangle = \langle b, a\rangle$. Consequently also β is symmetric: $\beta(1) = \sum_i a'_i \otimes a_i = \sum_i a_i \otimes a'_i$. Next, by the definition of the counit we have the following identity: $\langle x, 1\rangle = \varepsilon(x) = \mathrm{tr}_A(L_x) = \sum_i a_i^*(xa_i) = \sum_i \langle xa_i, a'_i\rangle = \langle x, \sum_i a_i a'_i\rangle$. Since this holds for all $x \in A$, we conclude

$$\sum_i a_i a'_i = 1\ . \tag{3.9}$$

For all $a, b \in A$ we have

$$\sum_i (aa'_i b) \otimes a_i = \sum_i a'_i \otimes (ba_i a) \quad \in A \otimes A\ ; \tag{3.10}$$

this can be proved by pairing both sides with an arbitrary x. On the left hand side, pairing with x produces $\sum_i \langle x, aa'_i b\rangle a_i = \sum_i \langle bxa, a'_i\rangle a_i = bxa$ by (3.8). The right hand side gives the same result.

If we interpret $\beta(1)$ as an element of the algebra $A^{\mathrm{op}} \otimes A$, the two equations above lead to the following result:

---

[9]The quotient $A/[A, A]$ is not necessarily isomorphic to $Z(A)$ – just take A to be the 3-dimensional algebra of upper triangular $2{\times}2$ matrices in which case $Z(A) \cong k$ while $A/[A, A] \cong k \oplus k$.

[10]Let $R_a(b) = ba$ denote the right multiplication by a and set $\varepsilon(a) = \mathrm{tr}_A(L_a)$, $\varepsilon'(a) = \mathrm{tr}_A(R_a)$. Then (A, ε) is a Frobenius algebra if and only if (A, ε') is a Frobenius algebra and in this case $\varepsilon(a) = \varepsilon'(a)$. For a proof see e.g. [**FRS1**, Lem. 3.9] in the special case that the tensor category is $\mathcal{V}ect_f(k)$. Note that if (A, ε) and (A, ε') are not Frobenius, the statement may be false; for example, if A is the 3-dimensional algebra of upper triangular $2{\times}2$ matrices and $a = \left(\begin{smallmatrix} 0 & 0 \\ 0 & 1 \end{smallmatrix}\right)$, then $\mathrm{tr}_A(L_a) = 1$ and $\mathrm{tr}_A(R_a) = 2$.

LEMMA 3.1. *In the algebra* $A^{\text{op}} \otimes A$ *we have*
(i) $\beta(1) \cdot (a \otimes 1) = \beta(1) \cdot (1 \otimes a)$ *and* $(a \otimes 1) \cdot \beta(1) = (1 \otimes a) \cdot \beta(1)$,
(ii) $\beta(1) \cdot \beta(1) = \beta(1)$.

PROOF. By definition, the product of $A^{\text{op}} \otimes A$ is $(a \otimes a') \cdot (b \otimes b') = (ba) \otimes (a'b')$. Statement (i) is nothing but (3.10), while (ii) follows from (i) and (3.9) via $\beta(1) \cdot \beta(1) = \sum_i \beta(1) \cdot (a_i' \otimes 1) \cdot (1 \otimes a_i) = \sum_i \beta(1) \cdot (1 \otimes a_i') \cdot (1 \otimes a_i) = \sum_i \beta(1) \cdot (1 \otimes (a_i' a_i)) = \beta(1) \cdot (1 \otimes 1)$. □

Consider a right A-module M and a left A-module N as above. For Frobenius algebras with trace pairing, the tensor product $M \otimes_A N$ can be canonically identified with a subspace of $M \otimes N$ as follows. The tensor product $M \otimes N$ carries an $A^{\text{op}} \otimes A$-action. Define the linear map $p_\otimes : M \otimes N \to M \otimes N$ to be the action of $\beta(1)$ on $M \otimes N$:

$$p_\otimes(m \otimes n) = \beta(1).(m \otimes n) = \sum_i (m.a_i') \otimes (a_i.n) \ . \tag{3.11}$$

LEMMA 3.2. *Let* A *be a Frobenius algebra with trace pairing.*
(i) $p_\otimes(m.a \otimes n) = p_\otimes(m \otimes a.n)$ *holds for all* $a \in A$, $m \in M$, $n \in N$.
(ii) $p_\otimes$ *is idempotent.*
(iii) $\text{im}(p_\otimes) = M \otimes_A N$.

PROOF. Parts (i) and (ii) are immediate consequences of Lemma 3.1. For part (iii), denote by

$$e_\otimes : \text{im}(p_\otimes) \longrightarrow M \otimes N \quad , \quad \pi_\otimes : M \otimes N \longrightarrow \text{im}(p_\otimes) \ , \tag{3.12}$$

the embedding of the image of $p_\otimes$ into $M \otimes N$ and the projection from $M \otimes N$ to the image. They satisfy

$$e_\otimes \circ \pi_\otimes = p_\otimes \quad , \quad \pi_\otimes \circ e_\otimes = \text{id}_{\text{im}(p_\otimes)} \ . \tag{3.13}$$

Consider the diagram

$$\begin{array}{ccccc} M \otimes A \otimes N & \xrightarrow{l-r} & M \otimes N & \xrightarrow{\pi_\otimes} & \text{im}(p_\otimes) \\ & & \downarrow f & \swarrow_{f \circ e_\otimes} & \\ & & V & & \end{array} \ . \tag{3.14}$$

Here V is a vector space and f satisfies $f \circ (l - r) = 0$. By part (i) the map $\pi_\otimes$ satisfies $\pi_\otimes \circ (l - r) = 0$. From $f(m.a \otimes n) = f(m \otimes a.n)$ it is easy to see that $f \circ p_\otimes = f$, and so $f = (f \circ e_\otimes) \circ \pi_\otimes$. Thus the above diagram commutes and we have verified the universal property of the cokernel. □

The construction of $p_\otimes$, $e_\otimes$, $\pi_\otimes$ works similarly for the $\circlearrowleft_A$ tensor product of an A-A-bimodule X. In this case,

$$p_\otimes(x) = \sum_i a_i.x.a_i' \ . \tag{3.15}$$

We use the same notation $e_\otimes : \circlearrowleft_A X \to X$ and $\pi_\otimes : X \to \circlearrowleft_A X$ as above. For multiple tensor products, the idempotents can be combined. For the state spaces of the TFT with defects to be constructed below, the maps

(3.16)
$$X_1 \otimes X_2 \otimes \cdots \otimes X_n \underset{e_\otimes}{\overset{\pi_\otimes}{\rightleftarrows}} \circlearrowleft_{A_{n,1}} X_1 \otimes_{A_{1,2}} X_2 \otimes_{A_{2,3}} \cdots \otimes_{A_{n-1,n}} X_n$$

will be useful. Here the X_i are bimodules with left/right actions of algebras as indicated. As above, the maps $e_\otimes$ and $\pi_\otimes$ satisfy $\pi_\otimes \circ e_\otimes = \mathrm{id}$ and $e_\otimes \circ \pi_\otimes = p_\otimes$. The projector $p_\otimes$ in this case is

$$(3.17)\quad p_\otimes(x_1 \otimes x_2 \otimes \cdots \otimes x_n) = \sum_{i_1,i_2,\dots,i_n} (a_{i_1}.x_1.a'_{i_2}) \otimes (a_{i_2}.x_2.a'_{i_3}) \otimes \cdots \otimes (a_{i_n}.x_n.a'_{i_1}) ,$$

where the a_{i_k} are bases of the corresponding algebras.

LEMMA 3.3. *Let A be a Frobenius algebra with trace pairing. Then*
$$Z(A) \cong \circlearrowleft_A A = A/[A,A] .$$

PROOF. Here $p_\otimes : A \to A$ is given by $p_\otimes(x) = \sum_i a_i x a'_i$. By the same reasoning as in (3.14) we conclude that the cokernel $A/[A,A]$ is isomorphic to $\mathrm{im}(p_\otimes)$. It remains to show that $\mathrm{im}(p_\otimes) = Z(A)$. For $z \in Z(A)$ we have $p_\otimes(z) = \sum_i a_i z a'_i = z(\sum_i a_i a'_i) = z$, where we used (3.9). Thus $Z(A) \subset \mathrm{im}(p_\otimes)$. Conversely, if $x \in \mathrm{im}(p_\otimes)$ then $x = \sum_i a_i x a'_i$. Then for all $y \in A$ we have

(3.18)
$$xy = \sum_i a_i x a'_i y = m \circ (\sum_i a_i \otimes (x a'_i y)) = m \circ (\sum_i (y a_i x) \otimes a'_i) = \sum_i y a_i x a'_i = yx ,$$

where we used (3.10) together with symmetry of β. Thus $\mathrm{im}(p_\otimes) \subset Z(A)$. □

REMARK 3.4. Recall that for a unital associative R-algebra (for R a commutative ring), the 0th Hochschild homology $HH_0(A)$ and cohomology $HH^0(A)$ are given by

$$(3.19)\qquad HH_0(A) = A/[A,A] \qquad \text{and} \qquad HH^0(A) = Z(A) ,$$

see [**Lo**, Sec. 1.1, 1.5]. Similarly, for an A-A-bimodule X one has $H_0(A,X) = \circlearrowleft_A X$ and $H^0(A,X) = \{x \in X \,|\, a.x = x.a \text{ for all } a \in A\}$. In the situation of Lemma 3.3, that is if A is a Frobenius algebra with trace pairing, one finds $H_0(A,X) \cong H^0(A,X)$. We will see in (3.49) below that the 0th Hochschild (co)homology provides the state space which the 2d lattice TFT with defects assigns to a circle with a single marked point.

3.4. Data for lattice TFT with defects. Fix sets D_2, D_1, D_0 with maps $s,t : D_1 \to D_2$ and a map j as in (2.2). The data that serves as input to the lattice TFT construction is as follows.

(1) For each $a \in D_2$ a Frobenius algebra A_a with trace pairing.
(2) For each $x \in D_1$ a finite-dimensional $A_{t(x)}$-$A_{s(x)}$-bimodule X_x.

There is also a piece of data associated to D_0, but we need a bit of preparation before we present it. For an A-B-bimodule X write $X^+ \equiv X$ and write X^- for the B-A-bimodule X^*. Recall the free category with conjugates $\mathbf{D} \equiv \mathbf{D}[D_2, D_1]$ defined in section 2.4. For $\underline{x} \in \mathbf{D}(a,b)$ we define the A_b-A_a-bimodule

$$(3.20)\qquad X_{\underline{x}} = X_{x_1}^{\varepsilon_1} \otimes_{A_{1,2}} X_{x_2}^{\varepsilon_2} \otimes_{A_{2,3}} \cdots \otimes_{A_{n-1,n}} X_{x_n}^{\varepsilon_n} ,$$

where $\underline{x} = ((x_1, \varepsilon_1), \ldots, (x_n, \varepsilon_n))$ and $A_{i,i+1}$ denotes the algebra which acts from the right on $X_i^{\varepsilon_i}$ and from the left on $X_{i+1}^{\varepsilon_{i+1}}$. Note that

$$X_{\underline{y} \circ \underline{x}} = X_{\underline{y}} \otimes_{A_b} X_{\underline{x}} \qquad \text{for} \quad c \xleftarrow{\underline{y}} b \xleftarrow{\underline{x}} a \; . \tag{3.21}$$

Let χ be an element of $D_1^{(n)}/C_n$ as in (2.2). Let $\underline{x} \in D_1^{(n)}$ be a representative of χ, i.e. $\chi = [\underline{x}]$, and let $\mathcal{O}_\chi = C_n.\underline{x}$ be the C_n orbit of $\underline{x}$ in $D_1^{(n)}$ (which is independent of the choice of $\underline{x}$). Let

$$\pi : J_\chi \to \mathcal{O}_\chi \tag{3.22}$$

be the vector bundle (with discrete base) whose fibres are given by the dual vector space

$$\pi^{-1}(\underline{y}) = \mathrm{Hom}_k(\circlearrowleft_{A_{s(\underline{y})=t(\underline{y})}} X_{\underline{y}}, k) \; , \tag{3.23}$$

where $\mathrm{Hom}_k(U, V)$ stands for the space of linear maps from U to V. Since $\underline{y} \in \mathbf{D}(s(\underline{y}), t(\underline{y}))$ is cyclically composable by assumption, we indeed have $s(\underline{y}) = t(\underline{y})$. An element σ of the cyclic group C_n acts on J_χ by taking a vector $\varphi \in \pi^{-1}(\underline{y})$ to $\varphi \circ \sigma^{-1} \in \pi^{-1}(\sigma(\underline{y}))$. By abuse of notation, here we also denoted by σ the linear isomorphism $\circlearrowleft_{A_{s(\underline{y})}} X_{\underline{y}} \to \circlearrowleft_{A_{s(\sigma\underline{y})}} X_{\sigma\underline{y}}$ obtained by shifting tensor factors. Denote by $\Gamma(J_\chi)^{\mathrm{inv}}$ the space of C_n-invariant sections of the bundle J_χ. The value of the section at $\underline{y}$ is then invariant under the action of the stabiliser of $\underline{y}$ in C_n.

For example, if $\chi = [\underline{x}]$ is such that all (x_i, ε_i) in $\underline{x}$ are mutually distinct, then the orbit has lengths n, all stabilisers are trivial, and $\Gamma(J_\chi)^{\mathrm{inv}}$ is isomorphic to any one of the fibres of J_χ. If all (x_i, ε_i) are identical, the orbit has length one and $\Gamma(J_\chi)^{\mathrm{inv}}$ consists of the C_n invariant vectors in $\mathrm{Hom}_k(\circlearrowleft_{A_{s(\underline{x})}} X_{\underline{x}}, k)$.

With this preparation, we can finally state the third piece of data for the lattice TFT construction.

3. For each $u \in D_0$, a vector $\varphi_u \in \Gamma(J_{j(u)})^{\mathrm{inv}}$.

This complicated construction will later ensure that the junction-condition is unchanged under 'rotations which leave the attached domain walls invariant', or in other words, it has no preferred 'starting edge'.

3.5. Functor on bordisms with cell decomposition. Fix D_i, $i = 0, 1, 2$ and the data described in the previous subsection. We proceed to define a symmetric monoidal functor

$$T^{\mathrm{cw}} : \mathrm{Bord}_{2,1}^{\mathrm{def,top,cw}} \longrightarrow \mathcal{V}ect_f(k) \; . \tag{3.24}$$

The action of T^{cw} on objects is as follows. Denote by O an object of the bordism category $\mathrm{Bord}_{2,1}^{\mathrm{def,top,cw}}$ consisting of a single S^1. Recall that $C_1(O)$ is the set of 1-cells (i.e. edges) of the cell decomposition of O. To each edge $e \in C_1(O)$ we assign the vector space

(3.25)

$$R_e = \begin{cases} A_a & ; \; e \text{ contains no marked point and carries label } a \in D_2, \\ X_x^\varepsilon & ; \; e \text{ contains a marked point with orientation } \varepsilon \text{ and label } x \in D_1, \end{cases}$$

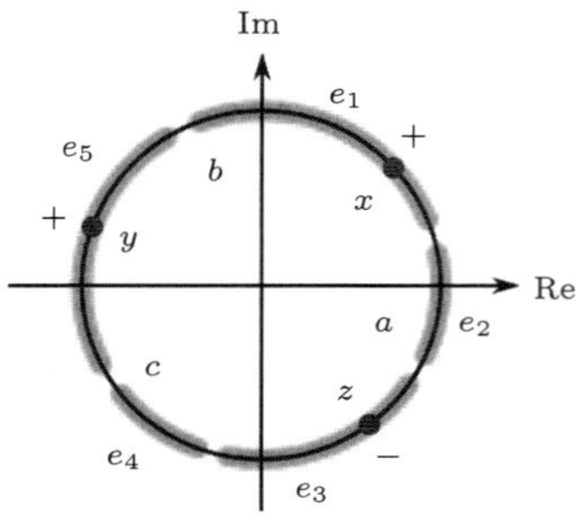

FIGURE 9. Assignment of vector spaces R_e to edges $e \in C_1(O)$: The figure shows a circle with three domains labelled $a, b, c \in D_2$ and three marked points labelled $x, y, z \in D_1$. The circle is decomposed into 5 edges, and the corresponding vector spaces are $R_{e_1} = X_x$, an A_b-A_a-bimodule; $R_{e_2} = A_a$; $R_{e_3} = X_z^*$, an A_a-A_c-bimodule (while X_z itself is an A_c-A_a-bimodule); $R_{e_4} = A_c$; $R_{e_5} = X_y$, an A_c-A_b-bimodule.

see figure 9 for an illustration. We set[11]

$$T^{\mathrm{cw}}(O) = \bigotimes_{e \in C_1(O)} R_e \ . \tag{3.26}$$

For an object $U = O_1 \sqcup O_2 \sqcup \cdots \sqcup O_n$ we take[12] $T^{\mathrm{cw}}(U) = T^{\mathrm{cw}}(O_1) \otimes \cdots \otimes T^{\mathrm{cw}}(O_n)$.

There are two types of morphisms in the bordism category: permutations of objects, and bordisms. A permutation $\sigma : U \to \sigma(U)$ is mapped by T^{cw} to the corresponding permutation of tensor factors. The description of T^{cw} for bordisms will take a little while. Given a bordism $M : U \to V$ (we assume that a representative of the equivalence class has been chosen and use the same symbol), we will write the functor as a composition of two linear maps

$$\begin{aligned} T^{\mathrm{cw}}(M) \ : \ T^{\mathrm{cw}}(U) &\xrightarrow{\mathrm{id}_{T^{\mathrm{cw}}(U)} \otimes P(M)} T^{\mathrm{cw}}(U) \otimes Q(M) \otimes T^{\mathrm{cw}}(V) \\ &\xrightarrow{E(M) \otimes \mathrm{id}_{T^{\mathrm{cw}}(V)}} T^{\mathrm{cw}}(V) \ . \end{aligned} \tag{3.27}$$

We will now describe the vector space $Q(M)$, and the maps $P(M)$ ('propagator') and $E(M)$ ('evaluation').

We start with the vector space $Q(M)$. Denote by $\partial_{\mathrm{in}} M$ the ingoing part of the boundary of M, that is, the part parametrised by U, and by $\partial_{\mathrm{out}} M$ the out-going part of the boundary of M, parametrised by V.

Consider triples (p, e, or), where $p \in C_2(M)$ is a 2-cell (i.e. an open polygon), $e \in C_1(M)$ is a 1-cell, and 'or' is an orientation of e. We only allow triples which

[11] Here, the tensor product stands for the tensor product over k of a family of vector spaces indexed by some set I (here $C_1(O)$). To define this tensor product it is not necessary to choose an ordering of I, i.e. a preferred way to write out the tensor product in a linear order. The same applies to similar tensor products below.

[12] Of course we could have taken the definition (3.26) also for a general object U instead of writing $T^{\mathrm{cw}}(U)$ as a tensor product with implied ordering of the factors. However, in this way it is easier to see that the functor is symmetric.

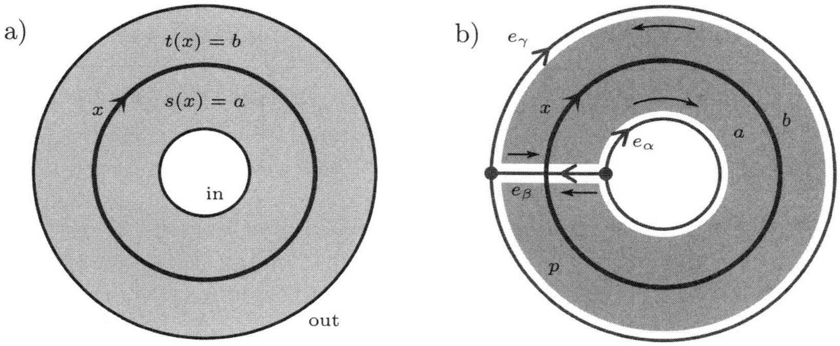

FIGURE 10. Denote by $O(a)$ the circle labelled by $a \in D_2$; figure a) shows a bordism $A(x) : O(a) \to O(b)$ with a single circular domain wall labelled x. Figure b) gives a possible cell decomposition of $A(x)$ into a 2-cell p, 1-cells e_α, e_β, e_γ and two unnamed 0-cells. The edges are oriented for convenience, so that we can describe the two possible orientations simply by $\pm$ (the orientation is not part of the data of the cell decomposition); the orientation of the boundary of p induced by the orientation of $A(x)$ is indicated by the arrows placed in p. There are 4 allowed triples: $(p, e_\alpha, +)$, $(p, e_\beta, +)$, $(p, e_\gamma, -)$, $(p, e_\beta, -)$. The corresponding vector spaces are $Q_{p,e_\alpha,+} = A_a$, $Q_{p,e_\beta,+} = X_x^*$, $Q_{p,e_\gamma,-} = A_b$, $Q_{p,e_\beta,-} = X_x$. As e_α lies on $\partial_{\text{in}} A(x)$, we have $Q(A(x)) = Q_{p,e_\beta,+} \otimes Q_{p,e_\gamma,-} \otimes Q_{p,e_\beta,-}$.

satisfy the following condition: the orientation of M also orients p, and this in turn induces an orientation of the boundary ∂p; we demand that (e, or) is part of ∂p as an oriented edge. This is illustrated in figure 10. To each allowed triple (p, e, or) we assign a vector space:

(3.28)

$$Q_{p,e,\text{or}} = \begin{cases} A_a & ; (e,\text{or}) \text{ does not intersect } M_1 \text{ and is in a component} \\ & \quad \text{of } M_2 \text{ labelled } a, \\ X_x & ; (e,\text{or}) \text{ intersects a component of } M_1 \text{ which is labelled } x \\ & \quad \text{and is oriented into the polygon } p, \\ X_x^* & ; (e,\text{or}) \text{ intersects a component of } M_1 \text{ which is labelled } x \\ & \quad \text{and is oriented out of the polygon } p. \end{cases}$$

Here, the pair (e, or) is understood as part of the boundary ∂p; this is important if the same edge e occurs twice in ∂p, see the example in figure 10. The vector space $Q(M)$ is given by

$$Q(M) = \bigotimes_{(p,e,\text{or}),\, e \notin \partial_{\text{in}} M} Q_{p,e,\text{or}} \quad , \tag{3.29}$$

where the tensor product is taken over all allowed triples (p, e, or) for which e does not lie in $\partial_{\text{in}} M$.

We now turn to the description of the map $P(M) : k \to Q(M) \otimes T^{\text{cw}}(V)$. Each edge $e \in C_1(M)$ in the interior of M occurs in two allowed triples, let us call them $(p(e)_1, e, \text{or}_1)$ and $(p(e)_2, e, \text{or}_2)$, where or_1 and or_2 are the two possible orientations of e, and $p(e)_i$ is the polygon which contains the oriented edge (e, or_i) in its boundary. Note that it may happen that $p(e)_1 = p(e)_2$, as it does in figure 10. For each interior edge e define the linear map

$$P_e : k \longrightarrow Q_{p(e)_1,e,\mathrm{or}_1} \otimes Q_{p(e)_2,e,\mathrm{or}_2} \tag{3.30}$$

according to the following two cases.

(1) If M_1 does not intersect e then according to (3.28) we have $Q_{p(e)_1,e,\mathrm{or}_1} = Q_{p(e)_2,e,\mathrm{or}_2} = A_a$, where a is the label of the component of M_2 containing e. We take $P_e = \beta_{A_a}$, with β the dual of the Frobenius pairing as in section 3.3. Since A_a has trace paring, β is symmetric and the map P_e is independent of the choice of order of $(p(e)_1, e, \mathrm{or}_1)$ and $(p(e)_2, e, \mathrm{or}_2)$.
(2) Suppose a component of M_1 labelled by x intersects e. Let u_i be a basis of X_x and let u_i^* be the dual basis of X_x^*. Choose the numbering '1' and '2' of $(p(e)_1, e, \mathrm{or}_1)$ and $(p(e)_2, e, \mathrm{or}_2)$ so that the orientation of the domain wall M_1 is such that it points into the polygon $p(e)_1$ at (e, or_1) and out of the polygon $p(e)_2$ at (e, or_2). Then $Q_{p(e)_1,e,\mathrm{or}_1} = X_x$ and $Q_{p(e)_2,e,\mathrm{or}_2} = X_x^*$ (see figure 10) and we set $P_e(\lambda) = \lambda \sum_i u_i \otimes u_i^*$.

If e is an edge on the out-going boundary $\partial_{\mathrm{out}} M$, i.e. the boundary component parametrised by V, then there is exactly one allowed triple which contains e. Let (p, e, or) be that triple. The parametrisation identifies e with an edge of $C(V)$ which we also call e. In this case P_e is defined as in (3.30), but with $Q_{p(e)_1,e,\mathrm{or}_1}$ and $Q_{p(e)_2,e,\mathrm{or}_2}$ replaced by $Q_{p(e),e,\mathrm{or}}$ and R_e. Comparing (3.25) and (3.28) (and using the conventions in figure 3), one checks that cases 1 and 2 above still apply.

Altogether, the map $P(M) : k \to Q(M) \otimes T^{\mathrm{cw}}(V)$ is defined as

$$P(M) = \bigotimes_{e \in C_1(M),\, e \notin \partial_{\mathrm{in}} M} P_e \,. \tag{3.31}$$

Finally, we need to define $E(M) : T^{\mathrm{cw}}(U) \otimes Q(M) \to k$. Note that $T^{\mathrm{cw}}(U) \otimes Q(M)$ contains one factor $Q_{p,e,\mathrm{or}}$ for each $p \in C_2(M)$ and $(e, \mathrm{or}) \in \partial p$, even if $e \subset \partial M$, such that

$$T^{\mathrm{cw}}(U) \otimes Q(M) = \bigotimes_{p \in C_2(M),\, (e,\mathrm{or}) \in \partial p} Q_{p,e,\mathrm{or}} \,. \tag{3.32}$$

For each polygon $p \in C_2(M)$ we define a linear map

$$E_p : \bigotimes_{(e,\mathrm{or}) \in \partial p} Q_{p,e,\mathrm{or}} \longrightarrow k \,. \tag{3.33}$$

Fix $p \in C_2(M)$. There are three cases to distinguish, depending on whether p intersects M_1 and/or M_0.

(1) Suppose p intersects neither M_1 nor M_0. Let a be the label of the component of M_2 containing p. Choose an edge $(e_1, \mathrm{or}_1) \in \partial p$ and denote by $(e_1, \mathrm{or}_1), (e_2, \mathrm{or}_2), \ldots, (e_m, \mathrm{or}_m)$ all oriented edges of ∂p in anti-clockwise ordering. Let further

$$q_1 \otimes q_2 \otimes \cdots \otimes q_m \in Q_{p,e_1,\mathrm{or}_1} \otimes Q_{p,e_2,\mathrm{or}_2} \otimes \cdots \otimes Q_{p,e_m,\mathrm{or}_m} \,. \tag{3.34}$$

Each Q_{p,e_i,or_i} is equal to A_a and we set $E_p(q_1 \otimes \cdots \otimes q_m) = \varepsilon_{A_a}(q_1 \cdots q_m)$, where ε_{A_a} is the counit of A_a. By symmetry of the pairing of the Frobenius algebra A_a, the result is independent of the choice of starting edge (e_1, or_1).

(2) Suppose p intersects M_1 but not M_0. In this case there is one oriented edge where M_1 leaves p, which we take to be (e_1, or_1). Then we order the oriented edges of ∂p anti-clockwise as in 1. Let (e_i, or_i) be the edge where M_1 enters p and let x be the label of the component of M_1 in p. In the notation from (3.34), we have $q_1 \in X_x^*$, $q_i \in X_x$, and $q_2, \dots, q_{i-1} \in A_{t(x)}$ and $q_{i+1}, \dots, q_m \in A_{s(x)}$. We set $E_p(q_1 \otimes \cdots \otimes q_m) = q_1\big((q_2 \cdots q_{i-1}).q_i.(q_{i+1} \cdots q_m)\big)$. Unlike case 1., case 2. did not involve an arbitrary choice, and there is no invariance condition to check.

(3) Suppose p contains a point u from M_0 of orientation $\nu_u \in \{\pm\}$ and with label $\hat{d}_0(u) = t \in D_0$. As in 1., we choose an arbitrary starting edge $(e_1, \mathrm{or}_1) \in \partial p$ and order the remaining edges anti-clockwise. Each edge e_i, $i = 1, \dots, m$ is transversed by a domain wall. For each $i = 1, \dots, m$ we thereby obtain a pair (x_i, ε_i) where $x_i \in D_1$ is the label of the domain wall crossing e_i, and $\varepsilon_i = +$ if this domain wall is oriented into the polygon at (e_i, or_i) and $\varepsilon_i = -$ otherwise. Let $\underline{x} = ((x_1, \varepsilon_1), \dots, (x_m, \varepsilon_m))$.

If $\nu_u = +$, the labelling has to satisfy $j(t) = [\underline{x}] =: \chi$. According to the construction in section 3.4, $\underline{x} \in \mathcal{O}_\chi$. Evaluating the section $\varphi_t \in \Gamma(J_{j(t)})^{\mathrm{inv}}$ at $\underline{x}$ gives an element $\psi \in \mathrm{Hom}_k(\circlearrowleft_{A_{s(\underline{x})=t(\underline{x})}} X_{\underline{x}}, k)$. Precomposing with the projection $\pi_\otimes : X_{\underline{x}} \to \circlearrowleft_{A_{s(\underline{x})=t(\underline{x})}} X_{\underline{x}}$ we obtain a linear form $\psi \circ \pi_\otimes : X_{x_1}^{\varepsilon_1} \otimes \cdots \otimes X_{x_m}^{\varepsilon_m} \to k$. We set $E_p(q_1 \otimes \cdots \otimes q_m) = \psi \circ \pi_\otimes(q_1 \otimes \cdots \otimes q_m)$. Independence of the choice of (e_1, or_1) follows since $\Gamma(J_{j(t)})^{\mathrm{inv}}$ consists of elements invariant under cyclic permutations.

If $\nu_u = -$, the labelling has to satisfy $j(t) = [\underline{x}^*]$, and the above construction is repeated with $\underline{x}^*$ instead of $\underline{x}$.

Figure 10 gives an example of case 2. There, $(e_1, \mathrm{or}_1) = (e_\beta, +)$, $(e_2, \mathrm{or}_2) = (e_\gamma, -)$, $(e_3, \mathrm{or}_3) = (e_\beta, -)$, $(e_4, \mathrm{or}_4) = (e_\alpha, +)$, and $E_p(q_1 \otimes q_2 \otimes q_3 \otimes q_4) = q_1(q_2.q_3.q_4)$, where $q_2.q_3.q_4$ is the left/right action of $q_2 \in A_b$ and $q_4 \in A_a$ on $q_3 \in X_x$.

Altogether, for $E(M)$ we take

$$E(M) = \bigotimes_{p \in C_2(M)} E_p \ . \tag{3.35}$$

This completes the definition of T^{cw}.

Let us briefly illustrate the construction in two related examples; more examples will be computed in section 3.7. For the bordism $A(x) : O(a) \to O(b)$ considered in figure 10, the composition of maps in (3.27) reads

(3.36)
$$T^{\mathrm{cw}}(A(x)) : T^{\mathrm{cw}}(O(a)) \xrightarrow{\mathrm{id}_{R_{e_\alpha}} \otimes P} R_{e_\alpha} \otimes Q_{p,e_\beta,+} \otimes Q_{p,e_\gamma,-} \otimes Q_{p,e_\beta,-} \otimes R_{e_\gamma} \xrightarrow{E \otimes \mathrm{id}_{R_{e_\gamma}}} T^{\mathrm{cw}}(O(b)) \ ,$$

where the edge on $O(a)$ is identified with e_α via the parametrisation, and the edge on $O(b)$ with e_γ. Substituting the definition of these vector spaces and maps gives

(3.37)
$$\begin{array}{rcccl} T^{\mathrm{cw}}(A(x)) : A_a & \xrightarrow{\mathrm{id}\otimes P} & A_a \otimes X_x^* \otimes A_b \otimes X_x \otimes A_b & \xrightarrow{E \otimes \mathrm{id}} & A_b \\ q & \longmapsto & \sum_{i,j} q \otimes u_i^* \otimes b_j' \otimes u_i \otimes b_j & \longmapsto & \sum_{i,j} u_i^*(b_j'.u_i.q)\, b_j \ . \end{array}$$

This map can be defined for any two Frobenius algebras with trace pairing A, B and a finite-dimensional B-A-bimodule X. One can check that the image of this

a)

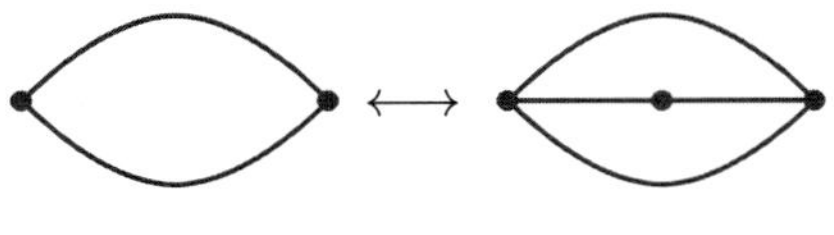

b)

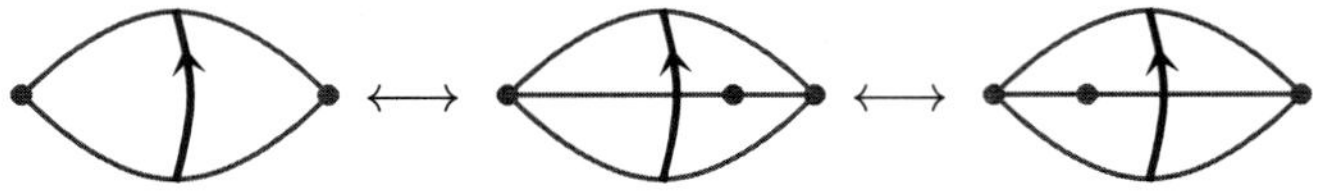

FIGURE 11. a) A local modification of the cell decomposition which adds an edge and a vertex turning a 2-gon into two triangles, or conversely. The two exterior vertices are allowed to be identical. b) The same in the presence of a domain wall; there are two modifications as the vertex can be added on either side of the domain wall.

map lies in $Z(B)$, and that the kernel of the projector $p_\otimes$ onto $Z(A)$ is contained in the kernel of $T(A(x))$. We therefore lose nothing if we restrict ourselves to $Z(A)$ and $Z(B)$ from the start:

$$D(X) : Z(A) \longrightarrow Z(B) \quad , \quad z \mapsto \sum_{i,j} u_i^*(b'_j.u_i.z)\, b_j \ . \tag{3.38}$$

This is an example of a defect operator, which we already briefly mentioned in Remark 2.3 (ii). Such defect operators have some nice properties[13]: if $X \cong X'$ as bimodules, then $D(X) = D(X')$; if Y is a C-B-bimodule, then $D(Y)D(X) = D(Y \otimes_B X)$; and for the A-A-bimodule A one has $D(A) = \mathrm{id}_{Z(A)}$.

A related example comes from the annulus as in figure 10, but without the domain wall x, so that necessarily $a = b$. The map in (3.37) specialises to $q \mapsto \sum_{i,j} \langle a'_i, a'_j a_i q \rangle\, a_j$. By (3.8), this is equal to $q \mapsto \sum_i a_i q a'_i = p_\otimes(q)$, cf. Lemma 3.3. Thus, T^{cw} maps the cylinder over $O(a)$ to the projector onto the centre of A_a.

It is fairly straightforward to see from the above construction that T^{cw} is compatible with composition and tensor products. Since we imposed that permutations of S^1-components in an object U of the bordism category get mapped to permutations of tensor factors in $T^{\mathrm{cw}}(U)$, the functor respects identities and is symmetric.

3.6. Independence of cell decomposition. In this subsection we abbreviate $\mathrm{Bord}^{\mathrm{cw}} \equiv \mathrm{Bord}_{2,1}^{\mathrm{def,top,cw}}$ and $\mathrm{Bord} \equiv \mathrm{Bord}_{2,1}^{\mathrm{def,top}}$. Objects and morphisms in $\mathrm{Bord}^{\mathrm{cw}}$ will be decorated by a tilde (e.g. $\tilde{U}$, $\tilde{M}$, ...). Recall the forgetful functor $F : \mathrm{Bord}^{\mathrm{cw}} \to \mathrm{Bord}$ from section 3.1. We will show that there exists a symmetric monoidal functor T making the diagram (3.3) commute (consequently, this functor is unique). This will be done in several steps, the key one being the following lemma.

[13]These properties are all easily checked directly with the methods of section 3.3. They have also been shown in arbitrary modular categories (instead of just the category $\mathcal{V}ect_f(k)$) in [**FRS3**, Lem. 2] and [**KR1**, Lem. 3.1].

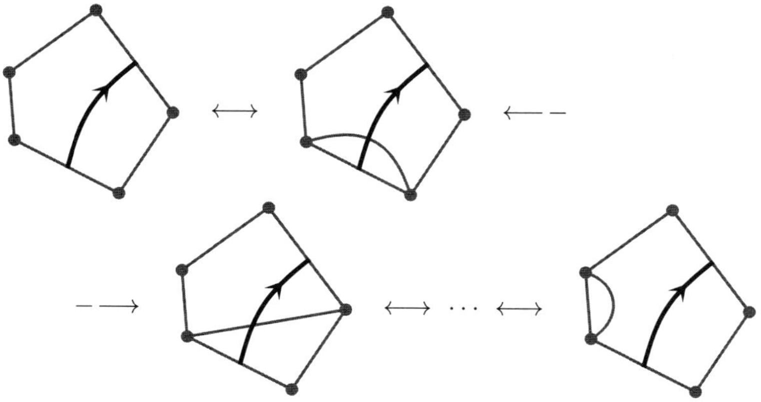

FIGURE 12. A local modification of the cell decomposition which adds a single edge to the interior of a 2-cell. The figure shows an exemplary situation. Alternatively, the 2-cell can be any n-gon with $n \geq 2$, the domain wall can run between other edges, or there could be no domain wall at all.

LEMMA 3.5. *Let $\tilde{U}, \tilde{V} \in \mathrm{Bord}^{\mathrm{cw}}$ and let $\tilde{M}, \tilde{M}' : \tilde{U} \to \tilde{V}$ be morphisms. If $\tilde{M}'$ is obtained from $\tilde{M}$ by one of the local modifications of the cell decomposition shown in figures* 11 *and* 12, *then $T^{\mathrm{cw}}(\tilde{M}) = T^{\mathrm{cw}}(\tilde{M}')$.*

PROOF. We will show the equality $T^{\mathrm{cw}}(\tilde{M}) = T^{\mathrm{cw}}(\tilde{M}')$ in the two cases displayed in figure 13; the remaining cases are treated analogously.

The part of the cell decomposition in figure 13 a) contributes the factor $E_p : X_x^* \otimes X_x \to k$, $\varphi \otimes x \mapsto \varphi(x)$ from (3.33) to the map E in (3.35) and (3.27). Figure 13 b) contributes the factors $E_{p_1} \otimes E_{p_2} : (X_x^* \otimes A_b \otimes X_x) \otimes (X_x^* \otimes A_b \otimes X_x) \to k$, $\varphi \otimes b \otimes x \otimes \varphi' \otimes b' \otimes x' \mapsto \varphi(b.x) \cdot \varphi'(b'.x')$ to E, and to P in (3.31) it contributes the factors $P_{e_3} \otimes P_{e_4} : k \to (A_b \otimes A_b) \otimes (X_x \otimes X_x^*)$, $1 \mapsto \sum_{i,j} b_i' \otimes b_i \otimes u_j \otimes u_j^*$. The composition of $P_{e_3} \otimes P_{e_4} \otimes \mathrm{id}_{X_x^*} \otimes \mathrm{id}_{X_x}$ and $E_{p_1} \otimes E_{p_2}$ (with the appropriate permutation of tensor factors) yields

$$\varphi \otimes x \mapsto \sum_{i,j} \varphi(b_i'.u_j) \cdot u_j^*(b_i.x) = \sum_i \varphi(b_i'.b_i.x) = \varphi(x) = E_p(\varphi \otimes x) \ . \tag{3.39}$$

Thus if $\tilde{M}$ and $\tilde{M}'$ differ only in one place as shown in figure 13 a,b), we still have $T^{\mathrm{cw}}(\tilde{M}) = T^{\mathrm{cw}}(\tilde{M}')$.

For figure 13 c,d) the argument is the same. The 2-cell p contributes the map $E_p : X_x^* \otimes A_b \otimes A_b \otimes X_x \otimes A_a \to k$, $\varphi \otimes b_1 \otimes b_2 \otimes x \otimes a \mapsto \varphi(b_1.b_2.x.a)$. The two cells p_1 and p_2 contribute $E_{p_1} \otimes E_{p_2} : (X_x^* \otimes A_b \otimes A_b \otimes X_x) \otimes (X_x^* \otimes X_x \otimes A_a) \to k$, $\varphi \otimes b_1 \otimes b_2 \otimes x' \otimes \varphi' \otimes x \otimes a \mapsto \varphi(b_1.b_2.x') \cdot \varphi'(x.a)$. The new edge e_6 gives the map $P_{e_6} : k \to X_x \otimes X_x^*$, $1 \mapsto \sum_j u_j \otimes u_j^*$. Composing the two as $(E_{p_1} \otimes E_{p_2}) \circ (\mathrm{id} \otimes P_{e_6} \otimes \mathrm{id})$ results in a map $X_x^* \otimes A_b \otimes A_b \otimes X_x \otimes A_a \to k$ which acts as

$$\begin{aligned} \varphi \otimes b_1 \otimes b_2 \otimes x \otimes a \ \mapsto \ & \textstyle\sum_j (E_{p_1} \otimes E_{p_2})(\varphi \otimes b_1 \otimes b_2 \otimes u_j \otimes u_j^* \otimes x \otimes a) \\ & = \textstyle\sum_j \varphi(b_1.b_2.u_j) \cdot u_j^*(x.a) = \varphi(b_1.b_2.x.a) \\ & = E_p(\varphi \otimes b_1 \otimes b_2 \otimes x \otimes a) \ . \end{aligned} \tag{3.40}$$

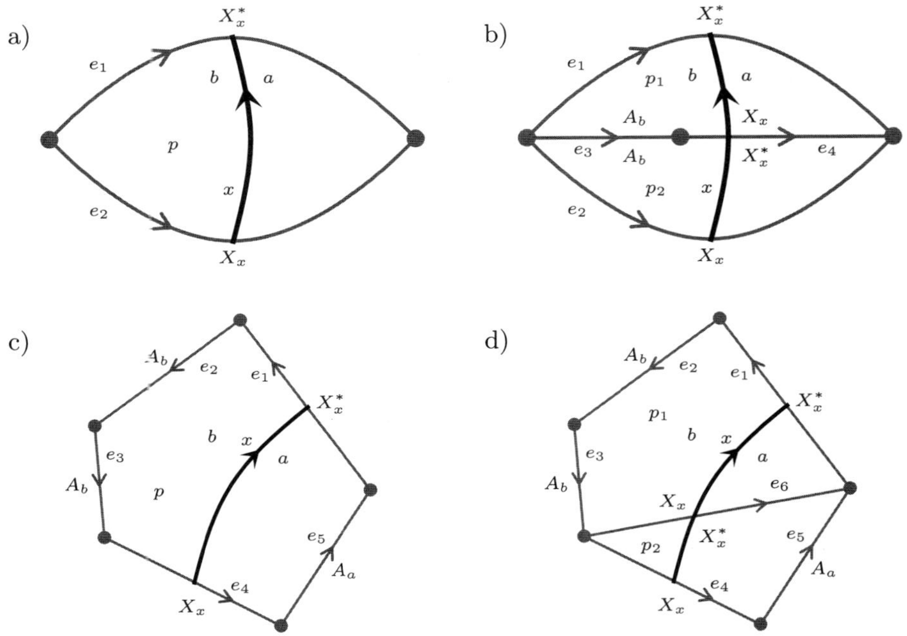

FIGURE 13. Two examples. Figures a,b: Adding two edges and a vertex; the vector spaces $Q_{p,e,\mathrm{or}}$ associated to triples (p, e, or) as in section 3.5 are also shown, e.g. $Q_{p_2,e_4,-} = X_x^*$ and $Q_{p_1,e_4,+} = X_x$. Figures c,d: Adding an edge to a 2-cell; also shown are the associated vector spaces.

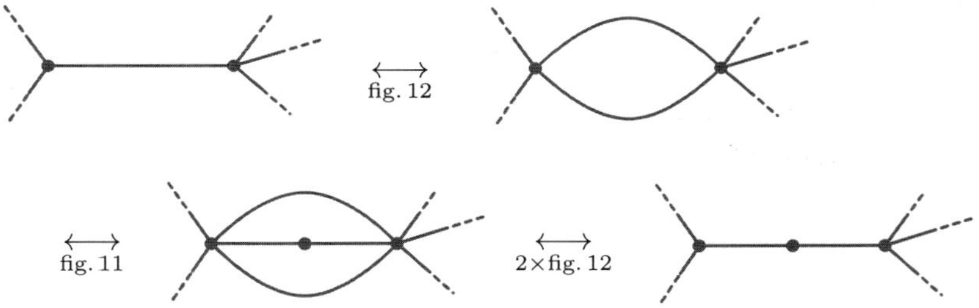

FIGURE 14. The moves in figures 11 and 12 allow one to split an edge by adding a new vertex.

Hence, if $\tilde{M}$ and $\tilde{M}'$ differ only in one place as shown in figure 13 c,d), we have $T^{\mathrm{cw}}(\tilde{M}) = T^{\mathrm{cw}}(\tilde{M}')$. □

One immediate consequence of the above lemma is that we can insert new vertices on edges which do not belong to the boundary of M via the sequence of moves in figure 14. (The cell-decomposition of the boundary is fixed by the parametrisation in terms of the source and target objects). In the absence of domain walls, the 'elementary subdivisions' (and their inverses) of a 2-cell (figure 12) and of a 1-cell (figure 14) have been shown in [**Ki**] to relate any two cell decompositions (more precisely: two PLCW-decompositions of a compact polyhedron, see [**Ki**] for details). The next lemma extends this to cell decomposition in the presence of domain walls and junctions; we will only sketch its proof.

LEMMA 3.6. *Let* $\tilde{U}, \tilde{V} \in \mathrm{Bord}^{\mathrm{cw}}$ *and let* $\tilde{M}, \tilde{M}' : \tilde{U} \to \tilde{V}$ *be two morphisms such that* $F(\tilde{M}) = F(\tilde{M}')$. *Then* $T^{\mathrm{cw}}(\tilde{M}) = T^{\mathrm{cw}}(\tilde{M}')$.

SKETCH OF PROOF. A 2-cell containing a junction is by construction the same in all cell decompositions, so we will remove such 2-cells from $\tilde{M}$ and $\tilde{M}'$ and treat their boundary edges as additional boundary components for the remaining cell complexes.

Next, use the moves in figures 12 and 14 to refine the cell decomposition of $\tilde{M}$ and $\tilde{M}'$ to a triangulation. The same moves allow us to make the part of the triangulation touched by the domain walls in $\tilde{M}$ and $\tilde{M}'$ agree: each component of the domain wall submanifold M_1 defines a string of triangles in $\tilde{M}$ and $\tilde{M}'$ and one can pass to a common refinement. (To achieve this refinement it is allowed to change the triangulation away from the domain walls.) We now remove all triangles containing domain walls from $\tilde{M}$ and $\tilde{M}'$, giving rise to yet more boundary components.

The triangulations of $\tilde{M}$ and $\tilde{M}'$ still remaining no longer contain any domain walls or junctions, and the standard proof of triangulation independence applies (see [**Ki**] or just check that figures 12 and 14 imply invariance under the Pachner moves). $\square$

The next step is to study the behaviour of T^{cw} on preimages of cylinders under F. For $U \in \mathrm{Bord}$ denote by $C_U : U \to U$ the morphism given by the cylinder over U. That is, $C_U = U \times [-1,1]$ with decomposition induced by that of U via $C_U = (C_U)_2 \cup (C_U)_1 \cup (C_U)_0$ with $(C_U)_i = U_{i-1} \times [-1,1]$, $i = 2,1$ and $(C_U)_0 = \emptyset$. Orientations and labellings are induced by U as well. Note that because morphisms of Bord are diffeomorphism classes, we have

$$C_U \circ C_U = C_U \tag{3.41}$$

as morphisms in Bord. Now pick objects $\tilde{U}, \tilde{U}'$ with $F(\tilde{U}) = F(\tilde{U}') = U$ and a morphism $\tilde{M} : \tilde{U} \to \tilde{U}'$ with $F(\tilde{M}) = C_U$. Define

$$\zeta_{\tilde{U}',\tilde{U}} := T^{\mathrm{cw}}(\tilde{M}) : T^{\mathrm{cw}}(\tilde{U}) \longrightarrow T^{\mathrm{cw}}(\tilde{U}') \ . \tag{3.42}$$

By Lemma 3.6, $\zeta_{\tilde{U},\tilde{U}'}$ is independent of $\tilde{M}$. As a consequence, given another preimage $\tilde{U}''$ of U, we have

$$\zeta_{\tilde{U}'',\tilde{U}'} \circ \zeta_{\tilde{U}',\tilde{U}} = \zeta_{\tilde{U}'',\tilde{U}} \ . \tag{3.43}$$

For an object $O \in \mathrm{Bord}$ consisting of a single component S^1 write $\underline{x}(O)$ for the list $((x_1,\varepsilon_1),\dots,(x_n,\varepsilon_n))$ of the marked points on O together with their orientation, ordered clockwise starting from the point $-1 \in S^1$. If $\tilde{O}$ is a preimage of O, denote by $e_1,\dots,e_m$ the 1-cells, again ordered clockwise starting from $-1 \in S^1$. For example, figure 9 shows a preimage $\tilde{O}$ with $m=5$ 1-cells, for O a circle with $\underline{x}(O) = ((y,+),(x,+),(z,-))$ so that $n=3$. Consider the maps

(3.44)

$$e(\tilde{O}) := \Big(\circlearrowleft_{A_{n,1}} X_{\underline{x}(O)} \xrightarrow{\sim} \circlearrowleft_{A_{m,1}} R_{e_1} \otimes_{A_{1,2}} R_{e_2} \otimes_{A_{2,3}} \cdots \otimes_{A_{m-1,m}} R_{e_m} \xrightarrow{e_\otimes} T^{\mathrm{cw}}(\tilde{O}) \Big)$$

and
(3.45)

$$\pi(\tilde{O}) := \Big(T^{\mathrm{cw}}(\tilde{O}) \xrightarrow{\pi_\otimes} \circlearrowleft_{A_{m,1}} R_{e_1} \otimes_{A_{1,2}} R_{e_2} \otimes_{A_{2,3}} \cdots \otimes_{A_{m-1,m}} R_{e_m} \xrightarrow{\sim} \circlearrowleft_{A_{n,1}} X_{\underline{x}(O)} \Big)$$

Here, $X_{\underline{x}}$ is the notation introduced in (3.20), R_e was defined in (3.25) and the intermediate algebras $A_{i,i+1}$ are as required by the bimodules. By (3.26), $T^{\mathrm{cw}}(\tilde{O})$ consists precisely of the tensor factors $R_{e_1} \otimes \cdots \otimes R_{e_m}$, and the maps $e_\otimes, \pi_\otimes$ are as in (3.16).

LEMMA 3.7. *Let $\tilde{O}$, $\tilde{O}'$ in* Bord^{cw} *be preimages of O. Then $\pi(\tilde{O}) \circ e(\tilde{O}) = \mathrm{id}_{\circlearrowleft_{A_{n,1}} X_{\underline{x}(O)}}$ and $e(\tilde{O}') \circ \pi(\tilde{O}) = \zeta_{\tilde{O}',\tilde{O}}$.*

PROOF. The first equality is the defining property of the maps $\pi_\otimes$ and $e_\otimes$ in (3.16).

Let $\tilde{C}_{\tilde{U}}$ be the cylinder over $\tilde{U}$ obtained by equipping C_U with the cell decomposition induced by that of $\tilde{U}$: each edge e of $\tilde{U}$ gets extended to the square 2-cell $e \times [-1, 1]$. If we apply the functor T^{cw} to $\tilde{C}_{\tilde{O}}$, a short calculation starting from the definition of T^{cw} (illustrated in the first example in section 3.7 below) shows that

$$\zeta_{\tilde{O},\tilde{O}} = p_\otimes \ , \tag{3.46}$$

where $p_\otimes$ is the idempotent on $X_{x_1}^{\varepsilon_1} \otimes \cdots \otimes X_{x_n}^{\varepsilon_n}$ whose image is $\circlearrowleft_{A_{n,1}} X_{\underline{x}(O)}$. Below we will furthermore check that

$$\pi(\tilde{O}') \circ \zeta_{\tilde{O}',\tilde{O}} \circ e(\tilde{O}) = \mathrm{id}_{\circlearrowleft_{A_{n,1}} X_{\underline{x}(O)}} \ . \tag{3.47}$$

Composing this from the left with $e(\tilde{O}')$ and from the right with $\pi(\tilde{O})$, and using $e_\otimes \circ \pi_\otimes = p_\otimes$ together with (3.46) and (3.43), proves the second equality of the lemma.

Let us now sketch the proof of (3.47). We identify $\circlearrowleft_{A_{n,1}} X_{\underline{x}(O)}$ and $\circlearrowleft_{A_{m,1}} R_{e_1} \otimes_{A_{1,2}} \cdots \otimes_{A_{m-1,m}} R_{e_m}$ with the images of the corresponding projectors $p_\otimes$ in $X_{x_1}^{\varepsilon_1} \otimes \cdots \otimes X_{x_n}^{\varepsilon_n}$ and $R_{e_1} \otimes \cdots \otimes R_{e_m} \equiv T^{\mathrm{cw}}(\tilde{O})$. Let $\sum_i x_1^{(i)} \otimes \cdots \otimes x_n^{(i)}$ be an element of $X_{x_1}^{\varepsilon_1} \otimes \cdots \otimes X_{x_n}^{\varepsilon_n}$ in the image of $p_\otimes$. The first arrow in (3.44) is the isomorphism mapping this to the element

$$v = p_\otimes \circ \Big(\sum_i 1_{A_{n,1}} \otimes \cdots \otimes x_1^{(i)} \otimes 1_{A_{1,2}} \otimes \cdots \otimes x_2^{(i)} \otimes 1_{A_{2,3}} \otimes \cdots \otimes x_n^{(i)} \otimes \cdots \otimes 1_{A_{n,1}} \Big) \tag{3.48}$$

of $R_{e_1} \otimes \cdots \otimes R_{e_m} \equiv T^{\mathrm{cw}}(\tilde{O})$. Here one unit element has been inserted for each factor R_{e_k} for which e_k does not contain a marked point (in this case $R_{e_k} = A_a$ for an appropriate a, cf. (3.25)). One can convince oneself that $\zeta_{\tilde{O}',\tilde{O}}$ maps v to an element v' of the same form in $T^{\mathrm{cw}}(\tilde{O})$ (i.e. v' has same factors $x_k^{(i)}$ but possibly a different number of factors $1_{A_{k,k+1}}$). We omit the details of this step. The final isomorphism in (3.45) maps v' back to $\sum_i x_1^{(i)} \otimes \cdots \otimes x_n^{(i)}$. □

In the last step, we define the sought-after functor T. On objects $O \in$ Bord with a single S^1 component, we set

$$T(O) := \circlearrowleft_{A_{m,1}} X_{\underline{x}(O)} \ . \tag{3.49}$$

For $U = O_1 \sqcup \cdots \sqcup O_n$, monoidality then requires $T(U) = T(O_1) \otimes \cdots \otimes T(O_n)$. For a bordism $M : U \to V$ in Bord pick a preimage $\tilde{M} : \tilde{U} \to \tilde{V}$ under the forgetful functor. Extend the definition of $e(\tilde{O})$ and $\pi(\tilde{O})$ to $\tilde{U}$ by taking tensor products. Define

$$T(M) := \Big(T(U) \xrightarrow{e(\tilde{U})} T^{\mathrm{cw}}(\tilde{U}) \xrightarrow{T^{\mathrm{cw}}(\tilde{M})} T^{\mathrm{cw}}(\tilde{V}) \xrightarrow{\pi(\tilde{V})} T(V)\Big) . \tag{3.50}$$

The first main result of this paper is:

THEOREM 3.8. *(i) $T(M)$ is independent of the choice of preimage $\tilde{M}$ of M.*
(ii) $T(C_U) = \mathrm{id}_{T(U)}$.
(iii) $T : \mathrm{Bord}_{2,1}^{\mathrm{def,top}} \to \mathcal{V}ect_f(k)$ is a symmetric monoidal functor.

PROOF. (i) Choose another preimage $\tilde{M}' : \tilde{U}' \to \tilde{V}'$ in $\mathrm{Bord}^{\mathrm{cw}}$ of $M : U \to V$, and choose preimages $\tilde{C}_U : \tilde{U} \to \tilde{U}'$ and $\tilde{C}_V : \tilde{V} \to \tilde{V}'$ of the cylinder C_U and C_V. Consider the diagram

$$\begin{array}{ccccccc} & & T^{\mathrm{cw}}(\tilde{U}) & \xrightarrow{T^{\mathrm{cw}}(\tilde{M})} & T^{\mathrm{cw}}(\tilde{V}) & & \\ & {}^{e(\tilde{U})}\nearrow & \downarrow{\scriptstyle \zeta_{\tilde{U}',\tilde{U}}} & & \downarrow{\scriptstyle \zeta_{\tilde{V}',\tilde{V}}} & \searrow^{\pi(\tilde{V})} & \\ \circlearrowleft_{A_{m,1}} X_{\underline{x}(U)} & & & & & & \circlearrowleft_{A_{n,1}} X_{\underline{x}(V)} \\ & {}_{e(\tilde{U}')}\searrow & T^{\mathrm{cw}}(\tilde{U}') & \xrightarrow{T^{\mathrm{cw}}(\tilde{M}')} & T^{\mathrm{cw}}(\tilde{V}') & \nearrow_{\pi(\tilde{V}')} & \end{array} \tag{3.51}$$

To see that the left triangle commutes, substitute $\zeta_{\tilde{U}',\tilde{U}} = e(\tilde{U}') \circ \pi(\tilde{U})$ and use that $\pi(\tilde{U}) \circ e(\tilde{U}) = \mathrm{id}$ (Lemma 3.7). Commutativity of the right triangle follows analogously. The following chain of equalities shows that also the central square commutes:
(3.52)

$$\zeta_{\tilde{V}',\tilde{V}} \circ T^{\mathrm{cw}}(\tilde{M}) \overset{(1)}{=} T^{\mathrm{cw}}(\tilde{C}_V) \circ T^{\mathrm{cw}}(\tilde{M}) \overset{(2)}{=} T^{\mathrm{cw}}(\tilde{C}_V \circ \tilde{M})$$
$$\overset{(3)}{=} T^{\mathrm{cw}}(\tilde{M}' \circ \tilde{C}_U) \overset{(4)}{=} T^{\mathrm{cw}}(\tilde{M}') \circ T^{\mathrm{cw}}(\tilde{C}_U) \overset{(5)}{=} T^{\mathrm{cw}}(\tilde{M}') \circ \zeta_{\tilde{U}',\tilde{U}} \,.$$

Step (1) is the definition of $\zeta_{\tilde{V}',\tilde{V}}$ in (3.42); step (2) is functoriality of T^{cw}; step (3) follows from Lemma 3.6 since $\tilde{C}_V \circ \tilde{M}$ and $\tilde{M}' \circ \tilde{C}_U$ are just different cell decompositions (but identical on the boundary) of the same bordism $\tilde{U} \to \tilde{V}'$; steps (4) and (5) are the same as (2) and (1). Thus the diagram (3.52) commutes, establishing (i).

(ii) By definition (3.50) and Lemma 3.7, $T(C_U) = \pi(\tilde{U}') \circ T^{\mathrm{cw}}(\tilde{C}_U) \circ e(\tilde{U}) = \pi(\tilde{U}') \circ \zeta_{\tilde{U}',\tilde{U}} \circ e(\tilde{U}) = \pi(\tilde{U}') \circ e(\tilde{U}') \circ \pi(\tilde{U}) \circ e(\tilde{U}) = \mathrm{id}$.

(iii) Let $U \xrightarrow{M} V \xrightarrow{N} W$ be two composable morphisms in Bord and choose a preimage $\tilde{U} \xrightarrow{\tilde{M}} \tilde{V} \xrightarrow{\tilde{N}} \tilde{W}$ in $\mathrm{Bord}^{\mathrm{cw}}$. To check compatibility with composition, we need to show $T(N \circ M) = T(N) \circ T(M)$. Inserting the definition, this amounts to

$$\pi(\tilde{W}) \circ T^{\mathrm{cw}}(\tilde{N} \circ \tilde{M}) \circ e(\tilde{U}) = \pi(\tilde{W}) \circ T^{\mathrm{cw}}(\tilde{N}) \circ e(\tilde{V}) \circ \pi(\tilde{V}) \circ T^{\mathrm{cw}}(\tilde{M}) \circ e(\tilde{U}) \tag{3.53}$$

That the two sides are indeed equal can be seen as follows. By Lemma 3.7, $e(\tilde{V}) \circ \pi(\tilde{V}) = \zeta_{\tilde{V},\tilde{V}}$, and, if $\tilde{C}$ is a preimage of C_V, by functoriality of T^{cw} the rhs is equal to $\pi(\tilde{W}) \circ T^{\mathrm{cw}}(\tilde{N} \circ \tilde{C} \circ \tilde{M}) \circ e(\tilde{U})$. But $F(\tilde{N} \circ \tilde{C} \circ \tilde{M}) = N \circ M$, so that by Lemma 3.6, the rhs is indeed equal to the lhs.

Monoidality and symmetry of T are implied by that of T^{cw}. □

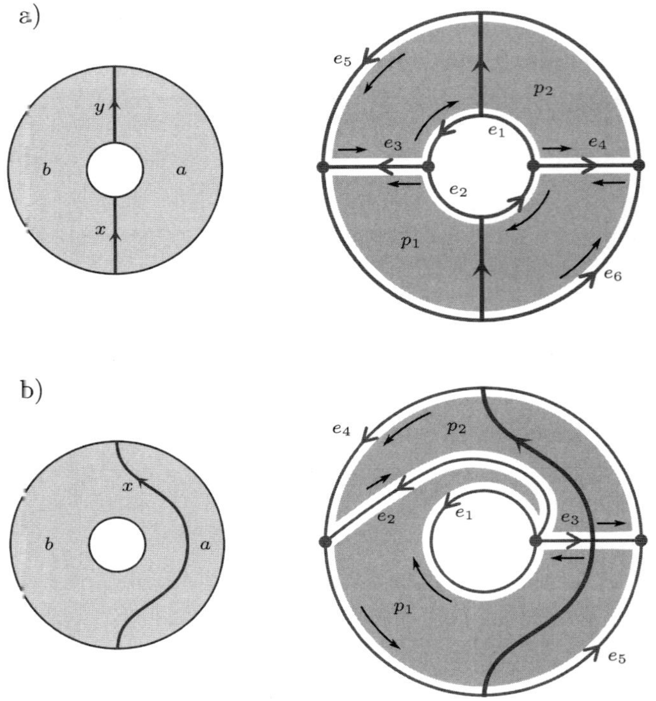

FIGURE 15. Two bordisms with defects together with a choice of cell-decomposition used in the sample computation. As in figure 10, the orientations of the edges are chosen for convenience and are not part of the data of the cell decomposition.

This concludes our construction of an example of a two-dimensional topological field theory with defects.

3.7. Some examples of amplitudes. Let us work through two more examples to see how the amplitude of a bordism

$$M : U \to V \tag{3.54}$$

in Bord $\equiv \mathrm{Bord}_{2,1}^{\mathrm{def,top}}$ is computed in lattice TFT. As in section 3.6, we denote by $\tilde{M} : \tilde{U} \to \tilde{V}$ a lift to $\mathrm{Bord}^{\mathrm{cw}} \equiv \mathrm{Bord}_{2,1}^{\mathrm{def,top,cw}}$.

The first example is shown in figure 15 a). Using the notation of section 2.4, let $U = V = O(y \circ x^*)$ be the object of Bord consisting of a single S^1 with two marked points $(x, -)$ and $(y, +)$. For $\tilde{U} = \tilde{V}$ we choose a decomposition with two 1-cells. The spaces R_e are:

e	e_1	e_2	e_5	e_6
R_e	X_y	X_x^*	X_y	X_x^*

The allowed triples are:

(p, e, or)	$(p_1, e_2, -)$	$(p_1, e_3, +)$	$(p_1, e_6, +)$	$(p_1, e_4, -)$
$Q_{p,e,\mathrm{or}}$	X_x^*	A_b	X_x	A_a

(p,e,or)	$(p_2,e_1,-)$	$(p_2,e_4,+)$	$(p_2,e_5,+)$	$(p_2,e_3,-)$
$Q_{p,e,\mathrm{or}}$	X_y	A_a	X_y^*	A_b

We now evaluate $T^{\mathrm{cw}}(\tilde{M})$ as given in (3.27), which is a linear map from $R_{e_1} \otimes R_{e_2}$ to $R_{e_5} \otimes R_{e_6}$, both of which spaces are equal to $X_y \otimes X_x^*$. The map $\mathrm{id} \otimes P$ in (3.27) maps the element $w \otimes \varphi \in R_{e_1} \otimes R_{e_2}$ to (not writing all '$\otimes$'-symbols)

$$\begin{array}{lcccccccc} \longrightarrow & & (R_{e_1} & R_{e_2}) & (Q_{p_1,e_3,+} & Q_{p_1,e_6,+} & Q_{p_1,e_4,-} & & \\ \longmapsto & \displaystyle\sum_{i,j,k,l} & (w \otimes & \varphi) \otimes & (b'_i \otimes & u_j \otimes & a'_k \otimes & & \\ & & & & Q_{p_2,e_4,+} & Q_{p_2,e_5,+} & Q_{p_2,e_3,-}) & (R_{e_5} & R_{e_6}) \\ & & & & a_k \otimes & v_l^* \otimes & b_i) & \otimes (v_l \otimes & u_j^*) \end{array}$$

which in turn gets mapped by $E \otimes \mathrm{id}$ to

$$\sum_{i,j,k,l} \varphi(b'_i.u_j.a'_k)\, v_l^*(b_i.w.a_k)\, v_l \otimes u_j^* \tag{3.55}$$

in $R_{e_5} \otimes R_{e_6}$. This can be simplified by carrying out the sum over the bases u_j, v_l and their duals, resulting in

$$T^{\mathrm{cw}}(\tilde{M})(w \otimes \varphi) = \sum_{i,k} (b_i.w.a_k) \otimes (a'_k.\varphi.b'_i)\ . \tag{3.56}$$

Comparing to the discussion in section 3.3, we see that this is nothing but the projector $p_\otimes$ on $X_y \otimes X_x^*$ whose image is $\circlearrowleft_{A_b} X_y \otimes_{A_a} X_x^*$. Combining this with the definition (3.50) of T, we see that $T(M) = \mathrm{id}$ on $\circlearrowleft_{A_b} X_y \otimes_{A_a} X_x^*$. This illustrates point (ii) of Theorem 3.8.

The following lemma provides a different point of view on this result which will be useful in understanding the 2-category of defect conditions. Denote by $\mathrm{ev}_V : V^* \otimes V \to k$ the evaluation of a vector space on its dual and write $\mathrm{Hom}_{A|B}(X,Y)$ for the space of bimodule maps between two A-B-bimodules X, Y.

LEMMA 3.9. *Let A, B be Frobenius algebras with trace pairing, and let X, Y be finite dimensional A-B-bimodules. The map $\phi : \circlearrowleft_A Y \otimes_B X^* \to \mathrm{Hom}_{A|B}(X,Y)$,*

$$\phi(\gamma) := \Big(X \xrightarrow{\gamma \otimes \mathrm{id}_X} (\circlearrowleft_A Y \otimes_B X^*) \otimes X \xrightarrow{e_\otimes \otimes \mathrm{id}_X} Y \otimes X^* \otimes X \xrightarrow{\mathrm{id}_X \otimes \mathrm{ev}_X} Y\Big) \tag{3.57}$$

is an isomorphism.

PROOF. Let $\gamma \in \circlearrowleft_A Y \otimes_B X^*$. That $\phi(\gamma)$ is a bimodule map is a straightforward calculation using $\mathrm{ev}_X(u \otimes (a.v.b)) = \mathrm{ev}_X((b.u.a) \otimes v)$ and $(e_\otimes \circ \gamma).a = a.(e_\otimes \circ \gamma)$. That ϕ is an isomorphism follows by the standard argument using the corresponding coevaluation map and the duality properties. □

Via this lemma we can also think of $T(O(y \circ x^*))$ as $\mathrm{Hom}_{A_b|A_a}(X_x, X_y)$. If we identify $X_y \otimes X_x^*$ with $\mathrm{Hom}(X_x, X_y)$ then $T^{\mathrm{cw}}(\tilde{M})$ becomes the projection from general linear maps to bimodule intertwiners.

The second example – figure 15 b) – is again an annulus with a domain wall, but this time the ingoing boundary sits entirely in one domain. Here, the source-object is $U = O(b)$, an S^1 with no marked points and labelled $b \in D_2$, and the target object is $V = O(x \circ x^*)$. The lift $\tilde{U}$ we chose contains a single edge, while $\tilde{V}$ contains two edges. The spaces R_e and $Q_{p,e,\mathrm{or}}$ in this case are:

e	e_1	e_4	e_5
R_e	A_b	X_x	X_x^*

(p, e, or)	$(p_1, e_1, -)$	$(p_1, e_2, +)$	$(p_1, e_5, +)$	$(p_1, e_3, -)$	$(p_2, e_2, -)$	$(p_2, e_3, +)$	$(p_2, e_4, +)$
$Q_{p,e,\text{or}}$	A_b	A_b	X_x	X_x^*	A_b	X_x	X_x^*

The map $\text{id} \otimes P$ takes $w \in A_b$ to

$$\longmapsto \sum_{i,j,k,l} \overset{R_{e_1}}{w} \otimes \Big(\overset{Q_{p_1,e_3,-}}{u_i^*} \otimes \overset{Q_{p_1,e_2,+}}{b_j'} \otimes \overset{Q_{p_1,e_5,+}}{u_k} \otimes \overset{Q_{p_2,e_4,+}}{u_l^*} \otimes \overset{Q_{p_2,e_2,-}}{b_j} \otimes \overset{Q_{p_2,e_3,+}}{u_i} \Big) \otimes \big(\overset{R_{e_4}}{u_l} \otimes \overset{R_{e_5}}{u_k^*} \big) .$$

This in turn is mapped to $\sum_{i,j,k,l} u_i^*((w \cdot b_j').u_k)\, u_l^*(b_j.u_i) \cdot u_l \otimes u_k^*$ by $E \otimes \text{id}$, which simplifies to

$$T^{\text{cw}}(\tilde{M})(w) = \sum_{j,k} \big((b_j w b_j').u_k\big) \otimes u_k^* = \sum_k (p_\otimes(w).u_k) \otimes u_k^* , \tag{3.58}$$

where $p_\otimes : A_b \to A_b$ is the projection to the centre of A_b, see Lemma 3.3. Accordingly, the resulting map for the bordism M is

$$T(M) : \begin{array}{ccc} Z(A_b) & \longrightarrow & \text{Hom}_{A_b|A_a}(X_x, X_x) \\ z & \longmapsto & (q \mapsto z.q) , \end{array} \tag{3.59}$$

where we have identified $\circlearrowleft_{A_b} X_x \otimes_{A_a} X_x^* \cong \text{Hom}_{A_b|A_a}(X_x, X_x)$ via Lemma 3.9.

3.8. Bicategory of algebras and 2-category of defect conditions. In the construction in sections 3.4–3.6, the domain wall conditions were given by bimodules. Bimodules naturally form a bicategory (see [**Be**, Sec. 2.5], [**Gr**, Sec. I.3] or [**ML**, Ch. XII.7]), and in this subsection we want to compare this bicategory to the 2-category of defect conditions described in section 2.4. Our conventions for bicategories can be found in appendix A.

DEFINITION 3.10. (i) The category $\mathcal{A}lg(k)$ has associative unital algebras over k as objects and (unital) algebra homomorphisms as morphisms.

(ii) The bicategory $\mathbf{Alg}(k)$ has associative unital algebras over k as objects. The morphism category $\mathbf{Alg}(k)(A, B)$ is given by the category of B-A-bimodules and bimodule intertwiners. The composition functor $\mathbf{Alg}(k)(B, C) \times \mathbf{Alg}(k)(A, B) \to \mathbf{Alg}(k)(A, C)$ is $(-) \otimes_B (-)$.

We will start with a small digression which is not restricted to Frobenius algebras with trace pairing. Namely, we will look at some properties of $\mathbf{Alg}(k)$.

Given a 1-category $\mathcal{C}$, we denote the bicategory obtained from $\mathcal{C}$ by adding only identity 2-morphisms again by $\mathcal{C}$. When comparing $\mathcal{A}lg(k)$ and $\mathbf{Alg}(k)$, we understand $\mathcal{A}lg(k)$ as a bicategory in this sense. For an algebra map $f : A \to B$ and a right B-module M, we denote by M_f the right A-module with action $(m, a) \mapsto m.f(a)$. In particular, B_f is a B-A-bimodule. The next lemma (following [**Be**, Sec. 5.7]) makes precise the idea that $\mathbf{Alg}(k)$ contains more 1- and 2-morphisms than $\mathcal{A}lg(k)$.[14]

[14]Note that while $\mathcal{A}lg(k)(A, B)$ is not additive (since $f+g$ is never an algebra homomorphism if f and g are), the category $\mathbf{Alg}(k)(A, B)$ has direct sums of 1-morphisms, so that we have added enough morphisms to 'linearise' $\mathcal{A}lg(k)$.

LEMMA 3.11. *(i) Let $A \xrightarrow{f} B \xrightarrow{g} C$ be algebra maps. The following map is well-defined and an isomorphism of C-A-bimodules:*

$$m_{g,f} : C_g \otimes_B B_f \longrightarrow C_{g\circ f} \quad , \quad c \otimes_B b \longmapsto c \cdot g(b) \ . \tag{3.60}$$

(ii) The assignment

$$i : \mathcal{A}lg(k) \longrightarrow \mathbf{Alg}(k) \ , \tag{3.61}$$

which is the identity on objects and which maps $A \xrightarrow{f} B$ to B_f, is a (non-lax) functor. The unit transformations are identities and the multiplication transformations are given by $m_{g,f}$.

(iii) Let $f, g : A \to B$ be algebra maps. Then $i(f)$ and $i(g)$ are 2-isomorphic in $\mathbf{Alg}(k)$ if and only if $f(-) = u \cdot g(-) \cdot u^{-1}$ for some $u \in B^\times$.

PROOF. Abbreviate $m \equiv m_{g,f}$.

(i) To see that m is well-defined, consider the map $\bar{m} : C_g \otimes B_f \to C_{g\circ f}$ given by $u \otimes v \mapsto ug(v)$. We verify the cokernel condition: for $b \in B$ we have $\bar{m}((u.b) \otimes v) = \bar{m}((ug(b)) \otimes v) = ug(b)g(v) = ug(bv) = \bar{m}(u \otimes (bv))$. Therefore, $\bar{m}$ induces a map $C_g \otimes_B B_f \to C_{g\circ f}$, which is precisely m. Since $c \mapsto c \otimes_B 1_B$ is an isomorphism from $C_{g\circ f}$ to $C_g \otimes_B B_f$, and since by composing with m one obtains the identity on C_g, it follows that m is an isomorphism. It is straightforward to check that m intertwines the C-A-bimodule structures.

(ii) We have to verify associativity and unit properties of the functor. We start with associativity. Given algebra maps $A \xrightarrow{f} B \xrightarrow{g} C \xrightarrow{h} D$, we must show commutativity of the diagram

$$\begin{array}{ccc}
(D_h \otimes_C C_g) \otimes_B B_f & \xrightarrow{\sim} & D_h \otimes_C (C_g \otimes_B B_f) \\
\downarrow{\scriptstyle m_{h,g} \otimes_B \mathrm{id}} & & \downarrow{\scriptstyle \mathrm{id} \otimes_C m_{g,f}} \\
D_{h\circ g} \otimes_B B_f & & D_h \otimes_C C_{g\circ f} \\
\downarrow{\scriptstyle m_{h\circ g,f}} & & \downarrow{\scriptstyle m_{h,g\circ f}} \\
D_{h\circ g\circ f} & \xrightarrow{=} & D_{h\circ g\circ f}
\end{array} \tag{3.62}$$

Acting on an element $d \otimes_C c \otimes_B b$, the left branch gives $d \cdot h(c) \cdot h(g(b))$ and the right branch gives $d \cdot h(c \cdot g(b))$. These are equal as h is an algebra map. The unit properties in turn amount to commutativity of the following two diagrams:

$$\begin{array}{ccc}
B \otimes_B B_f & \xrightarrow{\sim} & B_f \\
\downarrow{\scriptstyle \mathrm{id}} & & \uparrow{\scriptstyle \mathrm{id}} \\
B_{\mathrm{id}} \otimes_B B_f & \xrightarrow{m_{\mathrm{id},f}} & B_{\mathrm{id}\circ f}
\end{array} \quad , \quad
\begin{array}{ccc}
B_f \otimes_A A & \xrightarrow{\sim} & B_f \\
\downarrow{\scriptstyle \mathrm{id}} & & \uparrow{\scriptstyle \mathrm{id}} \\
B_f \otimes_A A_{\mathrm{id}} & \xrightarrow{m_{f,\mathrm{id}}} & B_{\mathrm{id}\circ f}
\end{array} \tag{3.63}$$

In the left diagram, both branches give the map $b \otimes_B b' \mapsto b \cdot b'$, and in the right diagram, both branches give $b \otimes_B a \mapsto b \cdot f(a)$.

Since by part (i) the m's are isomorphisms, we do indeed obtain a functor, not just a lax functor.

(iii) '$\Rightarrow$': Suppose that $\psi : B_f \to B_g$ is an isomorphism of B-A-bimodules. Then for all $x, b \in B$ and $a \in A$ we have $\psi(b \cdot x \cdot f(a)) = b \cdot \psi(x) \cdot g(a)$. From this we conclude that $f(a)\cdot\psi(1) = \psi(f(a)) = \psi(1)\cdot g(a)$. Since ψ is invertible, $\psi(1) \in B^\times$, and so $f(a) = \psi(1)\cdot g(a)\cdot \psi(1)^{-1}$.

'$\Leftarrow$': The isomorphism is given by $b \mapsto b\cdot u$. □

Recall the construction of the 2-category $\mathbf{D}[D_2, D_1; T]$ in (2.12), the assignment of algebras and bimodules to elements of D_2 and D_1 in the beginning of section 3.4, and the definition of the defect TFT T in Theorem 3.8. We want to define a functor

$$\Delta : \mathbf{D}[D_2, D_1; T] \longrightarrow \mathbf{Alg}(k) \ , \tag{3.64}$$

which on objects $a \in D_2$ acts as $\Delta(a) = A_a$ and on 1-morphisms $\underline{x} \in \mathbf{D}(a,b)$ as $\Delta(\underline{x}) = X_{\underline{x}}$, using the notation (3.20). The action on 2-morphisms will be described after the following remark.

REMARK 3.12. (i) By (2.9), the 2-morphism spaces of $\mathbf{D}[D_2, D_1; T]$ are given by $\mathbf{D}_2(\underline{x}, \underline{y}) := H^{\mathrm{inv}}(\underline{y} \circ \underline{x}^*)$. One may think that in a TFT all states are scale and translation invariant, and this is true but for one detail. Let $\underline{x} : a \to a$ and let $C_{O(\underline{x})}$ be the cylinder over $O(\underline{x})$. The defining property (2.7) of a scale and translation invariant family implies that all vectors $\psi_{\underline{x};r}$ lie in the image of the idempotent $T(C_{O(\underline{x})}) : T(O(\underline{x})) \to T(O(\underline{x}))$. Conversely, each vector in the image of $T(C_{O(\underline{x})})$ gives rise to a scale and translation invariant family. Indeed, for TFTs, $T(O(\underline{x})) \equiv T(O(\underline{x}; r))$ is independent of r, and so is the family $\psi_{\underline{x};r}$. We can therefore identify $H^{\mathrm{inv}}(\underline{x})$ with the image of the idempotent $T(C_{O(\underline{x})})$. For our lattice TFT construction, by Theorem 3.8 (iii) this does not make a difference, but for a general TFT, $T(C_{O(\underline{x})})$ may be different from the identity map on $T(O(\underline{x}))$.

(ii) Given a TFT for which the idempotents $T(C_U)$ for objects $U \in \mathrm{Bord}_{2,1}^{\mathrm{def,top}}$ are not always identity maps, one can define a new TFT T' in which one replaces all state spaces $T(U)$ by the image of the corresponding idempotent $T(C_U)$. The embedding of the image of $T(C_U)$ into $T(U)$ provides a monoidal natural transformation from T' to T. One can think of T' as the 'non-degenerate subtheory' of T, because an amplitude $T(U \xrightarrow{M} V)$ vanishes if its argument comes from the kernel of $T(C_U)$. In principle, one can always work with non-degenerate TFTs, but in some situations degenerate TFTs are useful as an intermediate step (such as in the orbifold construction of [**Frö2**], or in a sense also the construction in section 3.6, where T was defined precisely as the restriction of T^{cw} to images of idempotents).

According to part (i) of the above remark, in our lattice TFT example we have $H^{\mathrm{inv}}(\underline{x}) = T(O(\underline{x}))$. Substituting the definition of T on objects in (3.49), we see that for $\underline{x}, \underline{y} : a \to b$,

$$\mathbf{D}_2(\underline{x}, \underline{y}) = \circlearrowleft_{A_b} X_{\underline{y}} \otimes_{A_a} X_{\underline{x}^*} \ . \tag{3.65}$$

Using this and Lemma 3.9, we can finally state the action of the functor Δ on morphisms. Namely for $u \in \mathbf{D}_2(\underline{x}, \underline{y})$ we set $\Delta(u) = \phi(u) : X_{\underline{x}} \to Y_{\underline{y}}$.

We should now proceed to show that Δ thus defined is indeed a functor between bicategories, which in addition is locally fully faithful (since ϕ is an isomorphism). However, we will not go through these details and instead turn to the next topic, the relation between lattice TFT with defects and the centre of an algebra.

4. The centre of an algebra

The map which assigns to an algebra A its centre $Z(A)$ is not functorial, at least not in the obvious sense. Namely, given $A \in \mathcal{A}lg(k)$, then also $Z(A) \in \mathcal{A}lg(k)$, but for an algebra homomorphism $f : A \to B$ it is in general not true that $f|_{Z(A)}$ lands in $Z(B)$. For example, if A is the algebra of diagonal 2×2 matrices, if B is all 2×2 matrices and if f is the embedding map, then $Z(A) = A$, but $Z(B) = k\,\mathrm{id}$ which does not contain $f(Z(A))$.

For Frobenius algebras with trace pairing one could use the maps $e_\otimes$ and $\pi_\otimes$ between A and $Z(A) = \circlearrowleft_A A$ (cf. Lemma 3.3 – not true for general algebras) to map f to $\pi_\otimes \circ f \circ e_\otimes$, but this would in general not be compatible with composition and multiplication.

The main point of this section is to define a functorial version of the centre. This is done by first constructing a bicategory – or rather two versions thereof – whose objects are commutative algebras. The centre is then a lax functor into this bicategory; this functor will also be given in two versions (Theorem 4.12 and Remark 4.19). These constructions are motivated by 2-dimensional TFT with defects, so we begin the discussion by highlighting the relevant algebras and maps in the defect TFT.

4.1. Spaces and maps associated to defect TFTs. Let

$$T : \mathrm{Bord}_{2,1}^{\mathrm{def,top}}(D_2, D_1, D_0) \longrightarrow \mathcal{V}ect_f(k)$$

be a defect TFT (not necessarily obtained via lattice TFT). The functor T encodes an infinite number of state spaces and linear maps between them. In this subsection we will pick out some of the more fundamental ones and investigate their properties.

By Remark 3.12 (ii) we are entitled to assume that all idempotents $T(C_U)$ are in fact identity maps, and we will make this assumption for the rest of this subsection. Recall from section 2.4 the 2-category $\mathbf{D}[D_2, D_1; T]$ associated to a field theory with defects. By Remark 3.12 (i) and because of our assumption that $T(C_U) = \mathrm{id}_{T(U)}$, definition (2.9) of the 2-morphism spaces becomes

$$\mathbf{D}_2(\underline{x}, \underline{y}) = T(O(\underline{y} \circ \underline{x}^*)) \ . \tag{4.1}$$

Recall that the identity 1-morphism $\mathbf{1}_a : a \to a$, for $a \in D_2$, is the empty tuple $\mathbf{1}_a = ()$. Consider the space of 2-endomorphisms of $\mathbf{1}_a$, $\mathbf{D}_2(\mathbf{1}_a, \mathbf{1}_a) = T(O(a))$. This is an associative, commutative, unital algebra; the bordisms which give the multiplication and unit morphisms are those in figure 6 a,b), but without domain walls. Commutativity follows since precomposing the multiplication bordism with a transposition $\sigma : O(a) \sqcup O(a) \to O(a) \sqcup O(a)$ gives a diffeomorphic bordism. In fact, by the usual arguments, it is even a Frobenius algebra, and this Frobenius algebra defines the defect-free TFT given by $D_2 = \{a\}$ and $D_1 = D_0 = \emptyset$.

For an arbitrary 1-morphism $\underline{x} : a \to b$, the 2-endomorphisms $\mathbf{D}_2(\underline{x}, \underline{x})$ do still form an associative, unital algebra (even a Frobenius algebra), but this algebra need not be commutative. The horizontal composition functors for $(a \xrightarrow{\mathbf{1}_a} a \xrightarrow{\underline{x}} b) = a \xrightarrow{\underline{x}} b$ and $(a \xrightarrow{\underline{x}} b \xrightarrow{\mathbf{1}_b} b) = a \xrightarrow{\underline{x}} b$ give linear maps

$$\begin{aligned} \hat{R} &: \mathbf{D}_2(\underline{x}, \underline{x}) \otimes \mathbf{D}_2(\mathbf{1}_a, \mathbf{1}_a) \longrightarrow \mathbf{D}_2(\underline{x}, \underline{x}) \ , \\ \hat{L} &: \mathbf{D}_2(\mathbf{1}_b, \mathbf{1}_b) \otimes \mathbf{D}_2(\underline{x}, \underline{x}) \longrightarrow \mathbf{D}_2(\underline{x}, \underline{x}) \ . \end{aligned} \tag{4.2}$$

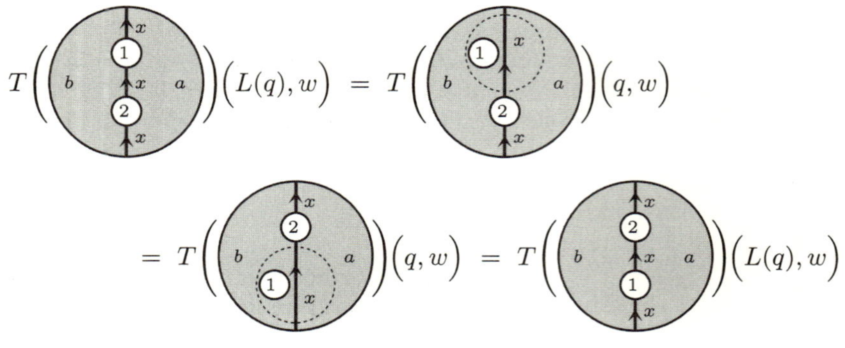

FIGURE 16. Manipulation of bordisms showing that $L \equiv \hat{L}(- \otimes \mathrm{id}_{\underline{x}}) : \mathbf{D}_2(\mathbf{1}_b, \mathbf{1}_b) \to \mathbf{D}_2(\underline{x}, \underline{x})$ maps to the centre of $\mathbf{D}_2(\underline{x}, \underline{x})$. Here $q \in \mathbf{D}_2(\mathbf{1}_b, \mathbf{1}_b)$ and $w \in \mathbf{D}_2(\underline{x}, \underline{x})$.

The bordisms for the maps $\hat{L}$ and $\hat{R}$ are as in figure 6 c), provided we specialise the latter to the case where only one of the two ingoing boundary circles has domain walls attached to it. If we insert the identity 2-morphism $\mathrm{id}_{\underline{x}}$, we obtain maps $R := \hat{R}(\mathrm{id}_{\underline{x}} \otimes -) : \mathbf{D}_2(\mathbf{1}_a, \mathbf{1}_a) \to \mathbf{D}_2(\underline{x}, \underline{x})$ and $L := \hat{L}(- \otimes \mathrm{id}_{\underline{x}}) : \mathbf{D}_2(\mathbf{1}_b, \mathbf{1}_b) \to \mathbf{D}_2(\underline{x}, \underline{x})$. The corresponding bordisms are obtained by gluing a disc as in figure 6 a) into the hole which has the domain walls attached. Figure 15 b) shows a bordism obtained in this way.

With the help of bordisms, it is easy to see that R and L are algebra homomorphisms whose images lie in the centre of $\mathbf{D}_2(\underline{x}, \underline{x})$. The bordism manipulations showing that the image of L lies in the centre are given in figure 16.[15]

Next, consider the space of 2-morphisms $\mathbf{D}_2(\underline{x}, \underline{y})$ between two 1-morphisms $\underline{x}, \underline{y} : a \to b$. This is the TFT state space for a circle with sequence of marked points $\underline{y} \circ \underline{x}^*$. By vertical composition, $\mathbf{D}_2(\underline{x}, \underline{y})$ is a right $\mathbf{D}_2(\underline{x}, \underline{x})$-module and a left $\mathbf{D}_2(\underline{y}, \underline{y})$-module. Using R to map $\mathbf{D}_2(\mathbf{1}_a, \mathbf{1}_a)$ into $\mathbf{D}_2(\underline{x}, \underline{x})$ and $\mathbf{D}_2(\underline{y}, \underline{y})$, we see that $\mathbf{D}_2(\underline{x}, \underline{y})$ is also a bimodule for $\mathbf{D}_2(\mathbf{1}_a, \mathbf{1}_a)$. However, by an argument analogous to that in figure 16 it is easy to check that the left and right action agree. Equally, L turns it into an $\mathbf{D}_2(\mathbf{1}_b, \mathbf{1}_b)$-bimodule with identical left and right action.

Let $f \in \mathbf{D}_2(\underline{x}, \underline{y})$ be a 2-morphism. Pre- and post-composing with f defines maps

$$f \circ (-) : \mathbf{D}_2(\underline{x}, \underline{x}) \to \mathbf{D}_2(\underline{x}, \underline{y}) \quad , \quad (-) \circ f : \mathbf{D}_2(\underline{y}, \underline{y}) \to \mathbf{D}_2(\underline{x}, \underline{y}) \tag{4.3}$$

Again by manipulating bordisms, one checks that $f \circ (-)$ intertwines the right $\mathbf{D}_2(\underline{x}, \underline{x})$ action and $(-) \circ f$ intertwines the left $\mathbf{D}_2(\underline{y}, \underline{y})$-action. All these maps are collected in figure 17.

REMARK 4.1. In conformal (and thus in particular in topological) field theory, one has the state-field correspondence, which says that the space of fields associated to a point on the world sheet is the same as the space of states on a small circle

[15] Manipulating such disk-shaped bordisms reminds one of the string-diagram notation for 2-categories [St]. Indeed, a string-diagram identity implies an identity for defect correlators on disks, but the converse is not true – the 2-category $\mathbf{D}[D_2, D_1; T]$ satisfies more conditions than a generic 2-category.

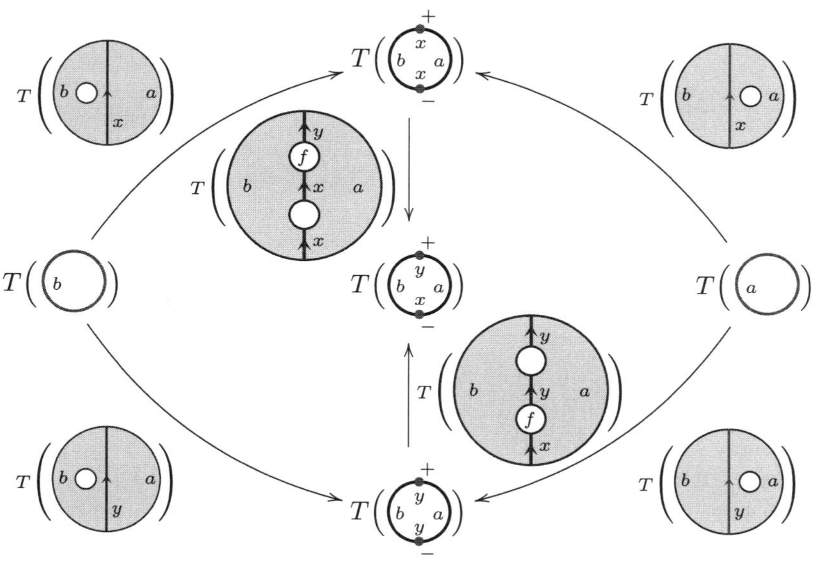

FIGURE 17. Summary of the state spaces and maps between them as described in section 4.1. Here only the case that $\underline{x} = ((x,+))$ and $\underline{y} = ((y,+))$ is shown. For tuples with more elements, the bordisms involve the corresponding sequences of parallel lines.

obtained by cutting out a small disc around this point; in fact, one can take this as a definition of what one means by a field. Then $\mathbf{D}_2(\mathbf{1}_a, \mathbf{1}_a)$ is the space of 'bulk fields' and $\mathbf{D}_2(\underline{x}, \underline{x})$ is the space of 'defect fields' supported on the defect x. An important notion in quantum field theory is the short distance expansion or operator product expansion (OPE). In TFT, of course, the distance between insertion points is immaterial. The above considerations isolate the three most important OPEs: the OPE of two bulk fields; the OPE of two defect fields; the expansion of a bulk field close to a defect line in terms of defect fields.

4.2. The bicategory of commutative algebras, version 1. In this subsection we will use some of the structure seen in 2d TFT with defects in the previous subsection to define a bicategory of commutative algebras in terms of cospans. All algebras will be unital, associative algebras over a field k.

DEFINITION 4.2. A *cospan between commutative algebras*, or *cospan* for short, is a tuple (A, α, T, β, B), where A, B are commutative algebras, T is an algebra, and α, β are algebra homomorphisms

$$\begin{array}{ccccc} & & T & & \\ & {\scriptstyle\alpha}\nearrow & & \nwarrow{\scriptstyle\beta} & \\ A & & & & B \end{array} \tag{4.4}$$

such that the images of α and β lie in the centre $Z(T)$ of T.

The definition has no preferred 'direction', but we will pick one anyway: we will think of (4.4) as going from B to A. The reason for this choice is that we will use the maps α, β to turn T into an A-B-bimodule and the composition of cospans (to be defined in more detail below) will be the tensor product of bimodules, just as in the bicategory $\mathbf{Alg}(k)$. In the latter, A-B-bimodules serve as 1-morphisms $B \to A$. We will write $T : B \to A$, or just T, to abbreviate the data in (4.4). Two different cospans $T, T' : B \to A$ can be compared via algebra homomorphisms $T \to T'$. This leads first to a category of cospans from B to A, and then to a bicategory $\mathbf{CAlg}(k)$ of cospans between commutative algebras. The construction is almost identical to the standard construction of the bicategory of spans for a given category with pullbacks (see [**Be**, Sec. 2.6] or [**Gr, ML**]), with the exception that not all three objects in (4.4) are taken from the same category (we use commutative algebras for the starting points of the cospan and not necessarily commutative algebras for the middle term)

DEFINITION 4.3. The category $\mathcal{C}osp(A, B)$ of cospans between commutative algebras from A to B is defined as follows.

- objects $T \in \mathcal{C}osp(A, B)$ are cospans $T : A \to B$.
- morphisms $f \in \mathcal{C}osp(A, B)(T, T')$ from $T : A \to B$ to $T' : A \to B$ are algebra maps $f : T \to T'$ such that the following diagram commutes:

(4.5)

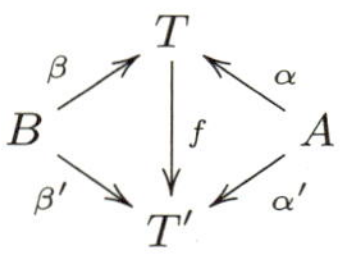

- the unit morphism in $\mathcal{C}osp(A, B)(T, T)$ is the identity map id_T, and composition of morphisms is composition of algebra maps.

The composition of two cospans $A \xrightarrow{T} B \xrightarrow{S} C$ is defined by the usual pushout square,

(4.6)

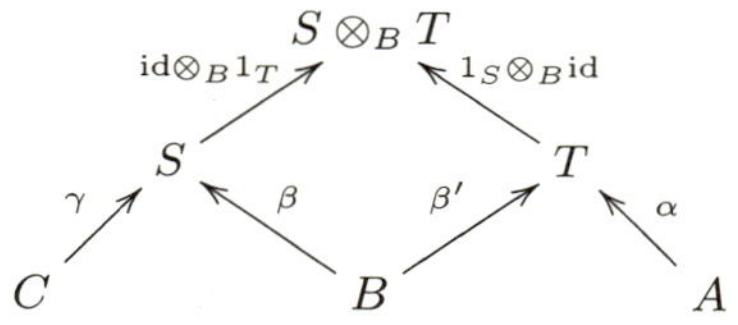

The algebra homomorphism β turns S into a right B-module, and β' turns T into a left B-module; these module structures are implied when writing $S \otimes_B T$. We still need to turn $S \otimes_B T$ into an algebra and show that A and C get mapped to the centre of this algebra. The multiplication on $S \otimes_B T$ is given by

$$(s \otimes_B t) \cdot (s' \otimes_B t') := (ss') \otimes_B (tt') \ . \tag{4.7}$$

To check that this is well-defined, we start with the map $\bar{m} : (S \otimes T) \otimes (S \otimes T) \to S \otimes_B T$, which takes $(s \otimes t) \otimes (s' \otimes t')$ to $(ss') \otimes_B (tt')$ and verify the cokernel condition. We present the calculation for the first factor, the one for the second factor is similar. With $b \in B$,

(4.8)

$$\begin{aligned} \bar{m}(s.b \otimes t \otimes s' \otimes t') &= (s\beta(b)s') \otimes_B (tt') \overset{(*)}{=} (ss'\beta(b)) \otimes_B (tt') \\ &= (ss').b \otimes_B (tt') = (ss') \otimes_B b.(tt') = (ss') \otimes_B (\beta'(b)tt') = \bar{m}(s \otimes b.t \otimes s' \otimes t') \end{aligned}$$

The step marked '$(*)$' uses that the image of the algebra homomorphism β is in the centre of S. This, by the way, is the reason not to allow general algebra homomorphisms in Definition 4.2: we want the tensor product over B to carry an induced algebra structure. Finally, it is straightforward to check that for all $a \in A$, $c \in C$, the elements $1_S \otimes_B \alpha(a)$ and $\gamma(c) \otimes_B 1_T$ are in the centre of $S \otimes_B T$.

The composition of cospans defined above forms part of a functor:

LEMMA 4.4. *The assignment*

$$\circledcirc_{C,B,A} : \mathcal{C}osp(B,C) \times \mathcal{C}osp(A,B) \longrightarrow \mathcal{C}osp(A,C)$$

$$\left(\begin{array}{ccc} & T & \\ C \overset{\gamma}{\nearrow} & \big\downarrow g & \overset{\beta_2}{\nwarrow} B \\ C \underset{\gamma'}{\searrow} & T' & \underset{\beta_2'}{\swarrow} B \end{array} , \begin{array}{ccc} & S & \\ B \overset{\beta_1}{\nearrow} & \big\downarrow f & \overset{\alpha}{\nwarrow} A \\ B \underset{\beta_1'}{\searrow} & S' & \underset{\alpha'}{\swarrow} A \end{array} \right) \longmapsto \begin{array}{ccc} & T \otimes_B S & \\ C \overset{\gamma\otimes_B 1}{\nearrow} & \big\downarrow g\otimes_B f & \overset{1\otimes_B \alpha}{\nwarrow} A \\ C \underset{\gamma'\otimes_B 1}{\searrow} & T' \otimes_B S' & \underset{1\otimes_B \alpha'}{\swarrow} A \end{array} \tag{4.9}$$

defines a functor.

PROOF. Note that the algebra map g is automatically a C-B-bimodule map. For example, $g(t.b) = g(t \cdot \beta_2(b)) = g(t) \cdot g(\beta_2(b)) = g(t) \cdot \beta_2'(b) = g(t).b$. Similarly, f is a B-A-bimodule map. Thus $g \otimes_B f$ is well-defined. That the two triangles on the rhs of (4.9) commute is immediate. Finally, functoriality of $\circledcirc_{C,B,A}$ amounts to the statement that

$$(g_2 \circ g_1) \otimes_B (f_2 \circ f_1) = (g_2 \otimes_B f_2) \circ (g_1 \otimes_B f_1) \ , \tag{4.10}$$

which is a property of the tensor product over B. □

We have now gathered the ingredients to define the first version of the bicategory of commutative algebras, which we denote by

$$\mathbf{CAlg}(k) \ . \tag{4.11}$$

Its objects are commutative algebras over k. Given two such algebras A, B, the category of morphisms from A to B is $\mathcal{C}osp(A,B)$. The identity in $\mathcal{C}osp(A,A)$ is

$$\begin{array}{ccc} & A & \\ \overset{\text{id}}{\nearrow} & & \overset{\text{id}}{\nwarrow} \\ A & & A \end{array} \ . \tag{4.12}$$

The composition functor is $\circledcirc_{C,B,A}$ from Lemma 4.4. The associativity and unit isomorphisms are just the natural isomorphisms $T \otimes_C (S \otimes_B R) \cong (T \otimes_C S) \otimes_B R$ and $R \otimes_A A \cong R \cong B \otimes_B R$, which we will not write out in the following. It is then clear that the coherence conditions of a bicategory – as listed in appendix A – are satisfied.

REMARK 4.5. For a category $\mathcal{C}$, we denote by $\mathcal{C}^{1,0}$ the subcategory containing only invertible morphisms. Similarly, given a bicategory $\mathbf{B}$, denote by $\mathbf{B}^{2,1}$ the bicategory obtained from $\mathbf{B}$ by restricting to invertible 2-morphisms, and by $\mathbf{B}^{2,0}$ the bicategory consisting only of invertible 1- and 2-morphisms (and $\mathbf{B} \equiv \mathbf{B}^{2,2}$; see [**Lu1**] for more on (m,n)-categories). If $\mathbf{B}$ is a k-linear bicategory, we obtain a lax functor

$$\mathbf{E} : \mathbf{B}^{2,1} \longrightarrow \mathbf{CAlg}(k) \ , \tag{4.13}$$

where '$\mathbf{E}$' stands for endomorphism. We will illustrate this functor in the case of the 2-category $\mathbf{B} \equiv \mathbf{D}[D_2, D_1; T]$ for a fixed 2d TFT T (which need not come from the lattice construction). On objects and 1-morphisms we set

(4.14)
$$\mathbf{E}(a) = \mathbf{B}(\mathbf{1}_a, \mathbf{1}_a) \quad , \quad \mathbf{E}(b \xleftarrow{\underline{x}} a) = \left(\mathbf{B}(\mathbf{1}_b,\mathbf{1}_b) \xrightarrow{L} \mathbf{B}(\underline{x},\underline{x}) \xleftarrow{R} \mathbf{B}(\mathbf{1}_a,\mathbf{1}_a) \right) \quad ;$$

the maps L and R have been given in section 4.1. To an invertible 2-morphism $u : \underline{x} \to \underline{y}$ we assign the algebra map $\mathbf{E}(u) : \mathbf{E}(\underline{x}) \to \mathbf{E}(\underline{y})$ given by conjugation with u. That is, $f : \underline{x} \to \underline{x}$ gets mapped to

$$\mathbf{E}(u)(f) = \left(\underline{y} \xrightarrow{u^{-1}} \underline{x} \xrightarrow{f} \underline{x} \xrightarrow{u} \underline{y} \right) . \tag{4.15}$$

We will omit the details of the proof that $\mathbf{E}$ is a lax functor. Note that, because (4.15) involves an inverse, $\mathbf{E}$ is only defined on $\mathbf{B}^{2,1}$. Nonetheless, $\mathbf{B}^{2,1}$ is not enough to define $\mathbf{E}$, instead one requires all of $\mathbf{B}$ so that $\mathbf{B}(\mathbf{1}_a, \mathbf{1}_a)$, etc., are indeed k-algebras. Also, even though the image $\mathbf{E}(u)$ of a 2-morphism is always invertible, it is not true that $\mathbf{E}$ is a functor to $\mathbf{CAlg}(k)^{2,1}$, because the associativity 2-morphism $\mathbf{E}(\underline{y}) \circledcirc \mathbf{E}(\underline{x}) \to \mathbf{E}(\underline{y} \circ \underline{x})$ is not necessarily invertible.

Denote by $\mathcal{A}lg(k)_{\text{com}}$ the full subcategory of commutative algebras in $\mathcal{A}lg(k)$. In the remainder of this subsection we illustrate that $\mathbf{CAlg}(k)$ enlarges the morphism spaces of $\mathcal{A}lg(k)_{\text{com}}$, but that it does not add new invertible morphisms.

LEMMA 4.6. *The assignment*

(4.16)
$$\begin{array}{cccc} I: & \mathcal{A}lg(k)_{\text{com}} & \longrightarrow & \mathbf{CAlg}(k)^{2,1} \\ & B \xleftarrow{f} A & \longmapsto & \left(B \xrightarrow{\text{id}} B \xleftarrow{f} A \right) \end{array}$$

defines a (non-lax) functor.

PROOF. Clearly, the identity gets mapped to the identity. Given two algebra homomorphisms $A \xrightarrow{f} B \xrightarrow{g} C$, one verifies that the map $m_{g,f}$ from Lemma 3.11 defines an isomorphism of cospans

(4.17)
$$\begin{array}{ccccc} & & C \otimes_B B & & \\ & {}^{\text{id}\otimes_B 1}\nearrow & \downarrow m_{g,f} & \nwarrow^{1\otimes_B f} & \\ C & & & & A \\ & {}_{\text{id}}\searrow & & \swarrow_{g\circ f} & \\ & & C & & \end{array} .$$

The verification of the associativity condition works along the same lines as the proof of Lemma 3.11 (ii). □

A 2-morphism between the cospans $I(f)$ and $I(g)$ would necessarily have to be the identity map in order to make the left triangle in the condition (4.5) commute. This then implies $f = g$. In particular, different algebra maps get mapped to non-2-isomorphic cospans. In this sense, I is faithful.

LEMMA 4.7. *A cospan*

(4.18)
$$B \xrightarrow{\beta} T \xleftarrow{\alpha} A$$

is invertible in $\mathbf{CAlg}(k)$ *if and only if* α *and* β *are isomorphisms.*

PROOF. '⇐': Suppose α and β are isomorphisms. Then

$$(4.19)\qquad \begin{array}{ccccc} & & T & & \\ & \overset{\beta}{\nearrow} & \big\downarrow{\scriptstyle \beta^{-1}} & \overset{\alpha}{\nwarrow} & \\ B & & & & A \\ & \underset{\mathrm{id}}{\searrow} & & \underset{\beta^{-1}\circ\alpha}{\swarrow} & \\ & & B & & \end{array}$$

is an isomorphism of cospans, i.e. $T \cong I(\beta^{-1} \circ \alpha)$. The latter cospan has inverse $I(\alpha^{-1} \circ \beta)$ by Lemma 4.6. Thus also T is invertible.

'⇒': Suppose $(A, \alpha', S, \beta', B)$ is a two-sided inverse of T. This means that there are isomorphisms f, g of cospans

$$(4.20)\qquad \begin{array}{ccccc} & & S\otimes_B T & & \\ & \overset{\alpha'\otimes_B 1}{\nearrow} & \big\downarrow{\scriptstyle f} & \overset{1\otimes_B \alpha}{\nwarrow} & \\ A & & & & A \\ & \underset{\mathrm{id}}{\searrow} & & \underset{\mathrm{id}}{\swarrow} & \\ & & A & & \end{array} \quad \text{and} \quad \begin{array}{ccccc} & & T\otimes_A S & & \\ & \overset{\beta\otimes_A 1}{\nearrow} & \big\downarrow{\scriptstyle g} & \overset{1\otimes_A \beta'}{\nwarrow} & \\ B & & & & B \\ & \underset{\mathrm{id}}{\searrow} & & \underset{\mathrm{id}}{\swarrow} & \\ & & B & & \end{array} .$$

Consider the left diagram. Since f is an isomorphism, it implies that also $f^{-1} = 1_S \otimes_B \alpha$ is an isomorphism. Thus we have the identities

$$(4.21)\qquad f \circ (1_S \otimes_B \alpha) = \mathrm{id}_A \quad , \quad (1_S \otimes_B \alpha) \circ f = \mathrm{id}_{S\otimes_B T} .$$

The first of these can be rewritten as $f \circ (1_S \otimes_B \mathrm{id}_T) \circ \alpha = \mathrm{id}_A$, showing that α has left-inverse $\hat{f} := f \circ (1_S \otimes_B \mathrm{id}_T) : T \to A$. An analogous argument gives the left inverse $\hat{g} := g \circ (1_T \otimes_A \mathrm{id}_S) : S \to B$ of β'.

Since $\beta' : B \to S$ is an algebra map and an intertwiner of right B-modules (by definition of the right B-action on S), we can write $1_S \otimes_B \alpha : A \to S \otimes_B T$ as $(\beta' \otimes_B \mathrm{id}_T) \circ (1_B \otimes_B \alpha)$. Inserting this into the second identity in (4.21) gives

$$(4.22)\qquad \mathrm{id}_{S\otimes_B T} = \Big(S \otimes_B T \xrightarrow{f} A \xrightarrow{\alpha} T \xrightarrow{1_B \otimes_B \mathrm{id}_T} B \otimes_B T \xrightarrow{\beta' \otimes_B \mathrm{id}_T} S \otimes_B T \Big)$$

We compose both sides with $\hat{g} \otimes_B \mathrm{id}_T$ and use that $\hat{g}$ is left-inverse to β'. This results in $\hat{g} \otimes_B \mathrm{id}_T = (1_B \otimes_B \mathrm{id}_T) \circ \alpha \circ f$. Finally, composing with $1_S \otimes_B \mathrm{id}_T$ from the left and using that $\hat{g}(1_S) = g(1_T \otimes_A 1_S) = 1_B$, shows that $1_B \otimes_B \mathrm{id}_T = (1_B \otimes_B \mathrm{id}_T) \circ \alpha \circ \hat{f}$. Since $1_B \otimes_B \mathrm{id}_T$ is an isomorphism, we see that $\hat{f}$ is also a right-inverse for α, and hence α is an isomorphism. That β is an isomorphism follows along the same lines. □

From the proof we see that the cospan T in (4.18) is 2-isomorphic to $I(\beta^{-1} \circ \alpha)$. Thus every 1-isomorphism lies in the essential image of I.

REMARK 4.8. An algebra isomorphism $f : A \to B$, when restricted to $Z(A)$, provides an isomorphism $f|_{Z(A)} : Z(A) \to Z(B)$. This gives a functor from $\mathcal{A}lg(k)^{1,0}$ to $\mathcal{A}lg(k)^{1,0}_{\text{com}}$, which, when composed with I, gives a functor

$$(4.23)\qquad \mathcal{A}lg(k)^{1,0} \xrightarrow{Z} \mathcal{A}lg(k)^{1,0}_{\text{com}} \xrightarrow{I} \mathbf{CAlg}(k)^{2,0} .$$

In Theorem 4.12 below, we will extend this beyond the groupoid case to a lax functor $Z : \mathcal{A}lg(k) \to \mathbf{CAlg}(k)$. As an aside, note that the composed functor in (4.23) is neither full nor faithful (isomorphic centres do not imply isomorphic algebras, and different algebra isomorphisms may restrict to the same map on the centre). However, $I : \mathcal{A}lg(k)^{1,0}_{\text{com}} \to \mathbf{CAlg}(k)^{2,0}$ *is* an equivalence since for invertible 1- and 2-morphisms, I is an equivalence on the morphism categories (as those morphism categories which lie in the image of I only contain identity 2-morphisms), and on objects it is just the identity.

4.3. Functorial centre, version 1. Given two not necessarily commutative algebras A, B and an algebra homomorphism $f : A \to B$, we define the *centraliser* $Z_{A,B}(f)$ to be the centraliser of the image of f in B,

$$Z_{A,B}(f) = \{\, b \in B \,|\, f(a)\, b = b\, f(a) \;\text{ for all } a \in A \,\} \,. \tag{4.24}$$

Let $\iota : Z(B) \to B$ be the embedding map and denote the restriction of f to $Z(A)$ also by f.

LEMMA 4.9. *Let A, B be algebras and $f : A \to B$ an algebra homomorphism. Then*

$$\begin{array}{ccccc} & & Z_{A,B}(f) & & \\ & \overset{\iota}{\nearrow} & & \overset{f}{\nwarrow} & \\ Z(B) & & & & Z(A) \end{array} \tag{4.25}$$

is a cospan of commutative algebras.

PROOF. Since $Z(B) \subset Z_{A,B}(f)$, ι is an algebra map which maps to the centre of $Z_{A,B}(f)$. Next we check that the image of $Z(A)$ under $f : A \to B$ lies in $Z_{A,B}(f)$. Given $z \in Z(A)$ set $b = f(z)$. For all $a \in A$ we have $f(a)b = f(a)f(z) = f(az) = f(za) = f(z)f(a) = bf(a)$. Thus $f(z) \in Z_{A,B}(f)$. It is then immediate that $f(Z(A))$ lies in the centre of $Z_{A,B}(f)$. ☐

REMARK 4.10. In the more restrictive setting of Frobenius algebras with trace pairing, the cospan (4.25) has actually already appeared in disguise in the lattice TFT construction. Consider the cospan forming the top of the diamond of maps in figure 17. By definition (3.49) and Lemma 3.3, $T(O(a)) = \circlearrowleft_{A_a} A_a = Z(A_a)$ and $T(O(b)) = Z(A_b)$. For the top entry we have $T(O(x^* \circ x)) = \circlearrowleft_{A_b} X_x \otimes_{A_a} X_x^* \cong \mathrm{Hom}_{A_b|A_a}(X_x, X_x)$, where we used Lemma 3.9. To make the connection to (4.25), take $A = A_a$, $B = A_b$ and let $f : A \to B$ be an algebra map. For X_x take the bimodule B_f defined in section 3.8. The map ϕ in

$$\begin{array}{ccccc} & & Z_{A,B}(f) & & \\ & \overset{\iota}{\nearrow} & & \overset{f}{\nwarrow} & \\ Z(B) & & \Big\downarrow \phi & & Z(A) \\ & \underset{\text{act}}{\searrow} & & \underset{\text{act}}{\swarrow} & \\ & & \mathrm{Hom}_{B|A}(B_f, B_f) & & \end{array} \quad , \tag{4.26}$$

defined as $\phi(b) := (u \mapsto u \cdot b)$, provides an isomorphism of cospans. Here 'act' refers to the map that takes $b \in Z(B)$ to the bimodule map $u \mapsto b.u$; $u \in B_f$ (resp. $a \in Z(A)$ to $u \mapsto u.a$). We omit the details.

LEMMA 4.11. *Let $A \xrightarrow{f} B \xrightarrow{g} C$ be algebra maps. The map*

$$m_{g,f} : Z_{B,C}(g) \otimes_{Z(B)} Z_{A,B}(f) \longrightarrow Z_{A,C}(g \circ f) \quad , \quad u \otimes_{Z(B)} v \longmapsto u \cdot g(v)$$

is a morphism of cospans

$$\begin{array}{ccccc} & & Z_{B,C}(g) \otimes_{Z(B)} Z_{A,B}(f) & & \\ & \overset{\iota \otimes_B 1}{\nearrow} & & \overset{1 \otimes_B f}{\nwarrow} & \\ Z(C) & & \Big\downarrow m_{g,f} & & Z(A) \\ & \underset{\iota}{\searrow} & & \underset{g \circ f}{\swarrow} & \\ & & Z_{A,C}(g \circ f) & & \end{array} \tag{4.27}$$

PROOF. Abbreviate $m \equiv m_{g,f}$ and $Y \equiv Z(B)$.

■ *m is well-defined:* That m gives a well-defined map to C is the same argument as in the proof of Lemma 3.11 (i). That the image of m lies in the centraliser $Z_{A,C}(g \circ f)$ amounts to, for all $a \in A$,

(4.28)
$$\begin{aligned} g(f(a)) \cdot m(u \otimes_Y v) &= g(f(a)) \cdot u \cdot g(v) \overset{(1)}{=} u \cdot g(f(a)) \cdot g(v) = u \cdot g(f(a)v) \\ &\overset{(2)}{=} u \cdot g(vf(a)) = u \cdot g(v) \cdot g(f(a)) = m(u \otimes_Y v) \cdot g(f(a)) \ , \end{aligned}$$

where (1) follows as $u \in Z_{B,C}(g)$ commutes with anything in the image of g, and (2) follows analogously from $v \in Z_{A,B}(f)$.

■ *m is an algebra map:* We have

(4.29)
$$\begin{aligned} m\big((u \otimes_Y v) \cdot (u' \otimes_Y v')\big) &= m\big((uu') \otimes_Y (vv')\big) = uu'g(vv') = uu'g(v)g(v') \\ &\overset{(*)}{=} ug(v)u'g(v') = m(u \otimes_Y v) \cdot m(u' \otimes_Y v') \ . \end{aligned}$$

The only perhaps not immediately obvious step is $(*)$, which follows since by definition for all $u' \in Z_{B,C}(g)$ and $v \in B$ we have $u'g(v) = g(v)u'$.

■ *The triangles commute:* Acting on arbitrary elements $c \in Z(C)$ and $a \in Z(A)$, commutativity of the two triangles amounts to the identities $c = m(c \otimes_Y 1_B)$ and $g(f(a)) = m(1_B \otimes_Y f(a))$, both of which are immediate upon substituting the definition of m. □

We have now collected the ingredients to state the second main result of this paper.

THEOREM 4.12. *The assignment*

$$Z : \quad \mathcal{A}lg(k) \longrightarrow \mathbf{CAlg}(k)$$

(4.30)
$$B \xleftarrow{f} A \quad \longmapsto \quad Z(B) \xrightarrow{\iota} Z_{A,B}(f) \xleftarrow{f} Z(A)$$

defines a lax functor. The unit transformations are identities and the multiplication transformations are given by $m_{g,f}$.

In other words, on objects the lax functor acts as $A \mapsto Z(A)$, on 1-morphisms $A \to B$ as $f \mapsto Z_{A,B}(f)$, and all 2-morphisms in $\mathcal{A}lg(k)$ are identities, which get mapped to identity 2-morphisms in $\mathbf{CAlg}(k)$.

PROOF. It remains to verify the associativity and unit properties. The argument is identical to that in the proof of Lemma 3.11 (ii). □

REMARK 4.13. (i) The map $m_{g,f}$ in Lemma 4.11 is typically not an isomorphism. For example, take $A = C = k \oplus k$ and B to be upper triangular $2{\times}2$ matrices. For the map f we take the diagonal embedding and for g the projection onto the diagonal part. By commutativity of the underlying algebra we see $Z(A) = Z(C) = Z_{B,C}(g) = Z_{A,C}(g \circ f) = k \oplus k$. The remaining algebras are $Z(B) \cong k$ (multiples of the identity matrix) and $Z_{A,B}(f) \cong k \oplus k$, the diagonal $2{\times}2$ matrices. Thus $Z_{B,C}(g) \otimes_{Z(B)} Z_{A,B}(f) \cong (k \oplus k) \otimes_k (k \oplus k)$ while $Z_{A,C}(g \circ f) = k \oplus k$. Therefore, we only have a lax functor.

(ii) If we restrict Z to commutative algebras $\mathcal{A}lg(k)_{\mathrm{com}}$ we obtain the functor I from Lemma 4.6 (since then $Z(B) = B$, $Z(A) = A$ and $Z_{A,B}(f) = B$ for all f). In this sense, Z is an extension of I to all algebras; the price to pay is that we have to work with bicategories and the functor becomes lax.

4.4. The bicategory of commutative algebras, version 2. Figure 17 suggests that there is an enlargement of $\mathbf{CAlg}(k)$, where the 2-morphisms are also replaced by cospans. The lattice TFT construction suggests that this enlargement becomes relevant if one wants to extend the centre functor from $\mathcal{A}lg(k)$ to $\mathbf{Alg}(k)$. This is the topic of the present subsection, as well as of the next one.

DEFINITION 4.14. (i) A *2-diagram* from a cospan $S : A \to B$ to $T : A \to B$ is a triple (g, M, f), where M is a T-S-bimodule, f is a right S-module map and g is a left T-module map such that the two squares in the diagram

$$\begin{array}{ccccc} & & S & & \\ & {\scriptstyle\beta_1}\nearrow & \downarrow{\scriptstyle f} & \nwarrow{\scriptstyle\alpha_1} & \\ B & & M & & A \\ & {\scriptstyle\beta_2}\searrow & \uparrow{\scriptstyle g} & \swarrow{\scriptstyle\alpha_2} & \\ & & T & & \end{array} \tag{4.31}$$

commute, and such that the induced left and right action of A on M agree, and those of B on M agree, i.e. that for all $a \in A$, $b \in B$, $m \in M$

$$\alpha_2(a).m = m.\alpha_1(a) \quad , \quad \beta_2(b).m = m.\beta_1(b) \ . \tag{4.32}$$

We will also abbreviate $M : S \to T$.

(ii) A *3-cell* between two 2-diagrams (g, M, f) and (g', M', f') is a T-S-bimodule map $\delta : M \to M'$ such that the following diagram commutes:

$$\begin{array}{ccccc} & & S & & \\ & {\scriptstyle f}\swarrow & & \searrow{\scriptstyle f'} & \\ M & & \xrightarrow{\ \delta\ } & & M' \\ & {\scriptstyle g}\nwarrow & & \nearrow{\scriptstyle g'} & \\ & & T & & \end{array} \tag{4.33}$$

(iii) The *category of 2-diagrams* $\mathcal{D}iag_{AB}(S, T)$ has 2-diagrams as objects and 3-cells as morphisms. The identity 3-cell for the object (g, M, f) is the identity map on M, the composition of 3-cells is given by composition of bimodule maps.

REMARK 4.15. (i) The category $\mathcal{D}iag_{AB}(S, T)$ can be used to define a bicategory $\mathbf{Cosp}(A, B)$ whose objects are cospans of commutative algebras from A to B and whose morphism categories are $\mathcal{D}iag_{AB}(S, T)$; this will be done in detail in [**DKR2**]. Conjecturally, there is a tricategory CALG(k) whose objects are commutative algebras and whose morphism bicategories are given by $\mathbf{Cosp}(A, B)$; we hope to return to this in the future.

(ii) The conjectural tricategory CALG(k) of part (i) is similar (but not equal) in structure to the tricategory of conformal nets described in [**BDH**]. In [**BDH**], the objects are conformal nets. An object in CALG(k), i.e. a commutative algebra, could be thought of as a 'topological net' which assigns the same algebra to every interval; this algebra is then necessarily commutative (but it is not a conformal net: the algebra does not have to be von Neumann and in general it violates the split property). According to [**BDH**, Def. 3], 1-morphisms are 'defects', i.e. conformal nets for bicoloured intervals. In the present language, this corresponds to the data of a cospan (A, α, T, β, B): evaluating the net on mono-coloured subintervals produces the two commutative algebras A, B, for a bicoloured interval one obtains T, and the inclusion of a mono– into a bicoloured interval provides the two maps α, β. In [**BDH**, Def. 4], 2-morphisms are sectors between the defect nets – this means a Hilbert space with compatible actions of all four conformal nets involved: the two defect nets and the two conformal nets which the defects go between. This provides all the data and constraints of a 2-diagram as in (4.31) *except* for the maps f and g, which are not part of the setting of [**BDH**]. We will need these two maps for the centre functor, see Lemma 4.18 below. The third level of categorical structure is constructed in [**BDH**] by making conformal nets a bicategory internal to symmetric monoidal categories.

Part (i) of the above remark motivates the notation $\underline{\mathbf{Cosp}}(A, B)$ for the category whose objects are cospans from A to B and whose morphisms are isomorphism classes of 2-diagrams. Composition in $\underline{\mathbf{Cosp}}(A, B)$ is given by

(4.34)

$$\odot : \ \underline{\mathbf{Cosp}}(A,B)(S,T) \times \underline{\mathbf{Cosp}}(A,B)(R,S) \longrightarrow \underline{\mathbf{Cosp}}(A,B)(R,T)$$

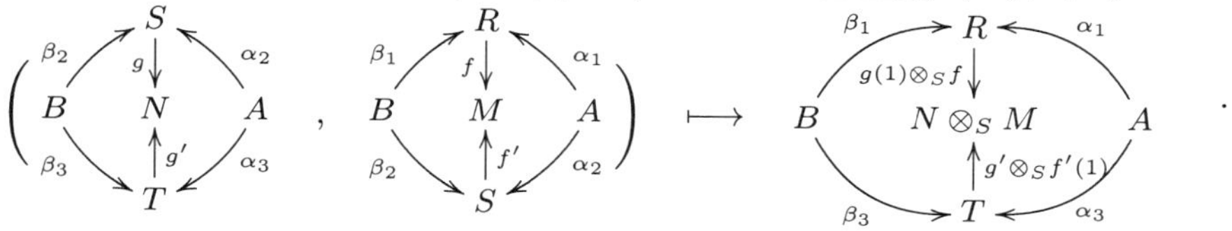

The right hand side is again a 2-diagram. Let us check explicitly the left square in (4.31) and the first condition in (4.32). For $b \in B$,

$$g(1) \otimes_S f(\beta_1(b)) \overset{(1)}{=} g(1) \otimes_S f'(\beta_2(b)) \overset{(2)}{=} g(1) \otimes_S \beta_2(b).f'(1)$$
(4.35)
$$\overset{(3)}{=} g(1).\beta_2(b) \otimes_S f'(1) \overset{(4)}{=} g(\beta_2(b)) \otimes_S f'(1) \overset{(5)}{=} g'(\beta_3(b)) \otimes_S f'(1) \ .$$

Here, (1) is commutativity of the left square in the 2-diagram (f', M, f), (2) is the fact that f' is left S-module map, (3) is the property of $\otimes_S$, (4) follows since g is a right S-module map, and finally (5) is commutativity of the left square in the 2-diagram (g', N, g). Next, that the left action of B on $N \otimes_S M$ agrees with the right action of B follows from

$$b.(n \otimes_S m) \ = (\beta_3(b).n) \otimes_S m = (n.\beta_2(b)) \otimes_S m$$
(4.36)
$$= n \otimes_S (\beta_2(b).m) = n \otimes_S (m.\beta_1(b)) = (n \otimes_S m).b$$

The unit morphism in $\underline{\mathbf{Cosp}}(A, B)(T, T)$ is

(4.37)

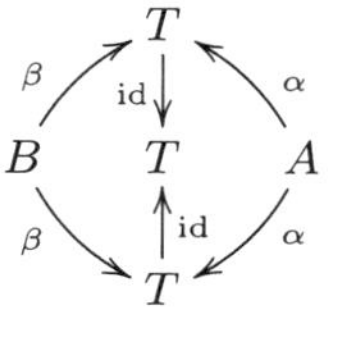

LEMMA 4.16. *A 2-diagram (g, M, f) between $S, T : A \to B$ is invertible if and only if both f and g are invertible.*

The proof is similar to that of Lemma 4.7 and we omit the details.

LEMMA 4.17. *The assignment*

$$\circledcirc_{C,B,A} : \underline{\mathbf{Cosp}}(B,C) \quad \times \quad \underline{\mathbf{Cosp}}(A,B) \quad \longrightarrow \quad \underline{\mathbf{Cosp}}(A,C)$$

$$(4.38) \qquad \left(\begin{array}{ccccc} & & T & & \\ & \overset{\gamma}{\nearrow} & \downarrow g & \overset{\beta_2}{\nwarrow} & \\ C & & N & & B \\ & \underset{\gamma'}{\searrow} & \uparrow g' & \underset{\beta_2'}{\swarrow} & \\ & & T' & & \end{array} , \begin{array}{ccccc} & & S & & \\ & \overset{\beta_1}{\nearrow} & \downarrow f & \overset{\alpha}{\nwarrow} & \\ B & & M & & A \\ & \underset{\beta_1'}{\searrow} & \uparrow f' & \underset{\alpha'}{\swarrow} & \\ & & S' & & \end{array} \right) \longmapsto \begin{array}{ccccc} & & T \otimes_B S & & \\ & \overset{\gamma \otimes_B 1}{\nearrow} & \downarrow g \otimes_B f & \overset{1 \otimes_B \alpha}{\nwarrow} & \\ C & & N \otimes_B M & & A \\ & \underset{\gamma' \otimes_B 1}{\searrow} & \uparrow g' \otimes_B f' & \underset{1 \otimes_B \alpha'}{\swarrow} & \\ & & T' \otimes_B S' & & \end{array}$$

defines a functor.

PROOF. It is evident that the functor maps a pair of identity cospans (4.37) to the identity cospan. To verify functoriality, choose another pair of 2-diagrams $N' : T' \to T''$ and $M' : S' \to S''$ in the product category. First composing in the product with $\odot \times \odot$ gives the pair $(N' \otimes_{T'} N, M' \otimes_{S'} M)$. Applying $\circledcirc$ yields $X := N' \otimes_{T'} N \otimes_B M' \otimes_{S'} M$. In the other order, by first applying $\circledcirc$, we obtain the two 2-diagrams $N' \otimes_B M' : T' \otimes_B S' \to T'' \otimes_B S''$ and $N \otimes_B M : T \otimes_B S \to T' \otimes_B S'$. Applying $\odot$ to this gives $Y := (N' \otimes_B M') \otimes_{T' \otimes_B S'} (N \otimes_B M)$. We claim that the isomorphism $N' \otimes N \otimes M' \otimes M \to N' \otimes M' \otimes N \otimes M$ given by permuting factors induces maps $X \to Y$ and $Y \to X$; these are then automatically inverse to each other. For example, the condition that the map $\phi : N' \otimes N \otimes M' \otimes M \to Y$ respects the tensor product over T' amounts to, for $t' \in T$,

$$(4.39) \qquad \begin{aligned} \phi\big((n'.t') \otimes n \otimes m' \otimes m\big) &= \big((n'.t') \otimes_B m'\big) \otimes_{T' \otimes_B S'} (n \otimes_B m) \\ &= \big((n' \otimes_B m').(t' \otimes_B 1)\big) \otimes_{T' \otimes_B S'} (n \otimes_B m) \\ &= \big(n' \otimes_B m'\big) \otimes_{T' \otimes_B S'} \big((t' \otimes_B 1).(n \otimes_B m)\big) \\ &= \big(n' \otimes_B m'\big) \otimes_{T' \otimes_B S'} \big((t'.n) \otimes_B m\big) \\ &= \phi\big(n' \otimes (t'.n) \otimes m' \otimes m\big) \ . \end{aligned}$$

The other tensor product cokernel conditions are checked similarly. That the induced isomorphism $X \to Y$ is a 3-cell is equally straightforward. □

We can now define the second version of the bicategory of commutative algebras, which we denote by

$$(4.40) \qquad \underline{\mathrm{CALG}}(k) \ ;$$

Objects and 1-morphisms are as in $\mathbf{CAlg}(k)$, but for 2-morphisms we take equivalence classes of 2-diagrams. In other words, the morphism category $A \to B$ is $\underline{\mathbf{Cosp}}(A,B)$. The composition functor is given in Lemma 4.17. The notation $\underline{\mathrm{CALG}}(k)$ is motivated by the conjectural tricategory of Remark 4.15 (i). The associativity and unit isomorphisms of $\underline{\mathrm{CALG}}(k)$ are just those of bimodules, and the required coherence conditions are satisfied for the same reason.

As compared to $\mathbf{CAlg}(k)$, the category $\underline{\mathrm{CALG}}(k)$ has more 2-morphisms. This is made precise by the observation that

$$\mathbf{I} : \quad \mathbf{CAlg}(k) \quad \longrightarrow \quad \underline{\mathrm{CALG}}(k)$$

$$(4.41) \qquad \begin{array}{ccccc} & & S & & \\ & \overset{\beta}{\nearrow} & & \overset{\alpha}{\nwarrow} & \\ B & & \downarrow f & & A \\ & \underset{\beta'}{\searrow} & & \underset{\alpha'}{\swarrow} & \\ & & T & & \end{array} \longmapsto \begin{array}{ccccc} & & S & & \\ & \overset{\beta}{\nearrow} & \downarrow f & \overset{\alpha}{\nwarrow} & \\ B & & T & & A \\ & \underset{\beta'}{\searrow} & \uparrow \mathrm{id} & \underset{\alpha'}{\swarrow} & \\ & & T & & \end{array}$$

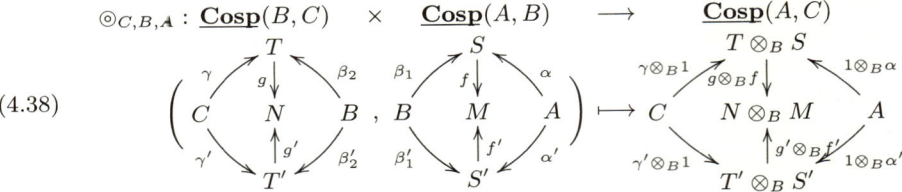

is a locally faithful functor from $\mathbf{CAlg}(k)$ to $\underline{\mathrm{CALG}}(k)$; we skip the details. By Lemma 4.16, each invertible morphism in $\underline{\mathbf{Cosp}}(A,B)(S,T)$ lies in the image of $\mathbf{I}$ (we again skip the details). Therefore, the restriction

$$\mathbf{I} : \mathbf{CAlg}(k)^{2,1} \xrightarrow{\sim} \underline{\mathrm{CALG}}(k)^{2,1} \tag{4.42}$$

is an equivalence of bicategories. In this sense, passing from $\mathbf{CAlg}(k)$ to $\underline{\mathrm{CALG}}(k)$ adds more non-invertible 2-morphisms.

4.5. Functorial centre, version 2. In this subsection we will try to extend the centre functor to $\mathbf{Alg}(k)$. We will see that $\underline{\mathrm{CALG}}(k)$ is not quite good enough as a target category, and we have to restrict ourselves to appropriate subcategories of $\mathbf{Alg}(k)$.

Let A, B be algebras and let X be a B-A-bimodule. Then

$$\begin{array}{ccccc} & & \mathrm{Hom}_{B|A}(X,X) & & \\ & \overset{\text{act}}{\nearrow} & & \overset{\text{act}}{\nwarrow} & \\ Z(B) & & & & Z(A) \end{array} \tag{4.43}$$

is a cospan of commutative algebras. As in Remark 4.10, 'act' refers to the map that takes $b \in Z(B)$ to the bimodule map $x \mapsto b.x$ (resp. $a \in Z(A)$ to $x \mapsto x.a$).

LEMMA 4.18. *Let A, B be algebras, let X, Y be A-B-bimodules and let $f : X \to Y$ be a bimodule homomorphism. Then*

$$\begin{array}{ccccc} & & \mathrm{Hom}_{A|B}(X,X) & & \\ & \overset{\text{act}}{\nearrow} & \downarrow{\scriptstyle f\circ(-)} & \overset{\text{act}}{\nwarrow} & \\ Z(A) & & \mathrm{Hom}_{A|B}(X,Y) & & Z(B) \\ & \underset{\text{act}}{\searrow} & \uparrow{\scriptstyle (-)\circ f} & \underset{\text{act}}{\swarrow} & \\ & & \mathrm{Hom}_{A|B}(Y,Y) & & \end{array} \tag{4.44}$$

is a 2-diagram.

PROOF. We need to verify the conditions in Definition 4.14. The composition of bimodule maps turns $\mathrm{Hom}_{A|B}(X,Y)$ into a right module over $\mathrm{Hom}_{A|B}(X,X)$ and a left module over $\mathrm{Hom}_{A|B}(Y,Y)$. The map $h \mapsto f \circ h$ from $\mathrm{Hom}_{A|B}(X,X)$ to $\mathrm{Hom}_{A|B}(X,Y)$ is a right module map (this translates into $f \circ (h \circ h') = (f \circ h) \circ h'$). Similarly, $(-) \circ f$ is a left module map. Commutativity of the left square amounts to equality of the two maps $x \mapsto f(a.x)$ and $x \mapsto a.f(x)$ for all $a \in Z(A)$, which follows since f is a bimodule map. That the right square commutes follows analogously. Finally, consider the two conditions in (4.32). The first condition amounts to equality of the two maps $x \mapsto g(x).b$ and $x \mapsto g(x.b)$ for all $b \in Z(B)$ and $g \in \mathrm{Hom}_{A|B}(X,Y)$, which holds since g is a bimodule map. The second condition can be checked similarly. □

As the constructions will now get somewhat technical, let us just outline in the remark below how the discussion continues from here, leaving the details to [**DKR2**].

REMARK 4.19. (i) The 2-diagram in (4.44) provides a lax functor

$$\mathbf{Z}_{A,B} : \mathbf{Alg}(k)(A,B) \longrightarrow \mathbf{Cosp}(Z(A), Z(B)) \ . \tag{4.45}$$

This functor is indeed lax for the following reason: The vertical composition (4.34) of two 2-diagrams of the form (4.44) belonging to bimodule maps $f : X \to Y$ and $g : Y \to Z$ yields a 2-diagram with central term

$$\mathrm{Hom}_{A|B}(Y,Z) \otimes_H \mathrm{Hom}_{A|B}(X,Y) \qquad \text{where } H \equiv \mathrm{Hom}_{A|B}(Y,Y) \ . \tag{4.46}$$

This space is in general not isomorphic to $\mathrm{Hom}_{A|B}(X,Z)$. So we cannot obtain a functor $\mathbf{Alg}(k)(A,B) \longrightarrow \underline{\mathbf{Cosp}}(Z(A),Z(B))$ in this way, and consequently not a – lax or otherwise – functor from $\mathbf{Alg}(k)$ to $\underline{\mathrm{CALG}}(k)$. However, we conjecture that the 2-diagram (4.44) does give rise to a lax functor $\mathbf{Z}$ from $\mathbf{Alg}(k)$ to the (conjectural) tricategory $\mathrm{CALG}(k)$.

(ii) If the maps f, g above are isomorphisms, the space (4.46) is isomorphic to $\mathrm{Hom}_{A|B}(X,Z)$. In this way, we at least obtain a functor $\mathbf{Z}_{A,B} : \mathbf{Alg}(k)(A,B)^{1,0} \longrightarrow \underline{\mathbf{Cosp}}(A,B)$ and with this also a lax functor

$$\mathbf{Z} : \mathbf{Alg}(k)^{2,1} \longrightarrow \underline{\mathrm{CALG}}(k) \ . \tag{4.47}$$

(iii) Denote by $\mathbf{F}$ the subcategory of $\mathbf{Alg}(k)$ consisting of Frobenius algebras with trace-pairing and finite-dimensional bimodules. One can show [**DKR2**] that the restriction

$$\mathbf{Z} : \mathbf{F}^{2,0} \longrightarrow \underline{\mathrm{CALG}}(k)^{2,0} \cong \mathbf{CAlg}(k)^{2,0} \cong \mathcal{A}lg(k)^{1,0}_{\mathrm{com}} \tag{4.48}$$

is locally fully faithful. This has the interpretation that all isomorphisms of lattice TFTs without defects (i.e. isomorphisms of Frobenius algebras with trace pairing) are implemented by invertible domain walls (i.e. bimodules inducing Morita equivalences).

REMARK 4.20. (i) There is a close link between the lattice TFTs with defects and the centre functor just defined. Let $T : \mathrm{Bord}^{\mathrm{def,top}}_{2,1}(D_2, D_1, D_0) \to \mathcal{V}ect_f(k)$ be a lattice TFT with defects as in Theorem 3.8, and let $\mathbf{D} \equiv \mathbf{D}[D_2, D_1; T]$ be the 2-category of defect conditions defined in section 2.4. Then we have the commuting square

$$\begin{array}{ccc} \mathbf{D}^{2,1} & \xrightarrow{\ \mathbf{E}\ } & \mathbf{CAlg}(k) \\ \big\downarrow{\scriptstyle \Delta} & & \big\downarrow{\scriptstyle \mathbf{I}} \\ \mathbf{Alg}(k)^{2,1} & \xrightarrow{\ \mathbf{Z}\ } & \underline{\mathrm{CALG}}(k) \end{array} \tag{4.49}$$

where the functor Δ was given in (3.64), $\mathbf{E}$ in (4.13), $\mathbf{I}$ in (4.41), and $\mathbf{Z}$ in (4.47). Indeed, evaluating the diagram on an invertible 2-morphism $f : \underline{x} \to \underline{y}$ for $\underline{x}, \underline{y} : a \to b$ gives for the upper path and lower path, in this order,

$$\begin{array}{ccccc} & & \mathrm{Hom}_{A|B}(X,X) & & \\ & \nearrow^{\mathrm{act}} & \big\downarrow{\scriptstyle f\circ(-)\circ f^{-1}} & \nwarrow^{\mathrm{act}} & \\ Z(A) & & \mathrm{Hom}_{A|B}(Y,Y) & & Z(B) \\ & \searrow_{\mathrm{act}} & \big\uparrow{\scriptstyle \mathrm{id}} & \swarrow_{\mathrm{act}} & \\ & & \mathrm{Hom}_{A|B}(Y,Y) & & \end{array} , \quad \begin{array}{ccccc} & & \mathrm{Hom}_{A|B}(X,X) & & \\ & \nearrow^{\mathrm{act}} & \big\downarrow{\scriptstyle f\circ(-)} & \nwarrow^{\mathrm{act}} & \\ Z(A) & & \mathrm{Hom}_{A|B}(X,Y) & & Z(B) \\ & \searrow_{\mathrm{act}} & \big\uparrow{\scriptstyle (-)\circ f} & \swarrow_{\mathrm{act}} & \\ & & \mathrm{Hom}_{A|B}(Y,Y) & & \end{array} , \tag{4.50}$$

These are isomorphic 2-diagrams, and thus equal in $\underline{\mathrm{CALG}}(k)$.

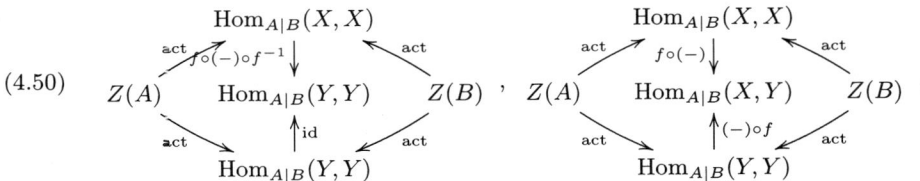

(ii) The commuting square (4.49) shows that the lattice construction of defect TFTs is an implementation of the centre functor. Conjecturally, the restriction to invertible 2-morphisms can be dropped if one replaces both bicategories on the right hand side with the (equally conjectural) tricategory CALG(k).

4.6. Generalisation motivated by 2d conformal field theory. Rational conformal field theories can be build in two steps. In the first step one starts from a rational vertex operator algebra V and finds its modules and the corresponding spaces of conformal blocks. The category $\mathcal{R}ep(V)$ of V-modules is a modular category in this case [**HL, Hu**].

The second step is combinatorial and consists of assigning a correlator to each world sheet, i.e. choosing a particular vector in the space of conformal blocks corresponding to the world sheet, such that the factorisation and locality constraints are satisfied. In the context of vertex operator algebras and for world sheets of genus zero and certain world sheets of genus one, such correlators were constructed in [**HK**] – see also the overview [**KR2**].

The second step can also be solved elegantly for world sheets of arbitrary genus with the help of three-dimensional topological field theory – provided one assumes that this 3d TFT correctly encodes the factorisation and monodromy properties of conformal blocks at arbitrary genus. The 3d TFT in question is the Reshetikhin-Turaev 3d TFT obtained from the modular category $\mathcal{R}ep(V)$. This combinatorial construction of CFT correlators in terms of 3d TFT was carried out in [**FRS1, Fj1**] – see also the overview [**RFFS**] – and in particular allows for a description of CFT correlators for world sheets with defects [**FRS1, Fröl**].

Generalising the considerations in section 4.1 from 2d TFT to 2d CFT suggests an interesting generalisation of the centre construction, which we now sketch.

Let us start with the (bi)categories $\mathcal{A}lg(k)$ and $\mathbf{Alg}(k)$. Instead of working with algebras and bimodules over a field k, that is, with algebras in the symmetric monoidal category of k-vector spaces, one considers algebras and bimodules in a general monoidal category $\mathcal{C}$ (in the CFT-context, this is the category $\mathcal{R}ep(V)$). In particular, we do not demand that $\mathcal{C}$ is symmetric or braided (though in the CFT context it is braided).

To generalise $\mathbf{CAlg}(k)$ and $\underline{\text{CALG}}(k)$, we need to be able to talk about commutative algebras, so here we consider cospans of commutative algebras in a braided monoidal category.

There is one major new ingredient when passing from vector spaces to more general categories, which is based on the following observation. For an algebra A in a general monoidal category $\mathcal{C}$ it makes no sense to talk about its centre as a subalgebra commuting with the entire algebra, because the formulation of this condition needs a braiding. A natural candidate to take the role of the centre in the case of general monoidal categories is the so-called *full centre* $Z(A)$ of A [**Fj2, Da**]. This is a commutative algebra which lives in the monoidal centre $\mathcal{Z}(\mathcal{C})$ (see [**JS**]) of the category $\mathcal{C}$. Since $\mathcal{Z}(\mathcal{C})$ is braided, we can talk about commutative algebras there. If $\mathcal{C}$ is the category of k-vector spaces, one has the degenerate situation that $\mathcal{Z}(\mathcal{C}) \cong \mathcal{C}$, and so many of the subtleties of the centre-construction are not visible. (In the context of rational CFT, and for modular categories in general, one has $\mathcal{Z}(\mathcal{C}) = \mathcal{C} \boxtimes \bar{\mathcal{C}}$, see [**Mü2**, Thm. 7.10].)

The constructions and results of sections 4.2–4.5 all have analogues in the more general setting of algebras in monoidal categories. For example, an instance of the equivalence (4.48), with the corresponding interpretation in terms of domain walls implementing equivalences of CFTs, has been found in [**DKR1**, Thm. 3.14]. More details will appear in [**DKR2**].

REMARK 4.21. As an aside, let us recall an observation from [**SFR**] which illustrates the usefulness of the 2-category of defect conditions defined in section 2.4 in the context of rational CFT. Namely, consider a fixed rational CFT (i.e. restrict your attention to only one world sheet phase) and consider only topological defects from this world sheet phase to itself, which in addition commute with the holomorphic and anti-holomorphic copy of the rational vertex operator algebra V. Then the 2-category $\mathbf{D}$ from section 2.4 has only one object (and so is a monoidal category). It turns out that $\mathbf{D}$ is Morita equivalent (in the sense of [**Mü1**, Def. 4.2]) to $\mathcal{R}ep(V)$; this follows since $\mathbf{D}$ is monoidally equivalent to the category of A-A-bimodules in $\mathcal{R}ep(V)$ for an appropriate Frobenius algebra A with trace pairing [**Frö1**, Sec. 2]. It also follows (from [**Sch**, Thm. 3.3]) that the monoidal centre $\mathcal{Z}(\mathbf{D})$ is braided monoidally equivalent to the monoidal centre $\mathcal{Z}(\mathcal{R}ep(V)) = \mathcal{R}ep(V) \boxtimes \overline{\mathcal{R}ep(V)}$. Thus, quite remarkably, if one knows the one-object 2-category of chiral symmetry preserving topological defects in a rational CFT, one obtains for free the braided monoidal category of representations of its chiral symmetry $V \otimes \bar{V}$.

5. Outlook

In this final section we would like to show some further directions that we find interesting and point out some open questions. From the perspective of this article, there are two evident problems which we left untouched:

(1) In the introduction we claimed that there are two natural ways in which higher categories arise in field theory: by demanding that the functor defining the field theory assigns data to manifolds of codimension larger than one, or by working with defects of various dimensions. Clearly, one should study these two constructions in unison. We are aware of three works in this direction: one in 2d TFT [**SP**], and two in arbitrary dimension – [**Lu1**, Sec. 4.3] and [**MW**, Sec. 6.7] – both 'extended down to points'. A better understanding of the relation between the two appearances of higher categories should allow one to make precise the idea that n-dimensional TFT extended down to points is in some sense dual to n-dimensional TFT which has defects in all dimensions.

(2) A symmetric monoidal functor defining a 2d TFT or 2d CFT without defects has a well-known presentation in terms of generators and relations which provides a link with Frobenius algebras [**So, Di, Ab, FRS1, HK**]. This connection has been useful in the construction of examples and in classification questions. For field theories with defects in 2d (let alone higher dimensions) such a generators and relations presentation is presently not known. Nonetheless, progress has been made on related questions: an algebraic description of 2d TFT with defects which extends down to points was presented in [**SP**], and for planar algebras, generators are given in [**KS**] and a construction in terms of a 1-morphism in a pivotal strict 2-category is presented in [**Go**]. For 2d homotopy TFTs over spaces with at least one of π_1 or π_2 trivial, a classification in terms of Frobenius algebras with extra structure is given in [**Tu, BT**].

Apart from these two points, let us list some further miscellaneous points to complement the material presented in this paper.

■ One nice application of quantum field theories with topological defects is the orbifold construction. Here, one introduces a domain wall which implements the 'averaging over the orbifold group', together with a selection of lower-dimensional junctions which allow one to glue these domain walls together. The orbifold theory is then defined in terms of a cell-decomposition of the original theory with the 'averaging domain wall' placed on the codimension-1 cells. The advantage of this point of view is that the 'averaging domain wall' need not actually be given by a sum over group symmetries, giving rise to a generalisation of the orbifold construction. In the case of 2d rational CFTs, this is described in [**Frö2**]. It is proved there that any two rational CFTs over the same left/right chiral symmetry algebra can be written as a generalised orbifold of one another.

■ In the application of field theories to questions in cohomology one considers field theories 'over a space X', see e.g. [**Tu, BT, ST**]. This means that objects and morphisms of the bordism category are in addition equipped with continuous maps to X. For each point $x \in X$, a field theory over X gives a field theory for undecorated bordisms by choosing these continuous maps to be constant with value x. The role of X is reminiscent of our D_n, the set labelling the top-dimensional domains M_n for an n-dimensional field theory with defects. However, in our setting the D_n label attached to a point in M_n is locally constant and may change only across M_{n-1}, and each such change has to be accompanied by specifying a domain wall which mediates this change. It would be interesting to have a continuous formulation of the framework presented here to be able to incorporate continuously changing domain conditions via 'smeared-out' domain walls and junctions.

■ While the general setup in section 2.1 allows for non-topological defects, in this paper we only studied the topological case. Theories with non-topological defects are much harder to treat and are much less studied. We mention here four examples in 2d CFT:

- There are only two 2d CFTs in which all conformally invariant domain walls (this includes the topological ones) from the CFT to itself are known[16]: the Lee-Yang model and the Ising model, see [**OA, QRW**].
- In [**QRW**], a transmission coefficient was introduced which measures the 'non-topologicality' of a domain wall.
- In [**BB**], the fusion product of certain non-topological domain walls in the free boson CFT (found in [**BBDO**]) was computed, showing that at least in these theories the notion of fusion makes sense for non-topological domain walls despite the short-distance singularities.
- In the operator-algebraic approach of [**BDH**], non-topological domain walls are included from their start and also their fusion is defined. The definition is via Connes' fusion of bimodules and does not involve a short-distance limit.

[16]Here 'known' means that one has a list of defect operators satisfying a selection of consistency conditions. Conjecturally, this uniquely specifies all conformally invariant defects, at least in 'semi-simple theories'. Only defects whose field content (the space $Q(O(x \circ x^*))$) has discrete $(L_0 + \bar{L}_0)$-spectrum are considered.

■ The centre of an algebra can be interpreted as a 'boundary-bulk map' in the following sense. When considering 2d TFT on surfaces with (unparametrised) boundaries, in addition to 'closed states' associated to circles there are 'open states' associated to intervals. The open states form a non-commutative Frobenius algebra and the closed states form a commutative Frobenius algebra, which one can take to be the centre of A (see e.g. [**Laz, MS**]). Thus, the centre defines a theory in one dimension higher (here in dimension two) for which the starting theory is a boundary theory (and the boundary is one-dimensional). The construction in section 4 can be understood as turning this boundary-bulk map into a functor. There are a number of situations in which such a boundary-bulk map occurs: In 2d rational CFT one finds that the boundary theory determines a unique bulk theory [**Fj2**]; algebraically this amounts to the construction of the full centre of an algebra in a monoidal category as briefly mentioned in section 4.6. In [**DKR2**] we will show that the bulk-boundary map is functorial also in the rational CFT case. For certain two-dimensional quantum spin-lattices (which are three-dimensional models because of the time direction), edge excitations determine the bulk excitations, see [**KK**]. Finally, an analogous (but much more general) result exists for $\mathbb{E}[k]$-algebras (related to algebras over the little-discs operad in k dimensions). Namely, in [**Lu2**] a construction is presented which assigns to an $\mathbb{E}[k]$-algebra (in a symmetric monoidal ∞-category) its centre, which is an $\mathbb{E}[k+1]$-algebra in the same category, see [**Lu2**, Cor. 2.5.13].[17] It would be interesting to understand the precise relation to the constructions presented here.

Acknowledgements: The authors would like to thank Nils Carqueville, Jens Fjelstad, Jürgen Fuchs, André Henriques, Chris Schommer-Pries, and especially Sebastian Novak for helpful discussions and comments on a draft of this article. AD would like to thank Tsinghua University, and IR the Beijing International Center for Mathematical Research, for hospitality while part of this work was completed.

Appendix A. Bicategories and lax functors

In this appendix we recall the definition of bicategories and related notions, see [**Be**] or [**Gr, Le**].

DEFINITION A.1. A bicategory $\mathbf{S}$ consists of a set of objects (in a given universe) and a category of morphisms $\mathcal{M}or(A,B)$ for each pair of objects A and B together with

(1) *identity morphism:* $\mathbf{1}_A : \mathbf{1} \to \mathcal{M}or(A,A)$ for all $A \in \mathbf{S}$, where $\mathbf{1}$ is a category with only one object and only the identity morphism. We will abbreviate $\mathbf{1}_A \equiv \mathbf{1}_A(\mathbf{1}) \in \mathcal{M}or(A,A)$,

(2) *composition functor:*

$$\circledcirc_{C,B,A} : \mathcal{M}or(B,C) \times \mathcal{M}or(A,B) \longrightarrow \mathcal{M}or(A,C) \quad , \quad (T,S) \longmapsto T \circ S \ ,$$

(3) *associativity isomorphisms:* for $A,B,C,D \in \mathbf{S}$, there is a natural isomorphism between functors $\mathcal{M}or(C,D) \times \mathcal{M}or(B,C) \times \mathcal{M}or(A,B) \to \mathcal{M}or(A,D)$:

$$\alpha : \circledcirc_{D,B,A} \circ (\circledcirc_{D,C,B} \times \mathrm{id}) \longrightarrow \circledcirc_{D,C,A} \circ (\mathrm{id} \times \circledcirc_{C,B,A}) \ ,$$

[17]IR would like to thank Owen Gwilliam for discussions on this point.

(4) *left and right unit isomorphisms:* for $A, B \in \mathbf{S}$ there are natural transformations between functors $\mathbf{1}\times\mathcal{M}or(A,B) \to \mathcal{M}or(A,B)$ and $\mathcal{M}or(A,B)\times\mathbf{1} \to \mathcal{M}or(A,B)$:

$$l : \circledcirc_{B,B,A} \circ (\mathbf{1}_B \times \mathrm{id}) \longrightarrow \mathrm{id} \quad , \quad r : \circledcirc_{B,A,A} \circ (\mathrm{id} \times \mathbf{1}_A) \longrightarrow \mathrm{id} \ ,$$

satisfying the following coherence conditions:

(1) *associativity coherence:*

(A.1)
$$\begin{array}{ccc}
((S\circ T)\circ U)\circ V & \xrightarrow{\alpha(S,T,U)\circ\mathrm{id}_V} & (S\circ(T\circ U))\circ V \\
\downarrow{\scriptstyle \alpha(S\circ T,U,V)} & & \downarrow{\scriptstyle \alpha(S,T\circ U,V)} \\
(S\circ T)\circ(U\circ V) & & S\circ((T\circ U)\circ V) \\
& \searrow{\scriptstyle \alpha(S,T,U\circ V)} \quad \swarrow{\scriptstyle \mathrm{id}_S\circ\alpha(T,U,V)} & \\
& S\circ(T\circ(U\circ V)) &
\end{array}$$

(2) *identity coherence:*

(A.2)
$$\begin{array}{ccc}
(S\circ\mathbf{1}_B)\circ T & \xrightarrow{\alpha(S,\mathbf{1}_B,T)} & S\circ(\mathbf{1}_B\circ T) \\
& \searrow{\scriptstyle r(S)\circ\mathrm{id}_T} \quad \swarrow{\scriptstyle \mathrm{id}_S\circ l(T)} & \\
& S\circ T &
\end{array}$$

DEFINITION A.2. Let $\mathbf{C}$ and $\mathbf{D}$ be two bicategories. A lax functor $\mathbf{F} : \mathbf{C} \to \mathbf{D}$ is a quadruple $\mathbf{F} = (F, \{\mathbf{F}_{(A,B)}\}_{A,B\in\mathbf{C}}, i, m)$ where

(1) F is a map of objects $X \mapsto F(X)$ for each object X in $\mathbf{C}$,
(2) $\mathbf{F}_{(A,B)} : \mathcal{M}or_{\mathbf{C}}(A,B) \to \mathcal{M}or_{\mathbf{D}}(F(A),F(B))$ is a functor for each pair of objects $A, B \in \mathbf{C}$,
(3) *unit transformation:* natural transformations $i_A : \mathbf{1}_{F(A)} \to \mathbf{F}_{(A,A)} \circ \mathbf{1}_A$ between two functors $\mathbf{1} \to \mathcal{M}or_{\mathbf{D}}(F(A),F(A))$ for all A,
(4) *multiplication transformation:* $m : \circledcirc_{\mathbf{D}} \circ (\mathbf{F}_{(B,C)} \times \mathbf{F}_{(A,B)}) \to \mathbf{F}_{(A,C)} \circ \circledcirc_{\mathbf{C}}$, i.e. a collection of morphisms $m_{S,T} : \mathbf{F}_{(B,C)}(S)\circ\mathbf{F}_{(A,B)}(T) \to \mathbf{F}_{(A,C)}(S\circ T)$ natural in $S \in \mathcal{M}or_{\mathbf{C}}(B,C), T \in \mathcal{M}or_{\mathbf{C}}(A,B)$,

satisfying the following commutative diagrams:

(1) *associativity:* for $S \in \mathcal{M}or_{\mathbf{C}}(C,D), T \in \mathcal{M}or_{\mathbf{C}}(B,C), U \in \mathcal{M}or_{\mathbf{C}}(A,B)$,

$$\begin{array}{ccc}
(\mathbf{F}_{(C,D)}(S)\circ\mathbf{F}_{(B,C)}(T))\circ\mathbf{F}_{(A,B)}(U) & \xrightarrow{\alpha_{\mathbf{D}}} & \mathbf{F}_{(C,D)}(S)\circ(\mathbf{F}_{(B,C)}(T)\circ\mathbf{F}_{(A,B)}(U)) \\
\downarrow{\scriptstyle m\circ\mathrm{id}} & & \downarrow{\scriptstyle \mathrm{id}\circ m} \\
\mathbf{F}_{(B,D)}(S\circ T)\circ\mathbf{F}_{(A,B)}(U) & & \mathbf{F}_{(C,D)}(S)\circ\mathbf{F}_{(A,C)}(T\circ U) \\
\downarrow{\scriptstyle m} & & \downarrow{\scriptstyle m} \\
\mathbf{F}_{(A,D)}((S\circ T)\circ U) & \xrightarrow{\mathbf{F}_{(A,D)}(\alpha_{\mathbf{C}})} & \mathbf{F}_{(A,D)}(S\circ(T\circ U))\ ,
\end{array}$$

(2) *unit properties:* for $S \in \mathcal{M}or_{\mathbf{C}}(A, B)$,

$$\begin{array}{ccc} \mathbf{1}_{F(B)} \circ \mathbf{F}_{(A,B)}(S) & \xrightarrow{l(F(S))} & \mathbf{F}_{(A,B)}(S) \\ \downarrow{\scriptstyle i_B \circ \mathrm{id}} & & \uparrow{\scriptstyle \mathbf{F}_{(A,B)}(l(S))} \\ \mathbf{F}_{(B,B)}(\mathbf{1}_B) \circ \mathbf{F}_{(A,B)}(S) & \xrightarrow{m} & \mathbf{F}_{(A,B)}(\mathbf{1}_B \circ S)\,, \end{array}$$

$$\begin{array}{ccc} \mathbf{F}_{(A,B)}(S) \circ \mathbf{1}_{F(A)} & \xrightarrow{r(F(S))} & \mathbf{F}_{(A,B)}(S) \\ \downarrow{\scriptstyle \mathrm{id} \circ i_B} & & \uparrow{\scriptstyle \mathbf{F}_{(A,B)}(r(S))} \\ \mathbf{F}_{(A,B)}(S) \circ \mathbf{F}_{(A,A)}(\mathbf{1}_A) & \xrightarrow{m} & \mathbf{F}_{(A,B)}(S \circ \mathbf{1}_A)\,. \end{array}$$

If we reverse all arrows, we obtain the notion of *oplax functor*. Given a lax functor $\mathbf{F}$, if the natural transformations i and m are actually isomorphisms, then $\mathbf{F}$ is called a functor.

Let P be a property of a functor between 1-categories like full, faithful, essentially surjective, etc. We say that a (lax, oplax or neither) functor is *locally* P if, for all objects A, B, the functors $\mathbf{F}_{(A,B)}$ have property P.

References

[Ab] L.S. Abrams, *Two-dimensional topological quantum field theories and Frobenius algebras*, J. Knot Theor. Ramifications **5** (1996) 569–587.

[At] M. Atiyah, *Topological Quantum Field Theories*, Inst. Hautes Etudes Sci. Publ. Math. **68** (1989) 175–186.

[BB] C. Bachas, I. Brunner, *Fusion of conformal interfaces*, JHEP **0802** (2008) 085 [0712.0076 [hep-th]].

[BBDO] C. Bachas, J. de Boer, R. Dijkgraaf and H. Ooguri, *Permeable conformal walls and holography*, JHEP **0206** (2002) 027 [hep-th/0111210].

[BD] J.C. Baez, J. Dolan, *Higher dimensional algebra and topological quantum field theory*, J. Math. Phys. **36** (1995) 6073–6105 [q-alg/9503002].

[BDH] A. Bartels, C.L. Douglas, A.G. Henriques, *Conformal nets and local field theory*, 0912.5307 [math.AT].

[Be] J. Bénabou, *Introduction to bicategories*, Reports of the Midwest Category Seminar, Lecture Notes in Math. 47 (1967) 1–77.

[BP] C. Bachas, P.M.S. Petropoulos, *Topological models on the lattice and a remark on string theory cloning*, Commun. Math. Phys. **152** (1993) 191–202 [hep-th/9205031].

[BT] M. Brightwell, P. Turner, *Representations of the homotopy surface category of a simply connected space*, J. Knot Theory Ramifications **9** (2000) 855–864 [math.AT/9910026].

[Ca] J.L. Cardy, *Boundary conditions, fusion rules and the Verlinde formula*, Nucl. Phys. B **324** (1989) 581–596.

[CR] N. Carqueville, I. Runkel, *Rigidity and defect actions in Landau-Ginzburg models*, 1006.5609 [hep-th].

[Da] A. Davydov, *Centre of an algebra*, Adv. Math. **225** (2010) 319–348 [math.CT/0908.1250].

[DGG] P. Das, S.K. Ghosh, V.P. Gupta *Perturbations of planar algebras*, 1009.0186 [math.QA].

[Di] R. Dijkgraaf, *A geometrical approach to two-dimensional conformal field theory*, Ph.D. thesis (Utrecht 1989).

[DKR1] A. Davydov, L. Kong, I. Runkel, *Invertible defects and isomorphisms of rational CFTs*, 1004.4725 [hep-th].

[DKR2] A. Davydov, L. Kong, I. Runkel, *in preparation.*

[FHK] M. Fukuma, S. Hosono, H. Kawai, *Lattice topological field theory in two-dimensions*, Commun. Math. Phys. **161** (1994) 157–176 [hep-th/9212154].

[Fj1] J. Fjelstad, J. Fuchs, I. Runkel, C. Schweigert, *TFT construction of RCFT correlators. V: Proof of modular invariance and factorisation*, Theo. Appl. Cat. **16** (2006) 342–433 [hep-th/0503194].

[Fj2] J. Fjelstad, J. Fuchs, I. Runkel, C. Schweigert, *Uniqueness of open/closed rational CFT with given algebra of open states*, Adv. Theor. Math. Phys. **12** (2008) 1283–1375 [hep-th/0612306].

[Fre] D.S. Freed, *Higher algebraic structures and quantization*, Commun. Math. Phys. **159** (1994) 343–398 [hep-th/9212115].

[Frö1] J. Fröhlich, J. Fuchs, I. Runkel, C. Schweigert, *Duality and defects in rational conformal field theory*, Nucl. Phys. B **763** (2007) 354–430 [hep-th/0607247].

[Frö2] J. Fröhlich, J. Fuchs, I. Runkel, C. Schweigert, *Defect lines, dualities, and generalised orbifolds*, conference proceedings 'XVI International Congress on Mathematical Physics' (Prague, August 2009) [0909.5013 [math-ph]].

[FRS1] J. Fuchs, I. Runkel, C. Schweigert, *TFT construction of RCFT correlators. I: Partition functions*, Nucl. Phys. B **646** (2002) 353–497 [hep-th/0204148].

[FRS2] J. Fuchs, I. Runkel, C. Schweigert, *TFT construction of RCFT correlators. IV: Structure constants and correlation functions*, Nucl. Phys. B **715** (2005) 539–638 [hep-th/0412290].

[FRS3] J. Fuchs, I. Runkel, C. Schweigert, *The fusion algebra of bimodule categories*, Appl. Cat. Str. **16** (2008) 123–140 [math.CT/0701223].

[FS] D. Friedan, S.H. Shenker, *The Analytic Geometry of Two-Dimensional Conformal Field Theory*, Nucl. Phys. B **281** (1987) 509–545.

[Go] S.K. Ghosh, *Planar algebras: a category theoretic point of view*, 0810.4186 [math.QA].

[Gr] J.W. Gray, *Formal Category Theory: Adjointness for 2-Categories*, Springer, 1974.

[HL] Y.-Z. Huang, J. Lepowsky, *Tensor products of modules for a vertex operator algebra and vertex tensor categories*, in: Lie Theory and Geometry, in honor of Bertram Kostant, ed. R. Brylinski, J.-L. Brylinski, V. Guillemin, V. Kac, Birkhäuser, Boston, 1994, 349–383 [hep-th/9401119].

[HK] Y.-Z. Huang, L. Kong, *Full field algebras*, Commun. Math. Phys. **272** (2007) 345–396 [math.QA/0511328].

[Hu] Y.-Z. Huang, *Rigidity and modularity of vertex tensor categories*, Commun. Contemp. Math. **10** (2008) 871–911 [math.QA/0502533].

[Jo] V.F.R. Jones, *Planar algebras, I*, math.QA/9909027.

[JS] A. Joyal, R. Street, *Braided tensor categories*, Adv. Math. **102** (1993) 20–78.

[Ka] A. Kapustin, *Topological Field Theory, Higher Categories, and Their Applications*, contribution to the ICM 2010, 1004.2307 [math.QA].

[Ki] A. Kirillov Jr, *On piecewise linear cell decompositions*, 1009.4227 [math.GT].

[KK] A. Kitaev, L. Kong, *Models for gapped boundaries and domain walls*, 1104.5047 [cond-mat.str-el].

[KR1] L. Kong, I. Runkel, *Morita classes of algebras in modular tensor categories*, Adv. Math. **219** (2008) 1548–1576 [0708.1897 [math.CT]].

[KR2] L. Kong, I. Runkel, *Algebraic structures in Euclidean and Minkowskian two-dimensional conformal field theory*, Noncommutative structures in Mathematics and Physics, 217–238, K. Vlaam. Acad. Belgie Wet. Kunsten (KVAB), Brussels, 2010 [0902.3829 [math.ph]].

[KPS] V. Kodiyalam, V. Pati, V.S. Sunder, *Subfactors and 1+1-dimensional TQFTs*, Int. J. Math. **18** (2007) 69–112 [math.QA/0507050].

[KS] V. Kodiyalam, V.S. Sunder, *On Jones' planar algebras*, J. Knot Theory and its Ramifications **13** (2004) 219–247.

[Law] R.J. Lawrence, *Triangulations, categories and extended topological field theories*, in *Quantum Topology*, World Scientific, Series on Knots and Everything **3** (1993) 191–208.

[Laz] C.I. Lazaroiu, *On the structure of open-closed topological field theory in two dimensions*, Nucl. Phys. B **603** (2001) 497–530 [hep-th/0010269].

[Le] T. Leinster, *Basic bicategories*, math.CT/9810017.

[Lo] J.-L. Loday, *Cyclic Homology*, Springer, 1992.

[Lu1] J. Lurie, *On the classification of topological field theories*, 0905.0465 [math.CT].

[Lu2] J. Lurie, *Derived Algebraic Geometry VI: $\mathbb{E}[k]$-Algebras*, 0911.0018 [math.AT].

[Ma] J.P. May, *The Geometry of Iterated Loop Spaces*, Springer, 1972.

[ML] S. Mac Lane, *Categories for the working mathematician*, 2nd ed., Springer, 1998.

[MN] D. McNamee, *On the mathematical structure of topological defects in Landau-Ginzburg models*, MSc Thesis, Trinity College Dublin, 2009.

[MS] G. Moore, G. Segal, *D-branes and K-theory in 2D topological field theory*, in 'Dirichlet Branes and Mirror Symmetry', Clay Mathematics Monographs **4** (2009) 27–108 [hep-th/0609042].

[Mü1] M. Müger, *From Subfactors to Categories and Topology I. Frobenius algebras in and Morita equivalence of tensor categories*, J. Pure Appl. Alg. **180** (2003) 81–157 [math.CT/0111204].

[Mü2] M. Müger, *From Subfactors to Categories and Topology II. The quantum double of tensor categories and subfactors*, J. Pure Appl. Alg. **180** (2003) 159–219 [math.CT/0111205].

[MW] S. Morrison, K. Walker, *The blob complex*, 1009.5025 [math.AT].

[OA] M. Oshikawa and I. Affleck, *Boundary conformal field theory approach to the critical two-dimensional Ising model with a defect line*, Nucl. Phys. B **495** (1997) 533–582 [cond-mat/9612187].

[PZ] V.B. Petkova, J.-B. Zuber, *Generalised twisted partition functions*, Phys. Lett. B **504** (2001) 157–164 [hep-th/0011021].

[QRW] T. Quella, I. Runkel, G.M.T. Watts, *Reflection and Transmission for Conformal Defects*, JHEP **0704** (2007) 095 [hep-th/0611296].

[RFFS] I. Runkel, J. Fjelstad, J. Fuchs, C. Schweigert, *Topological and conformal field theory as Frobenius algebras*, Contemp. Math. **431** (2007) 225–248 [math.CT/0512076].

[RS] I. Runkel, R.R. Suszek, *Gerbe-holonomy for surfaces with defect networks*, Adv. Theor. Math. Phys. **13** (2009) 1137–1219, [0808.1419 [hep-th]].

[Sch] P. Schauenburg, *The monoidal center construction and bimodules*, J. Pure. Appl. Alg. **158** (2001) 325–346.

[Se] G. Segal, *The definition of conformal field theory*, preprint 1988; also in: U. Tillmann (ed.), Topology, geometry and quantum field theory, London Math. Soc. Lect. Note Ser. **308** (2002) 421–577.

[SFR] C. Schweigert, J. Fuchs, I. Runkel, *Categorification and correlation functions in conformal field theory*, contribution to the ICM 2006, math.CT/0602079.

[So] H. Sonoda, *Sewing conformal field theories II*, Nucl. Phys. B **311** (1988) 417–432.

[SP] C. Schommer-Pries, *Topological Defects and Classifying Local Topological Field Theories in Low Dimensions*, presented at the workshop 'Strings, Fields and Topology' (Mathematisches Forschungsinstitut Oberwolfach, June 2009), slides on author's homepage.

[St] R. Street, *Categorical structures*, Handbook of Algebra **1** (1996) 529–577.

[ST] S. Stolz, P. Teichner, *Supersymmetric Euclidean field theories and generalized cohomology*, preprint 2008, available at second author's homepage.

[Tu] V. Turaev, *Homotopy field theory in dimension two and group algebras*, math.QA/9910010.

[Va] C. Vafa, *Conformal theories and punctured surfaces*, Phys. Lett. B **199** (1987) 195–202.

Dept. of Mathematics and Statistics, Univ. of New Hampshire, Durham NH, USA
E-mail address: alexei1davydov@gmail.com

Inst. for Advanced Study (Science Hall), Tsinghua Univ., Beijing 100084, China
E-mail address: kong.fan.liang@gmail.com

Fachbereich Mathematik, Univ. Hamburg, Bundesstr. 55, 20146 Hamburg, Germany
E-mail address: ingo.runkel@uni-hamburg.de

Quantization of Field Theories

Proceedings of Symposia in Pure Mathematics
Volume **83**, 2011

Homotopical Poisson reduction of gauge theories

Frédéric Paugam

ABSTRACT. The classical Poisson reduction of a given Lagrangian system with (local) gauge symmetries has to be done before its quantization. We propose here a coordinate free and self-contained mathematical presentation of the covariant Batalin-Vilkovisky Poisson reduction of a general gauge theory. It was explained in physical terms (DeWitt indices) in Henneaux and Teitelboim's book [**HT92**]. It was studied in coordinates using jet spaces by Barnich-Brandt-Henneaux [**BBH95**], Stasheff [**Sta98**], Fulp-Lada-Stasheff [**FLS02**], among others. The main idea of our approach is to use the functor of point approach to spaces of fields to gain coordinate free geometrical insights on the spaces in play, and to focus on the notion of Noether identities, that is a simple replacement of the notion of gauge symmetry, harder to handle algebraically. Our main results are a precise formulation and understanding of the optimal finiteness hypothesis necessary for the existence of a solution of the classical master equation, and an interpretation of the Batalin-Vilkovisky construction in the setting of homotopical geometry of non-linear partial differential equations.

CONTENTS

Introduction

This paper gives a self contained and coordinate free presentation of the Batalin-Vilkovisky formalism for homotopical Poisson reduction of gauge theories, in the setting of algebraic non-linear analysis, expanding on (and giving full proofs for) the very short presentation given in [**Pau10**], Section 4. We have tried to be as self-contained as possible so there may be some repetitions. We refer to loc. cit. for

1991 *Mathematics Subject Classification.* Primary 70S20, 14A20; Secondary 35G50, 70S10.
Key words and phrases. Algebraic analysis, gauge theory, Poisson reduction, homotopical geometry.
The author was supported by UPMC and CNRS.

further references on the various subjects treated here. We also refer to [**Pau11**] for a more complete and detailed account of this theory and of its applications in physics.

1. Lagrangian variational problems

For the reader's convenience, we recall shortly the formulation summed-up in [**Pau10**] and fully described in [**Pau11**] of general variational problems, and its grounding on functorial geometry. This is certainly useful, but not strictly necessary to understand our final results.

DEFINITION 1.0.1. A *Lagrangian variational problem* is made of the following data:

(1) A space M called the parameter space for trajectories,
(2) A space C called the configuration space for trajectories,
(3) A morphism $\pi : C \to M$ (often supposed to be surjective),
(4) A subspace $H \subset \Gamma(M, C)$ of the space of sections of π

$$\Gamma(M, C) := \{x : M \to C,\ \pi \circ x = \mathrm{id}\},$$

called the space of histories,
(5) A functional (partial function with a domain of definition)

$$S : H \to R$$

on histories with values in a space R usually given by $\mathbb{R}$ (or $\mathbb{R}[[\hbar]]$) called the action functional.

The main object of classical physics is the space

$$T = \{x \in H|\, d_x S = 0\}$$

of critical points of the action functional in histories.

Recall from loc. cit. that the word *space* of this definition means essentially a sheaf

$$X : \text{LEGOS}^{op} \to \text{SETS}$$

on a category LEGOS of geometrical building blocs equiped with a Grothendieck topology τ, also called a *space modeled on* (LEGOS, τ). Spaces that are locally representable are called *manifolds* or *varieties*. To present also higher gauge theory examples, one has to work with homotopical spaces, but we will not do that here.

We refer to Deligne-Morgan's lectures [**DM99**] and Manin's book [**Man97**] for an introduction to super-varieties. The reference [**Pau11**] gives a complete account of this in the functorial setting. We will work without further comments with the following types of spaces:

(1) Smooth spaces (also called diffeologies), modeled on the category LEGOS $=$ OPEN_{C^∞} of open subsets of $\mathbb{R}^n$ for varying n with smooth maps between them.
(2) Smoothly algebraic spaces, modeled on the category LEGOS $= \text{ALG}_{C^\infty}^{op}$ opposite to the category of Lawvere's smooth algebras (see [**MR91**] and [**Pau11**]), with its Zariski topology.
(3) Smooth super-spaces, modeled on the category LEGOS $= \text{OPEN}_{C^\infty}^{s}$ of smooth open subsets of the super affine space $\mathbb{R}^{n|m}$.

(4) Smoothly algebraic super-spaces, modeled on the category $\text{LEGOS} = \text{ALG}^s_{C^\infty}$ of smooth super-algebras, described in [**Pau11**].

All these types of spaces are useful (and actually necessary) to describe differential calculus on spaces of maps between smooth super-manifolds in a proper mathematical setting. If M and C are two varieties in the above sense, the space of maps

$$\underline{\text{Hom}}(M, C) : \text{LEGOS}^{op} \to \text{SETS}$$

is defined by

$$\underline{\text{Hom}}(M, C)(U) = \text{Hom}(M \times U, C).$$

If $\pi : C \to M$ is a morphism of varieties, the space $\underline{\Gamma}(M, C)$ is simply the corresponding subspace in $\underline{\text{Hom}}(M, C)$.

2. Algebraic analysis of partial differential equations

In this section, we present the natural coordinate free approach to partial differential equations, in the settings of $\mathcal{D}$-modules and $\mathcal{D}$-algebras. We refer to Schapira's survey [**Sch10**] for an efficient introduction to the general methods of linear algebraic analysis on varieties and to Beilinson and Drinfeld's book [**BD04**] for the non-linear setting.

This section expands on the article [**Pau10**], giving more details and explanations. In particular, we use systematically the functor of point approach to spaces of fields, as explained in Section 1 (see also [**Pau11**] for a complete treatment) without further comments. This means that spaces of superfunctions are treated essentially as usual spaces, and functionals are defined as partially defined functions between (functors of points of) usual spaces. We use Beilinson and Drinfeld's functorial approach [**BD04**] to non-linear partial differential equations, and we relate this approach to ours. We are also inspired by Vinogradov [**Vin01**] and Krasilshchik and Verbovetsky [**KV98**].

2.1. $\mathcal{D}$-modules and linear partial differential equations. We refer to Schneiders' review [**Sch94**] and Kashiwara's book [**Kas03**] for an introduction to $\mathcal{D}$-modules. We just recall here basic results, that are necessary for our treatment of non-linear partial differential equations in Section 2.3.

Let M be a smooth variety of dimension n and $\mathcal{D}$ be the algebra of differential operators on M. Recall that locally on M, one can write an operator $P \in \mathcal{D}$ as a finite sum

$$P = \sum_\alpha a_\alpha \partial^\alpha$$

with $a_\alpha \in \mathcal{O}_M$,

$$\partial = (\partial_1, \ldots, \partial_n) : \mathcal{O}_M \to \mathcal{O}_M^n$$

the universal derivation and α some multi-indices.

To write down the equation $Pf = 0$ with f in an $\mathcal{O}_M$-module $\mathcal{S}$, one needs to define the universal derivation $\partial : \mathcal{S} \to \mathcal{S}^n$. This is equivalent to giving $\mathcal{S}$ the structure of a $\mathcal{D}$-module. The solution space of the equation with values in $\mathcal{S}$ is then given by

$$\mathcal{S}ol_P(\mathcal{S}) := \{f \in \mathcal{S},\ Pf = 0\}.$$

Remark that

$$\mathcal{S}ol_P : \text{MOD}(\mathcal{D}) \to \text{VECT}_{\mathbb{R}_M}$$

is a functor that one can think of as representing the space of solutions of P. Denote $\mathcal{M}_P$ the cokernel of the $\mathcal{D}$-linear map

$$\mathcal{D} \xrightarrow{.P} \mathcal{D}$$

given by right multiplication by P. Applying the functor $\mathcal{H}om_{\text{MOD}(\mathcal{D})}(-,\mathcal{S})$ to the exact sequence

$$\mathcal{D} \xrightarrow{.P} \mathcal{D} \longrightarrow \mathcal{M}_P \to 0,$$

we get the exact sequence

$$0 \to \mathcal{H}om_{\text{MOD}(\mathcal{D})}(\mathcal{M}_P,\mathcal{S}) \to \mathcal{S} \xrightarrow{P.} \mathcal{S},$$

which gives a natural isomorphism

$$\mathcal{S}ol_P(\mathcal{S}) = \mathcal{H}om_{\text{MOD}(\mathcal{D})}(\mathcal{M}_P,\mathcal{S}).$$

This means that the $\mathcal{D}$-module $\mathcal{M}_P$ represents the solution space of P, so that the category of $\mathcal{D}$-modules is a convenient setting for the functor of point approach to linear partial differential equations.

Remark that it is even better to consider the derived solution space

$$\mathbb{R}\mathcal{S}ol_P(\mathcal{S}) := \mathbb{R}\mathcal{H}om_{\text{MOD}(\mathcal{D})}(\mathcal{M}_P,\mathcal{S})$$

because it encodes also information on the inhomogeneous equation

$$Pf = g.$$

Indeed, applying $\mathcal{H}om_{\mathcal{D}}(-,\mathcal{S})$ to the exact sequences

$$0 \to \mathcal{I}_P \to \mathcal{D} \to \mathcal{M}_P \to 0$$
$$0 \to \mathcal{N}_P \to \mathcal{D} \to \mathcal{I}_P \to 0$$

where $\mathcal{I}_P$ is the image of P and $\mathcal{N}_P$ is its kernel, one gets the exact sequences

$$0 \to \mathcal{H}om_{\mathcal{D}}(\mathcal{M}_P,\mathcal{S}) \to \mathcal{S} \to \mathcal{H}om_{\mathcal{D}}(\mathcal{I}_P,\mathcal{S}) \to \mathcal{E}xt^1_{\mathcal{D}}(\mathcal{M}_P,\mathcal{S}) \to 0$$
$$0 \to \mathcal{H}om_{\mathcal{D}}(\mathcal{I}_P,\mathcal{S}) \to \mathcal{S} \to \mathcal{H}om_{\mathcal{D}}(\mathcal{N}_P,\mathcal{S}) \to \mathcal{E}xt^1_{\mathcal{D}}(\mathcal{I}_P,\mathcal{S}) \to 0$$

If $Pf = g$, then $QPf = 0$ for $Q \in \mathcal{D}$ implies $Qg = 0$. The second exact sequence implies that this system, called the algebraic compatibility condition for the inhomogeneous equation $Pf = g$ is represented by the $\mathcal{D}$-module $\mathcal{I}_P$, because

$$\mathcal{H}om_{\mathcal{D}}(\mathcal{I}_P,\mathcal{S}) = \{g \in \mathcal{S},\ Q.g = 0,\ \forall Q \in \mathcal{N}_P\}.$$

The first exact sequence shows that $\mathcal{E}xt^1_{\mathcal{D}}(\mathcal{M},\mathcal{S})$ are classes of vectors $f \in \mathcal{S}$ satisfying the algebraic compatibility conditions modulo those for which the system is truly compatible. Moreover, for $k \geq 1$, one has

$$\mathcal{E}xt^k_{\mathcal{D}}(\mathcal{I}_P,\mathcal{S}) \cong \mathcal{E}xt^{k+1}_{\mathcal{D}}(\mathcal{M}_P,\mathcal{S})$$

so that all the $\mathcal{E}xt^k_{\mathcal{D}}(\mathcal{M}_P,\mathcal{S})$ give interesting information about the differential operator P.

Recall that the sub-algebra $\mathcal{D}$ of $\text{End}_{\mathbb{R}}(\mathcal{O})$, is generated by the left multiplication by functions in $\mathcal{O}_M$ and by the derivation induced by vector fields in Θ_M. There is a natural right action of $\mathcal{D}$ on the $\mathcal{O}$-module Ω^n_M by

$$\omega.\partial = -L_\partial \omega$$

with L_∂ the Lie derivative.

There is a tensor product in the category $\text{MOD}(\mathcal{D})$ given by

$$\mathcal{M} \otimes \mathcal{N} := \mathcal{M} \otimes_{\mathcal{O}} \mathcal{N}.$$

The $\mathcal{D}$-module structure on the tensor product is given on vector fields $\partial \in \Theta_M$ by Leibniz's rule

$$\partial(m \otimes n) = (\partial m) \otimes n + m \otimes (\partial n).$$

There is also an internal homomorphism object $\mathcal{H}om(\mathcal{M}, \mathcal{N})$ given by the $\mathcal{O}$-module $\mathcal{H}om_{\mathcal{O}}(\mathcal{M}, \mathcal{N})$ equipped with the action of derivations $\partial \in \Theta_M$ by

$$\partial(f)(m) = \partial(f(m)) - f(\partial m).$$

An important system is given by the $\mathcal{D}$-module of functions $\mathcal{O}$, that can be presented by the De Rham complex

$$\mathcal{D} \otimes \Theta_M \to \mathcal{D} \to \mathcal{O} \to 0,$$

meaning that $\mathcal{O}$, as a $\mathcal{D}$-module, is the quotient of $\mathcal{D}$ by the sub-$\mathcal{D}$-module generated by vector fields. The family of generators ∂_i of the kernel of $\mathcal{D} \to \mathcal{O}$ form a regular sequence, i.e., for every $k = 1, \dots, n$, ∂_k is not a zero divisor in $\mathcal{D}/(\partial_1, \dots, \partial_{k-1})$ (where $\partial_{-1} = 0$ by convention). This implies (see Lang [**Lan93**], XXI §4 for more details on Koszul resolutions) the following:

PROPOSITION 2.1.1. *The natural map*

$$\mathrm{Sym}_{(\mathrm{Mod}_{dg}(\mathcal{D}),\otimes)}([\mathcal{D} \otimes \Theta_M \to \mathcal{D}]) \longrightarrow \mathcal{O}$$

is a quasi-isomorphism of dg-$\mathcal{D}$-modules. The left hand side gives a free resolution of $\mathcal{O}$ as a $\mathcal{D}$-module called the universal Spencer complex.

PROPOSITION 2.1.2. *The functor*

$$\mathcal{M} \mapsto \Omega^n_M \otimes_{\mathcal{O}} \mathcal{M}$$

induces an equivalence of categories between the categories $\mathrm{Mod}(\mathcal{D})$ *and* $\mathrm{Mod}(\mathcal{D}^{op})$ *of left and right $\mathcal{D}$-modules whose quasi-inverse is*

$$\mathcal{N} \mapsto \mathcal{H}om_{\mathcal{O}_M}(\Omega^n_M, \mathcal{N}).$$

The monoidal structure induced on $\mathrm{Mod}(\mathcal{D}^{op})$ *by this equivalence is denoted* $\otimes^!$.

DEFINITION 2.1.1. Let $\mathcal{S}$ be a right $\mathcal{D}$-module. The *De Rham functor* with values in $\mathcal{S}$ is the functor

$$\mathrm{DR}_{\mathcal{S}} : \mathrm{Mod}(\mathcal{D}) \to \mathrm{Vect}_{\mathbb{R}_M}$$

that sends a left $\mathcal{D}$-module to

$$\mathrm{DR}_{\mathcal{S}}(\mathcal{M}) := \mathcal{S} \overset{\mathbb{L}}{\otimes}_{\mathcal{D}} \mathcal{M}.$$

The *De Rham functor* with values in $\mathcal{S} = \Omega^n_M$ is denoted DR and simply called the De Rham functor. One also denotes $\mathrm{DR}^r_{\mathcal{S}}(\mathcal{M}) = \mathcal{M} \overset{\mathbb{L}}{\otimes}_{\mathcal{D}} \mathcal{S}$ if $\mathcal{S}$ is a fixed left $\mathcal{D}$-module and $\mathcal{M}$ is a varying right $\mathcal{D}$-module, and $\mathrm{DR}^r := \mathrm{DR}^r_{\mathcal{O}}$.

PROPOSITION 2.1.3. *The natural map*

$$\begin{array}{ccc} \Omega^n_M \otimes_{\mathcal{O}} \mathcal{D} & \to & \Omega^n_M \\ \omega \otimes Q & \mapsto & \omega(Q) \end{array}$$

extends to a $\mathcal{D}^{op}$-linear quasi-isomorphism

$$\Omega^*_M \otimes_{\mathcal{O}} \mathcal{D}[n] \xrightarrow{\sim} \Omega^n_M.$$

PROOF. This follows from the fact that the above map is induced by tensoring the Spencer complex by Ω_M^n, and by the internal product isomorphism

$$\begin{array}{ccc} \mathrm{Sym}_{\mathrm{MOD}_{dg}(\mathcal{D})}([\underset{-1}{\mathcal{D}\otimes\Theta_M}\to\underset{0}{\mathcal{D}\otimes\mathcal{O}}])\otimes\Omega_M^n & \longrightarrow & (\Omega_M^*\otimes\mathcal{D}[n],d) \\ X\otimes\omega & \longmapsto & i_X\omega. \end{array}$$

□

We will see that in the super setting, this proposition can be taken as a definition of the right $\mathcal{D}$-modules of volume forms, called Berezinians.

The $\mathcal{D}$-modules we will use are usually not $\mathcal{O}$-coherent but only $\mathcal{D}$-coherent. The right duality to be used in the monoidal category $(\mathrm{MOD}(\mathcal{D}),\otimes)$ to get a biduality statement for coherent $\mathcal{D}$-modules is thus not the internal duality $\underline{\mathrm{Hom}}_{\mathcal{O}}(\mathcal{M},\mathcal{O})$ but the derived dual $\mathcal{D}^{op}$-module

$$\mathbb{D}(\mathcal{M}):=\mathbb{R}\mathcal{H}om_{\mathcal{D}}(\mathcal{M},\mathcal{D}).$$

The non-derived dual works well for projective $\mathcal{D}$-modules, but most of the $\mathcal{D}$-modules used in field theory are only coherent, so that one often uses the derived duality operation. We now describe the relation (based on biduality) between the De Rham and duality functors.

PROPOSITION 2.1.4. *Let $\mathcal{S}$ be a coherent $\mathcal{D}^{op}$-module and $\mathcal{M}$ be a coherent $\mathcal{D}$-module. There is a natural quasi-isomorphism*

$$\mathbb{R}\mathrm{Sol}_{\mathbb{D}(\mathcal{M})}(\mathcal{S}):=\mathbb{R}\mathcal{H}om_{\mathcal{D}^{op}}(\mathbb{D}(\mathcal{M}),\mathcal{S})\cong \mathrm{DR}_{\mathcal{S}}(\mathcal{M}),$$

where $\mathbb{D}(\mathcal{M}):=\mathbb{R}\mathrm{Hom}_{\mathcal{D}}(\mathcal{M},\mathcal{D})$ is the $\mathcal{D}^{op}$-module dual of $\mathcal{M}$.

The use of $\mathcal{D}$-duality will be problematic in the study of covariant operations (like Lie bracket on local vector fields). We will come back to this in Section 2.5.

2.2. Supervarieties and their Berezinians. We refer to Penkov's article [**Pen83**] for a complete study of the Berezinian in the $\mathcal{D}$-module setting and to Deligne-Morgan's lectures [**DM99**] and Manin's book [**Man97**] for more details on super-varieties. We also refer to [**Pau11**] for a treatment of smooth super-geometry making a systematic use of functors of points.

Let M be a super-variety of dimension $n|m$ and denote Ω_M^1 the $\mathcal{O}_M$-module of differential forms on M and Ω_M^* the super-$\mathcal{O}_M$-module of higher differential forms on M, defined as the exterior (i.e., odd symmetric) power

$$\Omega_M^*:=\wedge^*\Omega_M^1:=\mathrm{Sym}_{\mathrm{MOD}(\mathcal{O}_M)}\Omega_M^1[1].$$

Remark that Ω_M^* is strickly speaking a $\mathbb{Z}/2$-bigraded $\mathbb{R}$-module, but we can see it as a $\mathbb{Z}/2$-graded module because its diagonal $\mathbb{Z}/2$-grading identifies with $\mathrm{Sym}_{\mathrm{MOD}(\mathcal{O}_M)}T\Omega_M^1$, where $T:\mathrm{MOD}(\mathcal{O}_M)\to\mathrm{MOD}(\mathcal{O}_M)$ is the grading exchange. Thus from now on, we consider Ω_M^* as a mere $\mathbb{Z}/2$-graded module.

The super version of Proposition 2.1.3 can be taken as a definition of the Berezinian, as a complex of $\mathcal{D}$-modules, up to quasi-isomorphism.

DEFINITION 2.2.1. The *Berezinian* of M is defined in the derived category of $\mathcal{D}_M$-modules by the formula

$$\mathrm{Ber}_M:=\Omega_M^*\otimes_{\mathcal{O}}\mathcal{D}[n].$$

The *complex of integral forms* $I_{*,M}$ is defined by

$$I_{*,M}:=\mathbb{R}\mathrm{Hom}_{\mathcal{D}}(\mathrm{Ber}_M,\mathrm{Ber}_M).$$

The following proposition (see [**Pen83**], 1.6.3) gives a description of the Berezinian as a $\mathcal{D}$-module.

PROPOSITION 2.2.1. *The Berezinian complex is concentraded in degree* 0, *and equal there to*

$$\mathrm{Ber}_M := \mathcal{E}xt^n_{\mathcal{D}}(\mathcal{O}, \mathcal{D}).$$

It is moreover projective of rank 1 *over* $\mathcal{O}$.

PROOF. This follows from the fact that

$$\wedge^* \Theta_M \otimes_{\mathcal{O}} \mathcal{D}[-n] \to \mathcal{O}$$

is a projective resolution such that

$$\mathrm{Ber}_M := \Omega^*_M \otimes_{\mathcal{O}} \mathcal{D}[n] = \mathbb{R}\mathcal{H}om_{\mathcal{D}}(\wedge^* \Theta_M \otimes_{\mathcal{O}} \mathcal{D}[-n], \mathcal{D})$$

and this De Rham complex is exact (Koszul resolution of a regular module) except in degree zero where it is equal to

$$\mathcal{E}xt^n_{\mathcal{D}}(\mathcal{O}, \mathcal{D}).$$

□

PROPOSITION 2.2.2. *Suppose* M *is a super-variety of dimension* $m|n$. *The Berezinian is a locally free* $\mathcal{O}$*-module of rank* 1 *on* M *with generator denoted* $D(dx_1, \ldots, dx_m, d\theta_1, \ldots, d\theta_n)$. *It* $f : M \to M$ *is an isomorphism of super-varieties (change of coordinate) with local tangent map* $D_x f$ *described by the even matrix*

$$D_x f = \begin{pmatrix} A & B \\ C & D \end{pmatrix}$$

acting on the real vector space

$$T_x M = (T_x M)^0 \oplus (T_x M)^1,$$

the action of $D_x f$ *on* $D(dx_1, \ldots, dx_m)$ *is given by the Berezin determinant*

$$\mathrm{Ber}(D_x f) := \det(A - BD^{-1}C)\det(D)^{-1}.$$

PROOF. This is a classical result (see [**DM99**] or [**Man97**]). □

In the super-setting, the equivalence of left and right $\mathcal{D}$-modules is given by the functor

$$\mathcal{M} \mapsto \mathcal{M} \otimes_{\mathcal{O}} \mathrm{Ber}_M$$

that twists by the Berezinian right $\mathcal{D}$-module, which can be computed by using the definition

$$\mathrm{Ber}_M := \Omega^*_M \otimes_{\mathcal{O}} \mathcal{D}[n]$$

and passing to degree 0 cohomology.

A more explicit description of the complex of integral forms (up to quasi-isomorphism) is given by

$$I_{*,M} := \mathbb{R}\mathcal{H}om_{\mathcal{D}}(\mathrm{Ber}_M, \mathrm{Ber}_M) \cong \mathcal{H}om_{\mathcal{D}}(\Omega^*_M \otimes_{\mathcal{O}} \mathcal{D}[n], \mathrm{Ber}_M)$$

so that we get

$$I_{*,M} \cong \mathcal{H}om_{\mathcal{O}}(\Omega^*_M[n], \mathrm{Ber}_M) \cong \mathcal{H}om_{\mathcal{O}}(\Omega^*_M[n], \mathcal{O}) \otimes_{\mathcal{O}} \mathrm{Ber}_M$$

and in particular $I_{n,M} \cong \mathrm{Ber}_M$.

Remark that Proposition 2.1.3 shows that if M is a non-super variety, then Ber_M is quasi-isomorphic with Ω_M^n, and this implies that

$$I_{*,M} \cong \mathcal{H}om_{\mathcal{O}}(\Omega_M^*[n], \mathcal{O}) \otimes_{\mathcal{O}} \mathrm{Ber}_M \cong \wedge^* \Theta_M \otimes_{\mathcal{O}} \Omega_M^n[-n] \xrightarrow{i} \Omega_M^*,$$

where i is the internal product homomorphism. This implies the isomorphism

$$I_{*,M} \cong \Omega_M^*,$$

so that in the purely even case, integral forms essentially identify with ordinary differential forms.

The main use of the module of Berezinians is given by its usefulness in the definition of integration on super-varieties. We refer to Manin [**Man97**], Chapter 4 for the following proposition.

PROPOSITION 2.2.3. *Let M be a super-variety, with underlying variety $|M|$ and orientation sheaf* $\mathrm{or}_{|M|}$. *There is a natural integration map*

$$\int_M [dt^1 \dots dt^n d\theta^1 \dots d\theta^q] : \Gamma_c(M, \mathrm{Ber}_M \otimes \mathrm{or}_{|M|}) \to \mathbb{R}$$

given in a local chart (i.e., an open subset $U \subset \mathbb{R}^{n|m}$) for $g = \sum_I g_I \theta^I \in \mathcal{O}$ by

$$\int_U [dt^1 \dots dt^n d\theta^1 \dots d\theta^q] g := \int_{|U|} g_{1,\dots,1}(t) d^n t.$$

We finish by describing the inverse and direct image functors in the supergeometric setting, following the presentation of Penkov in [**Pen83**].

Let $g : X \to Y$ be a morphism of supermanifolds. Recall that for $\mathcal{F}$ a sheaf of $\mathcal{O}_Y$-modules on Y, we denote $g^{-1}\mathcal{F}$ the sheaf on X defined by

$$g^{-1}\mathcal{F}(U) := \lim_{g(U) \subset V} \mathcal{F}(V).$$

The $(\mathcal{D}_X, g^{-1}\mathcal{D}_Y)$ module of relative inverse differential operators is defined as

$$\mathcal{D}_{X \to Y} := \mathcal{O}_X \otimes_{g^{-1}\mathcal{O}_Y} g^{-1}\mathcal{D}_Y.$$

The $(g^{-1}\mathcal{D}_Y, \mathcal{D}_X)$ module of relative direct differential operators is defined as

$$\mathcal{D}_{X \leftarrow Y} := \mathrm{Ber}_X \otimes_{\mathcal{O}_X} \mathcal{D}_{X \to Y} \otimes_{g^{-1}\mathcal{O}_Y} (\mathrm{Ber}_Y^*).$$

The inverse image functor of $\mathcal{D}$-modules is defined by

$$g_{\mathcal{D}}^*(\_) := \mathcal{D}_{X \to Y} \overset{\mathbb{L}}{\otimes}_{g^{-1}\mathcal{O}_Y} g^{-1}(\_) : D(\mathcal{D}_Y) \to D(\mathcal{D}_X).$$

If $g : X \hookrightarrow Y$ is a locally closed embedding, the direct image functor is defined by

$$g_*^{\mathcal{D}}(\_) := \mathcal{D}_{Y \leftarrow X} \overset{\mathbb{L}}{\otimes}_{\mathcal{D}_X} (\_) : D(\mathcal{D}_X) \to D(\mathcal{D}_Y).$$

More generally, for any morphism $g : X \to Y$, one defines

$$g_*^{\mathcal{D}}(\_) := \mathbb{R}f_*(\_ \overset{\mathbb{L}}{\otimes}_{\mathcal{D}} \mathcal{D}_{X \to Y}).$$

2.3. Differential algebras and non-linear partial differential equations. In this section, we will use systematically the language of differential calculus in symmetric monoidal categories, and the functor of points approach to spaces of fields, described in [**Pau10**]. We restrict our presentation to polynomial partial differential equations with functional coefficients, but our results also apply to smooth partial differential equations (for a full treatment, that would be too long for this article, see [**Pau11**]). We specialize the situation to the symmetric monoidal category

$$(\mathrm{MOD}(\mathcal{D}_M), \otimes_{\mathcal{O}_M})$$

of left $\mathcal{D}_M$-module on a given (super-)variety M. Recall that there is an equivalence

$$(\mathrm{MOD}(\mathcal{D}_M), \otimes) \to (\mathrm{MOD}(\mathcal{D}_M^{op}), \otimes^!)$$

given by tensoring with the $(\mathcal{D}, \mathcal{D}^{op})$-modules Ber_M and $\mathrm{Ber}_M^{-1} = \mathcal{H}om_{\mathcal{O}}(\mathrm{Ber}_M, \mathcal{O})$. The unit objects for the two monoidal structures are $\mathcal{O}$ and Ber_M respectively. If $\mathcal{M}$ is a $\mathcal{D}$-module (resp. a $\mathcal{D}^{op}$-module), we denote $\mathcal{M}^r := \mathcal{M} \otimes \mathrm{Ber}_M$ (resp. $\mathcal{M}^\ell := \mathcal{M} \otimes \mathrm{Ber}_M^{-1}$) the corresponding $\mathcal{D}^{op}$-module (resp. $\mathcal{D}$-module).

Recall that if $P \in \mathbb{Z}[X]$ is a polynomial, one can study the solution space

$$\underline{\mathrm{Sol}}_{P=0}(A) = \{x \in A, \; P(x) = 0\}$$

of P with values in any commutative unital ring. Indeed, in any such ring, one has a sum, a multiplication, a zero and a unit that fulfill the necessary compatibilities to be able to write down the polynomial. One can thus think of the mathematical object given by the category of commutative unital rings as solving the mathematical problem of giving a natural setting for a coordinate free study of polynomial equations. This solution space is representable, meaning that there is a functorial isomorphism

$$\underline{\mathrm{Sol}}_{P=0}(-) \cong \mathrm{Hom}_{\mathrm{RINGS}_{cu}}(\mathbb{Z}[X]/(P), -).$$

This shows that the solution space of an equation essentially determine the equation itself. Remark that the polynomial P lives in the free algebra $\mathbb{Z}[X]$ on the given variable that was used to write it.

Suppose now given the bundle $\pi_1 : C = \mathbb{R} \times \mathbb{R} \to \mathbb{R} = M$ of smooth varieties. We would like to study an algebraic non-linear partial differential equation

$$F(t, \partial_t^i x) = 0$$

that applies to sections $x \in \Gamma(M, C)$, that are functions $x : \mathbb{R} \to \mathbb{R}$. It is given by a polynomial $F(t, x_i) \in \mathbb{R}[t, \{x_i\}_{i \geq 0}]$. The solution space of such an equation can be studied with values in any $\mathcal{O}$-algebra $\mathcal{A}$ equipped with an action of the differentiation ∂_t (that fulfills a Leibniz rule for multiplication), the basic example being given by the algebra $\mathrm{Jet}(\mathcal{O}_C) := \mathbb{R}[t, \{x_i\}_{i \geq 0}]$ above with the action $\partial_t x_i = x_{i+1}$. The solution space of the given partial differential equation is then given by the functor

$$\underline{\mathrm{Sol}}_{\mathcal{D},F=0}(\mathcal{A}) := \{x \in \mathcal{A}, \; F(t, \partial_t^i x) = 0\}$$

defined on all $\mathcal{O}_C$-algebras equipped with an action of ∂_t. To be more precise, we define the category of $\mathcal{D}$-algebras, that solves the mathematical problem of finding a natural setting for a coordinate free study of polynomial non-linear partial differential equations with smooth super-function coefficients.

DEFINITION 2.3.1. Let M be a variety. A *$\mathcal{D}_M$-algebra* is an algebra $\mathcal{A}$ in the monoidal category of $\mathcal{D}_M$-modules. More precisely, it is an $\mathcal{O}_M$-algebra equipped with an action

$$\Theta_M \otimes \mathcal{A} \to \mathcal{A}$$

of vector fields on M such that the product in $\mathcal{A}$ fulfills Leibniz's rule

$$\partial(fg) = \partial(f)g + f\partial(g).$$

Recall from [**Pau11**] that one can extend the jet functor to the category of smooth $\mathcal{D}$-algebras (and even to smooth super-algebras), to extend the forthcoming results to the study of non-polynomial smooth partial differential equations. The forgetful functor

$$\text{Forget} : \text{ALG}_{\mathcal{D}} \to \text{ALG}_{\mathcal{O}}$$

has an adjoint (free $\mathcal{D}$-algebra on a given $\mathcal{O}$-algebra)

$$\text{Jet} : \text{ALG}_{\mathcal{O}} \to \text{ALG}_{\mathcal{D}}$$

called the (infinite) jet functor. It fulfills the universal property that for every $\mathcal{D}$-algebra $\mathcal{B}$, the natural map

$$\text{Hom}_{\text{ALG}_{\mathcal{O}}}(\mathcal{O}_C, \mathcal{B}) \cong \text{Hom}_{\text{ALG}_{\mathcal{D}}}(\text{Jet}(\mathcal{O}_C), \mathcal{B})$$

induced by the natural map $\mathcal{O}_C \to \text{Jet}(\mathcal{O}_C)$ is a bijection.

Using the jet functor, one can show that the solution space of the non-linear partial differential equation

$$F(t, \partial_t^i x) = 0$$

of the above example is representable, meaning that there is a natural isomorphism of functors on $\mathcal{D}$-algebras

$$\underline{\text{Sol}}_{\mathcal{D},F=0}(-) \cong \text{Hom}_{\text{ALG}_{\mathcal{D}}}(\text{Jet}(\mathcal{O}_C)/(F), -)$$

where (F) denotes the $\mathcal{D}$-ideal generated by F. This shows that the jet functor plays the role of the polynomial algebra in the differential algebraic setting. If $\pi : C \to M$ is a bundle, we define

$$\text{Jet}(C) := \underline{\text{Spec}}(\text{Jet}(\mathcal{O}_C)).$$

One can summarize the above discussion by the following array:

Equation	Polynomial	Partial differential
Formula	$P(x) = 0$	$F(t, \partial^\alpha x) = 0$
Naive variable	$x \in \mathbb{R}$	$x \in \text{Hom}(\mathbb{R}, \mathbb{R})$
Algebraic structure	commutative unitary ring A	$\mathcal{D}_M$-algebra A
Free structure	$P \in \mathbb{R}[x]$	$F \in \text{Jet}(\mathcal{O}_C)$
Solution space	$\{x \in A, P(x) = 0\}$	$\{x \in A, F(t, \partial^\alpha x) = 0\}$

EXAMPLE 2.3.1. If $\pi : C = \mathbb{R}^{n+m} \to \mathbb{R}^n = M$ is a trivial bundle of dimension $m + n$ over M of dimension n, with algebra of coordinates $\mathcal{O}_C := \mathbb{R}[\underline{t}, \underline{x}]$ for $\underline{t} = \{t^i\}_{i=1,\ldots,n}$ and $\underline{x} = \{x^j\}_{j=1,\ldots,m}$ given in multi-index notation, its jet algebra is

$$\text{Jet}(\mathcal{O}_C) := \mathbb{R}[\underline{t}, \underline{x}_\alpha]$$

where $\alpha \in \mathbb{N}^m$ is a multi-index representing the derivation index. The $\mathcal{D}$-module structure is given by making $\frac{\partial}{\partial t^i}$ act through the total derivative

$$D_i := \frac{\partial}{\partial t^i} + \sum_{\alpha,k} x^k_{i\alpha} \frac{\partial}{\partial x^k_\alpha}$$

where $i\alpha$ denotes the multi-index α increased by one at the i-th coordinate. For example, if $\pi : C = \mathbb{R} \times \mathbb{R} \to \mathbb{R} = M$, one gets

$$D_1 = \frac{\partial}{\partial t} + x_1 \frac{\partial}{\partial x} + x_2 \frac{\partial}{\partial x_1} + \dots.$$

DEFINITION 2.3.2. Let $\pi : C \to M$ be a bundle. A *partial differential equation* on the space $\Gamma(M, C)$ of sections of π is given by a quotient $\mathcal{D}_M$-algebra

$$p : \mathrm{Jet}(\mathcal{O}_C) \twoheadrightarrow \mathcal{A}$$

of the jet algebra of the $\mathcal{O}_M$-algebra $\mathcal{O}_C$. Its *local space of solutions* is the $\mathcal{D}$-space whose points with values in $\mathrm{Jet}(\mathcal{O}_C)$-$\mathcal{D}$-algebras $\mathcal{B}$ are given by

$$\underline{\mathrm{Sol}}_{\mathcal{D},(\mathcal{A},p)} := \{x \in \mathcal{B} |\ f(x) = 0,\ \forall f \in \mathrm{Ker}(p)\}$$

The *non-local space of solutions* of the partial differential equation $(\mathcal{A}, p)$ is the subspace of $\underline{\Gamma}(M, C)$ given by

$$\underline{\mathrm{Sol}}_{(\mathcal{A},p)} := \{x \in \underline{\Gamma}(M, C) |\ (j_\infty x)^* L = 0 \text{ for all } L \in \mathrm{Ker}(p)\}$$

where $(j_\infty x)^* : \mathrm{Jet}(\mathcal{O}_C) \to \mathcal{O}_M$ is (dual to) the Jet of x. Equivalently, $x \in \underline{\mathrm{Sol}}_{(\mathcal{A},p)}$ if and only if there is a natural factorization

$$\begin{array}{ccc} \mathrm{Jet}(\mathcal{O}_C) & \xrightarrow{(j_\infty x)^*} & \mathcal{O}_M \\ & {\scriptstyle p}\searrow & \uparrow \\ & & \mathcal{A} \end{array}$$

of the jet of x through p.

2.4. Local functionals and local differential forms. The natural functional invariant associated to a given $\mathcal{D}$-algebra $\mathcal{A}$ is given by the De Rham complex

$$\mathrm{DR}(\mathcal{A}) := (I_{*,M} \otimes_{\mathcal{O}} \mathcal{D}[n]) \overset{\mathbb{L}}{\otimes}_{\mathcal{D}} \mathcal{A}$$

of its underlying $\mathcal{D}$-module with coefficient in the universal complex of integral forms $I_{*,M} \otimes_{\mathcal{O}} \mathcal{D}[n]$, and its cohomology $h^*(\mathrm{DR}(\mathcal{A}))$. We will denote

$$h(\mathcal{A}) := h^0(\mathrm{DR}(\mathcal{A})) = \mathrm{Ber}_M \otimes_{\mathcal{D}} \mathcal{A}$$

where Ber_M here denotes the Berezinian object (and not only the complex concentrated in degree 0). If M is a non-super variety, one gets

$$\mathrm{DR}(\mathcal{A}) = \Omega^n_M \overset{\mathbb{L}}{\otimes}_{\mathcal{D}} \mathcal{A} \qquad \text{and} \qquad h(\mathcal{A}) = \Omega^n_M \otimes_{\mathcal{D}} \mathcal{A}.$$

The De Rham cohomology is given by the cohomology of the complex

$$\mathrm{DR}(\mathcal{A}) = I_{*,M}[n] \otimes_{\mathcal{O}_M} \mathcal{A},$$

which gives

$$\mathrm{DR}(\mathcal{A}) = \wedge^* \Omega^1_M[n] \otimes_{\mathcal{O}_M} \mathcal{A}$$

in the non super case.

If $\mathcal{A}$ is a jet algebra of section of a bundle $\pi : C \to M$ with basis a classical manifold, the De Rham complex identifies with a sub-complex of the usual De Rham complex of $\wedge^*\Omega^1_{\mathcal{A}/\mathbb{R}}$ of $\mathcal{A}$ viewed as an ordinary ring. One can think of classes in $h^*(\mathrm{DR}(\mathcal{A}))$ as defining a special class of (partially defined) functionals on the space $\underline{\Gamma}(M, C)$, by integration along singular homology cycles with compact support.

DEFINITION 2.4.1. Let M be a super-variety of dimension $p|q$. For every smooth simplex Δ_n, we denote $\Delta_{n|q}$ the super-simplex obtained by adjoining q odd variables to Δ_n. The *singular homology of M with compact support* is defined as the homology $H_{*,c}(M)$ of the simplicial set

$$\mathrm{Hom}(\Delta_{\bullet|q}, M)$$

of super-simplices with compact support condition on the body and non-degeneracy condition on odd variables.

Recall that, following [**Pau11**], a functional $f \in \mathrm{Hom}(\underline{\Gamma}(M, C), \mathbb{A}^1)$ on a space of fields denotes in general, by definition, only a partially defined function (with a well-chosen domain of definition).

PROPOSITION 2.4.1. *Let $\pi : C \to M$ be a bundle and $\mathcal{A}$ be the $\mathcal{D}_M$-algebra* $\mathrm{Jet}(\mathcal{O}_C)$. *There is a natural integration pairing*

$$\begin{array}{rcl} H_{*,c}(M) \times h^{*-n}(\mathrm{DR}(\mathcal{A})) & \to & \mathrm{Hom}(\underline{\Gamma}(M, C), \mathbb{A}^1) \\ (\Sigma, \omega) & \mapsto & [x \mapsto \int_\Sigma (j_\infty x)^* \omega] \end{array}$$

where $j_\infty x : M \to \mathrm{Jet}(C)$ is the taylor series of a given section x. If $p : \mathrm{Jet}(\mathcal{O}_C) \to \mathcal{A}$ is a given partial differential equation (such that $\mathcal{A}$ is $\mathcal{D}$-smooth) on $\Gamma(M, C)$ one also gets an integration pairing

$$\begin{array}{rcl} H_{*,c}(M) \times h^{*-n}(\mathrm{DR}(\mathcal{A})) & \to & \mathrm{Hom}(\underline{\mathrm{Sol}}_{(\mathcal{A},p)}, \mathbb{A}^1) \\ (\Sigma, \omega) & \mapsto & S_{\Sigma,\omega} : [x(t, u) \mapsto \int_\Sigma (j_\infty x)^* \omega]. \end{array}$$

PROOF. Remark that the values of the above pairing are given by partially defined functions, with a domain of definition given by Lebesgue's domination condition to make $t \mapsto \int_\Sigma (j_\infty x_t)^* \omega$ a smooth function of t if x_t is a parametrized trajectory. The only point to check is that the integral is independent of the chosen cohomology class. This follows from the fact that the integral of a total divergence on a closed subspace is zero, by Stokes' formula (the super case follows from the classical one). □

DEFINITION 2.4.2. A functional $S_{\Sigma,\omega} : \underline{\Gamma}(M, C) \to \mathbb{A}^1$ or $S_{\Sigma,\omega} : \underline{\mathrm{Sol}}_{(\mathcal{A},p)} \to \mathbb{A}^1$ obtained by the above constructed pairing is called a *quasi-local functional.* We denote

$$\mathcal{O}^{qloc} \subset \mathcal{O} := \underline{\mathrm{Hom}}(\underline{\Gamma}(M, C), \mathbb{A}^1)$$

the space of quasi-local functionals.

Remark that for $k \leq n$, the classes in $h^{*-k}(\mathrm{DR}(\mathcal{A}))$ are usually called (higher) conservation laws for the partial differential equation $p : \mathrm{Jet}(\mathcal{O}_C) \to \mathcal{A}$.

If $\mathcal{A}$ is a $\mathcal{D}$-algebra, i.e., an algebra in $(\mathrm{MOD}(\mathcal{D}), \otimes_\mathcal{O})$, one defines, as usual, an $\mathcal{A}$-module of differential forms as an $\mathcal{A}$-module $\Omega^1_\mathcal{A}$ in the monoidal category $(\mathrm{MOD}(\mathcal{D}), \otimes)$, equipped with a ($\mathcal{D}$-linear) derivation $d : \mathcal{A} \to \Omega^1_\mathcal{A}$ such for every $\mathcal{A}$-module $\mathcal{M}$ in $(\mathrm{MOD}(\mathcal{D}), \otimes)$, the natural map

$$\mathrm{Hom}_{\mathrm{MOD}(\mathcal{A})}(\Omega^1_\mathcal{A}, \mathcal{M}) \to \mathrm{Der}_{\mathrm{MOD}(\mathcal{A})}(\mathcal{A}, \mathcal{M})$$

given by $f \mapsto f \circ d$ is a bijection.

Remark that the natural $\mathcal{O}$-linear map

$$\Omega^1_{\mathcal{A}/\mathcal{O}} \to \Omega^1_{\mathcal{A}}$$

is an isomorphism of $\mathcal{O}$-modules. The $\mathcal{D}$-module structure on $\Omega^1_{\mathcal{A}}$ can be seen as an Ehresman connection, i.e., a section of the natural projection

$$\Omega^1_{\mathcal{A}/\mathbb{R}} \to \Omega^1_{\mathcal{A}/M}.$$

EXAMPLE 2.4.1. In the case of the jet space algebra $\mathcal{A} = \mathrm{Jet}(\mathcal{O}_C)$ for $C = \mathbb{R}^{n+m} \to \mathbb{R}^n = M$, a basis of $\Omega^1_{\mathcal{A}/M}$ compatible with this section is given by the Cartan forms

$$\theta^i_\alpha = dx^i_\alpha - \sum_{j=1}^n x^i_{j\alpha} dt^j.$$

The De Rham differential $d : \mathcal{A} \to \Omega^1_{\mathcal{A}}$ in the $\mathcal{D}$-algebra setting and its De Rham cohomology (often denoted d^V in the literature), can then be computed by expressing the usual De Rham differential $d : \mathcal{A} \to \Omega^1_{\mathcal{A}/M}$ in the basis of Cartan forms.

As explained in Section 2.1, the right notion of finiteness and duality in the monoidal category of $\mathcal{D}$-modules is not the $\mathcal{O}$-finite presentation and duality but the $\mathcal{D}$-finite presentation and duality. This extends to the category of $\mathcal{A}$-modules in $(\mathrm{MOD}(\mathcal{D}), \otimes)$. The following notion of smoothness differs from the usual one (in general symmetric monoidal categories) because we impose the $\mathcal{A}[\mathcal{D}]$-finite presentation, where $\mathcal{A}[\mathcal{D}] := \mathcal{A} \otimes_{\mathcal{O}} \mathcal{D}$, to have good duality properties.

DEFINITION 2.4.3. The $\mathcal{D}$-algebra $\mathcal{A}$ is called *$\mathcal{D}$-smooth* if $\Omega^1_{\mathcal{A}}$ is a projective $\mathcal{A}$-module of finite $\mathcal{A}[\mathcal{D}]$-presentation in the category of $\mathcal{D}$-modules, and $\mathcal{A}$ is a (geometrically) finitely generated $\mathcal{D}$-algebra, meaning that there exists an $\mathcal{O}$-module $\mathcal{M}$ of finite type, and an ideal $\mathcal{I} \subset \mathrm{Sym}_{\mathcal{O}}(\mathcal{M})$ and a surjection

$$\mathrm{Jet}(\mathrm{Sym}_{\mathcal{O}}(\mathcal{M})/\mathcal{I}) \twoheadrightarrow \mathcal{A}.$$

PROPOSITION 2.4.2. *If $\mathcal{A} = \mathrm{Jet}(\mathcal{O}_C)$ for $\pi : C \to M$ a smooth map of varieties, then $\mathcal{A}$ is $\mathcal{D}$-smooth and the $\mathcal{A}[\mathcal{D}]$-module $\Omega^1_{\mathcal{A}}$ is isomorphic to*

$$\Omega^1_{\mathcal{A}} \cong \Omega^1_{C/M} \otimes_{\mathcal{O}_C} \mathcal{A}[\mathcal{D}].$$

In particular, if $\pi : C = \mathbb{R} \times M \to M$ is the trivial bundle with fiber coordinate u, one gets the free $\mathcal{A}[\mathcal{D}]$-module of rank one

$$\Omega^1_{\mathcal{A}} \cong \mathcal{A}[\mathcal{D}]^{(\{du\})}$$

generated by the form du.

2.5. Local vector fields and local operations. We refer to Beilinson-Drinfeld's book [**BD04**] for a complete and axiomatic study of general pseudo-tensor categories. We will only present here the tools from this theory needed to understand local functional calculus. Local operations are new operations on $\mathcal{D}$-modules, that induce ordinary multilinear operations on their De Rham cohomology. We start by explaining the main motivation for introducing these new operations when one does geometry with $\mathcal{D}$-algebras.

We now define the notion of local vector fields.

DEFINITION 2.5.1. Let $\mathcal{A}$ be a smooth $\mathcal{D}$-algebra. The $\mathcal{A}^r[\mathcal{D}^{op}]$-module of *local vector fields* is defined by

$$\Theta_{\mathcal{A}} := \mathcal{H}om_{\mathcal{A}[\mathcal{D}]}(\Omega^1_{\mathcal{A}}, \mathcal{A}[\mathcal{D}]),$$

where $\mathcal{A}^r[\mathcal{D}^{op}] := \mathcal{A}^r \otimes_{\mathrm{Ber}_M} \mathcal{D}^{op}$ acts on the right though the isomorphism

$$\mathcal{A}^r[\mathcal{D}^{op}] \cong (\mathcal{A} \otimes_{\mathcal{O}} \mathrm{Ber}_M) \otimes_{\mathrm{Ber}_M} \mathcal{D}^{op} \cong \mathcal{A}[\mathcal{D}^{op}].$$

Remark now that in ordinary differential geometry, one way to define vector fields on a variety M is to take the $\mathcal{O}_M$-dual

$$\Theta_M := \mathcal{H}om_{\mathcal{O}_M}(\Omega^1_M, \mathcal{O}_M)$$

of the module of differential forms. The Lie bracket

$$[.,.] : \Theta_M \otimes \Theta_M \to \Theta_M$$

of two vector fields X and Y can then be defined from the universal derivation $d : \mathcal{O}_M \to \Omega^1_M$, as the only vector field $[X,Y]$ on M such that for every function $f \in \mathcal{O}_M$, one has the equality of derivations

$$[X,Y].f = X.i_Y(df) - Y.i_X(df).$$

In the case of a $\mathcal{D}$-algebra $\mathcal{A}$, this construction does not work directly because the duality used to define local vector fields is not the $\mathcal{A}$-linear duality (because it doesn't have good finiteness properties) but the $\mathcal{A}[\mathcal{D}]$-linear duality. This explains why the Lie bracket of local vector fields and their action on $\mathcal{A}$ are new kinds of operations of the form

$$[.,.] : \Theta_{\mathcal{A}} \boxtimes \Theta_{\mathcal{A}} \to \Delta_*\Theta_{\mathcal{A}}$$

and

$$L : \Theta_{\mathcal{A}} \boxtimes \mathcal{A} \to \Delta_*\mathcal{A}$$

where $\Delta : M \to M \times M$ is the diagonal map and the box product is defined by

$$\mathcal{M} \boxtimes \mathcal{N} := p_1^*\mathcal{M} \otimes p_2^*\mathcal{N}$$

for $p_1, p_2 : M \times M \to M$ the two projections. One way to understand these construction is by looking at the natural injection

$$\Theta_{\mathcal{A}} \hookrightarrow \mathcal{H}om_{\mathcal{D}}(\mathcal{A}, \mathcal{A}[\mathcal{D}])$$

given by sending $X : \Omega^1_{\mathcal{A}} \to \mathcal{A}[\mathcal{D}]$ to $X \circ d : \mathcal{A} \to \mathcal{A}[\mathcal{D}]$. The theory of $\mathcal{D}$-modules tells us that the datum of this map is equivalent to the datum of a $\mathcal{D}^{op}_{M\times M}$-linear map

$$L : \Theta_{\mathcal{A}} \boxtimes \mathcal{A}^r \to \Delta_*\mathcal{A}^r.$$

Similarly, the above formula

$$[X,Y].f = X.i_Y(df) - Y.i_X(df)$$

of ordinary differential geometry makes sense in local computations only if we see $\Theta_{\mathcal{A}}$ as contained in $\mathcal{H}om_{\mathcal{D}}(\mathcal{A}, \mathcal{A}[\mathcal{D}])$ (and not in $\mathcal{H}om_{\mathcal{D}}(\mathcal{A}, \mathcal{A})$, contrary to what is usually done), so that we must think of the bracket as a morphism of sheaves

$$\Theta_{\mathcal{A}} \to \mathcal{H}om_{\mathcal{D}^{op}}(\Theta_{\mathcal{A}}, \Theta_{\mathcal{A}} \otimes \mathcal{D}^{op})^r.$$

This is better formalized by a morphism of $\mathcal{D}^{op}_{M\times M}$-modules

$$[.,.] : \Theta_{\mathcal{A}} \boxtimes \Theta_{\mathcal{A}} \to \Delta_*\Theta_{\mathcal{A}}$$

as above. Another way to understand these local operations is to make an analogy with multilinear operations on $\mathcal{O}_M$-modules. Indeed, if $\mathcal{F}$, $\mathcal{G}$ and $\mathcal{H}$ are three quasi-coherent $\mathcal{O}_M$-modules, one has a natural adjunction isomorphism

$$\mathcal{H}om_{\mathcal{O}_M}(\mathcal{F}\otimes\mathcal{G},\mathcal{H})\cong\mathcal{H}om_{\mathcal{O}_M}(\Delta^*(\mathcal{F}\boxtimes\mathcal{G}),\mathcal{H})\cong\mathcal{H}om_{\mathcal{O}_{M\times M}}(\mathcal{F}\boxtimes\mathcal{G},\Delta_*\mathcal{H}),$$

and local operations are given by a $\mathcal{D}$-linear version of the right part of the above equality. It is better to work with this expression because of finiteness properties of the $\mathcal{D}$-modules in play.

We now recall for the reader's convenience from Beilinson-Drinfeld [**BD04**] the basic properties of general local operations (extending straightforwardly their approach to the case of a base supervariety M).

DEFINITION 2.5.2. Let $(\mathcal{L}_i)_{i\in I}$ be a finite family of $\mathcal{D}^{op}$-modules and $\mathcal{M}$ be a $\mathcal{D}^{op}$-module. We define the *space of $*$-operations*

$$P_I^*(\{\mathcal{L}_i\},\mathcal{M}):=\mathrm{Hom}_{\mathcal{D}_{X^I}}(\boxtimes\mathcal{L}_i,\Delta_*^{(I)}\mathcal{M})$$

where $\Delta^{(I)}:M\to M^I$ is the diagonal embedding and $\boxtimes\mathcal{L}_i:=\otimes_{i\in I}p_i^*\mathcal{L}_i$ with $p_i:X^I\to X$ the natural projections. The datum $(\mathrm{MOD}(\mathcal{D}^{op}),P_I)$ is called the *pseudo tensor structure* on the category of $\mathcal{D}^{op}$-modules.

One can define natural composition maps of pseudo-tensor operations, and a pseudo-tensor structure is very similar to a tensor structure in many respect: one has, under some finiteness hypothesis, good notions of internal homomorphisms and internal $*$-operations.

The pseudo-tensor structure actually defines what one usually calls a colored operad with colors in $\mathrm{MOD}(\mathcal{D}^{op})$. It is very important because it allows to easily manipulate covariant objects like local vector fields or local Poisson brackets. The main idea of Beilinson and Drinfeld's approach to geometry of $\mathcal{D}$-spaces is that functions and differential forms usually multiply by using ordinary tensor product of $\mathcal{D}$-modules, but that local vector fields and local differential operators multiply by using pseudo-tensor operations. This gives the complete toolbox to do differential geometry on $\mathcal{D}$-spaces in a way that is very similar to ordinary differential geometry.

Another way to explain the interest of $*$-operations, is that they induce ordinary operations in De Rham cohomology. Since De Rham cohomology is the main tool of local functional calculus (it gives an algebraic presentation of local functions, differential forms and vector fields on the space $\underline{\Gamma}(M,C)$ of trajectories of a given field theory), we will make a systematic use of these operations. We refer to Beilinson-Drinfeld [**BD04**] for a proof of the following result, that roughtly says that the De Rham functor can be extended to respect the natural pseudo-tensor structures of its source and target categories. This means, in simpler terms, that local, i.e., $*$-operations can be used to define usual operations on quasi-local functionals.

PROPOSITION 2.5.1. *The central De Rham cohomology functor* $h:\mathrm{MOD}(\mathcal{D}^{op})\to\mathrm{MOD}(\mathbb{R}_M)$ *given by* $h(\mathcal{M}):=h^0(\mathrm{DR}^r(\mathcal{M})):=\mathcal{M}\otimes_{\mathcal{D}}\mathcal{O}$ *induces a natural map*

$$h:P_I^*(\{\mathcal{L}_i\},\mathcal{M})\to\mathrm{Hom}(\otimes_i h(\mathcal{L}_i),h(\mathcal{M}))$$

from $*$-operations to multilinear operations. At the level of complexes, the choice of a dg-$\mathcal{D}$-algebra resolution $\epsilon:\mathcal{P}\to\mathcal{O}$, that is flat as a dg-$\mathcal{D}$-module (i.e., $\otimes_{\mathcal{D}}\mathcal{P}$

transforms acyclic complexes in acyclic complexes), induces a natural morphism

$$\mathrm{DR} : P_I^*(\{\mathcal{L}_i\}, \mathcal{M}) \longrightarrow \mathbb{R}\mathcal{H}om(\overset{\mathbb{L}}{\otimes}_i \mathrm{DR}(\mathcal{L}_i), \mathrm{DR}(\mathcal{M})).$$

3. Gauge theories and the covariant phase space

We are inspired, when discussing the Batalin-Vilkovisky (later called BV) formalism, by a huge physical literature, starting with Peierls [**Pei52**] and De Witt [**DeW03**] for the covariant approach to quantum field theory, and with [**HT92**] and [**FH90**] as general references for the BV formalism. More specifically, we also use Stasheff's work [**Sta97**] and [**Sta98**] as homotopical inspiration, and [**FLS02**], [**Bar10**] and [**CF01**] for explicit computations.

3.1. A finite dimensional toy model. In this section, we will do some new kind of differential geometry on spaces of the form $X = \underline{\mathrm{Spec}}_{\mathcal{D}}(\mathcal{A})$ given by spectra of $\mathcal{D}$-algebras, that encode solution spaces of non-linear partial differential equations in a coordinate free fashion (to be explained in the next section).

Before starting this general description, that is entailed of technicalities, we present a finite dimensional analog, that can be used as a reference to better understand the constructions done in the setting of $\mathcal{D}$-spaces.

Let H be a finite dimensional smooth variety (analogous to the space of histories $H \subset \underline{\Gamma}(M, C)$ of a given Lagrangian variational problem) and $S : H \to \mathbb{R}$ be a smooth function (analogous to an action functional on the space of histories). Let $d : \mathcal{O}_H \to \Omega^1_H$ be the De Rham differential and

$$\Theta_H := \mathcal{H}om_{\mathcal{O}_H}(\Omega^1_H, \mathcal{O}_H).$$

There is a natural biduality isomorphism

$$\Omega^1_H \cong \mathcal{H}om_{\mathcal{O}_H}(\Omega^1_H, \mathcal{O}_H).$$

Let $i_{dS} : \Theta_H \to \mathcal{O}_H$ be given by the insertion of vector fields in the differential $dS \in \Omega^1_H$ of the given function $S : H \to \mathbb{R}$.

The claim is that there is a natural homotopical Poisson structure on the space T of critical points of $S : H \to \mathbb{R}$, defined by

$$T = \{x \in H,\ d_x S = 0\}.$$

Define the algebra of functions $\mathcal{O}_T$ as the quotient of $\mathcal{O}_H$ by the ideal $\mathcal{I}_S$ generated by the equations $i_{dS}(\vec{v}) = 0$ for all $\vec{v} \in \Theta_H$. Remark that $\mathcal{I}_S$ is the image of the insertion map $i_{dS} : \Theta_H \to \mathcal{O}_H$ and is thus locally finitely generated by the image of the basis vector fields $\vec{x}_i := \frac{\partial}{\partial x_i}$ that correspond to the local coordinates x_i on H. Now let $\mathcal{N}_S$ be the kernel of i_{dS}. It describes the relations between the generating equations $i_{dS}(\vec{x}_i) = \frac{\partial S}{\partial x_i}$ of $\mathcal{I}_S$.

The differential graded $\mathcal{O}_H$-algebra

$$\mathcal{O}_P := \mathrm{Sym}_{dg}([\underset{-1}{\Theta_H} \xrightarrow{i_{dS}} \underset{0}{\mathcal{O}_H}])$$

is isomorphic, as a graded algebra, to the algebra of multi-vectors

$$\wedge^*_{\mathcal{O}_H} \Theta_H.$$

This graded algebra is equipped with an odd, so called Schouten bracket, given by extending the Lie derivative

$$L : \Theta_H \otimes \mathcal{O}_H \to \mathcal{O}_H$$

and Lie bracket

$$[.,.] : \Theta_H \otimes \Theta_H \to \Theta_H$$

by Leibniz's rule.

PROPOSITION 3.1.1. *The Schouten bracket is compatible with the insertion map* i_{dS} *and makes* $\mathcal{O}_P$ *a differential graded odd Poisson algebra. The Lie bracket on* Θ_H *induces a Lie bracket on* $\mathcal{N}_S$.

Now let $\mathfrak{g}_S \to \mathcal{N}_S$ be a projective resolution of $\mathcal{O}_T$ as an $\mathcal{O}_H$-module. This graded module is the finite dimensional analog of the space of gauge (and higher gauge) symmetries. Suppose that $\mathcal{P}$ is bounded with projective components of finite rank.

DEFINITION 3.1.1. The *finite dimensional BV algebra* associated to $S : H \to \mathbb{R}$ is the bigraded $\mathcal{O}_H$-algebra

$$\mathcal{O}_{BV} := \mathrm{Sym}_{bigrad}\left(\begin{bmatrix} \mathfrak{g}_S[2] & \oplus & \Theta_H[1] & \oplus & \mathcal{O}_H \\ & & & & \oplus \\ & & & & {}^t\mathfrak{g}_S^*[-1] \end{bmatrix}\right),$$

where ${}^t\mathfrak{g}_S^*$ is the $\mathcal{O}_H$-dual of the graded module $\mathfrak{g}_S$ transposed to become a vertical ascending graded module.

The main theorem of the Batalin-Vilkovisky formalism, that is the aim of this section, is the following:

THEOREM 3.1.1. *There exists a non-trivial extension*

$$S_{cm} = S_0 + \sum_{i\geq 1} S_i \in \mathcal{O}_{BV}$$

of the classical function S that fulfills the classical master equation

$$\{S_{cm}, S_{cm}\} = 0.$$

The differential $D = \{S_{cm}, .\}$ *gives* $\mathcal{O}_{BV}$ *the structure of a differential graded odd Poisson algebra.*

The corresponding derived space $\mathbb{R}\underline{\mathrm{Spec}}(\mathcal{O}_{BV}, D)$ can be thought as a kind of derived Poisson reduction

$$\mathbb{R}\underline{\mathrm{Spec}}(\mathcal{O}_{BV}, D) \cong \mathbb{R}\underline{\mathrm{Spec}}(\mathcal{O}_H/\mathcal{I}_H)/_{\mathbb{R}}\mathcal{N}_S$$

that corresponds to taking the quotient of the homotopical critical space of S (cofibrant replacement of $\mathcal{O}_T = \mathcal{O}_H/\mathcal{I}_H$)) by the foliation induced by the Noether relations $\mathcal{N}_S$.

EXAMPLE 3.1.1. To be even more explicit, let us treat a simple example of the above construction. Let $H = \mathbb{R}^2$ be equipped with the polynomial function algebra $\mathcal{O}_H = \mathbb{R}[x, y]$. The differential one forms on H are given by the free $\mathcal{O}_H$-modules

$$\Omega^1_H = \mathbb{R}[x, y]^{(dx,dy)}.$$

Let $S \in \mathcal{O}_H$ be the function $F(x, y) = \frac{x^2}{2}$. One then has $dS = xdx$. The module Θ_H of vector fields is the free module

$$\Theta_H = \mathbb{R}[x, y]^{\left(\frac{\partial}{\partial x}, \frac{\partial}{\partial y}\right)}$$

and the insertion map is given by the $\mathbb{R}[x,y]$-module morphism

$$\begin{array}{cccc} i_{dS}: & \Theta_H & \to & \mathcal{O}_H \\ & \vec{v} & \mapsto & \langle dS, \vec{v}\rangle, \end{array}$$

and, in particular, $i_{dS}\left(\frac{\partial}{\partial x}\right) = x$, and $i_{dS}\left(\frac{\partial}{\partial y}\right) = 0$. The image of the insertion map is given by the ideal $\mathcal{I}_S = (x)$ in $\mathcal{O}_H = \mathbb{R}[x,y]$. The kernel $\mathcal{N}_S$ of the insertion map is the free submodule

$$\mathcal{N}_S = \mathfrak{g}_S = \mathbb{R}[x,y]^{\left(\frac{\partial}{\partial y}\right)}$$

of Θ_H. One then has a quasi-isomorphism

$$\mathrm{Sym}_{\mathcal{O}_H-dg}([\mathfrak{g}_S[2] \to \Theta_H[1] \xrightarrow{i_{dS}} \mathcal{O}_H]) \to \mathcal{O}_H/\mathcal{I}_S.$$

The obtained algebra is called the Koszul-Tate resolution of $\mathcal{O}_H/\mathcal{I}_S$. Now the bigraded BV algebra $\mathcal{O}_{BV}$ is given by

$$\mathcal{O}_{BV} := \mathrm{Sym}_{bigrad}\left(\left[\begin{array}{ccccc} \mathfrak{g}_S[2] & \oplus & \Theta_H[1] & \oplus & \mathcal{O}_H \\ & & & & \oplus \\ & & & & {}^t\mathfrak{g}_S^*[-1] \end{array}\right]\right).$$

It can be described more explicitely as

$$\mathcal{O}_{BV} = \mathrm{Sym}_{\mathcal{O}_H-bigrad}\left(\left[\begin{array}{ccccc} \mathcal{O}_H^{\left(\frac{\partial}{\partial y}\right)}[2] & \oplus & \mathcal{O}_H^{\left(\frac{\partial}{\partial x},\frac{\partial}{\partial y}\right)}[1] & \oplus & \mathcal{O}_H \\ & & & & \oplus \\ & & & & \mathcal{O}_H^{(dy)}[-1] \end{array}\right]\right).$$

The graded version is the $\mathcal{O}_H$-algebra

$$\mathcal{O}_{BV} = \mathrm{Sym}^*(\mathfrak{g}_S) \otimes \wedge^*\Theta_H \otimes \wedge^*\mathfrak{g}_S^*$$

on $4 = 1+2+1$ variables. The action map $\mathfrak{g}_S \otimes \mathcal{O}_H \to \mathcal{O}_H$ induces a differential operator

$$\mathcal{O}_H \to \mathrm{Hom}(\mathfrak{g}_S, \mathcal{O}_H)$$

whose extension to the algebra $\wedge^*\mathfrak{g}_S^*$ is the Chevalley-Eilenberg differential of the action. The extension of the natural differential on the horizontal part

$$\mathcal{O}_{KT} := \mathrm{Sym}^*(\mathfrak{g}_S) \otimes \wedge^*\Theta_H \subset \mathcal{O}_{BV}$$

of the BV algebra gives the Koszul-Tate differential. The BV formalism gives a way to combine the Koszul-Tate differential with the Chevalley-Eilenberg differential by constructing an $S_{cm} \in \mathcal{O}_{BV}$ such that some components of the bracket $\{S_{cm}, .\}$ induce both differentials on the corresponding generators of the BV algebra. In our case, the Koszul-Tate part of S_{cm} is defined by

$$S_{KT} = S + \frac{\partial}{\partial y}.dy,$$

so that

$$d_{KT} = \{S_{KT}, .\} : \mathcal{O}_{KT} \to \mathcal{O}_{KT}.$$

In this case, the expression S_{KT} is a solution S_{cm} of the classical master equation. This construction is not so interesting because the critical algebra $\mathcal{O}_T = \mathcal{O}_H/\mathcal{I}_S$ is simply $\mathbb{R}[x,y]/(x) \cong \mathbb{R}[y]$ and the foliation by gauge orbits on it is given by the free action of the vector field $\frac{\partial}{\partial y}$, so that the invariants $\mathcal{O}_T^{\mathfrak{g}_S}$ are simply the space $\mathbb{R}$ of functions on the critical space $T/\mathfrak{g}_S$, that is simply a point. The action being nice, there is no higher cohomology and the BV differential graded algebra $(\mathcal{O}_{BV}, D)$ is quasi-isomorphic to $\mathbb{R}$. In the case of spaces of fields and action functionals, the situation is not so simple, and the foliation by gauge orbits can be more singular on the critical space.

The aim of this section is to generalize the above construction to local variational problems, where $H \subset \underline{\Gamma}(M, C)$ is a space of histories (subspace of the space of sections of a bundle $\pi : C \to M$) and $S : H \to \mathbb{R}$ is given by the integration of a Lagrangian density. The main difficulties that we will encounter and overcome trickily in this generalization are that:

(1) The $\mathcal{D}$-module $\mathcal{D} \otimes_{\mathcal{O}} \mathcal{D}$ is not $\mathcal{D}$-coherent, so that a projective resolution of $\mathcal{N}_S$ will not be dualizable in practical cases. This will impose us to use finer resolutions of the algebra $\mathcal{O}_T$ of functions on the critical space.
(2) Taking the bracket between two *densities* of vector fields on the space H of histories is not an $\mathcal{O}_M$-bilinear operation but a new kind of operation, called a locally bilinear operation and described in Section 2.5.
(3) The ring $\mathcal{D}$ is not commutative. This will be overcome by using the equivalence between $\mathcal{D}$-modules and $\mathcal{D}^{op}$-modules given by tensoring by Ber_M.

Out of the above technical points, the rest of the constructions of this section are completely parallel to what we did on the finite dimensional toy model.

3.2. General gauge theories.

PROPOSITION 3.2.1. *Let M be a super-variety and $\mathcal{A}$ be a smooth $\mathcal{D}$-algebra. There is a natural isomorphism, called the local interior product*

$$\begin{array}{rccc} i: & h(\Omega^1_{\mathcal{A}}) := \mathrm{Ber}_M \otimes_{\mathcal{D}} \Omega^1_{\mathcal{A}} & \longrightarrow & \mathcal{H}om_{\mathcal{A}[\mathcal{D}]}(\Theta^{\ell}_{\mathcal{A}}, \mathcal{A}) \\ & \omega & \longmapsto & [X \mapsto i_X \omega]. \end{array}$$

PROOF. By definition, one has

$$\Theta_{\mathcal{A}} := \mathcal{H}om_{\mathcal{A}[\mathcal{D}]}(\Omega^1_{\mathcal{A}}, \mathcal{A}[\mathcal{D}])$$

and since $\mathcal{A}$ is $\mathcal{D}$-smooth, the biduality map

$$\Omega^1_{\mathcal{A}} \to \mathcal{H}om_{\mathcal{A}^r[\mathcal{D}^{op}]}(\Theta_{\mathcal{A}}, \mathcal{A}^r[\mathcal{D}^{op}])$$

is an isomorphism. Tensoring this map with Ber_M over $\mathcal{D}$ gives the desired result. □

DEFINITION 3.2.1. If $\mathcal{A}$ is a smooth $\mathcal{D}$-algebra and $\omega \in h(\Omega^1_{\mathcal{A}})$, the $\mathcal{A}[\mathcal{D}]$ linear map

$$i_\omega : \Theta^{\ell}_{\mathcal{A}} \to \mathcal{A}$$

is called the *insertion map*. Its kernel $\mathcal{N}_\omega$ is called the $\mathcal{A}[\mathcal{D}]$-module of *Noether identities* and its image $\mathcal{I}_\omega$ is called the *Euler-Lagrange ideal*.

If $\mathcal{A} = \mathrm{Jet}(\mathcal{O}_C)$ for $\pi : C \to M$ a bundle and $\omega = dS$, the Euler-Lagrange ideal $\mathcal{I}_{dS}$ is locally generated as an $\mathcal{A}[\mathcal{D}]$-module by the image of the local basis of vector fields in

$$\Theta^{\ell}_{\mathcal{A}} \cong \mathcal{A}[\mathcal{D}] \otimes_{\mathcal{O}_C} \Theta_{C/M}.$$

If M is of dimension n and the relative dimension of C over M is m, this gives n equations (indexed by $i = 1, \ldots, n$, one for each generator of $\Theta_{C/M}$) given in local coordinates by

$$\sum_{\alpha} (-1)^{|\alpha|} D_\alpha \left(\frac{\partial L}{\partial x_{i,\alpha}} \right) \circ (j_\infty x)(t) = 0,$$

where $S = [La^n t] \in h(\mathcal{A})$ is the local description of the Lagrangian density.

We now define the notion of local variational problem with nice histories. This type of variational problem can be studied completely by only using geometry of $\mathcal{D}$-spaces. This gives powerful finiteness and biduality results that are necessary to study conceptually general gauge theories.

DEFINITION 3.2.2. Let $\pi : C \to M$, $H \subset \underline{\Gamma}(M, C)$ and $S : H \to \mathbb{R}$ be a Lagrangian variational problem, and suppose that S is a local functional, i.e., if $\mathcal{A} = \mathrm{Jet}(\mathcal{O}_C)$, there exists $[L\omega] \in h(\mathcal{A}) := \mathrm{Ber}_M \otimes_{\mathcal{D}} \mathcal{A}$ and $\Sigma \in H_{c,n}(M)$ such that $S = S_{\Sigma, L\omega}$. The variational problem is called a *local variational problem with nice histories* if the space of critical points $T = \{x \in H,\ d_x S = 0\}$ identifies with the space $\underline{\mathrm{Sol}}(\mathcal{A}/\mathcal{I}_{dS})$ of solutions to the Euler-Lagrange equation.

The notion of variational problem with nice histories can be explained in simple terms by looking at the following simple example. The point is to define H by adding boundary conditions to elements in $\underline{\Gamma}(M, C)$, so that the boundary terms of the integration by part, that we do to compute the variation $d_x S$ of the action, vanish.

EXAMPLE 3.2.1. Let $\pi : C = \mathbb{R}^3 \times [0,1] \to [0,1] = M$, $\mathcal{A} = \mathbb{R}[t, x_0, x_1, \dots]$ be the corresponding $\mathcal{D}_M$-algebra with action of ∂_t given by $\partial_t x_i = x_{i+1}$, and $S = \frac{1}{2} m (x_1)^2 dt \in h(\mathcal{A})$ be the local action functional for the variational problem of newtonian mechanics for a free particle in $\mathbb{R}^3$. The differential of $S : \underline{\Gamma}(M, C) \to \underline{\mathbb{R}}$ at $u : U \to \underline{\Gamma}(M, C)$ along the vector field $\vec{u} \in \Theta_U$ is given by integrating by part

$$\langle d_x S, \vec{u} \rangle = \int_M \langle -m \partial_t^2 x, \frac{\partial x}{\partial \vec{u}} \rangle dt + \left[\langle \partial_t x, \frac{\partial x}{\partial \vec{u}} \rangle \right]_0^1 .$$

The last term of this expression is called the boundary term and we define nice histories for this variational problem by fixing the starting and ending point of trajectories to annihilate this boundary term:

$$H = \{x \in \underline{\Gamma}(M, C),\ x(0) = x_0,\ x(1) = x_1\}$$

for x_0 and x_1 some given points in $\mathbb{R}^3$. In this case, one has

$$T = \{x \in H,\ d_x S = 0\} \cong \underline{\mathrm{Sol}}(\mathcal{A}/\mathcal{I}_{dS})$$

where $\mathcal{I}_{dS}$ is the $\mathcal{D}$-ideal in $\mathcal{A}$ generated by $-m x_2$, i.e., by Newton's differential equation for the motion of a free particle in $\mathbb{R}^3$. The critical space is thus given by

$$T = \{x \in H,\ \partial_t x \text{ is constant on } [0,1]\},$$

i.e., the free particle is moving on the line from x_0 to x_1 with constant speed.

DEFINITION 3.2.3. A *general gauge theory* is a local variational problem with nice histories.

3.3. Regularity conditions and higher Noether identities. We now describe regularity properties of gauge theories, basing our exposition on the article [**Pau10**]. We will moreover use the language of homotopical and derived geometry in the sense of Toen-Vezzosi [**TV08**] to get geometric insights on the spaces in play in this section (See [**Pau11**] for an introduction and references). We denote $\mathcal{A} \mapsto Q\mathcal{A}$ a cofibrant replacement functor in a given model category. Recall that all differential graded algebras are fibrant for their standard model structure.

In all this section, we set $\pi : C \to M$, $H \subset \underline{\Gamma}(M, C)$, $\mathcal{A} = \mathrm{Jet}(\mathcal{O}_C)$ and $S \in h(\mathcal{A})$ a gauge theory. The kernel of its insertion map

$$i_{dS} : \Theta^{\ell}_{\mathcal{A}} \to \mathcal{A}$$

is called the space $\mathcal{N}_S$ of Noether identities. Its right version

$$\mathcal{N}_S^r = \mathrm{Ber}_M \otimes \mathcal{N}_S \subset \Theta_{\mathcal{A}}$$

is called the space of Noether gauge symmetries.

DEFINITION 3.3.1. The *derived critical space* of a gauge theory is the differential graded $\mathcal{A}$-space

$$\begin{array}{rccc} P := \mathrm{Spec}(\mathcal{A}_P): & dg-\mathcal{A}-\mathrm{ALG} & \to & \mathrm{SSETS} \\ & \mathcal{R} & \mapsto & s\mathrm{Hom}_{dg-Alg_{\mathcal{D}}}(\mathcal{A}_P, \mathcal{R}). \end{array}$$

whose coordinate differential graded algebra is

$$\mathcal{A}_P := \mathrm{Sym}_{dg}([\Theta^{\ell}_{\mathcal{A}}[1] \xrightarrow{i_{dS}} \mathcal{A}]).$$

A *non-trivial Noether identity* is a class in $H^1(\mathcal{A}_P, i_{dS})$.

We refer to Beilinson and Drinfeld's book [**BD04**] for the following proposition.

PROPOSITION 3.3.1. *The local Lie bracket of vector fields extends naturally to an odd local (so-called Schouten) Poisson bracket on the dg-$\mathcal{A}$-algebra $\mathcal{A}_P$ of coordinates on the derived critical space.*

The following corollary explains why we called $\mathcal{N}_S^r$ the space of Noether gauge symmetries.

COROLLARY 3.3.1. *The natural map*

$$\mathcal{N}_S^r \boxtimes \mathcal{N}_S^r \to \Delta_* \Theta_{\mathcal{A}}$$

induced by the local bracket on local vector fields always factors through $\Delta_* \mathcal{N}_S^r$ and the natural map

$$\mathcal{N}_S^r \boxtimes \mathcal{A}^r/\mathcal{I}_S^r \to \Delta_* \mathcal{A}^r/\mathcal{I}_S^r$$

is a local Lie $\mathcal{A}$-algebroid action.

The trivial Noether identities are those in the image of the natural map

$$\wedge^2 \Theta^{\ell}_{\mathcal{A}} \to \Theta_{\mathcal{A}},$$

and these usually don't give a finitely generated $\mathcal{A}[\mathcal{D}]$-module because of the simple fact that $\mathcal{D} \otimes_{\mathcal{O}} \mathcal{D}$ is not $\mathcal{D}$-coherent. This is a very good reason to consider only non-trivial Noether identities, because these can usually (i.e., in all the applications we have in mind) be given by a finitely generated $\mathcal{A}[\mathcal{D}]$-module.

DEFINITION 3.3.2. The *proper derived critical space* of a gauge theory is the (derived) space

$$\begin{array}{rccc} \mathbb{R}\mathrm{Spec}(\mathcal{A}/\mathcal{I}_S): & dg-\mathcal{A}-\mathrm{ALG} & \to & \mathrm{SSETS} \\ & \mathcal{R} & \mapsto & s\mathrm{Hom}_{dg-Alg_{\mathcal{D}}}(\mathcal{B}, \mathcal{R}). \end{array}$$

where $\mathcal{B} \xrightarrow{\sim} \mathcal{A}/\mathcal{I}_{dS}$ is a cofibrant resolution of $\mathcal{A}/\mathcal{I}_S$ as a dg-$\mathcal{A}$-algebra in degree 0.

From the point of view of derived geometry, differential forms on the cofibrant resolution $\mathcal{B}$ give a definition of the cotangent complex of the $\mathcal{D}$-space morphism

$$i : \underline{\mathrm{Spec}}_{\mathcal{D}}(\mathcal{A}/\mathcal{I}_S) \to \underline{\mathrm{Spec}}_{\mathcal{D}}(\mathcal{A})$$

of inclusion of critical points of the action functional in the $\mathcal{D}$-space of general trajectories. This notion of cotangent complex gives a well behaved way to study infinitesimal deformations of the above inclusion map i (see Illusie [**Ill71**]), even if it is not a smooth morphism (i.e., even if the critical space is singular).

It is easy to define a cofibrant resolution of the critical space related to Noether identities. If $\mathfrak{g}_S \to \mathcal{N}_S$ is a projective $\mathcal{A}[\mathcal{D}]$-resolution, the dg-algebra

$$\mathcal{B} = \mathrm{Sym}_{dg}([\mathfrak{g}_S[2] \to \Theta^{\ell}_{\mathcal{A}}[1] \to \mathcal{A}])$$

is a cofibrant resolution of $\mathcal{A}/\mathcal{I}_S$. An important problem with this simple construction is that the terms of the given resolution $\mathfrak{g}_S \to \mathcal{N}_S$ are usually *not finitely generated* as $\mathcal{A}[\mathcal{D}]$-modules. We will see how to define a finer, so-called Koszul-Tate resolution, that will have better finiteness properties, by using generating spaces of Noether identities. These can be defined by adapting Tate's construction [**Tat57**] to the local context. We are inspired here by Stasheff's paper [**Sta97**].

DEFINITION 3.3.3. A *generating space of Noether identities* is a tuple $(\mathfrak{g}_S, \mathcal{A}_n, i_n)$ composed of

(1) a negatively graded projective $\mathcal{A}[\mathcal{D}]$-module $\mathfrak{g}_S$,
(2) a negatively indexed family $\mathcal{A}_n$ of dg-$\mathcal{A}$-algebras with $\mathcal{A}_0 = \mathcal{A}$, and
(3) for each $n \leq -1$, an $\mathcal{A}[\mathcal{D}]$-linear morphism $i_n : \mathfrak{g}_S^{n+1} \to Z^n\mathcal{A}_n$ to the n-cycles of $\mathcal{A}_n$,

such that if one extends $\mathfrak{g}_S$ by setting $\mathfrak{g}_S^1 = \Theta^{\ell}_{\mathcal{A}}$ and if one sets

$$i_0 = i_{dS} : \Theta^{\ell}_{\mathcal{A}} \to \mathcal{A},$$

(1) one has for all $n \leq 0$ an equality

$$\mathcal{A}_{n-1} = \mathrm{Sym}_{\mathcal{A}_n}([\mathfrak{g}_S^{n+1}[-n+1] \otimes_{\mathcal{A}} \mathcal{A}_n \overset{i_n}{\to} \underset{0}{\mathcal{A}_n}]),$$

(2) the natural projection map

$$\mathcal{A}_{KT} := \varinjlim \mathcal{A}_n \to \mathcal{A}/\mathcal{I}_S$$

is a cofibrant resolution, called the *Koszul-Tate algebra*, whose differential is denoted d_{KT}.

We are now able to define the right regularity properties for a given gauge theory. These finiteness properties are imposed to make the generating space of Noether identities dualizable as an $\mathcal{A}[\mathcal{D}]$-module (resp. as a graded $\mathcal{A}[\mathcal{D}]$-module). Without any regularity hypothesis, the constructions given by homotopical Poisson reduction of gauge theories, the so-called derived covariant phase space, don't give $\mathcal{A}$-algebras, but only $\mathbb{R}$-algebras, that are too poorly behaved and infinite dimensional to be of any (even theoretical) use. We thus don't go through the process of their definition, that is left to the interested reader.

We now recall the language used by physicists (see for example [**HT92**]) to describe the situation. This can be useful to relate our constructions to the one described in physics books.

DEFINITION 3.3.4. A gauge theory is called *regular* if there exists a generating space of Noether identities $\mathfrak{g}_S$ whose components are finitely generated and projective. It is called *strongly regular* if this regular generating space is a bounded graded module. Suppose given a regular gauge theory. Consider the inner dual graded space (well-defined because of the regularity hypothesis)

$$\mathfrak{g}_S^\circ := \mathcal{H}om_{\mathcal{A}[\mathcal{D}]}(\mathfrak{g}_S, \mathcal{A}[\mathcal{D}])^\ell.$$

(1) The generators of $\Theta_\mathcal{A}^\ell$ are called *antifields* of the theory.
(2) The generators of $\mathfrak{g}_S$ of higher degree are called *antighosts*, or (non-trivial) *higher Noether identities* of the theory.
(3) The generators of the graded $\mathcal{A}^r[\mathcal{D}^{op}]$-module $\mathfrak{g}_S^r$ are called (non-trivial) *higher gauge symmetries* of the theory.
(4) The generators of the graded $\mathcal{A}[\mathcal{D}]$-module $\mathfrak{g}_S^\circ$ are called *ghosts* of the theory.

Remark that the natural map $\mathfrak{g}_S^{0,r} \to \mathcal{N}_S^r \subset \Theta_\mathcal{A}$ identifies order zero gauge symmetries with (densities of) local vector fields that induce tangent vectors to the $\mathcal{D}$-space $\underline{\mathrm{Spec}}_\mathcal{D}(\mathcal{A}/\mathcal{I}_S)$ of solutions to the Euler-Lagrange equation. This explains the denomination of higher gauge symmetries for $\mathfrak{g}_S^r$.

We now define an important invariant of gauge theories, called the Batalin-Vilkovisky bigraded algebra. This will be used in next section on the derived covariant phase space.

DEFINITION 3.3.5. Let $\mathfrak{g}_S$ be a regular generating space of the Noether gauge symmetries. The bigraded $\mathcal{A}[\mathcal{D}]$-module

$$\mathcal{V}_{BV} := \begin{bmatrix} \mathfrak{g}_S[2] & \oplus & \Theta_\mathcal{A}^\ell[1] & \oplus & 0 \\ & & & & \oplus \\ & & & & {}^t\mathfrak{g}_S^\circ[-1] \end{bmatrix},$$

where ${}^t\mathfrak{g}_S^\circ$ is the vertical chain graded space associated to $\mathfrak{g}_S^\circ$, is called the module of *additional fields*. The completed bigraded symmetric algebra

$$\hat{\mathcal{A}}_{BV} := \widehat{\mathrm{Sym}}_{\mathcal{A}\text{-bigraded}}(\mathcal{V}_{BV})$$

is called the *completed Batalin-Vilkovisky algebra* of the given gauge theory. The corresponding symmetric algebra

$$\mathcal{A}_{BV} := \mathrm{Sym}_{\mathcal{A}\text{-bigraded}}(\mathcal{V}_{BV})$$

is called the *Batalin-Vilkovisky algebra.*

In practical situations, physicists usually think of ghosts and antifields as sections of an ordinary graded bundle on spacetime itself (and not only on jet space). This idea can be formalized by the following.

DEFINITION 3.3.6. Let $\mathfrak{g}$ be a regular generating space of Noether symmetries for $S \in h(\mathcal{A})$. Suppose that all the $\mathcal{A}[\mathcal{D}]$-modules $\mathfrak{g}^i$ and $\Theta_\mathcal{A}^\ell$ are locally free on M. A *Batalin-Vilkovisky bundle* is a bigraded vector bundle

$$E_{BV} \to C$$

with an isomorphism of $\mathcal{A}[\mathcal{D}]$-modules

$$\mathcal{A}[\mathcal{D}] \otimes_{\mathcal{O}_C} \mathcal{E}_{BV}^* \to \mathcal{V}_{BV},$$

where $\mathcal{E}_{BV}$ are the sections of $E_{BV} \to C$. The sections of the graded bundle $E_{BV} \to M$ are called the *fields-antifields variables* of the theory.

Recall that neither $C \to M$, nor $E_{BV} \to M$ are vector bundles in general. To illustrate the above general construction by a simple example, suppose that the action $S \in h(\mathcal{A})$ has no non-trivial Noether identities, meaning that for all $k \geq 1$, one has $H^k(\mathcal{A}_P) = 0$. In this case, one gets

$$\mathcal{V}_{BV} = \Theta^\ell_{\mathcal{A}}[1]$$

and the relative cotangent bundle $E_{BV} := T^*_{C/M} \to C$ gives a BV bundle because

$$\Theta^\ell_{\mathcal{A}} \cong \mathcal{A}[\mathcal{D}] \otimes_{\mathcal{O}_C} \Theta_{C/M}.$$

The situation simplifies further if $C \to M$ is a vector bundle because then, the vertical bundle $VC \subset TC \to M$, given by the kernel of $TC \to \pi^*TM$, is isomorphic to $C \to M$. Since one has $T^*_{C/M} \cong (VC)^*$, one gets a natural isomorphism

$$E_{BV} \cong C \oplus C^*$$

of bundles over M. This linear situation is usually used as a starting point for the definition of a BV theory (see for example Costello's book [**Cos10**]). Starting from a non-linear bundle $C \to M$, one can linearize the situation by working with the bundle

$$C^{linear}_{x_0} := x_0^* T_{C/M} \to M$$

with $x_0 : M \to C$ a given solution of the equations of motion (sometimes called the vacuum).

PROPOSITION 3.3.2. *Let $E_{BV} \to C$ be a BV bundle. There is a natural isomorphism of bigraded algebras*

$$\mathrm{Jet}(\mathcal{O}_{E_{BV}}) \xrightarrow{\sim} \mathcal{A}_{BV} = \mathrm{Sym}_{bigrad}(\mathcal{V}_{BV}).$$

PROOF. Since $E_{BV} \to C$ is a graded vector bundle concentrated in non-zero degrees, one has

$$\mathcal{O}_{E_{BV}} = \mathrm{Sym}_{\mathcal{O}_C}(\mathcal{E}^*_{BV}).$$

The natural map

$$\mathcal{E}^*_{BV} \to \mathcal{V}_{BV}$$

induces a morphism

$$\mathcal{O}_{E_{BV}} = \mathrm{Sym}_{\mathcal{O}_C}(\mathcal{E}^*_{BV}) \to \mathcal{A}_{BV}.$$

Since $\mathcal{A}_{BV}$ is a $\mathcal{D}$-algebra, one gets a natural morphism

$$\mathrm{Jet}(\mathcal{O}_{E_{BV}}) \to \mathcal{A}_{BV} = \mathrm{Sym}_{bigrad}(\mathcal{V}_{BV}).$$

Conversely, the natural map $\mathcal{E}^*_{BV} \to \mathcal{O}_{E_{BV}}$ extends to an $\mathcal{A}[\mathcal{D}]$-linear map

$$\mathcal{V}_{BV} \to \mathrm{Jet}(\mathcal{O}_{E_{BV}}),$$

that gives a morphism

$$\mathcal{A}_{BV} \to \mathrm{Jet}(\mathcal{O}_{E_{BV}}).$$

The two constructed maps are inverse of each other. □

The main interest of the datum of a BV bundle is that it allows to work with non-local functionals of the fields and antifields variables. This is important for the effective renormalization of gauge theories, that involves non-local functionals.

DEFINITION 3.3.7. Let $E_{BV} \to C$ be a BV bundle. Denote $\mathbb{A}^1(A) := A$ the graded affine space. The space of *non-local functionals of the fields-antifields* is defined by

$$\mathcal{O}_{BV} := \underline{\mathrm{Hom}}(\underline{\Gamma}(M, E_{BV}), \mathbb{A}^1)$$

of (non-local) functionals on the space of sections of E_{BV}. The image of the natural map

$$h(\mathcal{A}_{BV}) \cong h(\mathrm{Jet}(\mathcal{O}_{E_{BV}})) \longrightarrow \mathcal{O}_{BV}$$

is called the space of *local functionals of the fields-antifields* and denoted $\mathcal{O}_{BV}^{qloc} \subset \mathcal{O}_{BV}$.

3.4. The derived covariant phase space. In all this section, we set $\pi : C \to M$, $H \subset \underline{\Gamma}(M, C)$, $\mathcal{A} = \mathrm{Jet}(\mathcal{O}_C)$ and $S \in h(\mathcal{A})$ a gauge theory. Suppose given a strongly regular generating space of Noether symmetries $\mathfrak{g}_S$ for S, in the sense of definitions 3.3.3 and 3.3.4.

The idea of the BV formalism is to define a (local and odd) Poisson dg-$\mathcal{A}$-algebra $(\mathcal{A}_{BV}, D, \{.,.\})$ whose spectrum $\mathbb{R}\underline{\mathrm{Spec}}_{\mathcal{D}}(\mathcal{A}_{BV}, D)$ can be though as a kind of homotopical space of leaves

$$\mathbb{R}\mathrm{Spec}(\mathcal{A}/\mathcal{I}_S)/\underset{\mathbb{R}}{} \mathcal{N}_S^r$$

of the foliation induced by the action (described in corollary 3.3.1) of Noether gauge symmetries $\mathcal{N}_S^r$ on the derived critical space $\mathbb{R}\underline{\mathrm{Spec}}_{\mathcal{D}}(\mathcal{A}/\mathcal{I}_S)$. It is naturally equipped with a homotopical Poisson structure, which gives a nice starting point for quantization. From this point of view, the above space is a wide generalization of the notion extensively used by DeWitt in his covariant approach to quantum field theory [**DeW03**] called the covariant phase space. This explains the title of this section.

We will first define the BV Poisson dg-algebra by using only a generating space for Noether identities, and explain in more details in the next section how this relates to the above intuitive statement.

PROPOSITION 3.4.1. *The local Lie bracket and local duality pairings*

$$[.,.] : \Theta_{\mathcal{A}} \boxtimes \Theta_{\mathcal{A}} \to \Delta_* \Theta_{\mathcal{A}} \quad \textit{and} \quad \langle .,. \rangle : (\mathfrak{g}_S^n)^r \boxtimes (\mathfrak{g}_S^{n\circ})^r \to \Delta_* \mathcal{A}^r,\ n \geq 0,$$

induce an odd local Poisson bracket

$$\{.,.\} : \hat{\mathcal{A}}_{BV}^r \boxtimes \hat{\mathcal{A}}_{BV}^r \to \Delta_* \hat{\mathcal{A}}_{BV}^r$$

called the BV-antibracket on the completed BV algebra

$$\hat{\mathcal{A}}_{BV} = \widehat{\mathrm{Sym}}_{bigrad}\left(\begin{bmatrix} \mathfrak{g}_S[2] & \oplus & \Theta_{\mathcal{A}}^{\ell}[1] & \oplus & \mathcal{A} \\ & & & & \oplus \\ & & & & {}^t\mathfrak{g}_S^{\circ}[-1] \end{bmatrix}\right)$$

and on the BV algebra $\mathcal{A}_{BV}$.

DEFINITION 3.4.1. Let $\mathfrak{g}_S$ be a regular generating space of Noether identities. A *formal solution to the classical master equation* is an $S_{cm} \in h(\hat{\mathcal{A}}_{BV})$ such that

(1) the degree $(0,0)$ component of S_{cm} is S,
(2) the component of S_{cm} that is linear in the ghost variables, denoted S_{KT}, induces the Koszul-Tate differential $d_{KT} = \{S_{KT}, .\}$ on antifields of degrees $(k, 0)$, and

(3) the *classical master equation*

$$\{S_{cm}, S_{cm}\} = 0$$

(meaning $D^2 = 0$ for $D = \{S_{cm}, .\}$) is fulfilled in $h(\hat{\mathcal{A}}_{BV})$.

A *solution to the classical master equation* is a formal solution that comes from an element in $h(\mathcal{A}_{BV})$.

The main theorem of homological perturbation theory, given in a physical language in Henneaux-Teitelboim [**HT92**], Chapter 17 (DeWitt indices), can be formulated in our language by the following.

THEOREM 3.4.1. *Let $\mathfrak{g}_S$ be a regular generating space of Noether symmetries. There exists a formal solution to the corresponding classical master equation, constructed through an inductive method. If $\mathfrak{g}_S$ is further strongly regular and the inductive method ends after finitely many steps, then there exists a solution to the classical master equation.*

PROOF. One can attack this theorem conceptually using the general setting of homotopy transfer for curved local L_∞-algebroids (see Schaetz's paper [**Sch09**] for a finite dimensional analog). We only need to prove the theorem when $\mathfrak{g}$ has all $\mathfrak{g}^i$ given by free $\mathcal{A}[\mathcal{D}]$-modules of finite rank since this is true locally on M. We start by extending S to a generator of the Koszul-Tate differential $d_{KT} : \mathcal{A}_{KT} \to \mathcal{A}_{KT}$. Remark that the BV bracket with S on $\mathcal{A}_{BV}$ already identifies with the insertion map

$$\{S, .\} = i_{dS} : \Theta^\ell_{\mathcal{A}} \to \mathcal{A}.$$

We want to define $S_{KT} := \sum_{k \geq 0} S_k$ with $S_0 = S$ such that

$$\{S_{KT}, .\} = d_{KT} : \mathcal{A}_{KT} \to \mathcal{A}_{KT}.$$

Let $C^*_{\alpha_i}$ be generators of the free $\mathcal{A}[\mathcal{D}]$-modules $\mathfrak{g}^i$ and C^{α_i} be the dual generators of the free $\mathcal{A}[\mathcal{D}]$-modules $(\mathfrak{g}^i)^\circ$. We suppose further that all these generators correspond to closed elements for the de Rham differential. Let $n_{\alpha_i} := d_{KT}(C^*_{\alpha_i})$ in $\mathcal{A}_{KT}$. Then setting $S_k = \sum_{\alpha_k} n_{\alpha_k} C^{\alpha_k}$, one gets

$$\begin{array}{rcl} \{S_i, C^*_{\alpha_i}\} & = & \{n_{\alpha_i} C^{\alpha_i}, C^*_{\alpha_i}\} \\ & = & n_{\alpha_i} \\ & = & d_{KT}(C^*_{\alpha_i}) \end{array}$$

so that $\{S_{KT}, .\}$ identifies with d_{KT} on $\mathcal{A}_{KT}$. Now let m_{α_j} denote the coordinates of n_{α_i} in the basis $C^*_{\alpha_i}$, so that

$$n_{\alpha_i} = \sum_j m_{\alpha_j} C^*_{\alpha_j}.$$

One gets in these coordinates

$$S_i = \sum_{\alpha_i, \alpha_j} C^*_{\alpha_j} m_{\alpha_j} C^{\alpha_i}.$$

The next terms in $S = \sum_{k \geq 0} S_k$ are determined by the recursive equation

$$2d_{KT}(S_k) + D_{k-1} = 0$$

where D_{k-1} is the component of Koszul-Tate degree (i.e., degree in the variables $C^*_{\alpha_i}$) $k-1$ in $\{R_{k-1}, R_{k-1}\}$, with

$$R_{k-1} = \sum_{j \leq k-1} S_j.$$

These equations have a solution because D_{k-1} is d_{KT}-closed, because of Jacobi's identity for the odd bracket $\{.,.\}$ and since d_{KT} is exact on the Koszul-Tate components (because it gives, by definition, a resolution of the critical ideal), these are also exact. If we suppose that the generating space $\mathfrak{g}_S$ is strongly regular (i.e., bounded) and the inductive process ends after finitely many steps, one can choose the solution S in $h(\mathcal{A}_{BV})$. □

Acknowledgements

We refer to the article [**Pau10**] for detailed acknowledgements and more references on this work, that is its direct continuation (with improvements and simplifications). Special thanks are due to Jim Stasheff for his detailed comments of loc. cit., to the referee, and to my students at IMPA and Jussieu's master class, that allowed me to improve both presentation and results. I also thank the editors, and in particular Urs Schreiber, for giving me the opportunity to publish in this book.

References

[Bar10] G. Barnich. A note on gauge systems from the point of view of lie algebroids. *ArXiv e-prints*, oct 2010.

[BBH95] Glenn Barnich, Friedemann Brandt, and Marc Henneaux. Local BRST cohomology in the antifield formalism. I. General theorems. *Comm. Math. Phys.*, 174(1):57–91, 1995.

[BD04] Alexander Beilinson and Vladimir Drinfeld. *Chiral algebras*, volume 51 of *American Mathematical Society Colloquium Publications*. American Mathematical Society, Providence, RI, 2004.

[CF01] Alberto S. Cattaneo and Giovanni Felder. Poisson sigma models and deformation quantization. *Modern Phys. Lett. A*, 16(4-6):179–189, 2001. Euroconference on Brane New World and Noncommutative Geometry (Torino, 2000).

[Cos10] Kevin Costello. Renormalization and effective field theory, 2010.

[DeW03] Bryce DeWitt. *The global approach to quantum field theory. Vol. 1, 2*, volume 114 of *International Series of Monographs on Physics*. The Clarendon Press Oxford University Press, New York, 2003.

[DM99] Pierre Deligne and John W. Morgan. Notes on supersymmetry (following Joseph Bernstein). In *Quantum fields and strings: a course for mathematicians, Vol. 1, 2 (Princeton, NJ, 1996/1997)*, pages 41–97. Amer. Math. Soc., Providence, RI, 1999.

[FH90] Jean M. L. Fisch and Marc Henneaux. Homological perturbation theory and the algebraic structure of the antifield-antibracket formalism for gauge theories. *Comm. Math. Phys.*, 128(3):627–640, 1990.

[FLS02] Ron Fulp, Tom Lada, and Jim Stasheff. Noether's variational theorem ii and the bv formalism. *arXiv*, 2002.

[HT92] Marc Henneaux and Claudio Teitelboim. *Quantization of gauge systems*. Princeton University Press, Princeton, NJ, 1992.

[Ill71] Luc Illusie. *Complexe cotangent et déformations. I*. Lecture Notes in Mathematics, Vol. 239. Springer-Verlag, Berlin, 1971.

[Kas03] Masaki Kashiwara. *D-modules and microlocal calculus*, volume 217 of *Translations of Mathematical Monographs*. American Mathematical Society, Providence, RI, 2003. Translated from the 2000 Japanese original by Mutsumi Saito, Iwanami Series in Modern Mathematics.

[KV98] Joseph Krasil'shchik and Alexander Verbovetsky. Homological methods in equations of mathematical physics. *arXiv*, 1998.

[Lan93] Serge Lang. *Algebra*. Addison-Wesley, New York, third edition, 1993.

[Man97] Yuri I. Manin. *Gauge field theory and complex geometry*, volume 289 of *Grundlehren der Mathematischen Wissenschaften [Fundamental Principles of Mathematical Sciences]*. Springer-Verlag, Berlin, second edition, 1997. Translated from the 1984 Russian original by N. Koblitz and J. R. King, With an appendix by Sergei Merkulov.

[MR91] Ieke Moerdijk and Gonzalo E. Reyes. *Models for smooth infinitesimal analysis*. Springer-Verlag, New York, 1991.

[Pau10] F. Paugam. Histories and observables in covariant field theory. *Journal of geometry and physics*, 2010.

[Pau11] F. Paugam. Towards the mathematics of quantum field theory, upmc/impa master course notes (book in preparation). *http://people.math.jussieu.fr/~fpaugam/*, 2011.

[Pei52] R. E. Peierls. The commutation laws of relativistic field theory. *Proc. Roy. Soc. London. Ser. A.*, 214:143–157, 1952.

[Pen83] I. B. Penkov. $\mathcal{D}$-modules on supermanifolds. *Invent. Math.*, 71(3):501–512, 1983.

[Sch94] Jean-Pierre Schneiders. An introduction to $\mathcal{D}$-modules. *Bull. Soc. Roy. Sci. Liège*, 63(3-4):223–295, 1994. Algebraic Analysis Meeting (Liège, 1993).

[Sch09] Florian Schätz. BFV-complex and higher homotopy structures. *Comm. Math. Phys.*, 286(2):399–443, 2009.

[Sch10] Pierre Schapira. Triangulated categories for the analysis. In *Triangulated categories*, volume 375 of *London Math. Soc. Lecture Note Ser.*, pages 371–388. Cambridge Univ. Press, Cambridge, 2010.

[Sta97] Jim Stasheff. Deformation theory and the Batalin-Vilkovisky master equation. In *Deformation theory and symplectic geometry (Ascona, 1996)*, volume 20 of *Math. Phys. Stud.*, pages 271–284. Kluwer Acad. Publ., Dordrecht, 1997.

[Sta98] Jim Stasheff. The (secret?) homological algebra of the Batalin-Vilkovisky approach. In *Secondary calculus and cohomological physics (Moscow, 1997)*, volume 219 of *Contemp. Math.*, pages 195–210. Amer. Math. Soc., Providence, RI, 1998.

[Tat57] John Tate. Homology of Noetherian rings and local rings. *Illinois J. Math.*, 1:14–27, 1957.

[TV08] Bertrand Toën and Gabriele Vezzosi. Homotopical algebraic geometry. II. Geometric stacks and applications. *Mem. Amer. Math. Soc.*, 193(902):x+224, 2008.

[Vin01] A. M. Vinogradov. *Cohomological analysis of partial differential equations and secondary calculus*, volume 204 of *Translations of Mathematical Monographs*. American Mathematical Society, Providence, RI, 2001. Translated from the Russian manuscript by Joseph Krasil′shchik.

Institut de Mathématiques de Jussieu, Université Pierre et Marie Curie, 75005 Paris
E-mail address: `frederic.paugam@upmc.fr`

Proceedings of Symposia in Pure Mathematics
Volume **83**, 2011

Orientifold Précis

Jacques Distler, Daniel S. Freed, and Gregory W. Moore

Abstract. We give a precise and concise formulation of the orientifold construction in Type II superstring theory. Our results include anomaly cancellation on the worldsheet and a K-theoretic computation of the background Ramond-Ramond charge.

Since its inception [**Sa, PS, Ho, DLP, BiS**] the orientifold construction of Type II superstring theory has proven useful both for the formal development of string theory and for its potential applications to phenomenology. For reviews see [**P, D, AnS, BH1, J, BH2, BHHW**]. Our work began many years ago when the first author showed the second the formula [**DJM, CR, MSS, S, SS**] for the Ramond-Ramond charge induced by an orientifold and inquired about its K-theoretic significance, especially in view of the unusual cousin of the Hirzebruch L-genus contained therein. In the process of our investigations we were led to more foundational questions about orientifolds and the Type II superstring. In this paper we describe some mathematical foundations we have developed to resolve these questions. In particular, we give careful definitions of both the worldsheet and spacetime fields of the Type II superstring, including orientifolds, with precise Dirac quantization conditions for the B-field and Ramond-Ramond (RR) field. The (Neveu-Schwarz)^2 =NSNS fields which appear in Definition 2 involve subtle topological structures which, for very different reasons, exactly fit what is needed in both the (short distance) worldsheet and (long distance) spacetime theories. The worldsheet fields are enumerated in Definition 4. A surprising challenge here is to define the integral of the B-field over the worldsheet. The lack of a proper

2010 *Mathematics Subject Classification.* 81T30, 83E30, 19L50.
The work of J.D. is supported by the National Science Foundation under grant PHY-0455649.
The work of D.S.F. is supported by the National Science Foundation under grant DMS-0603964.
The work of G.W.M. is supported by the DOE under grant DE-FG02-96ER40949.
We also thank the Aspen Center for Physics for providing a stimulating environment for many discussions related to this paper.
Report numbers: UTTG-04-09, TCC-019-09.

orientation leads to a novel prescription, one feature of which is that the B-field amplitude has a (classical) anomaly in the sense that it takes values in a complex line not canonically isomorphic to the complex numbers. It then cancels against the more standard (quantum) anomaly from the spinor field on the worldsheet, and the cancellation uses the "twisted" spin structure in spacetime.[1] The spacetime RR field is self-dual, so part of its formulation (Definition 6) involves a certain quadratic function. The general theory of self-dual fields defines an RR charge due to the orientifold background, which we compute with the prime 2 inverted, in which case it localizes to the orientifold fixed point set. Tensoring with the reals we recover the formula (8) which began this project.

Our definitions are new, even in the case of the Type I superstring, and offer some refinements of the standard (non-orientifold) Type II superstring. Taken together the orientifold data is an impressively tight structure, and leads to the most intricate matching we know between topological features in a short distance theory and its long distance approximation.

In recent years new abelian objects have entered differential geometry. Most familiar are "gerbes with connection" in various incarnations. Often these objects have cohomological significance and can be studied as part of generalized differential cohomology and twisted versions thereof. The abelian gauge fields in this paper are examples of such objects. Conversely, these geometrical constructs are precisely what we need to formulate orientifolds. The foundations of *equivariant* generalized differential cohomology have yet to be fully developed; see [**SV, O, BSc1**] for recent accounts of equivariant differential complex K-theory. On the other hand, our results about worldsheet anomalies and the RR background charge are purely topological and do not require these missing foundations.

We are writing longer accounts (e.g. [**DFM**]) which will on the one hand include detailed definitions, statements, and proofs; and on the other explain the results in a manner more accessible to physicists, including the relationship to previous approaches to orientifolds and several physical consequences. One of the virtues of Definition 2 is that it tames the zoo of orientifolds into a small set of data; in subsequent papers we will unpack this definition to recreate the zoo. Also, our formulation of the B-field allows us to state new consistency conditions for D-branes in orientifolds, and suggests the existence of new NS-branes with torsion charges. We have not investigated anomalies in the long distance spacetime theory, but have provided a framework in which to investigate them.

Previous work on K-theoretic interpretation of RR charge in orientifolds includes [**W2, G, Hor, BGH, OS, dBHKMMS, AH, GL, BGS, BS**]. The

[1] The twisted spin structure appears in Definition 2.

twisted KR-theory used here unifies all the various forms of K-theory in these papers.

We thank Michael Hopkins for topological consultations. We also thank Ilka Brunner, Frederik Denef, Emanuel Diaconescu, David Morrison, Graeme Segal, Constantin Teleman, and Edward Witten for useful discussions.

NSNS superstring backgrounds

We begin with some general concepts.

A *smooth orbifold* of dimension n is a space locally modeled on the quotient of $\mathbb{R}^n$ by the linear action of a finite group, and it has a smooth structure defined similarly to that of a smooth manifold; see [**ALR**] for a recent exposition. (Smooth manifolds are smooth orbifolds.) An important example to keep in mind is the case of a global quotient, where a finite group Γ acts on a smooth manifold Y; the quotient smooth orbifold is denoted $Y/\!/\Gamma$. Orbifolds are presented as groupoids, which may or may not be global quotients. A smooth orbifold can also be termed a 'smooth real Deligne-Mumford stack', and this leads to a more invariant description, but as that s-word conjures up demons, we avoid it here. A smooth orbifold has a geometric realization, or classifying space, which in the case of a global quotient $Y/\!/\Gamma$ is known as the Borel construction. We define the ordinary cohomology of an orbifold to be that of the geometric realization.

We make use of generalized cohomology theories [**A**], such as periodic KO-theory and its connective cover[2] ko, and also employ twisted versions. Twistings of a cohomology theory, E, are geometric objects. The most familiar example is a local system of abelian groups, which twists ordinary cohomology. Equivalence classes, of twistings of E, can be located in a cohomology theory, at least if E is suitably multiplicative (see [**ABGHR**] for a precise account). The cohomological degree is a particular type of twisting. For example, on an orbifold X the twistings of K-theory of interest here [**FHT**] are classified by the *set*

$$H^0(X;\mathbb{Z}/2\mathbb{Z}) \times H^1(X;\mathbb{Z}/2\mathbb{Z}) \times H^3(X;\mathbb{Z}). \tag{1}$$

Let R be the Postnikov section[3] $ko\langle 0\cdots 4\rangle$ of ko; it has homotopy groups concentrated in degrees $0,1,2$ and 4. The twistings of K-theory we use are classified by

[2]This is the real version of K-theory as developed in [**At**], before inverting the Bott element; see [**Se**] for another construction. For any space M the abelian groups $ko^q(M)$ vanish for $q>0$ and $ko^{-q}(M)\cong KO^{-q}(M)$ for $q\geqslant 0$.

[3]The 0-space of the spectrum R is constructed from the 0-space of the spectrum ko by attaching cells to kill all homotopy groups π_q for $q\geqslant 5$; see [**BT**, p. 250] for an exposition of this procedure. More precisely, we use the version of Postnikov truncation for connective E^∞-ring spectra [**Ba**] to construct R from ko.

the *abelian group* $R^{-1}(X)$, which as a set is isomorphic to (1). It is crucial for us that R is a *multiplicative* cohomology theory, more precisely an E^{∞} ring: we can multiply and integrate R-cohomology classes. Twistings and orientations of ko (or of the more familiar periodic version KO) induce twistings and orientations of R.

Abelian gauge fields and their associated currents are geometric objects which live in *differential* cohomology theories [**F1, F2, FMS2**]. The differential theory associated to a cohomology theory h is denoted $\check{h}$. A systematic development of some foundations is given in [**HS**]. A differential cohomology group is the set of equivalence classes in a groupoid.[4] For example, objects in degree two differential cohomology may be taken to be circle bundles with connection; the differential cohomology *group* $\check{H}^2(X)$ is the set of equivalence classes. We use the notation 'ob' below to emphasize that the fields are geometric objects in a groupoid, not equivalence classes in a group. There are different models for a given differential cohomology theory, and we needn't commit ourselves to a particular model. Twistings of differential theories are differential twists of the underlying topological theory. There is not yet a general theory of twisted differential cohomology theory in the literature, but see [**KV**] and [**BSc2**], §7.4.

With these preliminaries we can define the (Neveu-Schwarz)^2 sector of an orientifold background.

DEFINITION 2. *An* NSNS superstring background *consists of:*

(i) *a 10-dimensional smooth orbifold X together with Riemannian metric and real-valued function (the dilaton field);*
(ii) *a double cover $\pi\colon X_w \to X$;*
(iii) *a differential twisting $\check{\beta}$ of $\widecheck{KR}(X_w)$, the* B-field*;*
(iv) *and a* twisted spin structure $\kappa\colon \Re(\beta) \to \tau^{KO}(TX-10)$.

We call $\pi\colon X_w \to X$ the *orientifold double cover.* The symbol 'w' is used to denote the orientifold double cover as well as its equivalence class in $H^1(X;\mathbb{Z}/2\mathbb{Z})$. In the case of a global quotient $X = Y/\!/\Gamma$, the double cover may be specified by an index two subgroup $\Gamma_0 \subset \Gamma$: then $X_w = Y/\!/\Gamma_0$. In physics, one speaks of "the orientifold of the spacetime Y by the action of Γ." Our definition includes non-orientifold theories, by positing a trivialization of $X_w \to X$. Among those are the orbifold theories, where X is a global quotient. Underlying the B-field $\check{\beta}$ is a twisting β of $KR(X_w)$ whose equivalence class lies in $R^{w-1}(X)$, which is the twisted version of $R^{-1}(X)$ induced by the orientifold double cover. As a set $R^{w-1}(X)$ is isomorphic to the product of cohomology groups (1), but with the last factor replaced by $H^3(X;\mathbb{Z}_w)$ for $\mathbb{Z}_w$ the local coefficient system associated to $X_w \to X$. The "curvature" of $\check{\beta}$ is a closed twisted 3-form with integral periods, the 3-form field

[4] To obtain *local* objects, we need to consider higher groupoids, or spaces.

strength of the B-field. It lifts to an ordinary 3-form on X_w which is odd under the orientifold involution on X_w. What is new in our definition of the B-field, even in non-orientifold theories, is the inclusion of the H^0 and H^1 components in (1). The twisted spin structure in (iv) is an isomorphism of twistings of $KO(X)$: $\Re(\beta)$ is a lift of[5] $\bar{\beta}+\beta$ to a KO-twisting and $\tau^{KO}(V)$ denotes the KO-twisting determined by the real vector bundle V. In more down-to-earth terms, an ordinary spin structure on an n-dimensional Riemannian manifold (or orbifold) is a reduction of the principal O_n-bundle of orthonormal frames to a Spin_n-bundle. Note that Spin_n is a double cover of an index two subgroup of O_n. The twisted spin structure in (iv) is a similar reduction, but the double cover of the index two subgroup depends on the topological object β underlying the B-field; see [**DFM**] §6 for details. A twisted spin structure is a *discrete field* in the long distance supergravity theory; it has no differential form field strength.

The B-field is classified topologically by $R^{-1}(X)$ as in (1). The H^0 component t distinguishes between the usual Type B ($t=0$) and Type A ($t=1$). Its inclusion in the B-field incorporates the signs in the sum over spin structures of the worldsheet Type II theory [**SW, AgMV**] as part of the B-field amplitude (5), as we explain in [**DFM**] §2. Note that since $R^{-1}(X)$ is a group there is a distinguished zero, thus singling out the IIB superstring as "more fundamental." The H^1 component a is a further twisting of a Type II theory. We remark that the existence of a twisted spin structure implies the following relations between the Stiefel-Whitney classes of spacetime X and the B-field:

$$\begin{aligned} w_1(X) &= tw \\ w_2(X) &= tw^2 + aw. \end{aligned} \tag{3}$$

The Type I superstring on a spin 10-manifold Y is a special case of Definition 2. Then $X_w = Y$ with trivial involution and quotient X is a "non-effective" orbifold, presented as the groupoid $Y \times \mathrm{pt} /\!/ (\mathbb{Z}/2\mathbb{Z})$. The B-field reduces to a discrete field in $\mathrm{ob}\, H^2(Y;\mathbb{Z}/2\mathbb{Z})$—in particular, both t and a vanish—and the twisted spin structure is an ordinary spin structure on Y. The RR fields may be modeled as principal Spin_{32}-bundles "without vector structure" [**LMST, W3**]. As another special case, when $X_w = \mathbb{E}^{10}$ is flat Euclidean space and X the quotient by a reflection in $9-p$ directions, then the resulting constraint among t, a, and p reproduces the standard list of consistent orientifold projections [**P**].

[5] $\bar{\beta}$ is the complex conjugate of β, another KR-twisting.

Worldsheet theory

We now define a worldsheet in a given NSNS background (in the NSR formalism). Curiously, while low genus surfaces have been extensively investigated in the physics literature, a formulation valid for general worldsheets does not seem to be available, even for the Type I superstring.

DEFINITION 4. *Fix an NSNS superstring background as in Definition 2. Then a* worldsheet *consists of*

(i) *a compact smooth 2-manifold Σ (possibly with boundary) with Riemannian structure;*
(ii) *a spin structure α on the orientation double cover $\hat{\pi}\colon \widehat{\Sigma} \to \Sigma$ whose underlying orientation is that of $\widehat{\Sigma}$;*
(iii) *a smooth map $\phi\colon \Sigma \to X$;*
(iv) *an isomorphism $\phi^*w \to \hat{w}$, or equivalently a lift of ϕ to an equivariant map $\widehat{\Sigma} \to X_w$;*
(v) *a positive chirality spinor field ψ on $\widehat{\Sigma}$ with coefficients in $\hat{\pi}^*\phi^*(TX)$;*
(vi) *and a negative chirality spinor field χ on $\widehat{\Sigma}$ with coefficients in $T^*\widehat{\Sigma}$ (the gravitino).*

The orientation double cover $\widehat{\Sigma}$ carries a canonical orientation, used in (ii), whereas no orientation is assumed on Σ, which indeed may be nonorientable. We use '$\hat{w}$' to denote the orientation double cover of Σ. The spin structure α, which is a discrete field on Σ, is locally on Σ a choice of two spin structures with opposite orientations. The isomorphism in (iv) is also a discrete field on Σ. In case $X = Y/\!/\Gamma$ is a global quotient with double cover $X_w = Y/\!/\Gamma_0$, a map (iii) is given by a principal Γ-bundle $P \to \Sigma$ and a Γ-equivariant map $P \to Y$. Furthermore, P is oriented, elements of Γ_0 preserve the orientation, and elements of $\Gamma\backslash\Gamma_0$ reverse it. The spinor fields in (v) and (vi) use the spin structure α; the chirality refers to the canonical orientation of $\widehat{\Sigma}$.

Assume the boundary of Σ is empty. The exponentiated Euclidean action, after integrating out ψ and χ, has two factors on which we focus:

$$\exp\Bigl(2\pi i\int_\Sigma \check{\zeta}\cdot\phi^*\check{\beta}\Bigr)\cdot \operatorname{pfaff} D_{\widehat{\Sigma},\alpha}\bigl(\hat{\pi}^*\phi^*(TX) - T\widehat{\Sigma}\bigr), \tag{5}$$

the B-field amplitude and a Pfaffian. In the first factor the pullback $\phi^*\check{\beta}$ is an object in $\check{R}^{\hat{w}-1}(\Sigma)$, because of (iv) in Definition 4. Sadly, Σ is not endowed with an R-orientation[6] and so we have introduced a new object $\check{\zeta}$ in twisted differential

[6] An R-orientation is the same as an $\check{R}$-orientation. Because of the $\hat{w}$ twisting in the integrand we do not quite need an R-orientation, which is a spin structure on Σ, but rather a pin structure. Still, we do not have one.

R-theory in order to define the integral. We do not give here a precise definition, but remark that one ingredient is a pushforward of α to Σ, which measures the obstruction to refining α to a pin structure. Of utmost importance is that the first factor in (5) is not a number but rather an element in a certain Hermitian "B-line" L_B: the B-field amplitude is anomalous. Let S be a parameter space of non-fermionic worldsheet fields (Definition 4(i)–(iv)). In string theory one integrates over worldsheets, and therefore the exponentiated effective action after integrating out the fermions should be a *measure* on S.[7] The measure turns out to be a product of a measure which is manifestly well-defined on S and (5), whence the latter should be a function on S. The Pfaffian of the Dirac operator on $\widehat{\Sigma}$ has its usual anomaly: it takes values in a Hermitian "Pfaffian line" L_ψ. In a parametrized family both L_B and L_ψ are *flat* line bundles over S: these are global anomalies. The crucial result is that the tensor product $L_B \otimes L_\psi \to S$ has a *universal* trivialization, and so defines (5) as a function on S. This trivialization uses the twisted spin structure on spacetime X (see Definition 2(iv)); it is the manner in which that twisted spin structure enters the Lagrangian worldsheet theory. This anomaly cancellation—more precisely, this "setting of the quantum integrand" [**FM**]—is the most subtle example of its kind that we've seen. It applies as well in the non-orientifold case: then the B-field amplitude is a function, as the worldsheet carries a spin structure, and the trivialization of the Pfaffian line bundle $L_\psi \to S$ depends on the spin structure on spacetime.

We do not attempt to couple the worldsheet to the RR field or fermions on spacetime.

The RR field on spacetime

The RR field is self-dual, so we begin with some general remarks about self-dual fields. First, the cohomology theory used to quantize the self-dual charges and fluxes must itself be Pontrjagin self-dual [**FMS1**, Appendix B]. Furthermore, part of the definition of a self-dual field is a quadratic refinement of the pairing between electric and magnetic currents: this refinement is a topological datum. This quadratic function has a well-defined center of symmetry μ. We interpret $-\mu$ as the self-dual charge induced by the background. If μ is nonzero and X is compact, then there must be additional charged objects—D-branes—in the theory whose total charge j_{ext} equals μ. These charges have differential refinements[8]—the currents—and the self-dual gauge field is a (nonflat) isomorphism $\check{\mu} \to \check{j}_{\mathrm{ext}}$. See [**W1, F1, F2, HS, FMS1, BM**] for background on self-dual fields.

[7] In this paper we ignore all subtleties associated with the measure for supermoduli. We believe these are unrelated to the questions we address here.

[8] The lift of μ to a current may involve an additional topological choice.

Fix an NSNS superstring background. The RR charges and fluxes are quantized by a twisted form of periodic K-theory on X: the KR theory on X_w with its involution.[9] As X_w is an orbifold, not a manifold, we must specify what we mean by its KR-theory. Here we do not use the geometric realization or Borel construction, but rather use a geometric model which generalizes the equivariant vector bundles of Atiyah-Segal equivariant K-theory. (The analogous model for complex K-theory appears in [**FHT**, §3].) The quadratic function is a single topological choice which induces quadratic functions on families of manifolds of dimension $\leqslant 12$. Its simplest manifestation is integer-valued and occurs on a 12-dimensional orbifold M which has a double cover M_w, a B-field $\check{\beta}$, and a twisted spin structure κ.[10] One way to think of M is as a smooth 2-parameter family of 10-dimensional NSNS backgrounds.

DEFINITION 6. *Fix an NSNS superstring background as in Definition* 2. *Then*

(i) *an* RR current *is an object in* $\widetilde{KR}^{\check{\beta}}(X_w)$*;*

(ii) *the required quadratic function on a 12-manifold M is the composition*

(7)
$$\begin{array}{ccccccccc} KR^{\beta}(M_w) & \longrightarrow & KO^{\Re(\beta)}_{\mathbb{Z}/2\mathbb{Z}}(M_w) & \cong & KO^{\tau^{KO}(TM-4)}_{\mathbb{Z}/2\mathbb{Z}}(M_w) & \longrightarrow & KO^{-4}_{\mathbb{Z}/2\mathbb{Z}}(\mathrm{pt}) & \longrightarrow & \mathbb{Z} \\ j & \longrightarrow & & \kappa\bar{j}j & & \longmapsto & \displaystyle\int_{M_w} \kappa\bar{j}j & \longmapsto & \end{array}$$

ϵ-component

Notice that the B-field $\check{\beta}$ is used in the definition of the RR current to twist differential KR-theory. The twisted spin structure enters into the definition of the quadratic function at the second stage, and we have used Bott periodicity to adjust the degree. At the last stage of the quadratic function we identify the quaternionic representation group $KO^{-4}_{\mathbb{Z}/2\mathbb{Z}}(\mathrm{pt})$ of $\mathbb{Z}/2\mathbb{Z}$ with $\mathbb{Z}\oplus\mathbb{Z}\epsilon$, where ϵ is the sign representation. Our notation is schematic; details will appear in a subsequent paper. The quadratic function (7) generalizes that for non-orientifold Type II [**W4, FH1, DMW**] and for Type I [**MW, F1**].

One of our main results is the computation of the center μ of the quadratic function (7)—which equals *minus* the RR charge of the orientifold background—but only after inverting 2. Assume X_w is a 10-*manifold* with involution σ. Let $i\colon F \hookrightarrow X_w$ be the fixed point set of the involution and $\nu \to F$ the normal bundle. We apply the localization theorem in $\mathbb{Z}/2\mathbb{Z}$-equivariant KO-theory [**AS**] to compute the image of μ in a certain localization of $KR^{\beta}(X)$. The formula generalizes that in [**FH2**] for the special case of the Type I superstring. There a KO-theory analog

[9]Objects in $KR(X_w)$ carry a lift of the involution on X_w, so descend to X and should be regarded as living on X. But there is no standard notation for $KR(X_w)$ as a twisted K-group on X.

[10]The twisted spin structure on M is an isomorphism $\kappa\colon \Re(\beta) \to \tau^{KO}(TM-12)$.

of the Wu class appears, and in the general story it appears in a twisted form. Let $r = \operatorname{codim}_{X_w}(F)$. The RR charge in (possibly twisted) rational cohomology, obtained as a normalized[11] Chern character, is

$$(8) \qquad -\sqrt{\hat{A}(X)}\,\mathrm{ch}(\mu) = \pm\, 2^{5-r} i_* \left(\sqrt{\frac{L'(F)}{L'(\nu)}}\right), \qquad L'(V) = \prod \frac{x/4u}{\tanh x/4u},$$

where the L'-genus of a real vector bundle V is expressed as usual in terms of formal degree two classes, $\hat{A}$ is the A-hat genus, and u is the Bott generator of $K^2(\mathrm{pt})$. The L'-genus is reminiscent of Hirzebruch's L-genus, but the factors of 4 in the L'-genus are not seen in ordinary index theory. The sign in (8) (as well as omitted powers of the Bott elements in K and KR) depends on the twisting β. We remark that under the usual definition of $O^{\pm}$-planes the sign in (8) is $\pm$. We will give a precise formula for the sign, as well as a proof of (8) and its K-theory progenitor, in a subsequent paper.

Physical remarks

We conclude with some points where our work illuminates the physics.

(a) In the physics literature on global quotients one finds different species of "orientifold planes". Our work gives an intrinsic definition as well as global constraints on the distribution of orientifold planes. (Some examples of these constraints for toroidal orientifolds were investigated in [**dBHKMMS, BGS, GH**].)

(b) Our formula (8) for the RR charge induced by the orientifold is compatible with [**MSS, SS**] but not with formulæ in [**MS, HJ**].

(c) In the spacetime supergravity action, there are couplings between the spacetime fermions and the RR fields. Our definition of the twisted spin structure (for the fermions) and the twisting of differential KR-theory (for the RR fields) are such as to allow those couplings to be globally consistently defined in an orientifold background.

(d) The case $w = 0$ corresponds to Type II superstring theory, with no orientifold. In this case a twisted spin structure determines a spin structure on X. Moreover, the data (t, a), which correspond to a graded real line bundle on X, can be used to produce a second spin structure. One physical interpretation of this pair of spin structures is that they are the spin structures "seen" by the two gravitinos. (When $t = 0$ the two spin structures have the same underlying orientation, and when $t = 1$ they have opposite orientation.) Thus, introducing twistings with nonzero a incorporates and generalizes Scherk-Schwarz compactifications [**SSc1, SSc2, KM, H**].

---

[11] The normalization by $\sqrt{\hat{A}(X)}$ is discussed in [**GHM, CY, MM, F1**].

(e) For orbifolds which are global quotients $X = Y//\Gamma$ there is a subgroup of discrete B-fields classified by $H^3_\Gamma(\mathrm{pt};\mathbb{Z})$; these are known as "discrete torsion". There has been some controversy and confusion in the literature concerning the generalization to orientifolds. In our formulation the answer is clear: discrete torsion is classified by $R^{w-1}_\Gamma(\mathrm{pt})$. Whereas $H^3_\Gamma(\mathrm{pt};\mathbb{Z})$ classifies central extensions of Γ by the circle group $\mathbb{T}$, the group $R^{w-1}_\Gamma(\mathrm{pt})$ classifies non-central $\mathbb{Z}/2\mathbb{Z}$-*graded* extensions of Γ by $\mathbb{T}$, the action of Γ on $\mathbb{T}$ being determined by the orientifold. (See also [**BS**].)

(f) In the case that spacetime is a global quotient $Y//\Gamma$ there is a model for $KR^0(X_w)$ which makes contact with the tachyon field picture of K-theoretic charges [**W2**] and is a slight generalization of both standard equivariant K-theory and KR-theory. The Chan-Paton bundle of the unstable brane filling spacetime—an object in $KR^0(X_w)$—is a $\mathbb{Z}/2\mathbb{Z}$-graded complex vector bundle with the Γ action on Y lifted: elements of Γ_0 act $\mathbb{C}$-linearly and elements of $\Gamma - \Gamma_0$ act $\mathbb{C}$-*antilinearly.* The tachyon field is an odd endomorphism graded commuting with this Γ action.

(g) Suppose $W \subset X$ is the worldvolume of a D-brane. The open string field configurations on the D-brane include an object in differential KR-theory on W with twisting $\check{\tau}_W$. The induced RR current is computed by pushing forward under the inclusion $i\colon W \hookrightarrow X$, from which we deduce an important constraint relating the twisting class on W, the topology of W, and the B-field: namely, there must exist an isomorphism $\tau_W + \tau^{KR}(\nu) \cong i^*\beta$ in $R^{i^*w-1}(W)$, where $\nu \to W$ is the normal bundle. In the non-orientifold case this is the spacetime derivation of the anomaly derived in [**FW**] from the open string worldsheet. In case W is spin and coincides with an orientifold plane, we easily recover the standard rules for when a D-brane supports an orthogonal or symplectic gauge theory. There are many more possibilities in general. The need to view D-brane sources in orientifolds in a K-theoretic way in order to avoid paradoxes is central to the work of [**CDE**].

(h) Our work raises the possibility that there exist new solitonic objects in superstring theory. The magnetic NS current is an object in $\check{R}^{w+0}(X)$ while the electric NS current is an object in $\check{R}^{w+8}(X)$. If we consider NS-charged branes with constant charge density on their worldvolume, then the R-cohomology classification of NS currents recovers the fundamental string and the solitonic 5-brane. It also predicts the existence of an electrically charged particle with $\mathbb{Z}/2\mathbb{Z}$ torsion charge, as well as magnetically charged $\mathbb{Z}/2\mathbb{Z}$-torsion 7- and 8-branes. The 8-brane seems to be especially curious, being a domain wall between the type IIA and IIB theories.

(i) It would be interesting to reconcile our Dirac charge quantization condition for the B-field—which is different from that of the bosonic string—with that of the pure spinor formalism [**B**].

(j) We hope that our formulation of orientifold theory can help clarify some aspects of and prove useful to investigations in orientifold compactifications, especially in the applications to model building and the "landscape." In particular, our work suggests the existence of topological constraints on orientifold compactifications which have not been accounted for in the existing literature on the landscape.

References

[A] J. F. Adams, *Stable Homotopy and Generalised Homology*, Chicago Lectures in Mathematics, University of Chicago Press, Chicago, 1974.

[ABGHR] M. Ando, A. J. Blumberg, D. J. Gepner, M. J. Hopkins, and C. Rezk, *Units of ring spectra and Thom spectra*, arXiv:0810.4535.

[AgMV] Luis Àlvarez-Gaumé, Gregory W. Moore, and Cumrun Vafa, *Theta Functions, Modular Invariance, and Strings*, Commun. Math. Phys. **106** (1986), 1–40.

[AH] Michael Atiyah and Michael Hopkins, *A Variant of K-Theory: $K_\pm$*, London Math. Soc. Lecture Notes, vol. 308, pp. 5–17, Cambridge Univ. Press, Cambridge, 2004. arXiv:math/0302128.

[ALR] Alejandro Adem, Johann Leida, and Yongbin Ruan, *Orbifolds and stringy topology*, Cambridge Tracts in Mathematics, vol. 171, Cambridge University Press, Cambridge, 2007.

[AnS] Carlo Angelantonj and Augusto Sagnotti, *Open Strings*, Phys. Rept. **371** (2002), 1–150, arXiv:hep-th/0204089.

[AS] Michael F. Atiyah and Graeme B. Segal, *The index of elliptic operators. II*, Ann. of Math. (2) **87** (1968), 531–545.

[At] ———, *K-Theory*, second ed., Advanced Book Classics, Addison-Wesley, Redwood City, CA, 1989.

[B] Nathan Berkovits, *ICTP Lectures on Covariant Quantization of the Superstring*, arXiv:hep-th/0209059.

[Ba] M. Basterra, *André-Quillen cohomology of commutative S-algebras*, J. Pure Appl. Algebra **144** (1999), no. 2, 111–143.

[BGH] Oren Bergman, Eric G. Gimon, and Petr Hořava, *Brane Transfer Operations and T-Duality of Non-BPS States*, JHEP **04** (1999), 010, arXiv:hep-th/9902160.

[BGS] Oren Bergman, Eric G. Gimon, and Shigeki Sugimoto, *Orientifolds, RR Torsion, and K-Theory*, JHEP **05** (2001), 047, arXiv:hep-th/0103183.

[BH1] Ilka Brunner and Kentaro Hori, *Notes on Orientifolds of Rational Conformal Field Theories*, JHEP **07** (2004), 023, arXiv:hep-th/0208141.

[BH2] ———, *Orientifolds and Mirror Symmetry*, JHEP **11** (2004), 005, arXiv:hep-th/0303135.

[BHHW] Ilka Brunner, Kentaro Hori, Kazuo Hosomichi, and Johannes Walcher, *Orientifolds of Gepner Models*, JHEP **02** (2007), 001, arXiv:hep-th/0401137.

[BiS] Massimo Bianchi and Augusto Sagnotti, *On the Systematics of Open String Theories*, Phys. Lett. **B247** (1990), 517–524.

[BM] Dmitriy M. Belov and Gregory W. Moore, *Type II Actions from 11-Dimensional Chern-Simons Theories*, arXiv:hep-th/0611020.

[BS] Volker Braun and Bogdan Stefanski Jr., *Orientifolds and K-Theory*, arXiv:hep-th/0206158.

[BSc1] U. Bunke and T. Schick, *On the Topology of T-Duality*, Rev. Math. Phys. **17** (2005), 77–112.

[BSc2] U. Bunke and T. Schick, *Differential K-Theory. A Survey*, arXiv:1011.6663 [math.KT].

[BT] Raoul Bott and Loring W. Tu, *Differential forms in algebraic topology*, Graduate Texts in Mathematics, vol. 82, Springer-Verlag, New York, 1982.

[CDE] Andres Collinucci, Frederik Denef, and Mboyo Esole, *D-Brane Deconstructions in IIB Orientifolds*, JHEP **02** (2009), 005, arXiv:0805.1573 [hep-th].

[CR] Ben Craps and Frederik Roose, *Anomalous D-Brane and Orientifold Couplings from the Boundary State*, Phys. Lett. **B445** (1998), 150–159, arXiv:hep-th/9808074.

[CY] Yeuk-Kwan E. Cheung and Zheng Yin, *Anomalies, Branes, and Currents*, Nucl. Phys. **B517** (1998), 69–91, arXiv:hep-th/9710206.

[D] Atish Dabholkar, *Lectures on Orientifolds and Duality*, arXiv:hep-th/9804208.

[dBHKMMS] Jan de Boer et al., *Triples, Fluxes, and Strings*, Adv. Theor. Math. Phys. **4** (2002), 995–1186, arXiv:hep-th/0103170.

[DFM] G. W. Moore J. Distler, D. S. Freed, *Spin structures and superstrings*.

[DJM] Keshav Dasgupta, Dileep P. Jatkar, and Sunil Mukhi, *Gravitational Couplings and $\mathbb{Z}_2$ Orientifolds*, Nucl. Phys. **B523** (1998), 465–484, arXiv:hep-th/9707224.

[DLP] Jin Dai, R. G. Leigh, and Joseph Polchinski, *New Connections Between String Theories*, Mod. Phys. Lett. **A4** (1989), 2073–2083.

[DMW] Duiliu-Emanuel Diaconescu, Gregory Moore, and Edward Witten, *E_8 gauge theory, and a derivation of K-theory from M-theory*, Adv. Theor. Math. Phys. **6** (2002), no. 6, 1031–1134 (2003).

[F1] Daniel S. Freed, *Dirac Charge Quantization and Generalized Differential Cohomology*, Surveys in Differential Geometry, Int. Press, Somerville, MA, 2000, pp. 129–194. arXiv:hep-th/0011220.

[F2] ———, *K-theory in Quantum Field Theory*, Current Developments in Mathematics — 2001, Int. Press, Somerville, MA, 2002, pp. 41–87. arXiv:math-ph/0206031.

[FH1] Daniel S. Freed and Michael J. Hopkins, *On Ramond-Ramond Fields and K-Theory*, JHEP **05** (2000), 044, arXiv:hep-th/0002027.

[FH2] D. S. Freed and M. J. Hopkins, *KO and Anomalies in Type I*. Appendix B in [**F1**].

[FHT] D. S. Freed, M. J. Hopkins, and C. Teleman, *Loop Groups and Twisted K-Theory I*, 0711.1906.

[FM] Daniel S. Freed and Gregory W. Moore, *Setting the Quantum Integrand of M-Theory*, Commun. Math. Phys. **263** (2006), 89–132, arXiv:hep-th/0409135.

[FMS1] Daniel S. Freed, Gregory W. Moore, and Graeme Segal, *The Uncertainty of Fluxes*, Commun. Math. Phys. **271** (2007), 247–274, arXiv:hep-th/0605198.

[FMS2] ———, *Heisenberg Groups and Noncommutative Fluxes*, Annals Phys. **322** (2007), 236–285, arXiv:hep-th/0605200.

[FW] Daniel S. Freed and Edward Witten, *Anomalies in String Theory with D-Branes*, arXiv:hep-th/9907189.

[G] Sergei Gukov, *K-Theory, Reality, and Orientifolds*, Commun. Math. Phys. **210** (2000), 621–639, arXiv:hep-th/9901042.

[GH] Dongfeng Gao and Kentaro Hori, *On The Structure Of The Chan-Paton Factors For D-Branes In Type II Orientifolds*, 1004.3972.

[GHM] Michael B. Green, Jeffrey A. Harvey, and Gregory W. Moore, *I-Brane Inflow and Anomalous Couplings on D-Branes*, Class. Quant. Grav. **14** (1997), 47–52, arXiv:hep-th/9605033.

[GL] Hugo Garcia-Compean and Oscar Loaiza-Brito, *Branes and Fluxes in Orientifolds and K-Theory*, Nucl. Phys. **B694** (2004), 405–442, arXiv:hep-th/0206183.

[H] C. M. Hull, *Massive String Theories from M-Theory and F-Theory*, JHEP **11** (1998), 027, arXiv:hep-th/9811021.

[HJ] Pierre Henry-Labordère and Bernard Julia, *Gravitational Couplings of Orientifold Planes*, JHEP **01** (2002), 033, arXiv:hep-th/0112065.

[Ho] Petr Horava, *Strings on World Sheet Orbifolds*, Nucl. Phys. **B327** (1989), 461.

[Hor] Kentaro Hori, *D-Branes, T-Duality, and Index Theory*, Adv. Theor. Math. Phys. **3** (1999), 281–342, arXiv:hep-th/9902102.

[HS] M. J. Hopkins and I. M. Singer, *Quadratic Functions in Geometry, Topology, and M-Theory*, J. Diff. Geom. **70** (2005), 329–452, arXiv:math/0211216.

[J] Clifford V. Johnson, *D-Branes*, Cambridge Monographs on Mathematical Physics, Cambridge Univ. Press, Cambridge, 2003.

[KM] Nemanja Kaloper and Robert C. Myers, *The O(dd) Story of Massive Supergravity*, JHEP **05** (1999), 010, arXiv:hep-th/9901045.

[KV] A. Kahle and A. Valentino *T-Duality and Differential K-Theory*, arXiv:0912.2516 [math.KT].

[LMST] Wolfgang Lerche, Ruben Minasian, Christoph Schweigert, and Stefan Theisen, *A note on the geometry of CHL heterotic strings*, Phys. Lett. **B424** (1998), 53–59, arXiv:hep-th/9711104.

[MM] Ruben Minasian and Gregory W. Moore, *K-Theory and Ramond-Ramond Charge*, JHEP **11** (1997), 002, arXiv:hep-th/9710230.

[MS] Sunil Mukhi and Nemani V. Suryanarayana, *Gravitational Couplings, Orientifolds and M-Planes*, JHEP **09** (1999), 017, arXiv:hep-th/9907215.

[MSS] Jose F. Morales, Claudio A. Scrucca, and Marco Serone, *Anomalous Couplings for D-Branes and O-Planes*, Nucl. Phys. **B552** (1999), 291–315, arXiv:hep-th/9812071.

[MW] Gregory W. Moore and Edward Witten, *Self-Duality, Ramond-Ramond Fields, and K-Theory*, JHEP **05** (2000), 032, arXiv:hep-th/9912279.

[O] Michael L. Ortiz, *Differential Equivariant K-Theory*, arXiv:0905.0476 [math.AT].

[OS] Kasper Olsen and Richard J. Szabo, *Constructing D-Branes From K-Theory*, Adv. Theor. Math. Phys. **3** (1999), 889–1025, arXiv:hep-th/9907140.

[P] Joseph Polchinski, *String Theory. Volumes I,II*, Cambridge Monographs on Mathematical Physics, Cambridge University Press, Cambridge, 1998.

[PS] Gianfranco Pradisi and Augusto Sagnotti, *Open String Orbifolds*, Phys. Lett. **B216** (1989), 59.

[S] Bogdan Stefanski Jr., *WZ Couplings of D-Branes and O-Planes*, proceedings of NATO Advanced Study Institute: TMR Summer School on Progress in String Theory and M-Theory (Cargèse 99). arXiv:hep-th/9909105.

[Sa] Augusto Sagnotti, *Open Strings and their Symmetry Groups*, arXiv:hep-th/0208020.

[Se] Graeme Segal, *Categories and cohomology theories*, Topology **13** (1974), 293–312.

[SS] Claudio A. Scrucca and Marco Serone, *Anomaly Inflow and RR Anomalous Couplings*, arXiv:hep-th/9911223.

[SSc1] Joel Scherk and John H. Schwarz, *Spontaneous Breaking of Supersymmetry Through Dimensional Reduction*, Phys. Lett. **B82** (1979), 60.

[SSc2] ———, *How to Get Masses from Extra Dimensions*, Nucl. Phys. **B153** (1979), 61–88.

[SV] Richard J. Szabo and Alessandro Valentino, *Ramond-Ramond Fields, Fractional Branes and Orbifold Differential K-Theory*, Commun. Math. Phys. **294**, 647 (2010), arXiv:0710.2773 [hep-th].

[SW] Nathan Seiberg and Edward Witten, *Spin Structures in String Theory*, Nucl. Phys. **B276** (1986), 272.

[W1] Edward Witten, *Five-Brane Effective Action in M-Theory*, J. Geom. Phys. **22** (1997), 103–133, arXiv:hep-th/9610234.

[W2] ———, *D-Branes and K-Theory*, JHEP **12** (1998), 019, arXiv:hep-th/9810188.

[W3] ———, *Toroidal compactification without vector structure*, JHEP **02** (1998), 006, arXiv:hep-th/9712028.

[W4] ———, *Duality relations among topological effects in string theory*, J. High Energy Phys. (2000), no. 5, Paper 31, 31.

Theory Group, Department of Physics, and Texas Cosmology Center, University of Texas, 1 University Station C1600, Austin, TX 78712-0264

E-mail address: distler@golem.ph.utexas.edu

Department of Mathematics, University of Texas, 1 University Station C1200, Austin, TX 78712-0257

E-mail address: dafr@math.utexas.edu

NHETC and Department of Physics and Astronomy, Rutgers University, Piscataway, NJ 08855-0849

E-mail address: gmoore@physics.rutgers.edu

Two-Dimensional Quantum Field Theories

Proceedings of Symposia in Pure Mathematics
Volume **83**, 2011

Surface operators in 3d Topological Field Theory and 2d Rational Conformal Field Theory

Anton Kapustin and Natalia Saulina

Abstract. We study surface operators in 3d Topological Field Theory and their relations with 2d Rational Conformal Field Theory. We show that a surface operator gives rise to a consistent gluing of chiral and anti-chiral sectors in the 2d RCFT. The algebraic properties of the resulting 2d RCFT, such as the classification of symmetry-preserving boundary conditions, are expressed in terms of properties of the surface operator. We show that to every surface operator one may attach a Morita-equivalence class of symmetric Frobenius algebras in the ribbon category of bulk line operators. This provides a simple interpretation of the results of Fuchs, Runkel and Schweigert on the construction of 2d RCFTs from Frobenius algebras. We also show that every topological boundary condition in a 3d TFT gives rise to a commutative Frobenius algebra in the category of bulk line operators. We illustrate these general considerations by studying in detail surface operators in abelian Chern-Simons theory.

1. Introduction

There is a well-known relationship between 2d Rational Conformal Field Theory (RCFT) and 3d Topological Field Theory (TFT). Formally, this relationship arises from the fact that both conformal blocks of a chiral algebra of a 2d RCFT and bulk line operators in a semi-simple 3d TFT form a modular tensor category (i.e. a semi-simple ribbon category satisfying a certain non-degeneracy condition). More physically, one may consider a 3d TFT on a cylinder $D^2 \times \mathbb{R}$ and find [**Witten, EMSS**] that this is equivalent to a chiral sector of a 2d RCFT living on the boundary $S^1 \times \mathbb{R}$. This may be regarded as a toy model for holographic duality.

The simplest example of this relationship is probably the $U(1)$ Chern-Simons theory at level $k \in 2\mathbb{Z}$ which is holographically dual to a free boson RCFT with radius $R^2 = k$. A more complicated example is the duality between Chern-Simons gauge theory with a simple Lie group G and the Wess-Zumino-Witten model in 2d.

There is more to a 3d TFT than bulk line operators. For example, we may consider surface operators which are localized on 2d submanifolds. Surface operators form a 2-category whose 1-morphisms are line operators sitting at the junction of two surface operators, and whose 2-morphisms are local operators sitting at the

---

2010 *Mathematics Subject Classification.* Primary 81T40, 81T45.

junction of two line operators. The category of bulk line operators can be recovered from this 2-category: it is the category of 1-morphisms from the "invisible" surface operator to itself. The "invisible" surface operator is equivalent to no surface operator at all; more formally one may describe it by saying that the 2-category of surface operators has a monoidal structure given by fusing surface operators together, and the "invisible" surface operator is the identity object with respect to this monoidal structure.

In this note we discuss the following question: what is the meaning of surface operators in a 3d TFT from the point of view of the corresponding 2d RCFT? We will argue that surface operators in a given 3d TFT correspond to consistent gluings of chiral and anti-chiral sectors to form a modular-invariant 2d RCFT. More precisely, this interpretation applies to surface operators which do not admit nontrivial local operators. General surface operators correspond to consistent ways of gluing chiral and anti-chiral sectors to a 2d TFT. We illustrate the correspondence using the example of $U(1)_k$ current algebra.

One motivation for this work was the desire to give a more intuitive interpretation to some of the results of Fuchs, Runkel, Schweigert and collaborators on the algebraic construction of 2d RCFTs from modular tensor categories (see [**FRS, FRS2, FRS3, FRS4, FRS5, FFRS**] and references therein; an overview of this series of papers is contained in [**FRSO**]). These authors related 2d RCFTs with a fixed chiral algebra with Morita-equivalence classes of special symmetric Frobenius algebras in the corresponding modular tensor category. We will see that this result has a very natural explanation in terms of surface operators. A similar argument shows how to associate a commutative Frobenius algebra in the modular tensor category to any topological boundary condition in a 3d TFT. This can be regarded as a 3d analogue of the boundary state in 2d TFT. We describe this state in the case of $U(1)_k \times U(1)_{-k}$ Chern-Simons theory which admits many topological boundary conditions.

The authors are grateful to I. Runkel for comments on a preliminary draft, to A. Tsymbalyuk for the proof of the proposition presented in the appendix, and to Shlomit Wolf for drawing the more elaborate figures. A. K. would like to thank the Institute for Advanced Studies, Hebrew University, for hospitality during the completion of this work. The work of A. K. was supported in part by the DOE grant DE-FG02-92ER-40701.

2. Generalities

2.1. Surface operators and consistent gluings. Consider a 3d TFT $\mathcal{T}$ corresponding to a given chiral algebra $\mathcal{A}$. This means that the space of conformal blocks of $\mathcal{A}$ on a Riemann surface Σ is the space of states of the 3d TFT on Σ. A good example to keep in mind is Chern-Simons theory with gauge group G, in which case $\mathcal{A}$ is the affine Lie algebra corresponding to G. We will denote by $\mathcal{H}_\Sigma$ the space of conformal blocks on Σ; it is acted upon by the mapping class group of Σ. When we consider both chiral and anti-chiral sectors, the relevant representation of the mapping class group is $\mathcal{H}_\Sigma \otimes \mathcal{H}_\Sigma^*$. To obtain a well-defined 2d RCFT one needs to pick an invariant vector in all representation spaces $\mathcal{H}_\Sigma \otimes \mathcal{H}_\Sigma^*$.This choice must be compatible with cutting Σ into pieces, so that Segal's axioms [**Segal**] are satisfied.

From the 3d viewpoint the space $\mathcal{H}_\Sigma \otimes \mathcal{H}_\Sigma^*$ is the space of states of the 3d TFT $\mathcal{T} \otimes \bar{\mathcal{T}}$, where the bar denotes parity-reversal. There is a very natural way to get an invariant vector in this space. Consider the 3-manifold $\Sigma \times [0,1]$. Its boundary is a disjoint union of two copies of Σ. Let us pick a topological boundary condition in the theory $\mathcal{T} \otimes \bar{\mathcal{T}}$ and impose it on $\Sigma \times \{1\}$. The other copy of Σ is a "cut" boundary on which we do not impose boundary conditions. 3d TFT assigns to such a 3-manifold a diffeomorphism-invariant vector in $\mathcal{H}_\Sigma \otimes \mathcal{H}_\Sigma^*$. Compatibility with cutting and pasting is clearly ensured by the locality of the topological boundary condition imposed at $\Sigma \times \{1\}$.

By the usual folding trick we may reinterpret the boundary condition for the theory $\mathcal{T} \otimes \bar{\mathcal{T}}$ as a surface operator in the theory $\mathcal{T}$. Thus we consider the theory $\mathcal{T}$ on $\Sigma \times [0,2]$ with an insertion of a surface operator at $\Sigma \times \{1\}$. Chiral and anti-chiral degrees of freedom then live at $\Sigma \times \{0\}$ and $\Sigma \times \{2\}$, and the surface operator glues them together into a consistent 2d RCFT.

Among all surface operators there is always a trivial one equivalent to no surface operator at all. It corresponds to the diagonal gluing, and the corresponding vector in $\mathcal{H}_\Sigma \otimes \mathcal{H}_\Sigma^* = \operatorname{End} \mathcal{H}_\Sigma$ is $\mathbf{1}$. From an abstract viewpoint, surface operators in a 3d TFT form a monoidal 2-category (i.e. a category with an associative but not necessarily commutative tensor product), and the trivial surface operator is the identity object with respect to the product.

Given this correspondence between surface operators and consistent gluings we may interpret various properties of 2d RCFT in 3d terms. Recall that the main object of study in 2d RCFT is the modular tensor category $\mathcal{C}$ which from the 3d viewpoint is the category of bulk line operators in the theory $\mathcal{T}$. For example, if $\mathcal{T}$ is Chern-Simons theory, bulk line operators are Wilson lines labeled by integrable highest weight representations of the corresponding affine Lie algebra. Primary vertex operators in a 2d RCFT are labeled by a pair of objects W and W' of $\mathcal{C}$. From the 3d viewpoint a primary $V_{WW'}$ inserted at $p \in \Sigma$ is represented by a pair of bulk line operator inserted at $p \times [0,1]$ and $p \times [1,2]$ and labeled by W and W' respectively. These two bulk line operators meet at the point $p \times \{1\}$ which lies on the surface operator. Thus primaries labeled by W, W' are in 1-1 correspondence with local operators living on the surface operator which "convert" the bulk line operator W into the bulk line operator W'. The space of such local operators is the vector space which 3d TFT attaches to S^2 with the surface operator inserted at the equator and with W and W' inserted at the north and south poles (see Fig. 1).

If we take both W and W' to be trivial line operators (i.e. identity objects in the monoidal category $\mathcal{C}$), then we conclude that the space of primaries transforming in the vacuum representation of the chiral-anti-chiral algebra $\mathcal{A} \otimes \bar{\mathcal{A}}$ is isomorphic to the space of local operators on the surface operator (i.e. the space attached to S^2 with an insertion of the surface operator at the equator). One usually requires this space to be one-dimensional (this requirement is called uniqueness of the vacuum). To satisfy this axiom, the surface operator used for gluing should not admit any nontrivial local operators.

If a surface operator admits nontrivial local observables, then uniqueness of vacuum does not hold, i.e. the space of primaries in the vacuum representation has dimension greater than one. This space is actually a commutative algebra which we will denote A_S. The space of RCFT primaries transforming in the representation

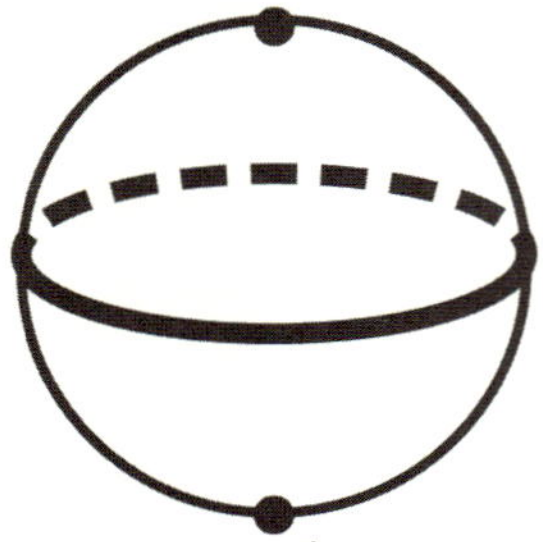

FIGURE 1. The space of RCFT primaries labeled by objects W, W' is the space of states of the 3d TFT on S^2 with insertions of line operators at the poles and the surface operator at the equator.

$W \otimes \bar{W}'$ is then a module over A_S. A trivial example of such a surface operator is obtained by tensoring an invisible surface operator with an arbitrary 2d TFT, so that this 2d TFT does not interact in any way with the bulk degrees of freedom. The algebra A_S in this case is simply the algebra of local operators in this 2d TFT. The corresponding 2d RCFT is the tensor product of the 2d RCFT corresponding to the diagonal gluing and the 2d TFT living on the surface operator. More generally, one can contemplate surface operators which carry a 2d TFT which interacts with bulk degrees of freedom; the corresponding gluing may be interpreted as a consistent way to couple chiral and anti-chiral sectors to a 2d TFT.

Another interesting question is the classification of boundary conditions for a 2d RCFT. For concreteness, let Σ be a disc. By definition, the *double* of such Σ is a pair of disks D^2_+ and D^2_- glued along their boundaries. We will identify the double with a unit sphere in $\mathbb{R}^3$, so that D^2_+ and D^2_- correspond to southern and northern hemispheres respectively. We wish to consider the 3d TFT on the ball D^3 which is the interior of the unit sphere and with an insertion of a surface operator at the equatorial plane (see Fig. 2). Chiral and anti-chiral degrees of freedom live on D^2_+ and D^2_- respectively. The support of the surface operator is a disk, so in order to specify the physics completely we need to specify how to terminate the surface operator. That is, we need to pick a line operator λ which sits at the boundary of the surface operator S. Such line operators can be thought of as 1-morphisms in the monoidal 2-category of surface operators (specifically, morphisms from the trivial surface operator $\mathbf{1}$ to the surface operator S). We conclude that boundary conditions for the 2d RCFT are labeled by such 1-morphisms.

In the special case $S = \mathbf{1}$ such line operators are the same as bulk line operators. Thus we end up with a familiar result that, in the case of diagonal gluing, boundary conditions preserving the full chiral algebra $\mathcal{A}$ are in 1-1 correspondence with representations of $\mathcal{A}$ [**Cardy, FFFS**].

Boundary-changing vertex operators can also be described in the language of surface operators. Let λ and λ' be two line operators on which a given surface operator S may terminate. As mentioned above they can be regarded as 1-morphisms from the trivial surface operator $\mathbf{1}$ to the surface operator S. Now, since we are dealing with a 2-category, we may consider morphisms between 1-morphisms, i.e.

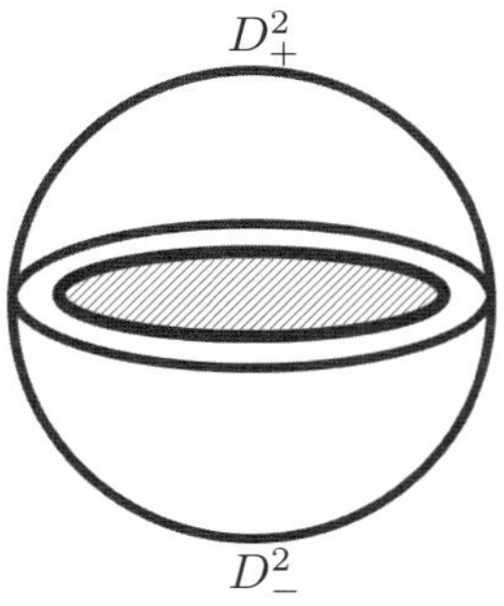

FIGURE 2. 3d TFT on a ball with an insertion of a surface operator (shaded region) on the equatorial plane. Chiral and anti-chiral sectors of the RCFT live on northern and southern hemispheres respectively. The surface operator terminates on a line operator denoted by a thick solid line.

2-morphisms. They form a vector space (the 2-category of surface operators is $\mathbb{C}$-linear) which from the viewpoint of the 3d TFT is the vector space attached to the picture shown in Fig. 3. In other words, this is the space of local operators which can be inserted at the junction of λ and λ'. Elements of this space label boundary-changing vertex operators which transform in the vacuum representation of the chiral algebra. In the special case of diagonal gluing where $S = \mathbf{1}$ and λ and λ' are bulk line operators, this is simply the space of morphisms between objects λ and λ' in the category $\mathcal{C}$.

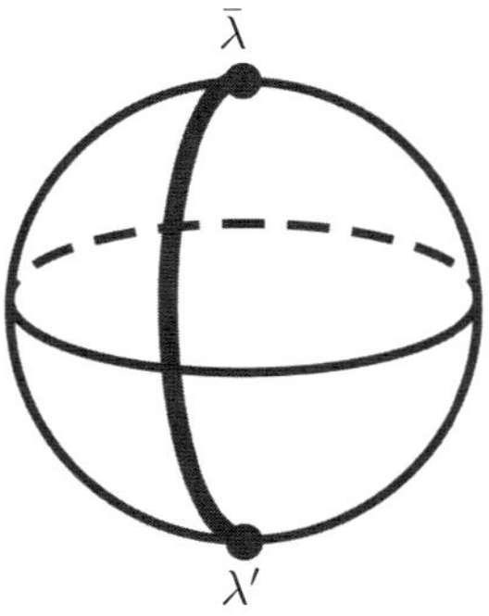

FIGURE 3. Boundary conditions are labeled by line operators on which the surface operator can terminate. Given two such line operators λ, λ', the space of boundary-changing operators in the vacuum representation of the chiral algebra is the space of states of the 3d TFT on S^2 with insertions of a surface operator (thick line) along a meridian semicircle and line operators $\bar{\lambda}$ and λ' at the poles.

More generally,, the space of boundary-changing primaries in the representation W is the vector space corresponding to the picture in Fig. 4. From the algebraic point of view, the category $\mathrm{Hom}(\mathbf{1}, S)$ is a module category over the braided monoidal category of bulk line operators, and the vector space corresponding to Fig. 4 is the space of 2-morphisms from λ to $\lambda' \otimes W$.

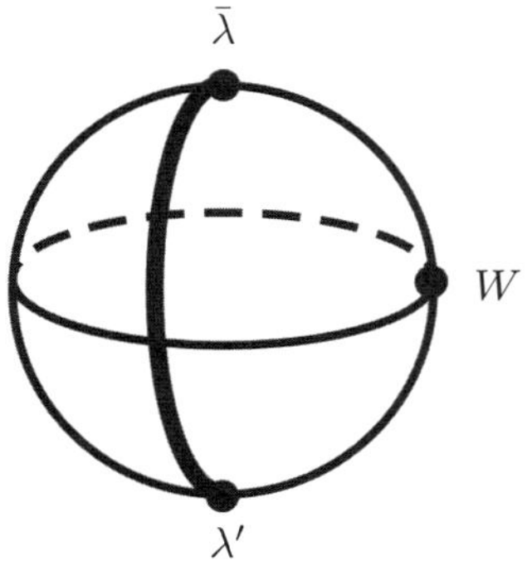

FIGURE 4. Computing the space of boundary-changing vertex operators in the representation W of the chiral algebra.

2.2. From surface operators to algebra-objects. Now let us make contact with the work of Fuchs, Runkel and Schweigert [**FRS**]. Let S be a surface operator and let us pick a line operator λ on which S can terminate (i.e. a 1-morphism from the trivial surface operator $\mathbf{1}$ to the surface operator S). Let $\bar{\lambda}$ be the line operator obtained from λ by orientation reversal; it can be regarded as a 1-morphism from S to the trivial surface operator. Consider now a long rectangular strip of S bounded on the two sides by λ and $\bar{\lambda}$ (see Fig. 5). Since we are dealing with a 3d TFT, the strip can be regarded as arbitrarily narrow and so is equivalent to a bulk line operator which we will denote $W(S, \lambda)$. This line operator comes with the natural framing[1]. More formally, if we regard λ as a 1-morphism from $\mathbf{1}$ to S and $\bar{\lambda}$ as a 1-morphism from S to $\mathbf{1}$, then their composition $W(S, \lambda) = \bar{\lambda} \circ \lambda$ is a 1-morphism from $\mathbf{1}$ to $\mathbf{1}$ and so is an object of the modular tensor category $\mathcal{C}$.

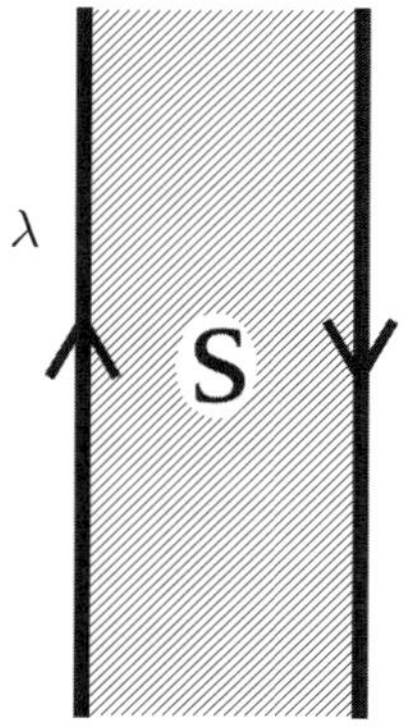

FIGURE 5. A strip of a surface operator S bounded from the left by a line operator λ and from the right by its orientation-reversal.

[1]Line operators are understood to have a framing as part of their data.

The object $W(S,\lambda)$ is very special: it is a symmetric Frobenius algebra in the modular tensor category $\mathcal{C}$. Let us remind what this means. First of all, an algebra A in a monoidal category $\mathcal{C}$ is an object A equipped with a morphism $m : \mathsf{A}\otimes\mathsf{A} \to \mathsf{A}$ (product) satisfying the associativity constraint expressed as the commutativity of the following diagram:

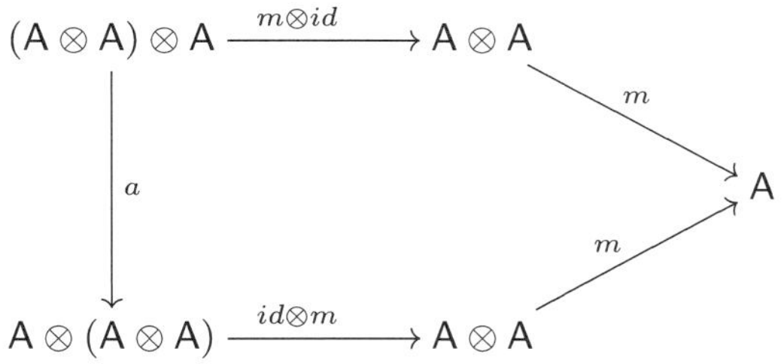

Instead of drawing commutative diagrams, it is convenient to use the graphical calculus for monoidal categories explained for example in [**FRS**] and [**BK**]. An object A of a monoidal category is then represented by an upward going line labeled by A, the tensor product of two objects is represented by drawing two lines going in the same direction, and a morphism from $A_1\otimes\ldots\otimes A_n$ to $B_1\otimes\ldots\otimes B_k$ is represented by a vertex with n incoming and k outgoing lines. Composition of morphisms corresponds to the concatenation of the diagrams in the vertical direction. The vertex corresponding to the identity morphism is usually omitted, so that we have

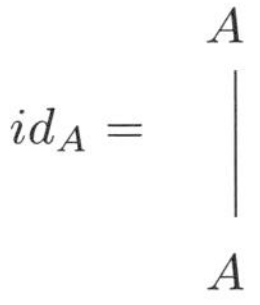

Then the associativity constraint on m is expressed at the equality of morphisms corresponding to the two diagrams shown below. We write associators only in formulas, but not in the graphical notation [2].

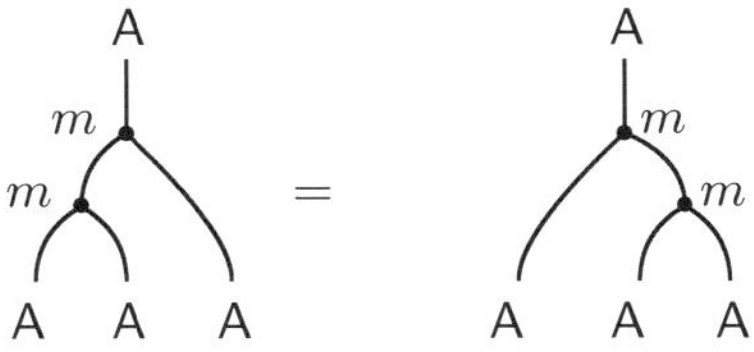

In the case of $W(S,\lambda)$ the morphism m arises from the "open-string vertex" shown in Fig. 6. The associativity constraint is satisfied because the two diagrams shown in Fig. 7 can be deformed into one another (i.e. they are related by a diffeomorphism connected to the identity diffeomorphism), and thus 3d TFT assigns the same morphism to them. Thus $W(S,\lambda)$ is an algebra in the category of bulk line operators.

Similarly, a coalgebra in $\mathcal{C}$ is an object A of $\mathcal{C}$ equipped with a morphism $\Delta : \mathsf{A} \to \mathsf{A}\otimes\mathsf{A}$ (coproduct) satisfying the coassociativity constraint expressed as the commutativity of the following diagram:

---

[2]All different ways to add associators to translate a picture into a formula describe the same morphism in the category by coherence.

$$\begin{array}{ccccc} & & \mathsf{A}\otimes\mathsf{A} & \xrightarrow{\Delta\otimes id} & (\mathsf{A}\otimes\mathsf{A})\otimes\mathsf{A} \\ & \overset{\Delta}{\nearrow} & & & \\ \mathsf{A} & & & & \downarrow a \\ & \underset{\Delta}{\searrow} & & & \\ & & \mathsf{A}\otimes\mathsf{A} & \xrightarrow{id\otimes\Delta} & \mathsf{A}\otimes(\mathsf{A}\otimes\mathsf{A}) \end{array}$$

In the case of $W(S,\lambda)$ the coproduct is given by the diagram shown in Fig. 8. The coassociativity follows from the fact that the two diagrams shown in Fig. 9 can be deformed into one another.

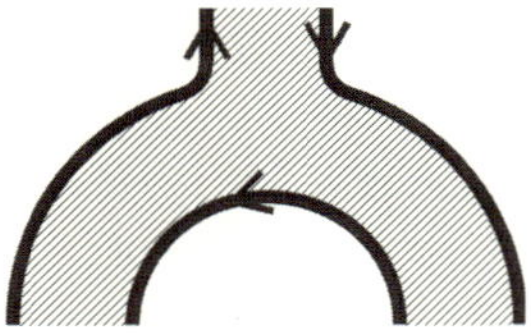

FIGURE 6. The product.

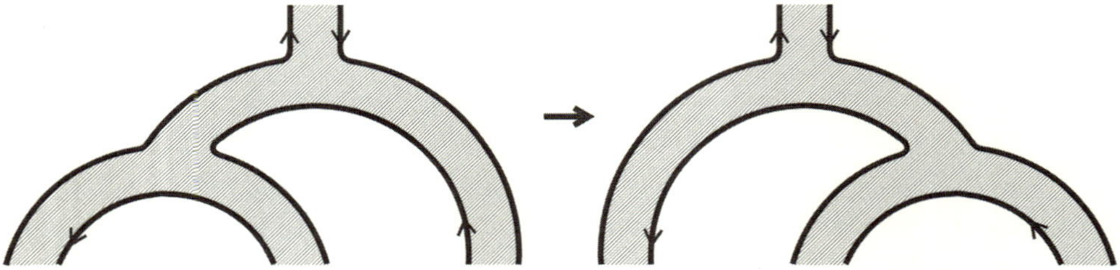

FIGURE 7. The associativity constraint. A horizontal arrow here and in other figures below indicates a diffeomorphism connected to the identity diffeomorphism.

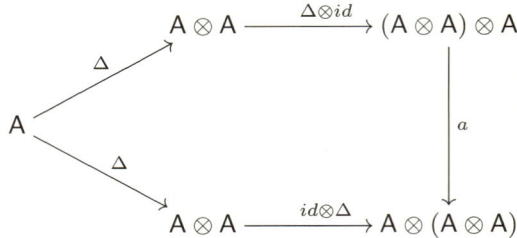

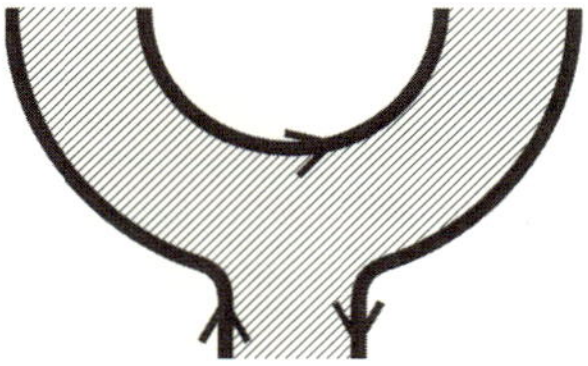

FIGURE 8. The coproduct.

An algebra A in a monoidal category $\mathcal{C}$ is unital algebra if one is given a morphism $\iota_{\mathsf{A}} : \mathbf{1} \to \mathsf{A}$ (called a unit) such that the following three pictures correspond to the same morphism:

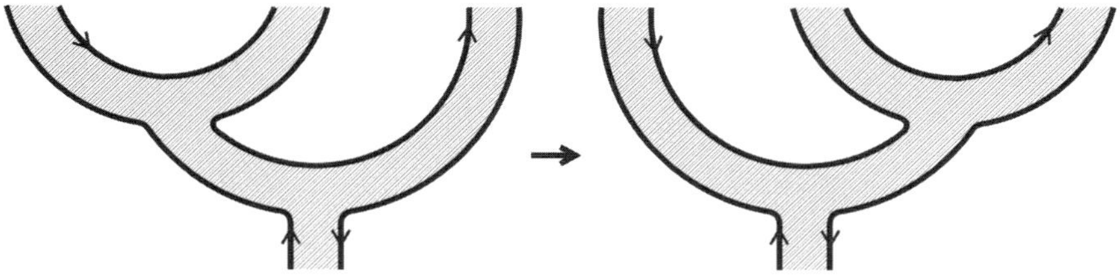

FIGURE 9. The coassociativity constraint.

$\iota_A =$

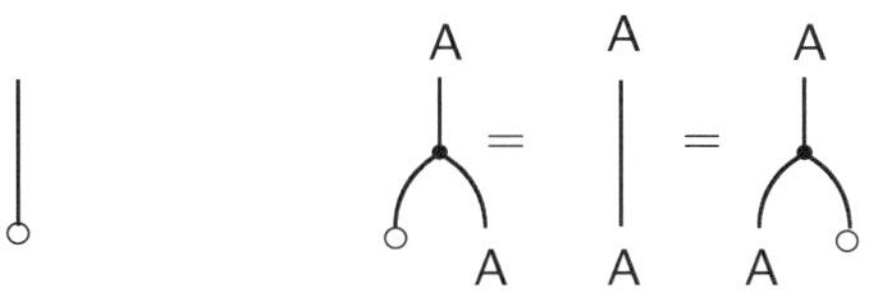

A coalgebra is a counital coalgebra if one is given a morphism $\epsilon_A : A \to \mathbf{1}$ (called counit) such that the the following three pictures correspond to the same morphism:

$\epsilon_A =$

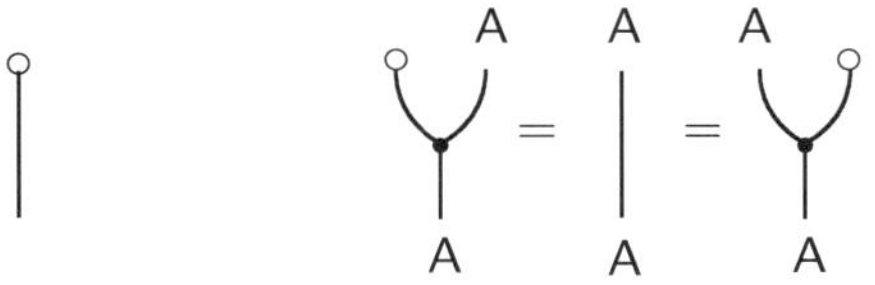

In the case of $W(S, \lambda)$ the unit and the counit are defined in the obvious way, see Fig. 10. Thus $W(S, \lambda)$ is simultaneosuly a unital algebra and a counital coalgebra in the category of bulk line operators.

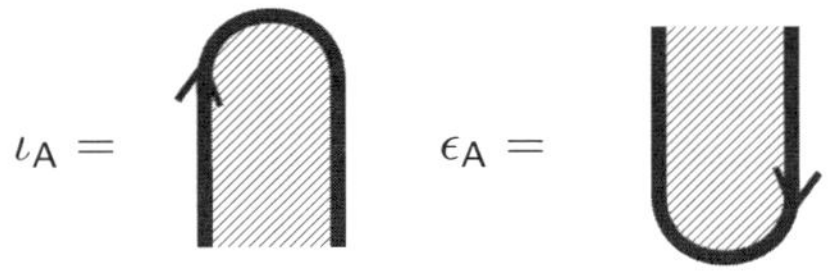

FIGURE 10. The unit and the counit.

A Frobenius algebra in $\mathcal{C}$ is an object A which is both a unital algebra and a counital coalgebra in $\mathcal{C}$ such that m and Δ are compatible in the sense that the following equalities hold:

$$(id_A \otimes m) \circ a_{AAA} \circ (\Delta \otimes id_A) = \Delta \circ m = (m \otimes id_A) \circ a^{-1}_{AAA} \circ (id_A \otimes \Delta)$$

Here $a_{AAA} : (A \otimes A) \otimes A \to A \otimes (A \otimes A)$ is the associator. The equalities can be represented graphically as follows (all vertices are labeled either by m or by Δ):

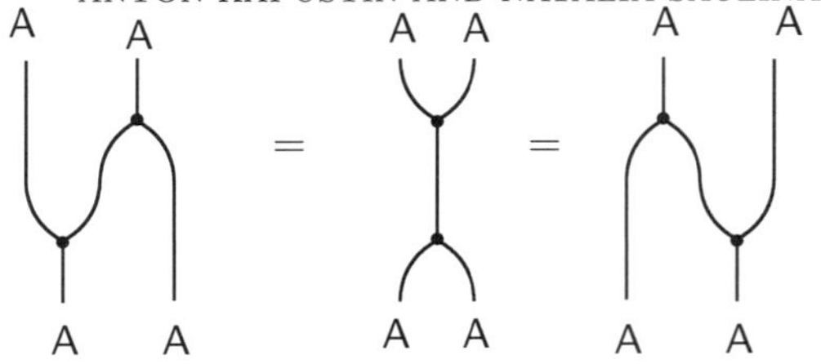

In the case of $W(S, \lambda)$ the compatibility of m and Δ follows from the fact that the three diagrams shown in Fig. 11 can be deformed into one another.

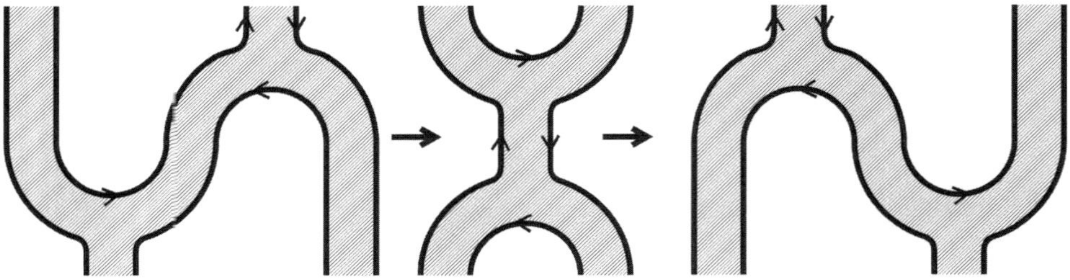

FIGURE 11. The Frobenius condition.

Recall that our category $\mathcal{C}$ is not only monoidal, but also rigid (because it is a ribbon category). In a rigid category every object A has a left-dual object ${}^*A$ and a right-dual object A^*; these objects are automatically unique up to isomorphism. In a ribbon category the left-dual is automatically isomorphic to the right-dual; a rigid monoidal category with this property is called sovereign.

From the physical viewpoint, the dual object is obtained by reversing the orientation of the bulk line operator. In the diagrammatic notation this is accounted for by placing arrows on all lines, so that a line labeled by A but with an arrow pointing down represents the object $A^* = {}^*A$. By the definition of a rigid category, for every A we have four morphisms $A \otimes A^* \to \mathbf{1}$, $\mathbf{1} \to A^* \otimes A$, $A^* \otimes A \to \mathbf{1}$ and $\mathbf{1} \to A \otimes A^*$ which are represented by the following four pictures:

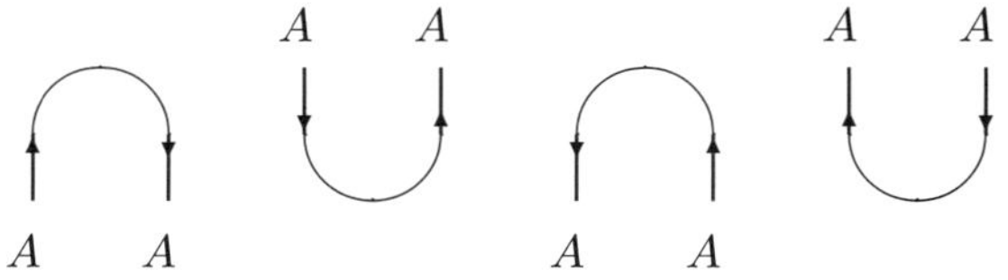

These four morphisms satisfy four obvious identities the first of which looks as follows:

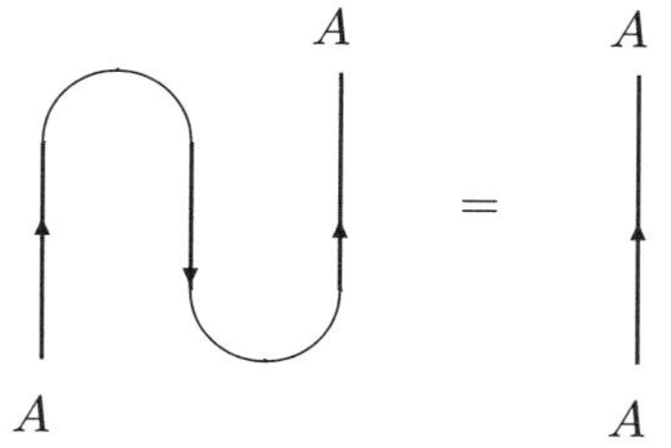

It is clear from Fig. 5 that the object $\mathsf{A} = W(S, \lambda)$ is self-dual: ${}^*\mathsf{A} = \mathsf{A}^* = \mathsf{A}$. This means that in all diagrams which involve only A we may drop the arrows on lines. Accordingly, two of the four duality morphisms become morphisms from $\mathsf{A} \otimes \mathsf{A}$ to $\mathbf{1}$, while the other two become morphisms from $\mathbf{1}$ to $\mathsf{A} \otimes \mathsf{A}$. It is easy to see that the former two are both given by $\epsilon_{\mathsf{A}} \circ m : \mathsf{A} \otimes \mathsf{A} \to \mathbf{1}$. Pictorially this is expressed as follows:

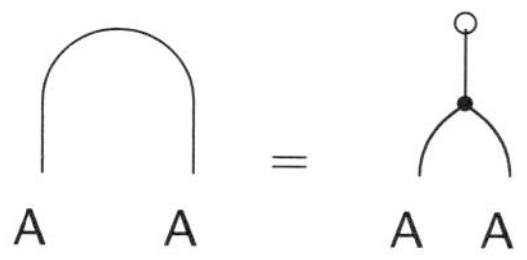

Similarly, the remaining two duality morphisms are both given by $\Delta \circ \iota_{\mathsf{A}} : \mathbf{1} \to \mathsf{A} \otimes \mathsf{A}$.

According to [**FRS**], a counital Frobenius algebra A in a sovereign monoidal category is called symmetric if the two morphisms from A to A^* shown below coincide:

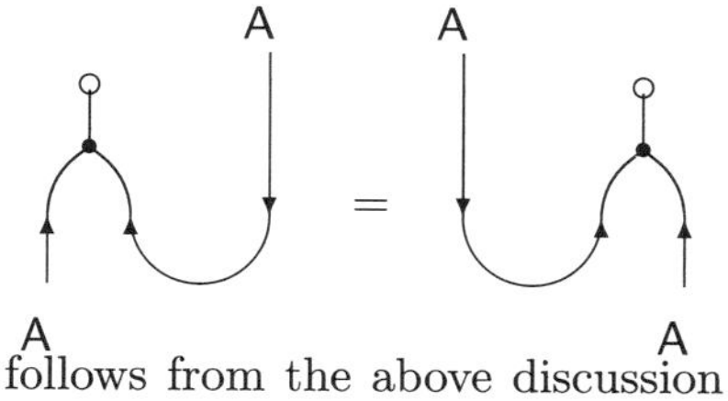

In the case of $W(S, \lambda)$ it follows from the above discussion that both morphisms are the identity morphism from A to $\mathsf{A}^* = \mathsf{A}$, so the Frobenius algebra $W(S, \lambda)$ is symmetric.

We are assuming all along that the identity object of $\mathcal{C}$ is simple, i.e. $\mathrm{Hom}(\mathbf{1}, \mathbf{1}) = \mathbb{C}$. Equivalently, the 3d TFT is assumed not to admit nontrivial local operators. Hence the composition of a unit and a counit of A is a number $\beta_1(\mathsf{A}) \in \mathbb{C}$. For the algebra-object $W(S, \lambda)$ this number has the following meaning: inserting a surface operator S shaped as a disk and bounded by the line operator λ is equivalent to mutliplying the correlator by $\beta_1(W(S, \lambda))$ (see Fig. 12).

Another natural operation is punching a hole in a surface operator S, so that the hole is bounded by the line operator λ. Such a hole is equivalent to a local operator on on S. If one imposes the uniqueness of vacuum axiom, S does not admit any nontrivial local operators, so this local operator must be the identity operator times a number $\beta_2(\mathsf{A}) \in \mathbb{C}$. Then punching a hole bounded by λ is equivalent to multiplying the correlator by $\beta_2(\mathsf{A})$ (see Fig. 13). Even if one does not impose uniqueness of vacuum, one may still require that a hole bounded by λ be equivalent to the identity operator times a complex number $\beta_2(\mathsf{A})$.

A Frobenius algebra A is called special [**FRS**] if this requirement is satisfied and both $\beta_1(\mathsf{A})$ and $\beta_2(\mathsf{A})$ are nonzero. Note that the numbers $\beta_1(\mathsf{A})$ and $\beta_2(\mathsf{A})$ for $\mathsf{A} = W(S, \lambda)$ are not independent but satisfy

$$\beta_1(\mathsf{A})\beta_2(\mathsf{A}) = \dim \mathsf{A},$$

where $\dim \mathsf{A}$ is the quantum dimension of A (the expectation value of the unknot labeled by the object A). Indeed, the unknot labeled by $W(S, \lambda)$ can be thought of as an annulus of S bounded by λ. Such an annulus can be thought of as the result of punching a hole in a disk-shaped S bounded by λ. Since the disk without a hole contributes $\beta_1(\mathsf{A})$ to the correlator and punching a hole multiplies the correlator by $\beta_2(\mathsf{A})$, we get the above identity.[3]

If A is a special Frobenius algebra, one can rescale ϵ by $\lambda \in \mathbb{C}^*$ and Δ by λ^{-1}, so that $\beta_1(\mathsf{A}) \mapsto \lambda\beta_1(\mathsf{A})$ and $\beta_2(\mathsf{A}) \mapsto \lambda^{-1}\beta_2(\mathsf{A})$ [**FRS, FFRS**]. The normalization condition adopted in [**FRS, FFRS**] is $\beta_1(\mathsf{A}) = \dim \mathsf{A}$, $\beta_2(\mathsf{A}) = 1$, so that punching a hole does not change the correlator. Another natural normalization condition is $\beta_1(\mathsf{A}) = \beta_2(\mathsf{A}) = \sqrt{\dim \mathsf{A}}$. It has the advantage of being invariant with respect to tensoring the line operator λ with a finite-dimensional vector space of dimension n (this procedure multiplies β_1 and β_2 by n and $\dim \mathsf{A}$ by n^2).

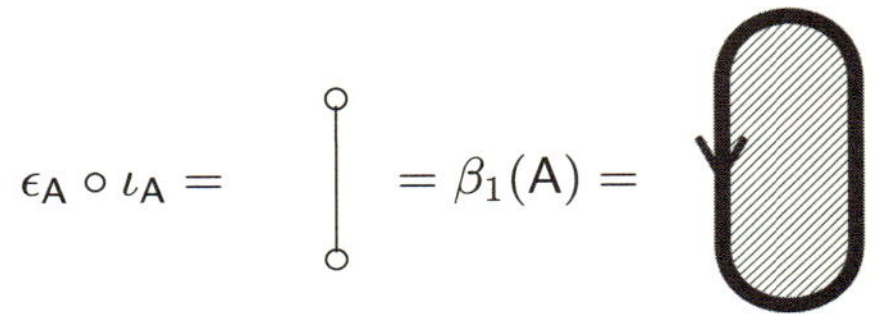

FIGURE 12. Composition of the unit and the counit.

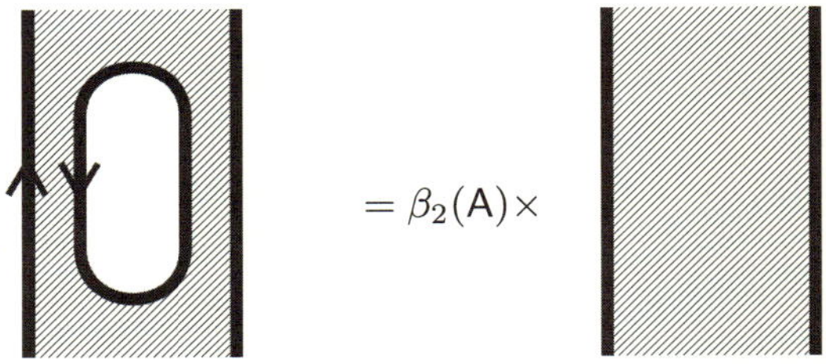

FIGURE 13. Punching a hole in a surface operator.

We have seen so far that to every surface operator S and a line operator λ on which it can terminate one can attach a symmetric Frobenius algebra in the category $\mathcal{C}$. On the other hand, the authors of [**FRS**] show that to any special symmetric Frobenius algebra one can associate a consistent gluing of chiral and anti-chiral sectors. Thus if S is special, we have two ways to associate a consistent gluing to the pair (S, λ): either we ignore λ and simply use the surface operator S to construct such a gluing, or we first compute the algebra $W(S, \lambda)$ in $\mathcal{C}$ and them

[3] One can show that this identity holds for an arbitrary special symmetric Frobenius algebra A [**FRS**].

construct the gluing following the recipe of [**FRS**]. It is not difficult to see that these two procedures give the same result, up to an overall scalar factor. Consider for simplicity the case when the Riemann surface Σ has empty boundary. According to the first approach, the gluing is given by a vector $v_S \in \mathcal{H}_\Sigma \otimes \bar{\mathcal{H}}_\Sigma$ which corresponds to the 3-manifold $\Sigma \times I$ with an insertion of the surface operator S at the midpoint of I. Using the fact that S is special, we may punch circular holes in S while changing v_S at most by a scalar factor. The surface operator S with holes in it can be deformed into a ribbon graph whose edges are labeled by objects $W(S,\lambda)$ and vertices correspond to the product morphism $m : W(S,\lambda) \otimes W(S,\lambda) \to W(S,\lambda)$. Thus we conclude that v_S can be obtained by considering $\Sigma \times I$ with an insertion of a ribbon graph. This is precisely the prescription of [**FRS**].

Note that the construction of [**FRS**] seems to depend not only on the surface operator S, but also on the choice of λ. The above discussion makes it clear that in fact v_S must be independent of λ, up to a scalar factor. This can be demonstrated as follows. First, we note that given a surface operator S and two line operators λ and λ' on which S can terminate, we have two canonical objects in the category $\mathcal{C}$: the strip of S bounded from the left by λ and from the right by $\bar{\lambda}'$, and the strip of S bounded from the left by λ' and from the right by $\bar{\lambda}$. We will denote them $W(S,\lambda,\lambda')$ and $W(S,\lambda',\lambda)$ respectively. It is easy to see that these are Morita-equivalence bimodules, so the algebras $W(S,\lambda)$ and $W(S,\lambda')$ are Morita-equivalent. Second, it is shown in [**FFRS**] that up to a scalar factor, the vector v_S depends only on the Morita-equivalence class of the algebra $W(S,\lambda)$.

3. Example: $U(1)$ current algebra

3.1. Surface operators in $U(1)$ Chern-Simons theory. To make our discussion more concrete, in the rest of this paper we will analyze the example of $U(1)$ Chern-Simons theory at level $k = 2N$. The corresponding chiral algebra $\mathcal{A}$ is the $U(1)$ current algebra at level $2N$. We are interested in classifying 2d RCFTs based on this current algebra. The category of conformal blocks (equivalently, the category of Wilson line operators in the $U(1)$ Chern-Simons theory) is a particular example of the following general construction. Let $\mathcal{D}$ be a finite abelian group. We will denote by $\mathsf{Vect}_\mathcal{D}$ the category of $\mathcal{D}$-graded finite-dimensional complex vector spaces. Its has an obvious monoidal structure given by the "convolution": the tensor product of a vector space E in degree $X \in \mathcal{D}$ and E' in degree $X' \in \mathcal{D}$ is a vector space $E \otimes E'$ in degree $X + X'$. The monoidal category $\mathsf{Vect}_\mathcal{D}$ has a deformation parameterized by an element $h \in H^3(\mathcal{D}, \mathbb{C}^*)$ which describes the associator morphism. In order for the deformation to be braided monoidal, h must have a special form: it must arise from a quadratic function $\mathsf{q} : \mathcal{D} \to \mathbb{R}/\mathbb{Z}$ [**JS**]. An explicit formula for h in terms of q can be found for example in [**S, KS**]. The braiding and the ribbon structure are also determined by q. We will denote by $\mathsf{Vect}^\mathsf{q}_\mathcal{D}$ the ribbon category parameterized by the quadratic function q.

In case of Chern-Simons theory with $G = U(1)$ at level $k = 2N$ we have $\mathcal{D} = \mathbb{Z}/2N\mathbb{Z}$; elements of $\mathcal{D}$ label Wilson line operators modulo those which can terminate on monopole operators (see [**KS**] for an explanation of this). The quadratic function $\mathsf{q} : \mathcal{D} \to \mathbb{R}/\mathbb{Z}$ has the form

$$\mathsf{q}(X) = \frac{X^2}{4N}, \quad X \in \mathbb{Z}/2N\mathbb{Z}. \tag{3.1}$$

Simple objects in $\mathsf{Vect}^{\mathsf{q}}_{\mathcal{D}}$ are labeled by elements of $\mathcal{D}$; we will denote by W_X the object labeled by $X \in \mathcal{D}$. The tensor product of simple objects W_X and W_Y is identified with W_{X+Y}; the braiding isomorphism can be thought of as a number

$$c_{X,Y} = \exp(\pi i \mathsf{g}(X,Y)),$$

where $\mathsf{g}(X,Y)$ is the $\mathbb{R}/\mathbb{Z}$-valued bilinear function on $\mathcal{D} \times \mathcal{D}$ derived from the quadratic function q:

$$\mathsf{g}(X,Y) = \mathsf{q}(X+Y) - \mathsf{q}(X) - \mathsf{q}(Y) = \frac{XY}{2N}.$$

Note that $c_{X,Y}$ is only well-defined up to a sign; to make it completely unambiguous one needs to choose particular representatives in $\mathbb{Z}$ for all elements $X \in \mathbb{Z}/2N\mathbb{Z}$. The physical reason for this is explained in [**KS**]. Usually one chooses representatives so that they lie in the interval from 0 to $2N-1$. We denote by W_i the Wilson line operator corresponding to an integer i in this interval.

Surface operators in this theory have been discussed in [**KS**]. It was shown there that they are labelled by divisors of N, i.e. $vm = N$ for some integer m. Let us recall how to construct a surface operator corresponding to a divisor v. We introduce a periodic scalar φ living on the support Σ of the surface operator located at $x^3 = 0$ and write the total action as:

$$S_{tot} = \frac{ik}{4\pi}\int_{x^3<0} A^{(+)}dA^{(+)} + \frac{ik}{4\pi}\int_{x^3>0} A^{(-)}dA^{(-)} + \frac{iv}{2\pi}\int_{\Sigma} \varphi(dA^{(+)} - dA^{(-)})$$

where $A^{(\pm)}$ are the limiting values of gauge fields to the left/right of Σ. The field φ should be viewed as a Lagrange multiplier field whose presence enforces the constraint that the connection 1-form $v(A^{(+)} - A^{(-)})$ is trivial on Σ. Note that if φ had been an ordinary non-periodic scalar, performing the functional integral over it would have produced the constraint $d(A^{(+)} - A^{(-)}) = 0$ on Σ, leaving the holonomy of the flat connection $A^{(+)} - A^{(-)}$ undetermined. The fact that φ takes values in $\mathbb{R}/2\pi\mathbb{Z}$ means that one also needs to perform a summation over its winding numbers (i.e. over the periods of the closed 1-form $d\varphi$). It is easy to show that this summation enforces the constraint that the holonomy of $A^{(+)} - A^{(-)}$ takes values in the group of vth roots of unity (*cf.* the discussion in section 2.2 of [**W2**]).

Let $G^{(\pm)}$ denote the groups of gauge transformations acting on $A^{(\pm)}$. The constraint

$$v(A^{(+)} - A^{(-)}) = 0$$

is invariant under the subgroup of $G^{(+)} \times G^{(-)}$ which consists of elements of the form

$$\left(\alpha, \alpha + \frac{2\pi\ell}{v}\right), \quad \alpha : \Sigma \to \mathbb{R}/2\pi\mathbb{Z}, \quad \ell \in \mathbb{Z}/v\mathbb{Z}.$$

This means that at the surface Σ the gauge group is broken down to $G_0 \simeq U(1) \times \mathbb{Z}_v$. The functional integral with an insertion of such a surface operator is gauge-invariant and diffeomorphism-invariant if we assign to φ charge $w = 2m$ with respect to the $U(1)$ factor in G_0 and charge m with respect to the $\mathbb{Z}_v$ factor in G_0 [**KS**]:

$$\varphi \mapsto \varphi + w\alpha, \quad \varphi \mapsto \varphi + m\frac{2\pi\ell}{v}.$$

The "invisible" surface operator is a special case corresponding to $v = 1$. Indeed, in this case the constraint reads $A^{(+)} = A^{(-)}$, which means that the gauge field is continuous on Σ.

A surface operator of the above kind may be characterized by the charges of Wilson lines which may meet at a point on the surface operator. That is, imagine a surface operator in $\mathbb{R}^3$ whose support Σ is a hyperplane dividing the space into left and right half-spaces, a Wilson line with charge X_1 coming to the surface from the left and a Wilson line with charge X_2 coming to the surface from the right, so that Wilson lines meet at $p \in \Sigma$. The net charge $X_1 + X_2$ need not be zero because one may insert a local operator of the form $\exp(i\nu\varphi)$. It was shown in section 6.2 of [**KS**] that for a fixed divisor v X_1 and X_2 must satisfy

$$X_1 - X_2 = 0 \bmod 2v, \quad X_1 + X_2 = 0 \bmod 2N/v. \tag{3.2}$$

Note that for $v = 1$ the second equation says that $X_1 = -X_2$ in $\mathcal{D}$, while the first equation is trivially satisfied. This means that the charge of the Wilson line does not jump at all, which is consistent with the fact that for $v = 1$ we are dealing with the "invisible" surface operator.

The set of pairs (X_1, X_2) satisfying the above equations defines a subgroup $\mathcal{L}_v$ of $\mathbb{Z}_{2N} \times \mathbb{Z}_{2N}$. More generally, a surface operator in any abelian Chern-Simons theory can be described by a subgroup $\mathcal{L}$ of $\mathcal{D} \times \mathcal{D}$ where the finite abelian group $\mathcal{D}$ classifies charges of bulk Wilson lines [**KS**]. This subgroup must be Lagrangian with respect to the quadratic form $\mathsf{q} \oplus (-\mathsf{q}) : \mathcal{D} \times \mathcal{D} \to \mathbb{R}/\mathbb{Z}$. Here "Lagrangian" means that the quadratic form vanishes when restricted to $\mathcal{L}$, and any element orthogonal to $\mathcal{L}$ (with respect to the bilinear form g derived from the quadratic form) is in $\mathcal{L}$. The Lagrangian subgroup $\mathcal{L}$ describes the charges of Wilson lines which can meet at the surface operator.

3.2. Algebra-objects corresponding to surface operators. Now let us compute the algebra-object $W(S_v, \lambda)$ corresponding to a surface operator S_v and a choice of a line operator λ on which S_v can terminate. Any object in the category of bulk line operators can be written in the form

$$\bigoplus_{i=0}^{2N-1} V_i \otimes W_i,$$

where the simple objects W_i have been defined above. The "expansion coefficients" V_i are finite-dimensional complex vector spaces. To compute V_i one needs to evaluate the vector space corresponding to $U(1)_{2N}$ Chern-Simons theory on $S^2 \times \mathbb{R}_t$ with an insertion of $W(S_v, \lambda)$ and the dual of W_i. Here $\mathbb{R}_t$ is the time coordinate. An insertion of $W(S_v, \lambda)$ can be thought of as an insertion of a surface operator S_v along a submanifold $I \times \mathbb{R}_t$, where I is a segment of the meridian of S^2, see Fig. 14.

It is easy to see that the vector space V_i is one-dimensional if the dual of W_i can cancel the charge carried by the Wilson lines on which S_v terminates, and is zero-dimensional otherwise. Thus all we need to determine is what bulk electric charge the surface operator bounded by λ and $\bar{\lambda}$ can carry. It is convenient to use the folding trick: squash the S^2 in Fig. 13 along the meridian on which S_v is placed. Then one gets $U(1)_{2N} \times U(1)_{-2N}$ Chern-Simons theory on a disk. The boundary of the disk is subdivided into two segments, and two different boundary conditions are imposed along these two segments. One of them is the boundary condition corresponding to the "invisible" surface operator, and the other one corresponds to the boundary condition S_v. At the two junctions of these segments one inserts line operators λ and $\bar{\lambda}$. We would like to determine which bulk line operators can

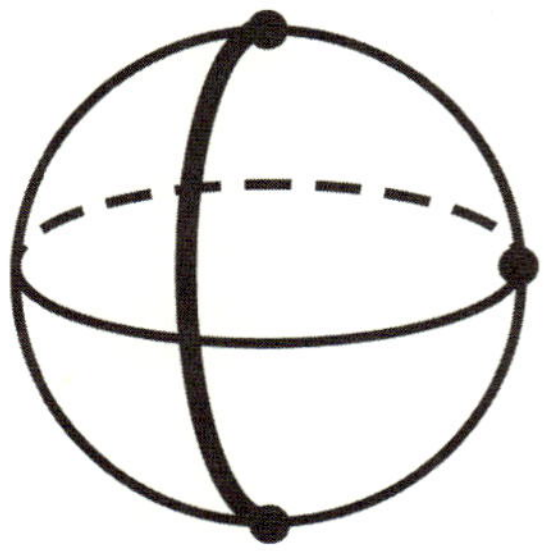

FIGURE 14. Computation of the algebra-object $W(S_v, \lambda)$. The surface operator S_v is placed along a meridian of S^2 and terminates on line operators λ and $\bar{\lambda}$. To get the "expansion coefficient" V_i one also needs to insert the bulk Wilson line dual to W_i.

be inserted in the interior of the disk to cancel the charge created by λ and $\bar{\lambda}$. Recall [**KS**] that boundary conditions in such a theory correspond to Lagrangian subgroups in the finite abelian group $\mathcal{D} \times \mathcal{D}$ equipped with the quadratic form $\mathsf{q} \oplus (-\mathsf{q})$, where elements of $\mathcal{D} \times \mathcal{D} \simeq \mathbb{Z}_{2N} \times \mathbb{Z}_{2N}$ label charges of bulk line operators. The surface operator S_v corresponds to the Lagrangian subgroup $\mathcal{L}_v$ defined by (3.2). The "invisible" surface operator is obtained by setting $v = 1$.

The meaning of this result is the following. The boundary gauge group is the subgroup of the bulk gauge group, and accordingly the group of boundary electric charges is the quotient of the group of bulk electric charges (i.e. it is a quotient of $\mathcal{D} \times \mathcal{D}$). The Lagrangian subgroup consists precisely of those bulk charges which project to zero in the group of boundary charges. Where boundary conditions corresponding to Lagrangian subgroups $\mathcal{L}_1$ and $\mathcal{L}_v$ meet along a line, the gauge group is broken further, and accordingly any bulk charge which belongs to $\mathcal{L}_1 + \mathcal{L}_v$ will be trivial as regards the unbroken gauge group at the junction. This has the following consequence: while the line operators sitting at the two ends of I have opposite charges with respect to the gauge group left unbroken at the junction, the net charge may be an arbitrary element of $\mathcal{L}_{tot} = \mathcal{L}_1 + \mathcal{L}_v$ when lifted to $\mathcal{D} \times \mathcal{D}$.

The subgroup $\mathcal{L}_{tot}$ consists of elements of the form

$$(n_1+x, n_2-x), \quad n_1+n_2 = 0 \bmod 2N/v, \quad n_1-n_2 = 0 \bmod 2v, \quad x = 0, \ldots, 2N-1.$$

We want to cancel this charge by a Wilson line whose charge after the folding trick has the form $(Y, 0)$, where Y is an arbitrary element of $\mathcal{D}$. Therefore we must require

$$n_2 = x, \quad Y = 2Nj/v,$$

where j is an arbitrary integer. Hence the algebra-object $W(S_v, \lambda)$ has the following expansion:

$$W(S_v, \lambda) = \bigoplus_{j=0}^{v-1} W_{jw},$$

where $w = 2N/v$. This is exactly the algebra-object denoted A_w in [**FRS**]. Thus we reproduced the classification of consistent RCFTs with a $U(1)_{2N}$ current algebra.[4] Note that all the symmetric Frobenius algebras A_w happen to be special.

3.3. Fusion of surface operators. Surface operators in a 3d TFT can be fused together; this gives rise to a monoidal structure (i.e. a kind of tensor product) on the 2-category of surface operators. In general, fusing two surface operators which do not admit local operators may produce a surface operator which does admit nontrivial local operators. In this subsection we discuss fusion of surface operators in the $U(1)_{2N}$ Chern-Simons theory and compare with analogous results for algebra-objects in [**FRS**]. A monoidal structure on a 2-category is a complicated notion which involves associator 1-morphisms, 2-associators between tensor products of associator 1-morphisms, and a coherence equation for 2-associators, see e.g. [**KV**] for a detailed discussion. Here we will only discuss how the tensor product works on the level of objects.

As discussed above, a surface operator may be characterized by charges of Wilson lines which may meet at it. These charge live in a Lagrangian subgroup $\mathcal{L}$ in $\mathcal{D} \times \mathcal{D}$. It is convenient to flip the orientation of the Wilson lines which comes in from the right, i.e. to think of it as an outgoing Wilson line instead of the incoming one. Thus instead of thinking about two Wilson lines coming to the surface from two different sides we will be thinking about a Wilson line which comes in from the left, pierces the surface operator and continues going in the same direction. A surface operator may cause a jump in the charge of a Wilson line from X_1 to X_2, and we will characterize a surface operator S_v by the set of the allowed jumps. Clearly, this set is a subset of $\mathcal{D} \times \mathcal{D}$ which may be obtained by taking $\mathcal{L}_v$ and flipping the sign of the second component. We will denote the resulting subset R_v and will think of it as a relation between $\mathcal{D}$ and $\mathcal{D}$.

The characterization of surface operators by relations is particularly useful when we consider the fusion of surface operators S_v and $S_{v'}$. Clearly, a Wilson line W_a may turn into a Wilson line W_c after meeting the composite surface operator $S_v \circ S_{v'}$ if and only if there exists $b \in \mathcal{D}$ such that[5] $(a, b) \in R_v$ and $(b, c) \in R_{v'}$. In other words, the composition of surface operators corresponds to the composition of relations.

It easy to see that the surface operator S_1 is the identity with respect to the fusion product, as expected. Indeed, S_1 is determined by the Lagrangian subgroup $\mathcal{L}_1$ consisting of elements of the form $(a, -a)$, $a \in \mathcal{D}$, so the corresponding relation R_1 is the diagonal in $\mathcal{D} \times \mathcal{D}$, i.e. the identity relation. Another interesting case is S_N; according to [**FRS**] the corresponding algebra-object A_2 is related to T-duality (i.e. the surface operator S_N is the T-duality wall). Let us compute the fusion product of S_v and S_N. The relation R_N corresponding to S_N is given by

$$a + b = 0, \quad a, b, \in \mathbb{Z}/2N\mathbb{Z}.$$

[4]In principle, we should also check that the Frobenius algebra structures agree. However, this is not really necessary since there is a unique (up to isomorphism) Frobenius algebra structure on the object A_w.

[5]If more than one such b exists for a given pair (a, c), then the composite surface operator is not merely $S_{v''}$ for some v'' but a sum of several such surface operators.

Composing it with R_v (in any order), we get a relation in $\mathcal{D} \times \mathcal{D}$ defined by pairs (a, b) such that

$$a - b = 0 \bmod 2v, \quad a + b = 0 \bmod 2N/v.$$

This is the relation $R_{N/v}$. Hence

$$S_v \circ S_N = S_N \circ S_v = S_{N/v}. \tag{3.3}$$

In particular, we see that the surface operator S_N is invertible: its inverse is S_N. This means that it implements a symmetry of the $U(1)_{2N}$ Chern-Simons theory. The relation (3.3) means that this symmetry maps the consistent gluing corresponding to S_v to the consistent gluing corresponding to $S_{N/v}$. This agrees with [**FRS**].

In general computing the composition of relations gives the following formula:

$$S_v \circ S_{v'} = g S_{v''},$$

where

$$g = \gcd(v, v', N/v, N/v'), \quad v'' = \operatorname{lcm}(\gcd(v, N/v'), \gcd(v', N/v)).$$

A proof (due to A. Tsymbalyuk) is given in the appendix. In particular we see that S_v is invertible if and only if $\gcd(v, N/v) = 1$, and that in this case the inverse is again S_v. These results agree with [**FRS**].

4. Boundary state map in 3d TFT

We have seen that to every surface operator S and a line operator λ on which it can terminate one can associate a symmetric Frobenius algebra in the ribbon category of bulk line operators. The proof made use of pictures which resemble open-string scattering diagrams. It is natural to ask if there is an analogous use for closed-string scattering diagrams in 3d TFT. In fact, closed-string scattering diagrams appear when one defines the 3d analogue of the boundary state familiar from 2d TFT. In the two-dimensional case for every object in the category of boundary conditions one can define a vector in the space of bulk local operators in the following way. Consider an annulus on whose interior boundary component one imposed the chosen boundary condition and whose exterior boundary component is a "cut" boundary. The path-integral of the 2d TFT on such a manifold defines a vector in the vector space associated to S^1, i.e. the space of local operators. From the categorical viewpoint, this space is the "derived center" (Hochschild cohomology) of the category of boundary conditions. This map from the set of boundary conditions to the space of bulk local operators is called the boundary state map.

If we consider the same annulus in a 3d TFT, it defines an object in the category associated to S^1. The latter category can be thought of either as the category of bulk line operators in the 3d TFT or as the Drinfeld center of the 2-category of boundary conditions. Thus we get a map from the set of boundary conditions to the set of bulk line operators. This is the 3d analogue of the boundary state map.

It turns out that the bulk line operator one gets from the boundary state map is very special: it is a commutative symmetric Frobenius algebra in the braided monoidal category of bulk line operators. This fact has no 2d analogue. To see why it is true one uses pictures resembling closed-string scattering diagrams. Let A be the bulk line operator corresponding to some boundary condition. Pictorially, it is represented by a hollow cylindrical tube in $\mathbb{R}^3$ on the boundary of which one imposes the boundary condition one is interested in. The product, coproduct, unit and counit arise from diagrams shown in Fig. 15. One should think about these diagrams as embedded in a three-dimensional ball, with the boundary circles placed on the boundary of the ball.

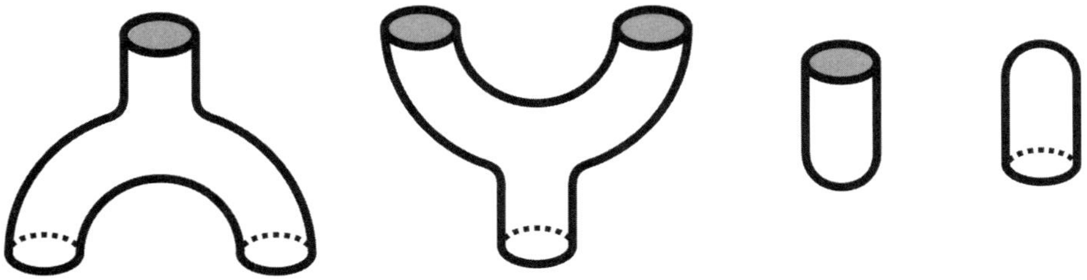

FIGURE 15. Product, coproduct, unit and counit for the algebra A corresponding to a boundary condition in a 3d TFT. The interior of the "termite holes" contains vacuum (trivial 3d TFT).

Commutativity means that the product morphism $m : \mathsf{A} \otimes \mathsf{A} \to \mathsf{A}$ is compatible with the self-braiding $c : \mathsf{A} \otimes \mathsf{A} \to \mathsf{A} \otimes \mathsf{A}$:

$$m \circ c = m.$$

The algebra A is commutative because the two diagrams shown in Fig. 16 can be deformed into one another.

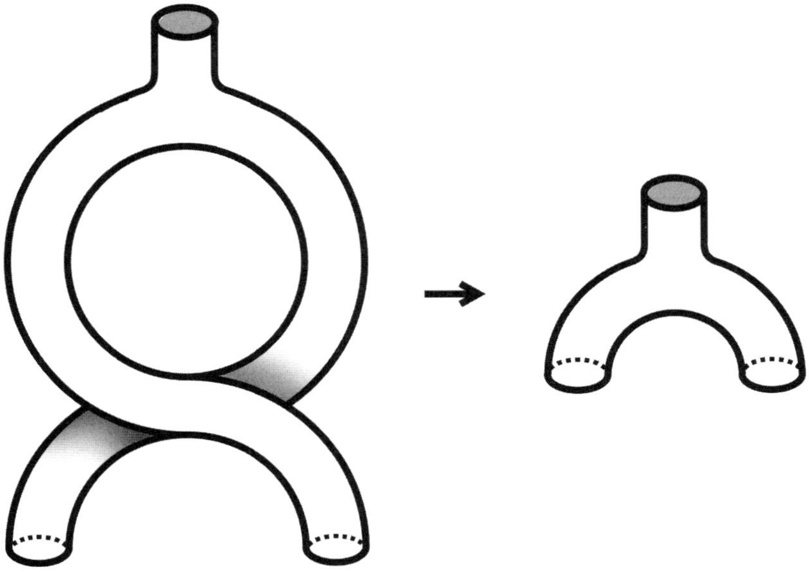

FIGURE 16. The commutativity constraint.

To give a concrete example, consider $U(1)_{2N} \times U(1)_{-2N}$ Chern-Simons theory. Boundary conditions in this theory are the same as surface operators in $U(1)_{2N}$ theory and, as explained in the previous section, are classified by divisors of N. Alternatively, they are classified by subgroups of $\mathbb{Z}_{2N} \times \mathbb{Z}_{2N}$ which are Lagrangian with respect to the quadratic function

$$\mathsf{q}(X_1, X_2) = \frac{1}{4N}(X_1^2 - X_2^2).$$

If v is a divisor of N, the corresponding Lagrangian subgroup $\mathcal{L}_v$ is given by (3.2). We denote by $W(\mathcal{L}_v)$ the bulk line operator corresponding to the Lagrangian subgroup $\mathcal{L}_v$ via the 3d boundary-bulk map.

We would like to compute the expansion of $W(\mathcal{L}_v)$ in terms of simple objects W_X where $X \in \mathbb{Z}_{2N} \times \mathbb{Z}_{2N}$ labels the bulk electric charge. The "expansion coefficient" is a vector space V_X which is the vector space corresponding to the disk with an insertion of a simple object $\overline{W}_X$ at the origin and with the boundary condition corresponding to $\mathcal{L}_v$ imposed on the boundary. It is easy to see that V_X is one-dimensional if X has vanishing charge with respect to the boundary gauge group and is zero-dimensional otherwise. Since $\mathcal{L}_v$ consists precisely of those bulk charges which are trivial with respect to the boundary gauge group, we conclude that $W(\mathcal{L}_v)$ is given by[6]

$$W(\mathcal{L}_v) = \bigoplus_{X \in \mathcal{L}_v} W_X.$$

The product morphism $m : W(\mathcal{L}_v) \otimes W(\mathcal{L}_v) \to W(\mathcal{L}_v)$ is essentially determined up to isomorphism by charge conservation. Indeed, charge conservation tells us that m maps the component $W_X \otimes W_Y = W_{X+Y}$ of $W(\mathcal{L}_v) \otimes W(\mathcal{L}_v)$ to the component W_{X+Y} of $W(\mathcal{L}_v)$. Assuming that this morphism is nonzero, we see that m is determined by a $\mathbb{C}^*$-valued function $m(X, Y)$ on $\mathcal{L}_v \times \mathcal{L}_v$. The associativity constraint reads

$$(\delta m)(X, Y, Z) = a(X, Y, Z), \quad X, Y, Z \in \mathcal{L}_v,$$

where δ is the differential in the standard cochain complex computing cohomology of $\mathcal{L}_v$ with coefficients in $\mathbb{C}^*$, and $a(X, Y, Z) \in \mathbb{C}^*$ is the restriction of the associator morphism to $\mathcal{L}_v$. Since $\mathcal{L}_v$ is isotropic (in fact, Lagrangian) with respect to the quadratic function q, the restriction of a to $\mathcal{L}_v$ is trivial [**KS**]. This means that $m(X, Y)$ is a 2-cocycle. Further, recall that $W(\mathcal{L}_v)$ is a commutative algebra. Since λ_v is isotropic with respect to the quadratic function q, for any two $X, Y \in \mathcal{L}_v$ the braiding is trivial. Hence the commutativity constraint becomes

$$m(X, Y) = m(Y, X).$$

Such a 2-cocycle is necessarily cohomologically trivial [**Brown**], i.e. there exists a function $n : \mathcal{L}_v \to \mathbb{C}^*$ such that

$$m(X, Y) = \frac{n(X + Y)}{n(X)n(Y)}.$$

[6]This is analogous to the following fact about Rozansky-Witten theory with target $\mathcal{X}$: the boundary state maps the boundary condition corresponding to a complex Lagrangian submanifold $\mathcal{Y}$ of $\mathcal{X}$ to the object of the derived category of $\mathcal{X}$ given by the push-forward of the structure sheaf of $\mathcal{Y}$ [**KRS**].

This means that $m(X, Y)$ can be made equal to 1 by replacing the algebra $(W(\mathcal{L}_v), m)$ with an isomorphic one, where the action of the isomorphism on the component W_X of $W(\mathcal{L}_v)$ is given by $n(X)$.

5. Concluding remarks

The main observation of this paper is that the classification of surface operators in a 3d TFT is equivalent to the classification of consistent gluings of chiral and anti-chiral sectors in the corresponding 2d RCFT. This observation led us to reinterpret the results of Fuchs, Runkel and Schweigert on the algebraic classification of consistent gluings in terms of surface operators. We considered a simple example of this relationship, $U(1)$ Chern-Simons theory, and showed that all 2d RCFTs with a $U(1)$ current algebra arise from surface operators. We note that the construction of gluing by means of surface operators seems to be more general than constructing of gluings by means of special symmetric Frobenius algebras. Indeed, while to any surface operator one can attach a symmetric Frobenius algebra (or rather, a Morita-equivalence class of symmetric Frobenius algebras), this algebra may fail to be special. In the case of $U(1)$ Chern-Simons theory all surface operators happen to yield special symmetric Frobenius algebras, so the two methods of constructing consistent gluings are equivalent.

Obviously, it would be desirable to classify surface operators in nonabelian Chern-Simons theory. However, the example of $SU(2)$ Chern-Simons theory indicates that this will not be easy. It is well known that 2d RCFTs with $SU(2)_k$ current algebra have ADE classification [**ADE**]. The A-type RCFTs exist for all levels k, the D-type RCFTs exist for all even k, while the E-type RCFTs exist for three "sporadic" values $k = 10, 16, 28$. This means that surface operators in $SU(2)$ Chern-Simons theory should have ADE classification as well. Clearly, the A-type surface operator must be the "invisible" surface operator. It is plausible that the D-type surface operator is defined by the following condition: if the $SU(2)$ gauge fields to the left and to the right of the surface operator are denoted $A^{(+)}$ and $A^{(-)}$, then the restrictions of $A^{(+)}$ and $A^{(-)}$ to the surface agree as $SO(3)$ gauge fields (but not necessarily as $SU(2)$ gauge fields). On the other hand, the E-type surface operators are much more mysterious, and the fact that they exist only for small enough k suggests that they cannot be understood using semiclassical reasoning at all.

It would also be interesting to extend our results to spin-topological 3d theories which depend on a choice of spin structure on a manifold. The simplest example of such a theory is $U(1)$ Chern-Simons theory at an odd level k. This theory describes Fractional Quantum Hall phases and therefore a classification of codimension-1 defects in this theory would be of considerable interest.

Appendix

PROPOSITION. (A. Tsymbalyuk)

Let N be a natural number, v be a divisor of N, and R_v be a relation (in fact, a subgroup) in $\mathbb{Z}_{2N} \times \mathbb{Z}_{2N}$ given by

$$a + b = 0 \bmod 2v, \quad a - b = 0 \bmod 2N/v.$$

Then for any two divisors v, v' we have

$$R_v \circ R_{v'} = gR_x,$$

where

$$g = \gcd(v, v', N/v, N/v'), \quad x = \operatorname{lcm}(\gcd(v, N/v'), \gcd(v', N/v)).$$

PROOF. Let a, b, c be integers such that $(a, b) \in R_v$ and $(b, c) \in R_{v'}$. Since

$$a + c = (a - b) + (b + c) = (a + b) - (b - c),$$

$a + c$ is divisible by $2\gcd(v', N/v)$ as well as by $2\gcd(v, N/v')$, and therefore is divisible by $2x$ where

$$x = \operatorname{lcm}(\gcd(v, N/v'), \gcd(v', N/v)).$$

Similarly, since

$$a - c = (a + b) - (b + c) = (a - b) + (b - c),$$

$a - c$ is divisible by $2y$ where

$$y = \operatorname{lcm}(\gcd(v, v'), \gcd(N/v, N/v').$$

To prove that $R_v \circ R_{v'} \subset R_x$ it is sufficient to show that $xy = N$. Let p be a prime factor of N, and let $\alpha = \operatorname{ord}_p N$, $i = \operatorname{ord}_p v$, $j = \operatorname{ord}_p v'$. Then

$$\operatorname{ord}_p x = \max(\min(i, \alpha - j), \min(\alpha - i, j)), \quad \operatorname{ord}_p y = \max(\min(i, j), \min(\alpha - i, \alpha - j)).$$

Since there is a symmetry exchanging v and v' and i and j, it is sufficient to consider the case $i \leq j$. Then there are four possible relative orderings of four numbers $i, j, \alpha - i, \alpha - j$. For each of them one can easily verify that $\operatorname{ord}_p x + \operatorname{ord}_p y = \operatorname{ord}_p N$. (For example, if $i \leq j \leq \alpha - j \leq \alpha - i$, then $\operatorname{ord}_p x = j$, $\operatorname{ord}_p y = \alpha - j$, and therefore $\operatorname{ord}_p x + \operatorname{ord}_p y = \alpha$). This proves that $xy = N$.

The opposite inclusion $R_v \circ R_{v'} \supset R_x$ follows from the Chinese remainder theorem. Therefore we have

$$R_v \circ R_{v'} = gR_x$$

for some natural number g.

Finally, let us compute g. Given a relation $R \subset \mathbb{Z}_{2N} \times \mathbb{Z}_{2N}$ we will say that $a \in \mathbb{Z}_{2N}$ is in the domain of R if there exists $b \in \mathbb{Z}_{2N}$ such that $(a, b) \in R$. It follows from the Chinese remainder theorem that for a fixed a in the domain of R_v there exist $\gcd(v, N/v)$ solutions of the condition $(a, b') \in S_v$. Hence if a is in the domain of $R_v \circ R_{v'}$, then there exist $\gcd(v, N/v) \cdot \gcd(v', N/v')$ solutions of the condition $(a, c') \in R_v \circ R_{v'}$. On the other hand, if $R_v \circ R_{v'} = gR_x$, this number must be equal to $g \cdot \gcd(x, N/x)$. Hence we have a formula for g:

$$g = \frac{\gcd(v, N/v) \cdot \gcd(v', N/v')}{\gcd(x, y)}.$$

The ratio on the right-hand-side of this formula is in fact equal to $\gcd(v, v', N/v, N/v')$. This is proved by analyzing the prime factorization of N as above. Let p be a prime factor of N, and $\alpha = \operatorname{ord}_p N$, $i = \operatorname{ord}_p v$, $\operatorname{ord}_p v' = j$. First of all, we have

$$\operatorname{ord}_p g = \min(i, \alpha - i) + \min(j, \alpha - j) - \min(\operatorname{ord}_p x, \operatorname{ord}_p y).$$

On the other hand, we have

$$\operatorname{ord}_p(gcd(v, v', N/v, N/v')) = \min(i, j, \alpha - i, \alpha - j).$$

Again we may assume without loss of generality that $i \leq j$, and then there are four cases to consider corresponding to four possible relative orderings of $i, j, \alpha - i, \alpha - j$. For example, if $i \leq j \leq \alpha - j \leq \alpha - i$, then $\operatorname{ord}_p g = i + j - j = i =$

$\text{ord}_p(gcd(v, v', N/v, N/v'))$. It is easy to verify that $\text{ord}_p g = \text{ord}_p(gcd(v, v', N/v, N/v'))$ in the other three cases as well. This concludes the proof.

□

References

[Witten] E. Witten, *Quantum field theory and the Jones polynomial*, Com. Math. Phys. **121**, p 351, 1989.

[EMSS] S. Elitzur, G. Moore, A. Schwimmer and N. Seiberg, *Remarks on the Canonical Quantization of the Chern-Simons-Witten Theory*, Nucl. Phys. B **326**, p 108, 1989.

[FRS] J. Fuchs, I. Runkel and C. Schweigert, *TFT construction of RCFT correlators. I: Partition functions*, Nucl. Phys. B **646**, p 353, 2002, [arXiv:hep-th/0204148].

[FRS2] J. Fuchs, I. Runkel and C. Schweigert, *TFT construction of RCFT correlators. II: Unoriented world sheets*, Nucl. Phys. B **678**, 511 (2004) [arXiv:hep-th/0306164].

[FRS3] J. Fuchs, I. Runkel and C. Schweigert, *TFT construction of RCFT correlators. III: Simple currents*, Nucl. Phys. B **694**, 277 (2004) [arXiv:hep-th/0403157].

[FRS4] J. Fuchs, I. Runkel and C. Schweigert, *TFT construction of RCFT correlators. IV: Structure constants and correlation functions*, Nucl. Phys. B **715**, 539 (2005) [arXiv:hep-th/0412290].

[FRS5] J. Fjelstad, J. Fuchs, I. Runkel and C. Schweigert, *TFT construction of RCFT correlators. V: Proof of modular invariance and factorisation*, Theor. Appl. Categor. **16**, 342 (2006) [arXiv:hep-th/0503194].

[FFRS] J. Frohlich, J. Fuchs, I. Runkel and C. Schweigert, *Duality and defects in rational conformal field theory*, Nucl. Phys. B **763**, 354 (2007) [arXiv:hep-th/0607247].

[FRSO] J. Fuchs, I. Runkel and C. Schweigert, *Conformal correlation functions, Frobenius algebras and triangulations*, Nucl. Phys. B **624**, 452 (2002) [arXiv:hep-th/0110133].

[Segal] G. Segal, *The definition of conformal field theory" in Topology, geometry and quantum field theory*, London Math. Soc. Lecture Note Ser., **308**, p 423, Cambridge Univ. Press, Cambridge, 2004.

[Cardy] J. Cardy, *Boundary Conditions, Fusion Rules and the Verlinde Formula*, Nucl. Phys. B **324**, (1989) p. 581.

[FFFS] G. Felder, J. Frohlich, J. Fuchs, C. Schweigert, *Correlation functions and boundary conditions in RCFT and three- dimensional topology*, Compos. Math. **131** (2002) p. 189.

[BK] B. Bakalov, A. Kirillov, *Lectures on tensor categories and modular functors* American Mathematical Society, 2000.

[JS] A. Joyal, R. Street, *Braided tensor categories*,Adv. Math. **102**, p 20, 1983.

[S] S. Stirling, *Abelian Chern-Simons theory with toral gauge group, modular tensor categories, and group categories*,[arXiv:0807.2857].

[KS] A. Kapustin, N. Saulina, *Topological boundary conditions in abelian Chern-Simons theory*, [arXiv:1008.0654].

[W2] E. Witten, *SL(2,Z) action on three-dimensional conformal field theories with Abelian symmetry*, In Shifman, M. (ed.) et al., *From fields to strings*, vol. **2**, pp. 1173-1200.

[KV] M. Kapranov, and V. Voevodsky, *2-categories and Zamolodchikov tetrahedra equations*, in *Algebraic groups and their generalizations: quantum and infinite-dimensional methods (University Park, PA, 1991)*, Proc. Sympos. Pure Math. **56**, p. 177.

[KRS] A. Kapustin, L. Rozansky, N. Saulina, *Three-dimensional topological field theory and symplectic algebraic geometry I*, Nucl. Phys. B **816**, p 295, 2009, [arXiv:0810.5415].

[Brown] K. S. Brown, *Cohomology of groups* (Springer Verlag, Berlin 1982).

[ADE] A. Cappelli, C. Itzykson, and J. -B. Zuber, *A-D-E Classification of Conformal Field Theories,* Commun. Math. Phys. **113**, p 1, 1987.

California Institute of Technology, 1200 E. California Blvd., Pasadena, CA 91125, United States

Current address: California Institute of Technology, 1200 E. California Blvd., Pasadena, CA 91125, United States

E-mail address: kapustin@theory.caltech.edu

Perimeter Institute, 31 Caroline Street North, Waterloo, Ontario, N2L2Y5, Canada

Current address: Perimeter Institute, 31 Caroline Street North, Waterloo, Ontario, N2L2Y5, Canada

E-mail address: saulina@theory.caltech.edu

Proceedings of Symposia in Pure Mathematics
Volume **83**, 2011

Conformal field theory and a new geometry

Liang Kong

Abstract. This paper is a review of open-closed rational conformal field theory (CFT) via the theory of vertex operator algebras (VOAs), together with a proposal of a new geometry based on CFTs and D-branes. We will start with an outline of the idea of the new geometry, followed by some philosophical background behind this vision. Then we will review a working definition of CFT slightly modified from Segal's original definition and explain how VOA emerges from it naturally. Next, using the representation theory of rational VOAs, we will discuss a classification result of open-closed rational CFTs, from which some basic properties of a rational CFT, such as the Holographic Principle, can be derived. They will also serve as supporting evidences for the vision of a new geometry. In the end, we briefly discuss the connection between our vision of a new geometry and other topics.

To see a world in a grain of sand,
And a heaven in a wild flower,
Hold infinity in the palm of your hand,
And eternity in an hour.

— William Blake "Auguries of Innocence"

1. Introduction and summary

This paper has grown out of many talks I have given in 2009 and 2010 on a new geometry based on conformal field theory (CFT). Although the idea is rooted in many works in both physics and mathematics, as far as I know, it has never been clearly stated and emphasized. It is not surprising that this sounds quite shocking and alien to some of my audiences, even to myself when I first realized it in May 2007. So in this paper, I will try to cover some philosophical aspects of this new geometry which are impossible to cover in my talks. Besides this vision of a new geometry, this paper is pretty much a review of open-closed CFT from the point of view of vertex operator algebra (VOA). To avoid any confusion, CFTs in this paper are all 2-dimensional. Moreover, a CFT in Section 1 and 2 means a CFT with or without supersymmetry, but for later sections, we

1991 *Mathematics Subject Classification.* Primary: 81T40; Secondary: 17B69, 18D10, 18D50, 18R10.

Key words and phrases. Conformal field theory, D-branes, vertex operator algebra, stringy algebraic geometry.

The author is supported in part by the Basic Research Young Scholars Program of Tsinghua University, Tsinghua University independent research Grant No. 20101081762 and by NSFC Grant No. 20101301479.

will only discuss CFT without supersymmetry because the mathematical foundations of open-closed supersymmetric CFT are still lacking.

String theory is proposed to be a theory of quantum gravity. By Einstein, gravity is nothing but geometry. Therefore, it is equally good to say that string theory is a proposal for a theory of *quantum geometry*. One of the aims of string theory is to answer how geometry emerges. All classical notions in geometry, such as points, lines, surfaces, dimensions and metrics, should all be derived concepts. They should arise as certain derived quantities, moduli or invariants from the abstract mathematical structures of string theory. By giving up all familiar geometric notions, string theory challenges us to find a radical new point of view of geometry.

It is impossible to know how to write a new geometry from nothing. Any theoretical speculation needs hints from experimental data. One of the testing grounds or laboratories of this new quantum geometry is provided by the so-called non-linear sigma models. These models connect mathematical structures of string theory directly to notions in classical geometry. The outcomes of this laboratory are many "phenomenological theories" manifesting themselves as new topics in geometry. To name a few: mirror symmetry, quantum cohomology, Gromov-Witten theory, topological string, elliptic cohomology, etc. What is the secret structure behind of all these clues?

Perhaps we can ask an easier question. What is the new message from string theory that is so distinguishable that it alone can explain why we missed so many new insights before the advent of string theory? There are perhaps many answers to this question. One answer is that string theory seems to emphasize a point of view from loop space which has richer structures than the original manifold. If we want to take this message seriously, we would immediately ask the question: what are the structures on loop space? Again, there are perhaps many answers to this question. One answer suggested by string theory is that the free loop space, or rather the space of functions on the free loop space, has a natural algebraic structure which is called closed conformal field theory. One should take the term "loop space" and "CFT" in a flexible way. They can have different meanings in different contexts. For example, loop space can be defined in both the topological context and the (derived-)algebro-geometric context.

CFT is a formidable subject and unfamiliar to most mathematicians. In spite of many works on this subject, its full nature is still very mysterious even to experts. Ironically, within the frameworks of known definitions of CFT, we cannot even claim to have a single nontrivial example! Nevertheless, many ingredients of CFT such as vertex operator algebra, modular functor, modular tensor category, etc. are explicitly constructed and relatively well understood. Moreover, many variants of CFT or VOA such as string topology [**ChSu**], topological CFT (TCFT) [**Cos**], homological CFT [**Go**], ..., etc. have been constructed. Very often, for convenience, we will pretend to talk about a CFT while we are only talking about its substructures such as a VOA or its variants such as a TCFT or string topology.

Inspired by the suggestion that the free loop space is a closed CFT, a few groups of mathematicians initiated programs to realize this picture. We will list some constructions by these mathematicians. The list is not complete due to my limited knowledge.

- Malikov, Schechtman and Vaintrob constructed the Chiral de Rham complex [**MSV**] which is a sheaf of vertex operator algebras (VOA) on a smooth manifold. It can be viewed as a shadow of certain structure on formal loop space.
- Kapranov and Vasserot [**KV1**] constructed an algebro-geometric version of free loop space and showed that it has a natural structure of factorization monoid, which can be viewed as a non-linear version of factorization algebra that contains VOA as local data.

- Chas and Sullivan constructed the closed string topology [**ChSu**], which is an algebraic structure on the homology of the free loop space. A closed string topology can be viewed as a homological closed conformal field theory [**Go**].
- Ben-Zvi, Francis and Nadler [**BFN**] showed that the stable symmetric monoidal ∞-category of quasi-coherent sheaves on the loop space of a perfect stack has the structure of a 2-dimensional topological field theory (TFT) [**Lu2**] which is a categorified analogue of a TCFT [**Cos**].

By all these works, we can be more confident about identifying a closed CFT as an algebraic model for a free loop space, at least for our philosophical discussion. More importantly, as a theory of quantum geometry, string theory demands us to forget about the loop space in the first place because it can only be a derived concept. Instead, we shall take a closed CFT as initial data to rebuild the entire geometry.

One convenient approach to see the connection between CFT and geometry is to work only with certain substructures of nonlinear sigma models. For example, substructures such as superconformal algebras, chiral rings, A-branes, B-branes, partition functions, etc., are directly linked to mirror symmetry, quantum cohomology, Fukaya category, bounded derived category of coherent sheaves, elliptic genus, respectively. This approach has been very successful so far. But the disadvantage of it is that we might lose the global picture (if there is any) and perhaps some deeper revelations.

In recent years, there have been some developments on the mathematical foundations of open-closed rational CFTs without supersymmetry[1] via the representation theory of VOAs [**HK1**][**HK2**][**Ko1**][**Ko2**][**Ko3**], and another successful and nearly complementary program led by Fuchs, Runkel and Schweigert [**FS**][**FRS1**][**FRS3**][**FRS4**][**FjFRS1**] in which an open-closed rational CFT is studied as a holographic boundary of a 3-dimensional topological field theory. These two approaches provide compatible results [**FRS1**][**KR2**] and can be unified [**KR3**]. Together, they give a rather complete[2] picture of rational open-closed CFTs. Perhaps, it is dangerous to use the word "complete" because we are always subject to certain axiom systems. But I believe that the system is rich enough to reveal many global features of CFT and to recover or link to other missing aspects. As we will review in later sections, these developments, in particular, give the following two results:

(1) A closed CFT is indeed a stringy generalization of a commutative ring. As we will show in Section 3.5, a closed CFT is a full field algebra satisfying a stringy commutativity condition, and also a commutative associative algebra in a braided tensor category [**Ko1**] (see Theorem 3.11).
(2) For a closed CFT $A_{\rm cl}$, its D-branes can be parametrized by modules over an open CFT compatible with $A_{\rm cl}$. It will become clear later that these modules are automatically *chiral modules*[3] over $A_{\rm cl}$ (see Remark 4.25, 5.4).

I believe that this relation between a closed CFT and its D-branes is a stringy analogue of that between a commutative ring and its prime ideals. Moreover, it is well known to physicists that D-branes, as boundary conditions for open strings, indeed behave like generalized points or submanifolds[4], e.g. X and Y in a target manifold M in Figure 1. This fact has been used by physicists to probe the geometry of the target manifold. What

---

[1]I believe that supersymmetric CFTs share similar properties with CFTs without supersymmetry. We will not distinguish them in our philosophical discussion in this section.

[2]High genus theory for rational CFTs is still not available. But they are uniquely determined by genus-0 theory and there are good reasons to believe that they indeed exist.

[3]The word "chiral" refers to the fact that it is a module over only one copy of the Virasoro algebra (see Remark 4.8).

[4]Usually, such a submanifold is equipped with additional structures, such as a bundle over it and a connection.

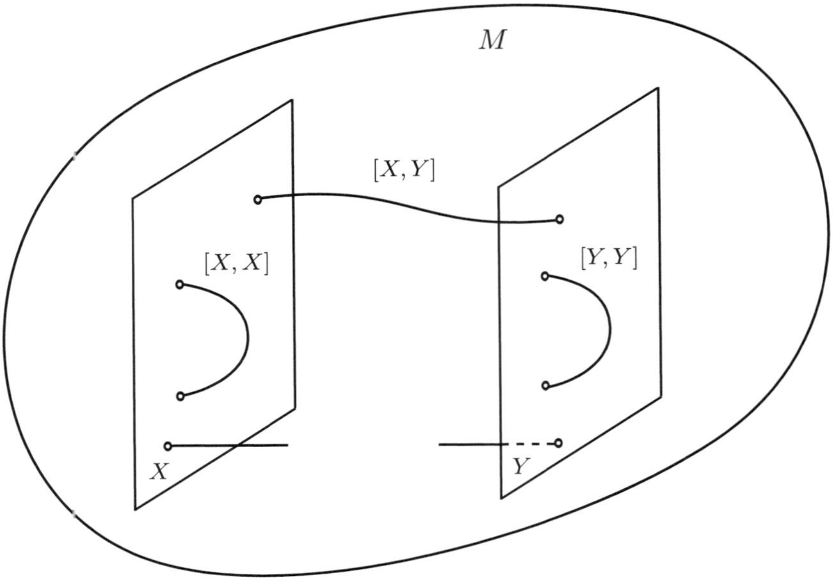

FIGURE 1. Geometric intuition of the category of D-branes associated to a target manifold M: a closed CFT A_{cl} models the free loop space; a boundary condition, i.e. a chiral A_{cl}-module X, might be able to model the space of $[0, 1)$-paths starting from a submanifold associated to X (also denoted by X); an open CFT $[X, X]$ models the space of paths from X to X; and the $[Y, Y]$-$[X, X]$-bimodule $[X, Y]$ models the space of paths from X to Y.

is suggested by above results is a surprisingly simple picture: CFT provides an immediate stringy generalization of classical algebraic geometry!

In order to get a more precise picture of this analogy, we have to take a closer look at the nature of the "set" of D-branes. Because the operator product expansion (OPE) of fields on a D-brane provides a stringy algebra structure, a module X over an open CFT A does not contain the information of fields on a D-brane in general. It is only a boundary condition literally. This fact becomes rather clear in Fuchs-Runkel-Schweigert's state-sum construction of rational CFTs via 3-d TFT techniques [**FRS1**], in which boundary edges of a triangulation of a surface are labeled by modules over an open CFT A which is a separable symmetric Frobenius algebra in a modular tensor category $\mathcal{C}$ (see Section 3.7, 4.6). As we will show later in this case, the closed CFT is given by the center $Z(A)$ of A. An A-module is automatically a chiral module over the closed CFT. Moreover, the category of A-modules is an indecomposable component of the category of chiral modules over $Z(A)$ (see Theorem 5.3). Very often, *boundary conditions* and *D-branes* are used synonymously in the literature. In this work, we will distinguish them. For us, a boundary condition is just an A-module X for an open CFT A and a chiral module over the closed CFT. Since the geometric intuition behind a closed CFT A_{cl} is given by the free loop space LM, I believe that the geometric intuition behind a chiral A_{cl}-modules X is the space of $[0, 1)$-paths starting from a submanifold associated to X[5] (see Figure 1). I believe that one can find a precise meaning of it in the framework of string topology. We

[5]A $[0, 1)$-path starting from a submanifold X of M is a continuous map $\gamma : [0, 1) \to M$ such that $\gamma(0) \in X$.

further assume that the open CFT A can be chosen to be commutative in a stringy sense[6] so that the category of boundary conditions $\mathcal{C}_A$ is a monoidal category and we can talk about internal homs[7]. In Section 4.6, we will give an example of such situation which is called Cardy case.

A D-brane is a physical (or dynamical) object, the physical content of which lies in its interaction with other D-branes (including itself) and the bulk theory. From the intuition of the loop space, the information of the "self-interaction" of a D-brane is contained in the fusion of all open strings with two ends ending on it. For example, when the boundary condition is just a point, all the open strings ending on it form a based loop space. The algebraic model for the fusion of open strings is given by an open CFT compatible with the given closed CFT. By "compatible" we mean that together they form an open-closed CFT (see Section 4.1 for a precise definition). As we will show later on, in a rational CFT, for a boundary condition X, the corresponding open CFT is given by the internal hom $[X, X]$ in $\mathcal{C}_A$. In this case, the closed CFT is also given by the center of $[X, X]$. For irrational CFTs, due to the existence of OPE, it is reasonable to believe that such internal-hom type of constructions still work. Geometrically, it is perhaps also true that one can view the space of paths from a submanifold X to X as some kind of "internal endo-hom" of the space of $[0, 1)$-paths starting from X. More general "interaction" between two different D-branes X, Y is given by the space of paths between them. Its algebraic model is given by a $[Y, Y]$-$[X, X]$-bimodule $[X, Y]$ which is also an internal hom in $\mathcal{C}_A$. In the end, we obtain a category of boundary conditions but enriched by internal homs[8]. We will call this enriched category *the category of D-branes*[9]. The geometric intuition behind the category of D-branes is depicted in Figure 1. A D-brane can then be defined by the functor $[X, -]$. Sometimes, it is also convenient, by only looking at its relation to the bulk theory, to define a D-brane by a pair $(X, [X, X])$ or just the open CFT $[X, X]$. In the rest of this work, we will ignore the ambiguity in the meaning of a D-brane, and leave readers to figure out which one is appropriate in the context. For a given boundary condition X, the map $Y \mapsto [X, Y]$ defines a functor from the category of boundary conditions to the category of right $[X, X]$-modules. This functor is an equivalence for all X in a rational CFT. Nothing is known for irrational CFTs, but we assume that the equivalence holds at least for some X. That is why we often see in the literature that the category of boundary conditions is defined to be the category of modules over an open CFT.

As we have mentioned, objects in the category of D-branes can be viewed as stringy primary ideals of the closed CFT which is a stringy analogue of a commutative ring. A chiral module over a closed CFT is a module over only one copy of the Virasoro algebra (or a superconformal algebra in a superconformal field theory) which can be roughly viewed as the Laplacian (or the Dirac operator) on loop space or path spaces. It means that the hom spaces of the category of D-branes are physical spectra. Therefore, the category of D-branes is naturally a combination of the algebro-geometric spectrum (boundary conditions) and physical spectra (internal homs). We will take it to be a stringy analogue of the spectrum of commutative rings[10] in usual algebraic geometry. This suggests to lift the

---

[6]I believe that such stringy commutative open CFTs exist in most non-linear sigma models.

[7]For irrational CFTs, a tensor product may not exist. Nevertheless, due to the existence of the OPE, at least certain weak version of a tensor product should exist so that we can still talk about internal homs.

[8]It is a standard construction from a module $\mathcal{M}$ over a tensor category $\mathcal{C}$ to obtain a category $\mathcal{M}^{\mathcal{C}}$ enriched by internal homs in $\mathcal{C}$.

[9]Notice that the category of D-brane is not much different from a single open CFT because the direct sum (or a coend in general) of the hom spaces of the category of D-branes can gives again an open CFT.

[10]Of course, it makes little difference to use the category of boundary conditions instead to define the spectrum because the relation between these two categories is clear and standard.

usual algebraic description of geometry as familiar from the field of algebraic geometry to a *stringy algebraic geometry* (SAG), which is summarized in the following table:

Classical algebraic geometry	Stringy algebraic geometry
A commutative ring R	a closed CFT A_{cl}
$\mathrm{Spec}(R)$ = the set of prime ideals of R	$\mathrm{Spec}(A_{\mathrm{cl}})$ = the category of D-branes

If we agree that such a new geometry indeed exists, the SAG will have the following new features that are different from classical algebraic geometry:

(1) CFT emphasizes a bordism point of view towards the notion of space instead of the usual sheaf-theoretical point of view. It regards a space as a network of subspaces linked by paths instead of a union of open charts. In particular, the spectrum is a category instead of a set.

(2) If we take a closed CFT loosely as an algebraic model for free loop space and an open CFT as an model for path space with a given boundary condition, we see immediately that this stringy geometry has the so-called holographic phenomenon. For example, a D-brane associated to a point is just the based loop space which contains the information of the entire space. It is very different from the usual sheaf-theoretical geometry in which a single chart does not contain any global information. This aspect is somewhat similar to that of noncommutative geometry [**Con**], where the noncommutativity encodes the information of gluing data[11].

(3) It includes spectral geometry as an ingredient. The information of "distance" and "intersection" between two D-branres X and Y should be all encoded in the internal hom $[X, Y]$ which is the spectrum of Laplacian or Dirac operator. Be aware that the usual notion of distance only makes sense in the low energy limit. At high energy, different points are highly entangled via large loops or paths. Therefore, we expect a radical new point of view of metric in SAG.

We will further discuss the naturalness of these categorical, holographic and spectral nature of SAG in Section 2 in order to support our vision. We now summarize these properties of SAG and their classical counterparts in the following table:

Classical geometry	Classical algebra	Stringy geometry	Stringy algebra
Affine scheme M	comm. ring R	loop space LM	closed CFT
points $x, y \in M$	prime ideals of R	submanifolds of $X, Y \subset M$	boundary conditions X, Y.
		the path space between X and Y	internal hom $[X, Y]$
		Laplacian (Dirac) operators on LM or path spaces	Virasoro algebra (superconformal algebra)

From the content and the properties of SAG, it is fair to say that SAG is a natural combination of algebraic geometry and spectral geometry.

It is tempting to enrich the "stringy geometry" column by including "higher paths" and higher bordisms between higher paths all the way to infinity to make it an ∞-category [**Lu1**]. We stop at level 2 because of the very nature of string theory and 2-dimensional CFT. There is no higher data on the CFT side to encode the higher bordisms. If one insists to consider higher paths beyond level 2, one need consider higher-dimensional QFTs, which provide higher categorical spectra. It is unclear how a CFT can be extended to a higher dimensional theory. But this might be due to our limited understanding of the situation.

[11] I want to thank Matilde Marcolli for clarifying this point.

Recently, Morrison and Walker proposed a scheme to extend lower dimensional QFTs to higher dimensional QFTs [**MW**]. I hope to come back to this issue in the future.

The classical notion of manifold should emerge in the low energy regime as moduli spaces of certain substructures or invariants of the spectrum of SAG. For example, the target manifold can be emergent as the moduli space of certain 0-dimensional objects in the category of D-branes [**As**]; the time evolution might come from the moduli space of certain defects (see Section 5). Another way to let a manifold emerge from CFTs is to take the so-called large volume limit. It was studied at a mathematical level by Roggenkamp-Wendland [**RW1, RW2**] and Kontsevich-Soibelman [**KS**] (see also Soibelman's contribution [**Soi**] in this book). In [**KS, Soi**], it was shown that collapsing a family of unitary CFTs leads to Riemannian manifolds (possible singular) with non-negative Ricci curvature.

As many differences between SAG and the notion of scheme are noticeable, one can question the necessity of the notion of spectrum of this new geometry. Even if we take it for granted, one can still question the naturalness of such a definition of spectrum[12]. Without any real progress in examples of this new geometry, it is hard to tell what is the natural thing to do at the current stage. An ideal attitude is to keep an open mind. I hope that such vagueness will not be viewed as an evidence against its existence. In Section 2, we will try to analyze the issue from the historical, philosophical and physical point of view in order to provide some support for our vision of SAG.

To the best of my knowledge, above simple picture has never been clearly stated in the literature. It is, however, certainly not new to physicists, and has been actively pursued by many string theorists. In the years before my sudden enlightenment in May, 2007 during a talk given by Tom Bridgeland in Max-Planck Institute for Mathematics at Bonn, I was very much influenced by Connes' noncommutative geometry [**Con**], Douglas' D-geometry [**D**], Aspinwall's stringy geometry as the moduli space of D0-branes [**As**], the geometric interpretation of D-branes in non-linear Sigma models [**FFFS**][**BDLR**][**MMS**] and Kapranov-Vasserot's program on infinite dimensional algebraic geometry [**KV1**][**Ka2**][**KV2**].

The idea that the rich structure of a CFT can be reduced to produce a spectral triple [**Con**] goes back to Fröhlich and Gawedzki [**FG**], later developed by Roggenkamp and Wendland [**RW1**], and more recently it was shown that a proper completion of a super-VOA leads to a spectral triple [**CHKL**][13]. But by doing so, some stringy features of CFT are lost. I believe that there should be an intrinsic stringy geometry associated to CFT. Such a picture was suggested by Kontsevich and Soibelman. They proposed to look at a CFT as a stringy generalization of a spectral triple [**KS**] (see Soibelman's contribution [**Soi**] in this book). For this reason, SAG can very well be called *2-spectral geometry*[14]. So our proposal on SAG is not new. But Kontsevich-Soibelman's proposal seems to emphasize different aspects of the same picture.

---

[12]Another category plays an important role in CFT is the bicategory with open CFTs as objects and the bimodules over two open CFTs as 1-cells and bimodule maps as 2-cells. Equivalently, one can define an object by the category of modules (also called boundary conditions) over an open CFT, and define a hom category by the category of functors. This bicategory appears often in the study of extended TFTs. In an n-dimensional TFT, it becomes the n-category of $(n-1)$-categories of boundary conditions. Comparing to the category of D-branes, the essential new things in this bicategory is the inclusion of all bimodules. There are physical motivations to include more general bimodules. They are called defects in physics. Some of them encode the information of dualities [**FrFRS**] of the bulk theory (see Section 5.2).

[13]Recently, Carpi, Hillier, Kawahigashi, Longo and Xu [**CHKLX**] connect subfactor theory and noncommutative geometry through the conformal net approach towards superconformal field theory. D-branes are not mentioned in their approach. I hope that a program of new geometry parallel to SAG can be developed in the conformal-net approach to CFT.

[14]This name was suggested by Urs Schreiber. I would also like to thank him for pointing out the works [**KS**][**Soi**] to me after I sent him the first draft of this paper.

SAG proposed here has a strong commutative flavor because a closed CFT is stringy commutative. Perhaps that is why the low energy effective theory of closed string theory can describe classical gravity, which is a commutative geometry and refuses to be quantized in any naive way[15]. The physics on a D-brane, as governed by an open CFT, is noncommutative in general even in the stringy sense. Thus we expect its low energy effective theory to be quantum in nature.

During my preparation of this paper, I came across an interesting article "*What is a brane?*" by Gregory Moore [**Mo2**], in which he wrote:

> However, a common theme in the study of D-branes has been the idea that in fact, the (string) field theory on the brane is the primary concept, whereas the spacetime itself is a secondary, derived, concept. This notion has been given some degree of precision in the so-called Matrix theory formulation of M-theory. A rough analogy of what physicists expect may be described in the context of purely topological branes, where the field theory on a brane is described in terms of a noncommutative Frobenius algebra, and the spacetime in which it propagates is derived from the Hochschild cohomology of that algebra. These ideas might ultimately lead to a profound revision of the way we regard spacetime.

As we will show in later sections that a closed rational CFT is a commutative Frobenius algebra in a braided tensor category and a rational D-brane (an open CFT) is a noncommutative Frobenius algebra in a tensor category. Moreover, the closed CFT very often appears as the center of the open CFT. Meanwhile, Hochschild cohomology is nothing but a notion of derived center. Therefore, the proposal of SAG can be viewed as a relatively detailed explanation of Moore's prediction in the framework of CFT from my own point of view.

There are other attempts in proposing new geometries associated to D-branes. One influential approach is the so-called derived algebraic geometry [**DAG**], which was also influenced at its early stage by the intuition from CFT through Kontsevich. In a nutshell, derived algebraic geometry replaces a commutative ring by the Hochschild cohomology $\mathrm{HH}^\bullet(A)$ of an associative algebra A, which carries a structure of topological conformal field theory [**Cos**]. Therefore, it can be viewed as a parallel program to our stringy algebraic geometry. In general, one can replace the associative algebra by an A_∞-algebra, or a $\otimes_\infty$-category, or an E_n-algebra, etc. DAG is a very rich theory, but it contains no information on the metric. Another algebraic geometry approach towards D-branes was taken by Liu and Yau [**LY1, LY2**], they proposed an Azumaya non-commutative geometry formulation of D-branes. This approach is sheaf-theorectical and largely motivated by formulating new moduli problems. The relation between D-branes and schemes was also discussed by Gomez and Sharpe [**GS**].

Perhaps it is wise to remind readers again that it is still premature to say what this new geometry really is because of the lack of examples. Nevertheless, it is important to initiate the discussion on a larger scale about the general philosophy of this topic as such philosophical inquiries have already generated new questions and inspired new works including some of my own works on CFT. As we will shown in Section 6, it is perhaps also interesting to explore the possible application of such philosophy to other subjects such as string topology, geometric Langlands correspondence, etc. The SAG proposed here is a step towards such goal. Similar to Connes' noncommutative geometry, working with examples is crucial in this program. We will pursue it in our future publications.

---

[15]Quantizing gravity is not necessary the right question to ask in the first place [**Hu**].

The layout of this paper: in Section 2, we will discuss some historical, philosophical and physical backgrounds of SAG; in Section 3, we will review the mathematical foundations of closed CFTs; in Section 4, we will review the mathematical foundations of open-closed CFTs, and give a classification of open-closed CFTs over a rational VOA; in Section 5, we will discuss a few properties (relevant to SAG) of open-closed rational CFTs; in Section 6, we will briefly discuss the connection between SAG and other topics.

Convention of notations: $\mathbb{N}, \mathbb{Z}, \mathbb{Z}_+, \mathbb{R}, \mathbb{R}_+, \mathbb{C}$ denote the set of natural numbers, integers, positive integers, real numbers, positive real numbers, complex numbers, respectively. We set $\mathbb{H} := \{z \in \mathbb{C} | \mathrm{Im}\, z > 0\}$, $\overline{\mathbb{H}} := \{z \in \mathbb{C} | \mathrm{Im}\, z < 0\}$. The ground field is always assumed to be $\mathbb{C}$.

Acknowledgement: I would like to thank Tom Bridgeland whose lecture delivered in Max-Planck Institute for Mathematics at Bonn in May 2007 triggered a quantum leap in my own understanding of this subject. I thank Hao Zheng for many inspiring conversations during my preparation of this paper, Fong-Ching Chen for his interesting lecture on Newton, Matilde Marcolli for clarifying my confusions on noncommutative geometry, Yi-Zhi Huang, Antun Milas, Ingo Runkel and Wang Zhong for many valuable comments on earlier drafts of this paper. I want to thank Urs Schreiber and referee whose comments on an earlier draft leads to a significant revision of it. I want to thank Arthur Greenspoon who has corrected a lot of my English grammar mistakes. I also want to thank Yi-Zhi Huang, James Lepowsky, Christoph Schweigert, Matilde Marcolli and Dennis Sullivan for their constant support for many years. This work was supported in part by the Basic Research Young Scholars Program of Tsinghua University, Tsinghua University independent research Grant No. 20101081762 and by NSFC Grant No. 20101301479.

2. Geometry and physics

In this section, we will discuss the naturalness of the categorical, holographic and spectral properties of SAG from a historical, philosophical and physical point of view. The central question in this endeavor is: what is a space or geometry?

> It is known that geometry assumes, as things given, both the notion of space and the first principles of constructions in space. She gives definitions of them which are merely nominal, while the true determinations appear in the form of axioms. The relation of these assumptions remains consequently in darkness; we neither perceive whether and how far their connection is necessary, nor a priori, whether it is possible.
>
> — Bernhard Riemann

The categorical nature of SAG is not only natural but also childlike. Indeed, our physical world as perceived from the eyes of a human being consists of various physical objects, such as trees, houses, furniture, mountains, etc, and their interrelations. Namely, our innocent view towards reality is already categorical. It is natural for us to think likewise in our search for a definition of physical space.

Geometry is an old science grown out of the practice of "earth-measuring". Instead of a set-theorectical foundation, geometry in ancient times regards geometrical objects, such as squares, triangles and cubes, as independent objects rather than the combinations of their points. For example, in Euclid's "Elements", the notion of line is independent of that of point. It is even possible that the notion of point came later than that of line if they did not appear at the same time. This possibility seems natural to me because a point (or rather its approximation) rarely exists alone in nature. Instead, it appears more often as an end of a line-like object, such as a rope, or a corner of a 3-dimensional body, or an intersection of two lines.

Unfortunately, this intuitive and innocent point of view towards physical space did not lead us very far. The real revolution happened after René Descartes' invention of Cartesian coordinates. It provided a foundation of analytic geometry, in which a geometric object, such as a line or a surface, can be defined by the coordinates of its points. This is perhaps the beginning of a set-theorectial point of view of geometry, which prevailed in modern geometry in the 20th century. In particular, a space is made of points, which are ideal and abstract and have no internal structures. The content of geometry lies in how these ideal points are synthesized together. However, we cannot really glue points together due to the lack of internal structure. What one can do is only to parametrize them by known mathematical structures such as the real numbers $\mathbb{R}$ which is a mathematical model for continuum space.

The birth of calculus further strengthened this set-theorectial point of view of geometry. The success of calculus was largely due to its applications in physics, such as in the study of the orbits of celestial bodies. The key to these applications is a deep observation that many physical processes have little to do with the internal structure of the physical objects involved in this process at least in the first order of approximation. Therefore, to study the dynamics only, one can ignore the internal structure of these physical objects. In many cases, by doing so, these physical objects can be simplified to ideal points. Their dynamics, as a consequence, can be translated into a problem of solving differential equations in calculus. This radical simplification has a deep influence on our perception of geometry. The notion of a space in geometry gradually lost its meaning as a physical space. It is treated more and more as a parameter space of a physical process or a moduli space of certain mathematical structures. For this reason, I would like to borrow terminology from computer science and call such a geometry as a *process oriented geometry*.

An extreme point of view along this line is to accept the notion of Cartesian spacetime as an absolute physical existence which is independent of matter living in it. This is Newton's idea of an absolute spacetime, which had a big influence on later physics and geometry. Newton's idea was not accepted without resistance. For example, Leibniz argued against it by showing that the physical existence of such absolute space is contradictory to the principle of sufficient reason and the identity of indiscernibles [**Lb**]. As we know, this absolute-space point of view was shattered by Einstein's theory of general relativity in which physical matter and geometry are indispensable to each other. Nevertheless, Einstein's theory, as a physical theory of geometry at large length scales, preserves many local properties of Cartesian space so that we can still apply calculus at least locally. Thus it is a global generalization of calculus. This fact is manifest in the modern definition of *manifold* which is locally given by open sets in $\mathbb{R}^n$. So it is fair to say that modern differential geometry, as a global generalization of calculus, is still a process oriented geometry. In hindsight, it is hard to tell if it is the most natural thing to do or just a historical convenience due to our strong reluctance to give up calculus as a powerful tool. Because of the success of the theory of manifolds in physics, after a few generations of physicists being educated in this framework, it is very difficult to imagine that there might be other possibilities.

Our main purpose in introducing the terminology of process oriented geometry is to introduce a different type of geometry which can very well be called *object oriented geometry*. In contrast to Newton, Leibniz described a space as a network of relation between physical objects, and it can not exist alone without the existence of physical objects. It is more intuitive than Newtonian space because it is exactly how we perceive the world from our daily life. But this familiar experience is not very helpful because it is too complex to be formalized. All physical objects, even atoms in modern physics, contain structures too rich to be handled easily. But complexity is not the real difficulty. There are more serious problems in this approach. Before the advent of Newtonian Gravity, a popular natural philosophy in the 17th century was the mechanical philosophy, a main contributor to which was René Descartes. He believed that there is no vacuum, everything physical is made of tiny "corpuscles" of matter, and force is possible only through the collision of adjacent corpuscles. Leibniz's network view also suggests something similar to the nature of force. From these points of view, it is hard to understand Newton's gravity because it acts at a distance. Therefore, it is a real challenge for any successful theory from Leibniz's school to understand Newton's gravity. It perhaps had to wait until modern condensed matter physics to suggest a possible answer: gravity is emergent [**Sak**].

A real breakthrough for this object oriented point of view of geometry happened in the 1960s when Grothendieck and his school established a complete new foundation of algebraic geometry. Although the notion of scheme is still set-theoretical, Grothendieck emphasized that it should be understood as a representable functor from the category of schemes to the category of sets. In other words, a scheme can be defined by the its relation to other schemes including those that cannot be viewed as its subschemes. This is a true object oriented geometry. Even in retrospect, it is still hard to understand how Grothendieck made this breakthrough, since it was as if the historical burden had nearly no effect on him. He simply started everything from scratch. The success of Grothendieck's algebraic geometry in mathematics created a wave of replacing the set-theoretical foundation of mathematics by a categorical one. In the 1980s and 1990s, this wave found its new driving force from another field of science: physics.

The ultimate source of information on what a space or geometry really is should come from the observation of our physical universe. That is the field of physics. As we mentioned before, from a childlike viewpoint, the world is obviously a network of physical objects and their interactions. A physical object can only be understood through its interaction with other physical objects, beyond which there is no further reality. Only interaction is directly observable. For example, what one observes in a cloud chamber that connected to a particle accelerator is not a particle itself but its interaction with other particles. It is similar to the slogan in category theory that morphisms are more important than objects. But, for most working physicists, such a naive viewpoint is not enough to convince them to accept categories as a new universal language. They need something workable and computable. Therefore, a better way is to look at the categorical roots in modern physical theories.

Physics in the last century has gone through a quantum revolution. As we dig deeper and deeper into the microscopic world, we see a rather strange universe. Cartesian geometric intuition loses its meaning at short length scales and becomes only a low-energy illusion. It turns out that physics at the atomic scale is governed by quantum mechanics and quantum field theory (QFT). Although they are very successful theories, they are not only difficult for beginners to study but also often puzzling to experts. The key reason, as pointed out by Alain Connes, is that they lack geometric foundations. Connes went ahead to develop such a geometric foundation which is now called Noncommutative Geometry [**Con**]. It was inspired by the Heisenberg picture of quantum mechanics. In particular, it starts with an operator algebra $\mathcal{A}$ acting on a Hilbert space $\mathcal{H}$. When enriched by a so-called Dirac operator D, the triple $(\mathcal{A}, \mathcal{H}, D)$ (called a spectral triple) determines a new geometry. In the case that the operator algebra $\mathcal{A}$ is commutative, the spectral triple, equipped with additional natural structures and satisfying certain natural conditions, recovers the Riemannian geometry. This is a spectral geometry by its nature.

A very different approach to quantum theory is the path integral approach. It is the most popular and productive approach to study quantum field theory. The key to its success lies in the fact that much classical geometric intuition is preserved in this formulation. Indeed, superficially, there is no non-commutative variable appearing in the path integral. Of course, it is not really classical. Its hidden noncommutativity can be seen from the fact that the free algebra of a few noncommutative variables can be viewed as coordinate functions on the path space of a lattice [**Ka1**]. Nevertheless, the path integral approach carries a lot of commutative flavors. As we will show in Section 3, a closed

CFT is not commutative in the classical sense but commutative in a stringy sense. More precisely, a closed CFT is a commutative associative algebra in a braided tensor category obtained from the representation theory of a VOA. Similar commutativity also appeared in higher dimensional topological field theories [**Lu1**]. In string theory, we know that the low energy effective theory of a closed string theory reproduces Einstein's theory of gravity. I believe that this correspondence between stringy commutativity and the commutative geometry underlying Einstein's General Relativity is not accidental. Perhaps it is this hidden commutativity which forbids any naive way to quantize gravity.

Another interesting feature of the path integral is its categorical nature. Although a proper measure for the path integral is very difficult to construct[16], the formal properties of the path integral are rather clear. It has been formalized by Segal [**Se**], Atiyah [**At**] and Kontsevich as a (projective) symmetric monoidal functor from a geometric category, where the compositions of morphisms are given by gluing world-sheets, to the category of vector spaces. This gave birth to an unprecedented participation of mathematicians in the study of QFT. Category has become a common language in the study of QFT among mathematicians since then. It is important to notice that this categorical aspect of QFT is on the level of the world sheet. It turns out that the path integral also suggests to view the target space, which is our main focus, as a network of interesting subspaces connected by bordisms. This is manifest in the study of D-branes in non-linear sigma models. Geometrically, a D-brane, as a boundary condition of open strings, can be wrapped around a subspace of the target manifold. So it can be viewed as a generalized subspace of the target manifold or simply a generalized point. What lies between two such subspaces are paths from one subspace to the other. Pictorially, we perceive the target manifold as a network of subspaces linked by paths. An algebraic model for this network is nothing but the category of D-branes introduced in Section 1. The geometric content of the space is completely encoded in this category. For example, the information of the intersection and the distance of a pair of D-branes should be extracted from the Hom space between these two D-branes. Therefore, it is reasonable to define our physical space (consisting of all observables) by the category of D-branes. The classical geometric intuition should be emergent from this mathematical structures of observables. We want to generalize this picture obtained from non-linear sigma models to all CFTs. Namely, we will forget about the target manifold and start from a closed CFT and its category of D-branes. It is just our SAG introduced in Section 1. Since the category of D-branes is itself a physical space, it is reasonable to define it to be the spectrum[17] of SAG. Moreover, as we discussed in Section 1, the category of D-branes is also a natural combination of the algebro-geometric spectrum (boundary conditions) and physical spectra (of Laplacians).

Taking a closer look at this spectrum, we see that the endo-hom spaces are open CFTs compatible with a given closed CFT. We will show in Section 4 that these open CFTs are given by certain open-string vertex operator algebras. The Hom space between a pair of D-branes is given by a particular bimodule over two open-string vertex operator algebras. As we will show in later sections, open CFTs, closed CFTs and their (bi-)modules are all constructible via the representation theory of vertex operator algebra. Therefore, all these data are well-defined, constructible and computable. Moreover, sometimes these data can be reduced to more familiar mathematical structures. For example, in certain twisted theories of some supersymmetric CFTs, by considering only zero modes on D-branes, the

[16]There is a nice construction of such measure for scalar fields by Doug Pickrell [**Pi**].

[17]One can ask why not define the spectrum to be the category of D0-branes. According to our philosophy of object oriented geometry, a point loses its priority as a basic building block of other objects, such as lines, surfaces and even a two-point set. Instead, a point can be viewed as an intersection of two lines for example. The relative relation between all D-branes is the full content of the geometry. Therefore, we prefer to have all D-branes in our categorical spectrum from which all kinds of moduli spaces, such as the moduli spaces of Dp-branes for $p \geq 0$, can emerge.

category of D-branes can be reduced to the bounded derived category of coherent sheaves on the target manifold [**Sh**][**D**].

The category of D-branes provides us a new way to look at geometry. From this new point of view, the classical set-theoretical geometries are not fundamental and should be viewed as emergent phenomena. For example, a set-theoretical target manifold can emerge as the moduli space of D0-branes [**As**]. Moreover, restricting ourselves to zero modes on D-branes, it is known that sometimes the original target manifold (or scheme) can be reconstructed from the bounded derived category of coherent sheaves and the group of autoequivalences [**BO**]. This fits into our philosophy that manifolds as a concept in process oriented geometry can emerge from the moduli spaces of certain mathematical structures.

We have seen that the categorical nature of the path integral approach to CFT naturally leads us to SAG. But perhaps there are deeper reasons for us to follow this route. In retrospect, the success of QFT is based on a rather strange logical order. Namely, we start with a very abstract notion of spacetime and rebuild everything on it. For example, the Hamiltonian is the dual of time, and fundamental particles are time-invariant states. In other words, physical matter, which lies at the heart of our sense of existence, is even a derived concept of the spacetime. Obviously, it is the Newtonian point of view lying at the core of this logic. It is rather strange from the eyes of a child or anyone belonging to the Leibniz school. Let's leave the debate between Newton and Leibniz aside. Remember, we are now looking for a theory of spacetime. It means that we have to look for some concepts that are even more basic than that of spacetime. Unfortunately, the Newtonian point of view provides no clue to such a question. But the Leibniz school of thought suggests that at least you can reverse the order of the logic by assuming that the existence of physical objects is more basic than that of spacetime. On the practical side, to work out this dual picture is certainly a way to enrich and deepen our understanding of the situation. On the philosophical side, I do believe that the question on the existence of something is deeper than the notion of spacetime. What lies at the heart of the former question is the notion of information [**Bek**], which can tell an observer that there is something. On the one hand, the total information of an observable physical object should be nicely structured. On the other hand, information itself should be some kind of structure that is observable to other structures. Therefore, we can simply define a physical object as a nicely structured stack of information. As a consequence, our physical space is nothing but a network of structured stacks of information, from which spacetime can emerge. If all the observable information is spectral, this network is already very close to our proposal of the spectrum of SAG.

We have argued that the categorical feature of SAG is natural and in some sense a return to innocence. More importantly, it is computable. But it still might not be sufficient to convince us that it is the right track to follow. In physics, the search of a new theory is often guided by principles, which arise from experimental or theoretical observations but are excluded from existing theories. Such principles are required to lie at the core of any successful theory, and serve as severe constraints in new theories. In the search for a theory of spacetime, many physicists believe that a fundamental guiding principle is the so-called Holographic Principle[18] (see [**Bou**] for a review), which says that the boundary of the universe contains all the information of the universe. Rooted in black hole thermodynamics, the Holographic Principle was first proposed by Gerardus 't Hooft [**tH**] and given a precise string-theory interpretation by Leonard Susskind [**Sus**]. There are several different ways to formalize the idea of Holographic Principle. We will use a particular one in the context of CFT. It says that a boundary theory determines the bulk theory uniquely. Ideally, we would like to check this Holographic Principle in SAG for

[18]It is perhaps better to call it holographic hypothesis at the current stage.

the most general D-branes, the so-called conformally invariant D-branes (see Definition 4.9). Unfortunately, such a result is still not available. What we can do now is to test it in rational CFTs for D-branes respecting a chiral symmetry given by a rational VOA V, also called V-invariant D-branes (see Definition 4.9). In these cases, the Holographic Principle indeed holds [**FjFRS2**][**KR2**]. More precisely, the closed CFT can be obtained by a given open CFT (over V) A_{op} by taking the center $Z(A_{\mathrm{op}})$ of A_{op} (see Theorem 5.1). Moreover, such holographic phenomena also appear in string topology [**ChSu**][**BCT**], TCFT [**Cos**][**BFN**], extended Turaev-Viro topological field theories [**KK**] and topological field theories in general [**Lu1**][**Lu2**]. Having partially passed the test, we conjecture that the Holographic Principle holds for many sufficiently nice conformally invariant D-branes. In other words, our SAG should be automatically holographic. A single nice D-brane is enough to determine the bulk theory which further determines the entire geometry. In the light of William Blake's poem, we will be able to see the world in a grain of D-brane.

In physics, a strong support for the Holographic Principle comes from the AdS/CFT correspondence [**Ma**] which also suggests that the Holographic Principle can provide a natural mechanism for gravity to emerge (see also for example [**Ve**]). This is also compatible with our philosophy. Indeed, the Holographic Principle can be viewed as only one direction of boundary-bulk duality. The inverse statement of Holographic Principle says that a bulk theory does not determine the boundary theory uniquely. The ambiguity of non-uniqueness is nothing but the spectrum of an entirely new geometry SAG, from which gravity or Riemannian geometry can emerge.

That SAG incorporates the Holographic Principle as an intrinsic property might not sound exciting at all to mathematicians. In mathematics, a good program is often judged by its richness and its power in solving old problems. Although one can get a glimpse of the power of CFT from its application in many topics, such as mirror symmetry, low dimensional topology and the geometric Langlands correspondence, etc, its full power in solving problems in other fields of mathematics including number theory has not been sufficiently disclosed yet. We will choose to comment on its richness instead.

Metric is a geometric notion that plays an important role in physics. It is also considered by Riemann as a necessary ingredient of the doctrine of space. Modern algebraic geometry was developed by Grothendieck's school without any influence from physics. Its interaction with physics came rather late. This delay is partly due to the lack of metric in algebraic geometry. The recent marriage of algebraic geometry and physics via string theory perhaps is simply a sign for a possible unification of algebraic geometry with the metric. In CFT, the information of metric is encoded in the modules of the Virasoro algebra or its supersymmetric counterpart. Such modules can be roughly viewed as the spectrum of the Laplacian or Dirac operator on the loop space or path spaces. One can extract metric information from the spectrum of the Laplacian or Dirac operator similar to Connes' spectral triple. But one should keep an open mind. The intrinsic notion of metric in this 2-spectral geometry might not be the one familiar to us. One might expect a hierarchy of metrics or even more exotic notion of metric. The possible exotic behavior can be seen from our loop space intuition, which suggests that space is highly entangled at high energy through the channels of long paths. The usual intuition on distance between objects loses its pertinence at high energy. Therefore, we expect a radical new and rich notion of 2-metric in this 2-spectral geometry. In this sense, a closed CFT, as a "stringy commutative ring" over a "stringy ground field" constructed from the Virasoro algebra (or super-conformal algebra), provides a natural unification of algebraic geometry with the metric.

Unification of this kind is quite inspiring if you are a believer in the unity of mathematics. In addition to it, CFT has revealed much more unifying power in mathematics. Very often, seemingly unrelated mathematical structures appear in the same CFT. For example, the first discovered CFT in mathematics was the Monster moonshine CFT [**FLM1**].

It contains the tensor-product of 48 Ising models [**DMZ**], its automorphism group is the Monster group [**Bor**][**FLM2**] and its partition function is the j-function which is the generator of all modular functions [**FLM1**]. What underlies all these miracles is a very deep and simple reason: QFT is a theory for physical systems with infinitely many degrees of freedom! The very existence of a mathematical structure (QFT) that is rich enough to model these infinite systems is already a miracle. The mathematical structures appearing in QFT are often infinite dimensional. It is not surprising that one can discover many familiar but seemingly unrelated mathematical structures from a single QFT through different finite or infinite dimensional windows. The price we pay is that to construct such a rich structure of QFT explicitly is often a difficult problem.

SAG is perhaps just what many physicists had in mind and has already been actively pursued for many years. Many aspects of SAG are known to physicists. Unfortunately, these results are scattered like leaves of a tree, but float in the air without knowing its roots. For example, the supposed-to-be-derived notions such as points, lines, paths and surfaces are unavoidably used in a lot of discussion on the emergence of spacetime in a lot of physical literature as something fundamental. How mathematicians can help is to construct the roots and provide a new calculus so that notions like points, line, surfaces, space, time and causal structure can all be derived from deeper structures and principles.

Before we leave this section, we would like to point out that 2-dimensional CFTs are certainly not the only source for such object oriented geometries. Notice that the key ingredients of the above SAG is the boundary-bulk duality. This duality also exists in higher dimensional TFTs perhaps in other QFTs as well. For example, for an n-dimensional TFT, we will have an $(n-1)$-category of boundary conditions. In many cases, boundary theories are again obtained by the internal homs of boundary conditions, and the bulk theory is obtained by taking the center of a boundary theory [**Lu1**][**Lu2**]. So I believe that it is possible to define a new geometry for many higher dimensional QFTs. The richer the structure of a QFT is, the richer the corresponding geometry is. Unfortunately, QFTs other than TFTs and 2-dimensional CFTs are still far from being understood. The structure of a CFT is certainly richer than that of a TFT. For example, there is not any spectral geometry ingredient in a TFT. Therefore, SAG based on 2-dimensional CFT is a good place to invest our labor.

3. Closed conformal field theories

In this section, we will provide the mathematical foundations of closed CFTs without supersymmetry. From now on, by a CFT we always mean a theory without supersymmetry. A systematic study of CFT in physics start from Belavin, Polyakov and Zamolodchikov's seminal paper [**BPZ**] in 1984. It was followed by a flood of physics papers on CFT. Some of them made deep impacts on mathematics. But the list is too long to be given here. Readers can get some physical ideas behind the materials presented in this review from [**BPZ**][**FriS**][**Va**][**MSei**][**Wi**]. See also Francesco, Mathieu and Senechal's book [**FMS**] and references therein. Now we shift our attention to the mathematical side of the story.

3.1. Basic definitions. Around 1987 Segal [**Se**] and Kontsevich independently gave a mathematical definition of two-dimensional closed conformal field theory (closed CFT or just CFT). This definition become very important for the development of CFT in mathematics. Segal defines a closed CFT as a projective symmetric monoidal functor $\mathcal{F} : \mathcal{RS}^b \to \mathcal{TV}$ between two symmetric monoidal categories $\mathcal{RS}^b$ and $\mathcal{TV}$. The objects of the category $\mathcal{RS}^b$ are finitely ordered sets and the morphism set $\mathrm{Mor}(A, B)$ for $A, B \in \mathcal{RS}^b$ is the set of conformal equivalence classes of Riemann surfaces with $|A|$ ordered negatively oriented parametrized boundaries and $|B|$ ordered positively oriented parametrized boundaries, where $|A|$ and $|B|$ denote the cardinalities of the set A and B

respectively. A negatively (positively) oriented parametrized boundary component is a boundary component equipped with a germ of conformal map from a neighborhood of the boundary to an inside (outside) neighborhood of the unit circle in complex plane such that its restriction to the boundary is real analytic[19]. The composition of morphisms is defined as sewing of Riemann surfaces along oppositely oriented parametrized boundaries. The compositions of morphisms are clearly associative. Notice that this category is not a unital category. The category $\mathcal{TV}$ is the category of locally convex complete topological vector spaces with continuous linear maps as morphisms. The projectivity of $\mathcal{F}$ means that $\mathcal{F}$, as a map on each set of morphisms, is only well-defined up to constant scalars, i.e. the image of $\mathcal{F}$ lies in the projectivization of the space of morphisms in $\mathcal{TV}$.

Segal's beautiful definition of CFT immediately leads to some interesting structures of conformal field theory such as the modular functor. However, it also has some inconveniences.

(1) A quantum field $\phi(x)$ in physics is defined for a point x on a worldsheet. But world sheets with punctures do not live in Segal's definition. Correlation functions of CFTs in physics are usually defined as sections of certain bundles on the moduli space of Riemann surfaces with punctures [**FriS**]. This suggests working with Riemann surfaces with punctures instead of boundaries. A geometric formulation of CFT in physics was given by Vafa [**Va**] as sewing of Riemann surfaces with punctures and local coordinates.
(2) Physicists usually do not work with the entire Hilbert space. Instead they often only work with a graded and dense subspace, which has already revealed a rich algebraic structure. For example, the chiral algebra (equivalent to VOA in mathematics) and its modules as crucial ingredients of a CFT are graded vector spaces. To complete properly the dense subspace to a complete topological vector space is a difficult problem.

In order to take advantage of the power of the algebraic method (via chiral algebras or VOAs) widely used by physicists, we would like to modify accordingly Segal's definition.

In physics, Vafa [**Va**] proposed a definition of CFT via sewing of closed Riemann surfaces with punctures after Segal. In mathematics, Igor Frenkel [**F**] started a program to study CFTs in physics via the theory of vertex operator algebras even before the appearance of Segal's definition of closed CFT. One of the highlights of this program is Huang's thesis [**H1**] (see also [**H2**][**H4**]) in which Huang established the precise geometric meaning of VOA as the sewing of Riemann surfaces with punctures. A VOA is a structure defined on a $\mathbb{Z}$-graded vector space over $\mathbb{C}$. Huang's result suggests that it is natural to consider graded vector spaces instead of complete topological vector spaces when one studies CFTs on surfaces with punctures. Vafa and Huang's works implicitly suggest a convenient and workable definition of a CFT by modifying Segal's definition only slightly. Before we do that, we need first to introduce two categories $\mathcal{RS}^p$ and $\mathcal{GV}$ below.

The category $\mathcal{RS}^p$: We first introduce some terminology. By *a puncture in a Riemann surface* Σ, we mean a marked point p in Σ equipped with a local coordinate, which is a germ of conformal map f_p from a neighborhood of p to a neighborhood of $0 \in \mathbb{C}$ such that $f_p(p) = 0$, and an orientation, which is a label $\varepsilon_p = \pm$ on p. It is called positively oriented if $\varepsilon_p = +$, and negatively oriented if $\varepsilon_p = -$. Such a puncture is denoted by (p, f_p, ε_p). A Riemann surface (not necessarily connected) with k ordered positively oriented punctures and l ordered negatively oriented punctures will be called a (k,l)-surface and will be denoted by

$$(\Sigma | (p_1, f_{p_1}, +), \ldots, (p_k, f_{p_k}, +); (q_1, f_{q_1}, -), \ldots, (q_l, f_{q_l}, -)) \ . \tag{3.1}$$

[19]Segal's definition is more general. We only care about this restricted case here.

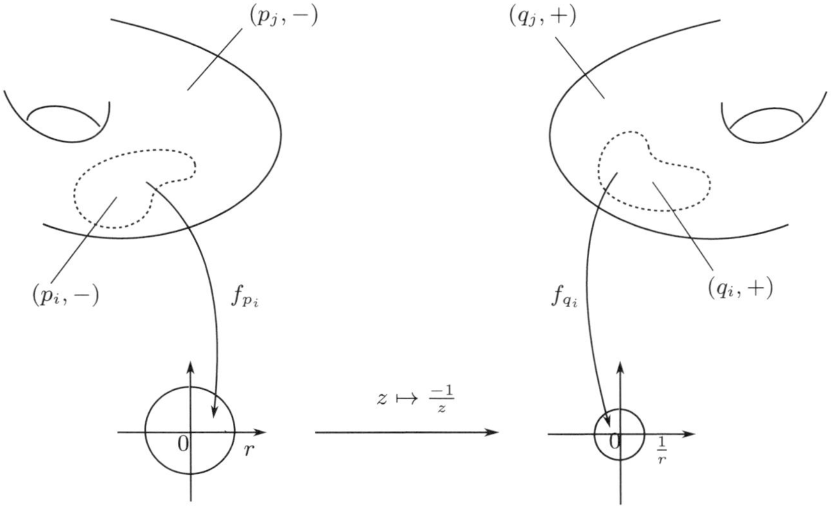

FIGURE 2. Sewing a negatively oriented puncture $(p_i, f_{p_i}, -)$ to a positively oriented puncture $(q_i, f_{q_i}, +)$.

or Σ for simplicity. Two (k,l)-surfaces are conformally equivalent if there is a biholomorphic map between them that preserves the orders, the orientations and local coordinates of the punctures. We denote the conformal equivalence class of Σ by $[\Sigma]$ and the moduli space of (k,l)-surfaces of genus g by $\mathcal{E}^g_{(k,l)}$. We also set $\mathcal{E}^g := \cup_{k,l}\mathcal{E}^g_{(k,l)}$, $\mathcal{E}_{(k,l)} := \cup_g \mathcal{E}^g_{(k,l)}$ and $\mathcal{E} := \cup_g \mathcal{E}^g$. These moduli spaces naturally has a structure of infinity dimensional complex manifolds [**H4**]. We denote the space of germs of conformal maps f_p by F_p. Notice that the F_p are isomorphic to F_0 which is the space of germs of conformal maps from a neighborhood of $0 \in \mathbb{C}$ to $0 \in \mathbb{C}$. If we denote the moduli space of Riemann surfaces with marked points by $\mathcal{M}^g_{(k,l)}$, then $\mathcal{E}^g_{(k,l)}$ can be viewed as a bundle over $\mathcal{M}^g_{(k,l)}$ with fiber given by F_0^{k+l}. We set $\mathcal{M} := \cup_{g,k,l}\mathcal{M}^g_{(k,l)}$.

Now we discuss how to sew a (k,l)-surface to an (l,m)-surface. Let Σ_1 be a (k,l)-surface with l ordered negatively oriented punctures $(p_i, f_{p_i}, -), i = 1, \ldots, l$, and let Σ_2 be an (l,m)-surface with l ordered positively oriented punctures $(q_i, f_{q_i}, +), i = 1, \ldots, l$. Let $D(r, 0)$ denote the closed disk in $\mathbb{C}$ centered at 0 with radius $r > 0$. We say that p_i can be sewn to q_i for all $i = 1, \ldots, l$ at the same time if there exist $r_i > 0, i = 1, \ldots, l$ such that $f_{p_i}^{(-1)}(D(r_i, 0))$ in Σ_1 and $f_{q_i}^{-1}(D(1/r_i, 0))$ in Σ_2 are all disjoint. Depicted in Figure 2, the sewing operation is defined by first cutting off the interiors of the disks $f_{p_i}^{(-1)}(D(r_i, 0))$ and $f_{q_i}^{-1}(D(1/r_i, 0))$, then identifying their boundaries via the maps $f_{q_i}^{-1} \circ J \circ f_{p_i}$ for $i = 1, \ldots, l$, where $J : \mathbb{C}^\times \to \mathbb{C}^\times$ is given by $w \mapsto \frac{-1}{w}$. When such r_i do not exist, a sewing operation cannot be defined. Notice that we defined the sewing operation by a choice of $r_i, i = 1, \ldots, l$. For different choices, the surfaces obtained after sewing operations are all conformally equivalent. Thus the sewing operations are well-defined on $\mathcal{E}$ (independent of choices). Notice that a sewing operation between a pair of oppositely oriented punctures can always be defined if we rescale f_{p_i} or f_{q_i}.

Let $\mathcal{RS}^p$ be a partial category of finite ordered sets, and $\mathrm{Mor}_{\mathcal{RS}^p}(A, B) := \mathcal{E}_{(|A|,|B|)}$ for $A, B \in \mathcal{RS}^p$, and the compositions of morphisms be given by sewing operations. The word "partial" refers to the fact that the sewing operations are not always well-defined. The compositions of morphisms are associative in the sense that $[\Sigma_1] \circ ([\Sigma_2] \circ [\Sigma_3]) = ([\Sigma_1] \circ [\Sigma_2]) \circ [\Sigma_3]$ whenever both sides are well-defined. The symmetric monoidal structure on $\mathcal{RS}^p$ is given by disjoint union. It is not hard to see that $\mathcal{RS}^p$ is actually a unital category.

The category $\mathcal{GV}$: By *a graded vector space* H in this work, we always mean a vector space H over $\mathbb{C}$ graded by an abelian group G_H, i.e. $H = \oplus_{n\in G_H} H_{(n)}$, with finite dimensional homogeneous spaces and a weak topology induced from the restricted dual space $H' := \oplus_{n\in G_H} H^*_{(n)}$. For the purpose of this work, the abelian group G_H can be taken to be $\mathbb{Q}^i$ or $\mathbb{R}^j$ for some $i, j \in \mathbb{N}$. We will use $\overline{H}$ to denote the algebraic completion of H, i.e. $\overline{H} = \prod_{n\in G_H} H_{(n)}$. We define a non-associative partial category $\mathcal{GV}$. The objects in $\mathcal{GV}$ are graded vector spaces. For $H_1, H_2 \in \mathcal{GV}$, the morphism set $\mathrm{Mor}_{\mathcal{GV}}(H_1, H_2)$ is defined to be $\mathrm{Hom}_{\mathbb{C}}(H_1, \overline{H_2})$ instead of $\mathrm{Hom}_{\mathbb{C}}(H_1, H_2)$! Notice that the compositions of morphisms in $\mathcal{GV}$ are not well-defined a priori. Let $f \in \mathrm{Hom}_{\mathbb{C}}(H_1, \overline{H_2})$ and $g \in \mathrm{Hom}_{\mathbb{C}}(H_2, \overline{H_3})$. If the sum $\sum_{n\in G_{H_2}} g(P_n f(u))$, where $P_n : \overline{H_2} \to (H_2)_{(n)}$ is the projection operator, is absolutely convergent for all $u \in H_1$, then we say that $g\circ f$ is well-defined and $g \circ f(u) := \sum_{n\in G_{H_2}} g(P_n f(u))$. The composition is non-associative. For any triple of morphisms h, g, f, $h\circ(g\circ f) \neq (h\circ g)\circ f$ in general even if both sides are well-defined. The symmetric monoidal structure on $\mathcal{GV}$ is the usual one on the category of vector spaces.

The moduli space $\mathcal{M}$ are naturally equipped with coordinates which can be expressed by a sequence of complex variables $(z_1, z_2, \dots)$ [**H4**]. Any functor defined on $\mathcal{RS}^p$ depending on each z_i smoothly, holomorphically and real-analytically will be called smooth, holomorphic and real-analytic, respectively. In particular, $\mathcal{F}$ being real-analytic means that it has the following expansion:

$$\mathcal{F}(z_i) = \sum_{m,n\in\mathbb{R}} F_{mn} z_i^m \bar{z}_i^n, \qquad i = 1, 2, 3, \cdots . \tag{3.2}$$

For many algebraic constructions of CFTs, it is enough to assume that $\mathcal{F}$ is real-analytic. $\mathcal{F}$ is called holomorphic if $\mathcal{F}(z_i) = \sum_{n\in\mathbb{Z}} F_n z_i^n$.

In this work, we will take the following working definition of CFT.

Definition 3.1. A (closed) CFT is a real-analytic projective symmetric monoidal functor $\mathcal{F} : \mathcal{RS}^p \to \mathcal{GV}$.

Note that such a functor $\mathcal{F}$ requires that $\mathcal{F}([\Sigma_1]) \circ \mathcal{F}([\Sigma_2])$ is well-defined whenever the composition $[\Sigma_1] \circ [\Sigma_2]$ is well-defined in $\mathcal{RS}^p$ and

$$\mathcal{F}([\Sigma_1]) \circ (\mathcal{F}([\Sigma_2]) \circ \mathcal{F}([\Sigma_3])) = (\mathcal{F}([\Sigma_1]) \circ \mathcal{F}([\Sigma_2])) \circ \mathcal{F}([\Sigma_3])$$

whenever $[\Sigma_1] \circ ([\Sigma_2] \circ [\Sigma_3])$ and $([\Sigma_1] \circ [\Sigma_2]) \circ [\Sigma_3]$ are both well-defined. These are highly nontrivial conditions. For rational CFTs, one can see that the convergence domains of genus-0 correlation functions exactly match with the domains in $\mathcal{M}$ where the sewing operations are well-defined [**H4**][**Ko1**].

Remark 3.2. It is possible to construct some CFTs that are smooth instead of real-analytic by some functional-analytic methods, for example via some rigorous constructions of path integral (see for example [**Pi**]). But for algebraic constructions considered in this article, it is enough to restrict to real-analytic CFTs.

Now we discuss a possible relation between Segal CFTs and the CFTs in Definition 3.1. On the one hand, using parametrization, one can glue open disks containing the closed unit disk centered at 0 or ∞ in $\hat{\mathbb{C}} := \mathbb{C} \cup \{\infty\}$ to a Riemann surface with parametrized boundaries. One obtains a closed Riemann surface with ordered oriented marked points (0 or ∞ of the unit disk) and local coordinates, each of which maps a neighborhood of the point to a neighborhood of the closed unit disk centered at 0 or ∞ depending on its orientation. Since the local coordinates also contain the information of orientations, we can get rid of this redundancy by requiring that all local coordinates map to neighborhoods of $0 \in \mathbb{C}$. In this way, we obtain an embedding of categories $\mathcal{RS}^b \hookrightarrow \mathcal{RS}^p$. We denote the subspace of $\mathcal{E}$ consisting of elements from the morphisms in $\mathcal{RS}^b$ by $\mathcal{E}_{\text{Segal}}$. Therefore, from a Segal CFT one cannot automatically obtain a CFT in Definition 3.1 a priori. On the other hand, Huang proposed how to obtain a Segal CFT from a CFT in Definition 3.1 by properly completing the graded vector spaces. Huang showed explicitly in [**H6**][**H7**] how to do this on a substructure of genus-0 CFTs (see Theorems 3.3 and 3.5 below).

3.2. Operads and Huang's Theorems. We will show how a VOA naturally appears as an ingredient of a CFT via a theorem of Huang. We will use the terminology of (partial) operad freely. The reader can find an introduction to the notion of operad in Ittay Weiss' contribution [**We**] to this book.

Let $\{\text{pt}\}$ be a one-point set. Since $\mathcal{F}$ is monoidal, for a finite ordered set $A \in \mathcal{RS}^p$, $\mathcal{F}(A) \cong \mathcal{F}(\{\text{pt}\})^{\otimes|A|}$. For a morphism $f \in \text{Mor}_{\mathcal{RS}^p}(A,B)$, $\mathcal{F}(f)$ can be viewed as a morphism $\mathcal{F}(\{\text{pt}\})^{\otimes|A|} \to \overline{\mathcal{F}(\{\text{pt}\})^{\otimes|B|}}$. Therefore, the structure of a CFT can be transported to a structure on $\mathcal{F}(\{\text{pt}\})$. Namely, a CFT is nothing but a graded vector space

$$V_{\text{cl}} := \mathcal{F}(\{\text{pt}\})$$

equipped with multilinear maps $\mathcal{F}([\Sigma]) : V_{\text{cl}}^{\otimes k} \to \overline{V_{\text{cl}}^{\otimes l}}$ for all (k,l)-surfaces Σ and $k,l \in \mathbb{N}$ satisfying all the conditions required by the functoriality of $\mathcal{F}$. For this reason, we will also denote a CFT by

$$(V_{\text{cl}}, \{\mathcal{F}([\Sigma])\}_{[\Sigma]\in\mathcal{E}}).$$

This notation $(V_{\text{cl}}, \{\mathcal{F}([\Sigma])\}_{[\Sigma]\in\mathcal{E}})$ for CFT is very convenient for us to address substructures of a CFT. For example, the space $\mathcal{E}$ equipped with sewing operations is just a partial PROP. The CFT structure $(H, \{\mathcal{F}([\Sigma])\}_{[\Sigma]\in\mathcal{E}})$ is nothing but a projective algebra over this partial PROP. The sewing operations restricted to $\mathcal{E}^0$ induce the structure of a partial dioperad [**Ga**] on $\mathcal{E}^0$ called the sphere partial dioperad in [**Ko3**]. Therefore, a genus-0 CFT, or equivalently the structure $(V_{\text{cl}}, \{\mathcal{F}([\Sigma])\}_{[\Sigma]\in\mathcal{E}^0})$ where all sewing operations resulting in higher genus surfaces are not allowed, is nothing but a smooth projective algebra over the sphere partial dioperad $\mathcal{E}^0$. A genus-0,1 CFT is defined to be the structure $(V_{\text{cl}}, \{\mathcal{F}([\Sigma])\}_{\Sigma\in\mathcal{E}^0\cup\mathcal{E}^1})$. We will also be interested in the space $K := \cup_{k\in\mathbb{N}}\mathcal{E}^0_{(k,1)}$ which has the structure of a partial operad called the sphere partial operad in [**H4**]. Then the structure $(V_{\text{cl}}, \{\mathcal{F}([\Sigma])\}_{[\Sigma]\in K})$ is nothing but a smooth projective K-algebra.

With all this terminology, we can state Huang's fundamental result.

Theorem 3.3 ([**H4**]). *A VOA V canonically gives a holomorphic projective K-algebra or equivalently a structure $(V_{\text{cl}}, \{\mathcal{F}([\Sigma])\}_{[\Sigma]\in K})$ for $V_{\text{cl}} = V$ as a substructure of a CFT.*

We will illustrate the main structures of a VOA in Section 3.3 by the above theorem. Namely, a VOA is a vector space V equipped with multilinear maps $\{\mathcal{F}([\Sigma]) : V^n \to \overline{V} | \Sigma \in \mathcal{E}^0_{(n,1)}\}_{n=0}^{\infty}$ satisfying conditions required by the functoriality of $\mathcal{F}$.

Remark 3.4. Huang's original theorem says that the category of VOAs with central charge c is isomorphic to the category of holomorphic algebras (not projective!) over a certain partial-operad extension of K involving the determinant line bundle over K and satisfying additional natural properties. Therefore, VOA should be viewed as a natural ingredient of holomorphic CFT with natural properties. To avoid technical details, we are satisfied with the weak version stated in Theorem 3.3 for the purpose of this paper. For more details, one should consult Huang's book [**H4**].

Similarly, we can also denote a Segal CFT by $(H_{\text{Segal}}, \{\mathcal{F}_{\text{Segal}}([\Sigma])\}_{[\Sigma]\in\mathcal{E}_{\text{Segal}}})$, a genus-0 Segal CFT as $(H_{\text{Segal}}, \{\mathcal{F}_{\text{Segal}}([\Sigma])\}_{[\Sigma]\in\mathcal{E}^0_{\text{Segal}}})$ where H_{Segal} is a locally convex complete topological vector space and $\mathcal{E}^0_{\text{Segal}} := \mathcal{E}_{\text{Segal}} \cap \mathcal{E}^0$. Let $K_{\text{Segal}} := \mathcal{E}_{\text{Segal}} \cap K$. Sewing operations are always defined on K_{Segal}. Thus K_{Segal} is an operad instead of a partial operad. The structure $(H_{\text{Segal}}, \{\mathcal{F}_{\text{Segal}}([\Sigma])\}_{[\Sigma]\in K_{\text{Segal}}})$ is nothing but a smooth K_{Segal}-aglebra structure on H_{Segal}. Huang also showed how to obtain a smooth K_{Segal}-aglebra by properly completing a finitely generated VOA.

Theorem 3.5 ([**H6**][**H7**]). *By properly completing a finitely generated VOA V to a locally convex complete topological vector space $\tilde{V}$, one can obtain a holomorphic projective K_{Segal}-algebra on $H_{\text{Segal}} = \tilde{V}$.*

Remark 3.6. Note that $V \subset \tilde{V} \subset \overline{V}$. The topology on $\tilde{V}$ in Huang's construction is even nuclear [**CS**]. Such a projective K_{Segal}-algebra $(H_{\text{Segal}}, \{\mathcal{F}_{\text{Segal}}([\Sigma])\}_{[\Sigma]\in K_{\text{Segal}}})$ gives a substructure of a Segal CFT.

3.3. Vertex operator algebras. In this paper, we will not give the standard definition of VOA in terms of formal variables because it is not very illuminating for our purpose. Intrigued readers can consult many reviews of VOAs (see for example [**LL**]). Instead, we will introduce all ingredients of a VOA $V_{\text{cl}} = V$ from the structure $(V, \{\mathcal{F}([\Sigma])\}_{\Sigma\in K})$ of a holomorphic projective K-algebra by Huang's theorem 3.3, and summarize it into a working definition of VOA (see Definition 3.7). To avoid the technicalities associated to the word "projective", let us pretend that a VOA V gives a K-algebra instead of a projective K-algebra.

Building blocks of a VOA:

(1) *Grading properties*: A VOA V is a $\mathbb{Z}$-graded vector space such that $\dim V_{(n)} < \infty$ and the grading is bounded from below, i.e. $V = \oplus_{n\in\mathbb{Z}} V_{(n)}$ and $V_{(n)} = 0$ for $n << 0$. For $u \in V_{(n)}$, n is also called *conformal weight* of u, denoted by $\text{wt}\, u$. The restricted dual space is defined by $V' := \oplus_n (V_{(n)})^*$.

(2) *Unit*: Recall (3.1) and consider the element $\Sigma_{\mathbf{1}} := [(\hat{\mathbb{C}}, (\infty, f_\infty, -))] \in \mathcal{E}^0_{(0,1)} \subset K$ where $f_\infty : w \mapsto \frac{-1}{w}$. This element[20] in K can be sewn with any positively oriented puncture on any sphere in K and the surface thus obtained is simply a sphere with one fewer positively oriented punctures but with all other data unchanged. This gives rise to a distinguished element $\mathcal{F}([(\hat{\mathbb{C}}, (\infty, f_\infty, -))])$ in $\overline{V}$, denoted by $\mathbf{1}$. In the case of VOA, $\mathbf{1} \in V_{(0)}$.

(3) In the case of VOA, the $\mathcal{F}$ image of the element $\Sigma_{\text{id}} := [(\hat{\mathbb{C}}, (0, f_0, +), (\infty, f_\infty, -))] \in \mathcal{E}^0_{(1\ 1)}$, where $f_0 : w \to w$ and $f_\infty : w \mapsto \frac{-1}{w}$, is nothing but id_V. Moreover, for $f_0 : w \to aw$, $a \in \mathbb{C}^\times$ and $f_\infty : w \mapsto \frac{-1}{w}$, $\mathcal{F}[(\hat{\mathbb{C}}, (0, f_0, +), (\infty, f_\infty, -))]$ acts on $V_{(n)}$ as a^{-n}.

(4) *Vertex operator*: Consider the following element:

$$P(z) := [(\hat{\mathbb{C}}, (z, f_z, +), (0, f_0, +), (\infty, f_\infty, -))] \in \mathcal{E}^0_{(2,1)}$$

where $z \in \mathbb{C}^\times, f_z : w \to w - z, f_0 : w \to w, f_\infty : w \to \frac{-1}{w}$. The linear map $\mathcal{F}(P(z))$ gives the vertex operator: $Y(\cdot, z)\cdot := \mathcal{F}(P(z)) : V \otimes V \to \overline{V}$ for $z \in \mathbb{C}^\times$. For $u, v \in V$, we have

$$u \otimes v \mapsto Y(u, z)v = \sum_{n\in\mathbb{Z}} u_n v z^{-n-1},$$

where $u_n : V \to \overline{V}$. In the case of VOA, $u_n \in \text{End}(V)$. Moreover, if $u \in V_{(m)}$, then $u_n : V_{(k)} \to V_{(k+m-n-1)}$. Therefore, $u_n v = 0$ for $n >> 0$.

(5) *Unit properties*: If we sew the puncture at ∞ on $[\Sigma_{\mathbf{1}}]$ to the puncture at z on $P(z)$, we obtain the sphere $[\Sigma_{\text{id}}]$. This sewing identity is transported by $\mathcal{F}$ to the identity $Y(\mathbf{1}, z) = \text{id}_V$, which is called the *left unit property* of the VOA. If we first sew the puncture on $[\Sigma_{\mathbf{1}}]$ to the puncture at 0 in $P(z)$, then

[20] We choose to use a convention that is slightly different from the one used in Huang's book [**H4**], where f_∞ is chosen to be $w \mapsto \frac{1}{w}$.

take the limit $z \to 0$, we obtain $[\Sigma_{\mathrm{id}}]$ again. Using $\mathcal{F}$, it gives the identity: $\lim_{z\to 0} Y(\cdot, z)\mathbf{1} = \mathrm{id}_V$, which is called the *right unit property* of the VOA.

(6) *Convergence property I*: The puncture at 0 on $P(z_1)$ can be sewn to that at ∞ on $P(z_2)$ if and only if $|z_1| > |z_2| > 0$. By the axioms of CFT, this sewability implies the following convergence property: the sum

$$\langle u', Y(u_1, z_1)Y(u_2, z_2)u_3\rangle := \sum_{n\in\mathbb{Z}} \langle u', Y(u_1, z_1)P_n Y(u_2, z_2)u_3\rangle$$

is absolutely convergent when $|z_1| > |z_2| > 0$ for all $u_1, u_2, u_3 \in V$ and $u' \in V'$. Actually, a VOA satisfies a stronger *convergence property I*: the sum

$$\begin{aligned}&\langle u', Y(u_1, z_1)\cdots Y(u_n, z_n)v\rangle \\ &\qquad = \sum_{k_1,\cdots,k_n\in\mathbb{Z}} \langle u', Y(u_1, z_1)P_{k_1} Y(u_2, z_2)\cdots P_{k_n} Y(u_n, z_n)v\rangle\end{aligned}$$

is absolutely convergent when $|z_1| > \cdots > |z_n| > 0$ for all $u_1, \cdots, u_n, v \in V$ and $u' \in V'$.

(7) *Convergence property II*: The puncture at z_2 on $P(z_2)$ can be sewn to that at ∞ on $P(z_1 - z_2)$ if and only if $|z_2| > |z_1 - z_2| > 0$. This sewability implies the following convergence property: the sum

$$\langle u', Y(Y(u_1, z_1 - z_2)u_2, z_2)u_3\rangle := \sum_{n\in\mathbb{Z}} \langle u', Y(P_n Y(u_1, z_1 - z_2)u_2, z_2)u_3\rangle$$

is absolutely convergent when $|z_2| > |z_1 - z_2| > 0$ all $u_1, u_2, u_3 \in V$ and $u' \in V'$.

(8) *Associativity*: The two sewing operations defined in the convergence property I and II are both well-defined when $|z_1| > |z_2| > |z_1 - z_2| > 0$. They give the same surface. By applying $\mathcal{F}$, we obtain the associativity of VOA:

$$\langle u', Y(u_1, z_1)Y(u_2, z_2)u_3\rangle = \langle u', Y(Y(u_1, z_1 - z_2)u_2, z_2)u_3\rangle$$

when $|z_1| > |z_2| > |z_1 - z_2| > 0$ for all $u_1, u_2, u_3 \in V$ and $u' \in V'$. It is nothing but the operator product expansion (OPE) in physics.

(9) *Commutativity*: By sewing the 0 on $P(z_1)$ to ∞ on $P(z_2)$ when $|z_1| > |z_2| > 0$ and sewing the 0 on $P(z_2)$ to ∞ on $P(z_1)$ when $|z_2| > |z_1| > 0$, we obtain two different elements in $\mathcal{E}^0_{(3,1)}$ which are path connected. By the fact that $\mathcal{F}$ is holomorphic and monodromy-free[21], we see that the $\mathcal{F}$ images of these two elements in $\mathcal{E}^0_{(3,1)}$ must be the unique analytic continuations of each other. This implies that

$$\langle u', Y(u_1, z_1)Y(u_2, z_2)u_3\rangle, \qquad |z_1| > |z_2| > 0,$$

and

$$\langle u', Y(u_2, z_2)Y(u_1, z_1)u_3\rangle, \qquad |z_2| > |z_1| > 0,$$

are analytic continuations of each other for all $u_1, u_2, u_3 \in V$ and $u' \in V'$. This property is called the commutativity or the locality of VOA.

(10) *Conformal properties*: Consider the element $\Sigma_\omega := [(\hat{\mathbb{C}}, (\infty, f_\infty, -))] \in \mathcal{M}^0_{(0,1)}$ with $f_\infty : w \to \frac{-1}{w-\varepsilon}$. It is clear that $\mathcal{F}(\Sigma_\omega)$ is an element in $\overline{V}$. Then $\omega := \frac{d}{d\varepsilon}|_{\varepsilon=0}\mathcal{F}(\Sigma_\omega)$ gives a distinguished element in $\overline{V}$. In the case of VOA, $\omega \in V_{(2)}$ and we have

$$Y(\omega, z) = \sum_{n\in\mathbb{Z}} L(n)z^{-n-2}, \tag{3.3}$$

[21] $\mathcal{F}$ is well-defined on the space $\mathcal{E}^0_{(3,1)}$, which is not simply connected.

where $L(n), n \in \mathbb{Z}$, can be shown by the sewing properties of spheres in $\mathcal{E}_{(1,1)}$, to generate a Virasoro Lie algebra (see the proof of [**H4**, Prop. 5.4.4]), i.e.

$$[L(m), L(n)] = (m-n)L(m+n) + \frac{c}{12}(m^3 - m)\delta_{m+n,0}\ , \qquad \forall m, n \in \mathbb{Z}, \tag{3.4}$$

where $c \in \mathbb{C}$ is called the central charge[22]. Moreover, $\omega = L(-2)\mathbf{1}$ and $L(0)$ acts on V as the grading operator. We also have

$$[L(-1), Y(u,z)] = Y(L(-1)u, z) = \frac{d}{dz}Y(u,z) \tag{3.5}$$

which can be easily derived by considering sewing the puncture on Σ_ω to the puncture at z on $P(z)$. Using (3.5) or sewing the puncture on $\Sigma_{\mathbf{1}}$ to the puncture at 0 on $P(z)$, one can derived the following identity:

$$Y(u,z)\mathbf{1} = e^{zL(-1)}u$$

which is sometimes called the creation property.

We can summarize some of the properties of VOA listed above into a working definition of VOA: Definition 3.7. This definition is equivalent to the usual definition of VOA [**H5**, Prop. 1.7].

Definition 3.7. A VOA is a quadruple $(V, Y, \mathbf{1}, \omega)$ where $V = \oplus_{n\in\mathbb{Z}} V_{(n)}$ is a $\mathbb{Z}$-graded vector space with grading operator $L(0)$, $\mathbf{1} \in V_{(0)}$, $\omega \in V_{(2)}$ and Y is a vertex operator:

$$\begin{array}{rcl} Y(\cdot, z)\cdot : V \otimes V & \to & \overline{V} \\ u \otimes v & \mapsto & Y(u,z)v = \sum_n u_n v z^{-n-1}, \quad u_n \in \mathrm{End} V, \end{array}$$

satisfying the grading properties, the unit properties, the convergence property I and II, the associativity and the commutativity listed above and the identity (3.3), (3.4), (3.5).

It is straight forward to define the notion of a module over a VOA. We will not do it here. Readers can consult [**LL**]. The tensor product theory of modules over a VOA is rather complicated. A relative easier notion is the so-called intertwining operator [**FHL**], which is used to define the tensor product. We will not recall this notion here either. Readers who are interested in these topics should consult [**FHL**][**HL1**][**HL2**][**HL3**][**H3**].

Notice that we have only used spheres $\Sigma_{\mathbf{1}}, \Sigma_{\mathrm{id}}, P(z), \Sigma_\omega$ and spheres in $\mathcal{E}_{(1,1)}$ to define the data in a VOA. But these spheres generate the entire sphere partial operad K by sewing operations. Therefore, the data in a VOA is enough to construct a projective K-algebra $(V, \{\mathcal{F}([\Sigma])\}_{\Sigma\in K})$. An explicit formula for $\mathcal{F}([\Sigma])$ for a generic element $[\Sigma] \in K$ in terms of the data of a VOA is known [**H4**, eq. (5.4.1)]. Intrigued readers should consult Huang's book [**H4**].

3.4. Holomorphic CFTs. Since one cannot obtain all elements in $\mathcal{E}^0$ from those in K by sewing operations, a VOA is not enough to construct a genus-0 CFT. It was proved in [**Ko1**] that a VOA V equipped with a non-degenerate invariant bilinear form on V is enough to give a holomorphic genus-0 closed CFT $(V_{\mathrm{cl}}, \{\mathcal{F}([\Sigma])\}_{\Sigma\in\mathcal{E}^0})$ by taking $V_{\mathrm{cl}} = V$. The notion of an invariant bilinear form on a VOA was first introduced in [**FHL**]. More precisely, a bilinear form $(\cdot,\cdot) : V \otimes V \to \mathbb{C}$ is called invariant if the following identity:

$$(w, Y(u,z)v) = (Y(e^{-zL(1)}z^{-2L(0)}u, -z^{-1})w, v)$$

holds for $u, v, w \in V$ and $z \in \mathbb{C}^\times$. This definition of invariant bilinear form is slightly different from that given in [**FHL**]. We follow the convention taken in [**Ko1**] in order to

[22] We cheat here. This is the place we cannot ignore the "projectivity". In order to obtain the Virasoro algebra with a nontrivial central charge, we need extend the sphere partial operad K by the $\frac{c}{2}$-power of determinant line bundle over K (see [**H4**] for more details).

obtain simpler categorical formulation later. Such an invariant bilinear form is automatically symmetric. For simplicity, a VOA equipped with a non-degenerate invariant bilinear form will be called a self-dual VOA.

Note that all elements in $\mathcal{E}$ can be obtained by applying sewing operations on elements in $\mathcal{E}^0$ repeatedly. Therefore, if a genus-0 CFT is extendable to all higher genera, then it uniquely determines the entire CFT. Higher genus theories provide no new data but only some compatibility conditions. Therefore, Huang's theorem 3.3 suggests that one might be able to obtain a CFT from a self-dual VOA. If this is indeed possible, then this CFT will be holomorphic. A CFT requires $\mathcal{F}$ to be well-defined on $\mathcal{E}$. Equivalently, $\mathcal{F}$ is invariant under the actions of all mapping class groups. In the case of holomorphic CFTs, this invariance property puts serious constrain on the VOAs. So far, it is not known which VOAs satisfy all the necessary properties. But if we restrict to genus-0,1 CFT, the answer is known. The examples of such VOAs are the Monstrous Moonshine VOA $V^\natural$ [**FLM1**][**Bor**][**FLM2**] and the VOAs constructed from self-dual positive even lattices [**FLM2**]. Since such VOAs are candidates to construct holomorphic CFTs, they are called holomorphic VOAs.

3.5. Full field algebras. In general, we cannot expect CFTs to be holomorphic. In this work, we are interested in real-analytic CFTs (recall Remark 3.2). Again, we first restrict our attention to K. Can we obtain a real-analytic K-algebra from VOAs? The answer is yes. Given a pair of VOAs $(V^L, Y_{V^L}, \mathbf{1}_{V^L}, \omega_{V^L})$ and $(V^R, Y_{V^R}, \mathbf{1}_{V^R}, \omega_{V^R})$, one can combine them in a way to yield another interesting structure. More precisely, we set $V_{\text{cl}} = V^L \otimes V^R$,

$$\mathbf{1}_{\text{cl}} = \mathbf{1}_{V^L} \otimes \mathbf{1}_{V^R}, \qquad \omega_{\text{cl}} = \omega_{V^L} \otimes \mathbf{1}_{V^R} + \mathbf{1}_{V^L} \otimes \omega_{V^R},$$

and

$$Y_{\text{cl}}(u^L \otimes u^R; z, \bar{z})(v^L \otimes v^R) = Y_{V^L}(u^L, z)v^L \otimes Y_{V^R}(u^R, \bar{z})v^R \,, \tag{3.6}$$

for $u^L, v^L \in V^L, u^R, v^R \in V^R$. Such a quadruple $(V_{\text{cl}}, Y_{\text{cl}}, \mathbf{1}_{\text{cl}}, \omega_{\text{cl}})$ is not a VOA, but a so-called full field algebra introduced in [**HK2**]. This quadruple is not much different from a VOA. In particular, if one replaces the $\bar{z}$ in (3.6) by z, one obtains again a VOA [**FHL**]. Thus the tensor product $V^L \otimes V^R$ has both a VOA structure and a full field algebra structure depending on how the vertex operator is defined. The latter structure is interesting because the quadruple $(V_{\text{cl}}, Y_{\text{cl}}, \mathbf{1}_{\text{cl}}, \omega_{\text{cl}})$ gives another projective K-algebra which is not holomorphic but real-analytic. A full field algebra can be viewed as a real-analytic analogue of a VOA.

In general, if V^L and V^R are not holomorphic VOAs, one cannot construct the entire CFT structure on a full field algebra $V^L \otimes V^R$ due to the lack of the modular invariance property on $V^L \otimes V^R$. However, if a VOA V is rational[23], it was proved by Huang (see Theorem 3.9) that $\mathcal{C}_V$, the category of V-modules, is a modular tensor category [**RT**][**T**] on which all the mapping class groups act. If V^L and V^R are rational, then the VOA $V^L \otimes V^R$ is also rational [**DMZ**][**HK2**]. In this case, it is reasonable to expect that we might be able to obtain a CFT of all genera by extending the full field algebra $V^L \otimes V^R$ by adding modules over $V^L \otimes V^R$ viewed as a VOA. Namely, we will look for a CFT extension V_{cl} of $V^L \otimes V^R$ as a $V^L \otimes V^R$-module. Such a CFT, if it exists, will be called a CFT over $V^L \otimes V^R$. We will denote the canonical embedding $V^L \otimes V^R \hookrightarrow V_{\text{cl}}$ by ι_{cl}. In this case, we can use the powerful tools of tensor category to give a classification of CFTs over $V^L \otimes V^R$. A full field algebra extension of $V^L \otimes V^R$ is called a full field algebra over $V^L \otimes V^R$.

We will not give a precise definition of a full field algebra. Being a projective K-algebra, a full field algebra has properties similar to those of a VOA. In particular, it also satisfies a certain associativity and a commutativity. The difference lies only in their

[23] V is rational if it satisfies the conditions in Theorem 3.9.

analyticity properties. We will describe only a few crucial ingredients and properties of a full field algebra that are important for the purpose of this paper. See [**HK2**][**Ko1**] for more details.

(1) *Grading properties*: $V_{\rm cl}$ is a $\mathbb{R}\times\mathbb{R}$-graded vector space, i.e. $V_{\rm cl} = \oplus_{m,n\in\mathbb{R}}(V_{\rm cl})_{(m,n)}$, such that $\dim V_{(m,n)} < \infty$ and $V_{(m,n)} = 0$ if $m-n \notin \mathbb{Z}$ or $m << 0$ or $n << 0$.
(2) *Unit*: $\mathbf{1}_{\rm cl} \in (V_{\rm cl})_{(0,0)}$.
(3) *Vertex operator*: for $u, v \in V_{\rm cl}$ and $z, \zeta \in \mathbb{C}^\times$ with branching cut: $-\pi < \operatorname{Arg}(z) \leq \pi$, $-\pi \leq \operatorname{Arg}(\zeta) < \pi$;
$$\begin{array}{rcl} Y_{\rm cl}(\cdot; z, \zeta) : V_{\rm cl} \otimes V_{\rm cl} & \to & \overline{V_{\rm cl}} \\ u \otimes v & \mapsto & Y_{\rm cl}(u; z, \zeta)v = \displaystyle\sum_{m,n\in\mathbb{R}} u_{(m,n)} v z^{-m-1}\zeta^{-n-1}, \end{array}$$
where for $u \in (V_{\rm cl})_{(a,b)}$, $u_{(m,n)} : (V_{\rm cl})_{(k,l)} \to (V_{\rm cl})_{(k+a-m-1, l+b-n-1)}$.
(4) *Unit property*: $Y_{\rm cl}(\mathbf{1}; z, \zeta) = \mathrm{id}_{V_{\rm cl}}$.
(5) *Convergence property I*: Let $V_{\rm cl}'$ be the restricted dual space of $V_{\rm cl}$, i.e. $V_{\rm cl}' = \oplus_{m,n\in\mathbb{R}}(V_{\rm cl})^*_{(m,n)}$. For $u_1, u_2, v \in V_{\rm cl}, v' \in V_{\rm cl}'$, the following sum
$$\langle v', Y_{\rm cl}(u_1; z_1, \zeta_1) Y_{\rm cl}(u_2; z_2, \zeta_2)v\rangle = \sum_{m,n} \langle v', Y_{\rm cl}(u_1; z_1, \zeta_1) P_{(m,n)} Y_{\rm cl}(u_2; z_2, \zeta_2)v\rangle$$
where $P_{(m,n)} : \overline{V_{\rm cl}} \to (V_{\rm cl})_{(m,n)}$ is the projection operator, is absolutely convergent when $|z_1| > |z_2| > 0$ and $|\zeta_1| > |\zeta_2| > 0$.
(6) *Convergence property II*: For $u_1, u_2, v \in V_{\rm cl}, v' \in V_{\rm cl}'$, the following sum
$$\begin{aligned} &\langle v', Y_{\rm cl}(Y_{\rm cl}(u_1; z_1 - z_2, \zeta_1 - \zeta_2)u_2; z_2, \zeta_2)\rangle \\ &\quad = \sum_{m,n} \langle v', Y_{\rm cl}(P_{(m,n)} Y_{\rm cl}(u_1; z_1 - z_2, \zeta_1 - \zeta_2)u_2; z_2, \zeta_2)\rangle \end{aligned}$$
is absolutely convergent when $|z_2| > |z_1 - z_2| > 0$ and $|\zeta_2| > |\zeta_1 - \zeta_2| > 0$.
(7) *Associativity*: For $u_1, u_2, v \in V_{\rm cl}, v' \in V_{\rm cl}'$,
$$\begin{aligned} &\langle v', Y_{\rm cl}(u_1; z_1, \bar{z}_1) Y_{\rm cl}(u_2; z_2, \bar{z}_2)v\rangle \\ &\quad = \langle v', Y_{\rm cl}(Y_{\rm cl}(u_1; z_1 - z_2, \bar{z}_1 - \bar{z}_2)u_2; z_2, \bar{z}_2)\rangle \end{aligned}$$
when $|z_1| > |z_2| > |z_1 - z_2| > 0$. This associativity is also called the OPE of bulk fields in physics.
(8) *Commutativity*: For $u_1, u_2, v \in V_{\rm cl}, v' \in V_{\rm cl}'$
$$\langle v', Y_{\rm cl}(u_1; z_1, \zeta_1) Y_{\rm cl}(u_2; z_2, \zeta_2)v\rangle$$
in the domain $\{|z_1| > |z_2| > 0, |\zeta_1| > |\zeta_2| > 0\}$ is the analytic continuation of
$$\langle v', Y_{\rm cl}(u_2; z_2, \zeta_2) Y_{\rm cl}(u_1; z_1, \zeta_1)v\rangle$$
in the domain $\{|z_2| > |z_1| > 0, |\zeta_2| > |\zeta_1| > 0\}$ along a path of z_2 (fixing z_1) running clockwisely around z_1 with zero winding number and a path of ζ_2 (fixing ζ_1) running counter-clockwisely around ζ_1 with zero winding number. Both paths do not cross the branching cuts.
(9) *Left Virasoro element*: $\omega_L \in (V_{\rm cl})_{(2,0)}$; *right Virasoro element*: $\omega_R \in (V_{\rm cl})_{(0,2)}$. We have
$$\begin{aligned} Y_{\rm cl}(\omega_L; z, \zeta) &= \sum_{n\in\mathbb{Z}} L^L(n) z^{-n-1}, \\ Y_{\rm cl}(\omega_R; z, \zeta) &= \sum_{n\in\mathbb{Z}} L^R(n) \zeta^{-n-1}, \end{aligned}$$
where $\{L^L(n)\}$ generate a Virasoro algebra of central charge $c^L \in \mathbb{C}$ and $\{L^R(n)\}$ generate a Virasoro algebra of central charge $c^R \in \mathbb{C}$. $[L^L(m), L^R(n)] =$

0 for $m, n \in \mathbb{Z}$. Moreover, $L^L(0)$ and $L^R(0)$ are left and right grading operators, i.e. $L^L(0)|_{(V_{\mathrm{cl}})_{(m,n)}} = m\,\mathrm{id}_{(V_{\mathrm{cl}})_{(m,n)}}$ and $L^R(0)|_{(V_{\mathrm{cl}})_{(m,n)}} = n\,\mathrm{id}_{(V_{\mathrm{cl}})_{(m,n)}}$.

Remark 3.8. Among of all these properties, the commutativity is especially interesting to us. It is nothing but the stringy commutativity of a closed CFT mentioned in the Introduction. In Section 3.7, we will give a categorial formulation of a full field algebra over $V^L \otimes V^R$ where both V^L and V^R are rational. In this categorical formulation, the commutativity listed above reduces to the usual commutativity of an algebra in a braided tensor category [**Ko1**](see Theorem 3.11).

For readers who are interested in how a full field algebra can produce a projective K-algebra, a construction and a detailed proof can be found in [**Ko1**, Thm. 1.19]. To lift a projective K-algebra V_{cl} to a genus-0 CFT, we need a bilinear form on V_{cl}. Let's define $Y_{\mathrm{cl}}^f(u; x, y)v = \sum_{(m,n)} u_{(m,n)} v x^{-n-1} y^{-n-1}$ where x, y are formal variables. A bilinear form $(\cdot,\cdot)_{\mathrm{cl}} : V_{\mathrm{cl}} \otimes V_{\mathrm{cl}} \to \mathbb{C}$ is called invariant [**Ko1**, eq. (2.1)]) if the following identity:

$$(w, Y_{\mathrm{cl}}^f(u; x, y)v)_{\mathrm{cl}} = (Y_{\mathrm{cl}}^f(e^{-xL^L(1)-yL^R(1)} x^{-2L^L(0)} y^{-2L^R(0)} u, e^{\pi i} x^{-1}, e^{-\pi i} y^{-1})w, v)_{\mathrm{cl}}$$

holds for $u, v, w \in V_{\mathrm{cl}}$. Such an invariant bilinear form is automatically symmetric [**Ko1**, Prop. 2.3]. A full field algebra equipped with a non-degenerate invariant bilinear form is called self-dual. It was proved in [**Ko1**, Thm. 2.7] that a self-dual full field algebra canonically gives a genus-0 CFT.

3.6. Modular tensor categories. From now on, we will assume that both VOAs V^L and V^R are rational. We will first recall the notion of a modular tensor category [**RT**][**T**].

Let $\mathcal{C}$ be a tensor category. We set the notations: $\mathbf{1}$ for the unit object; $\otimes : \mathcal{C} \times \mathcal{C} \to \mathcal{C}$ for the tensor bifunctor; $l_A : \mathbf{1} \otimes A \cong A$ and $r_A : A \otimes \mathbf{1} \cong A$ for $A \in \mathcal{C}$ for unit isomorphisms; $\alpha_{A,B,C} : (A \otimes B) \otimes C \cong A \otimes (B \otimes C), \forall A, B, C \in \mathcal{C}$ for the associator. If $\mathcal{C}$ is braided, we use $c_{A,B} : A \otimes B \to B \otimes A$, for $A, B \in \mathcal{C}$, to denote the braiding isomorphisms.

If $\mathcal{C}$ is rigid, each object U is equipped with a left dual ${}^\vee U$ and a right dual $U^\vee$. A ribbon category is a rigid braided tensor category with a twist isomorphism $\theta_U : U \to U$ for each $U \in \mathcal{C}$ satisfying certain balancing properties. In particular, in a ribbon category, one can take ${}^\vee U = U^\vee$. In this case, we will write the dualities as

$= d_U : U^\vee \otimes U \to \mathbf{1}$,

$= \tilde{d}_U : U \otimes U^\vee \to \mathbf{1}$,

$= b_U : \mathbf{1} \to U \otimes U^\vee$,

$= \tilde{b}_U : \mathbf{1} \to U^\vee \otimes U$.

A modular tensor category $\mathcal{C}$ is a semisimple abelian finite $\mathbb{C}$-linear ribbon category with a simple unit $\mathbf{1}$, $\mathrm{End}(\mathbf{1}) = \mathbb{C}$ and satisfying an additional non-degeneration condition on braiding (given below). We denote the set of equivalence classes of simple objects in $\mathcal{C}$ by I, elements in I by $i, j, k \in I$ and their representatives by U_i, U_j, U_k. We define $\mathrm{Dim}(\mathcal{C}) := \sum_i (\dim U_i)^2$. It is known that $\mathrm{Dim}\,\mathcal{C} \neq 0$ [**ENO1**]. We also set $U_0 = \mathbf{1}$. We define numbers $s_{i,j} \in \mathbb{C}$ by

$s_{i,j} =$

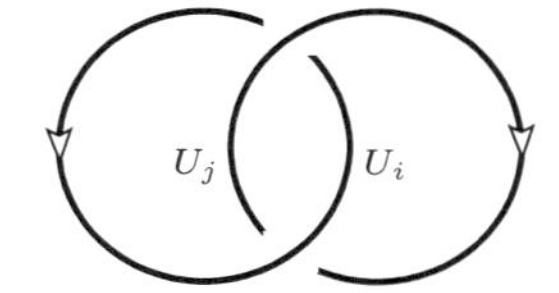

.

We also define $\dim(U) := \tilde{d}_U \circ b_U \in \mathbb{C}$. We have $s_{i,j} = s_{j,i}$ and $s_{0,i} = \dim U_i$. The non-degeneracy condition on the braiding of a modular tensor category is that the $|I|\times|I|$-matrix s is invertible.

For $A \in \mathcal{C}$, we also choose a basis $\{b_A^{(i;\alpha)}\}$ of $\mathrm{Hom}_{\mathcal{C}}(A, U_i)$ and the dual basis $\{b^A_{(i;\beta)}\}$ of $\mathrm{Hom}_{\mathcal{C}}(U_i, A)$ for $i \in I$ such that $b_A^{(i;\alpha)} \circ b^A_{(i;\beta)} = \delta_{\alpha\beta}\,\mathrm{id}_{U_i}$. We use the graphical notation

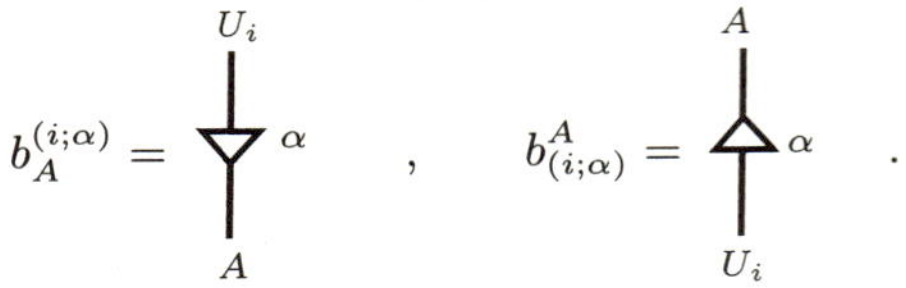

Based on Huang and Lepowsky's earlier works on the tensor product of modules over a VOA [**HL1**][**HL2**] [**HL3**][**H3**][**HL4**], Huang proved the following theorem which is very important for us.

Theorem 3.9 ([**H11**]). *If V is a simple VOA satisfying*

(1) $V_{(n)} = 0$ *for* $n < 0$, $V_{(0)} = \mathbb{C}\mathbf{1}$ *and* $V' \cong V$ *as* V*-modules,*
(2) *Every* $\mathbb{N}$*-gradable weak* V*-module is completely reducible,*
(3) V *is* C_2*-cofinite.*

Then we say V is rational. Moreover, the category $\mathcal{C}_V$ of V-modules is a modular tensor category.

3.7. Non-holomorphic CFTs over $V^L \otimes V^R$. An algebra in $\mathcal{C}$ or a $\mathcal{C}$-algebra is a triple $A = (A, m, \eta)$ where A is an object of $\mathcal{C}$, m (the multiplication) is a morphism $A \otimes A \to A$ such that $m \circ (m \otimes \mathrm{id}_A) \circ \alpha_{A,A,A} = m \circ (\mathrm{id}_A \otimes m)$, and η (the unit) is a morphism $\mathbf{1} \to A$ such that $m \circ (\mathrm{id}_A \otimes \eta) = \mathrm{id}_A \circ r_A$ and $m \circ (\eta \otimes \mathrm{id}_A) = \mathrm{id}_A \circ l_A$. An algebra A is called commutative if $m_A \circ c_{A,A} = m_A$. Similarly, one can define a coalgebra $A = (A, \Delta, \varepsilon)$ where $\Delta : A \to A \otimes A$ and $\varepsilon : A \to \mathbf{1}$ obey a coassociativity and a counit conditions.

Definition 3.10. A Frobenius algebra $A = (A, m, \eta, \Delta, \varepsilon)$ is an algebra and a coalgebra such that the coproduct is an intertwiner of A-bimodules, i.e.

$$(\mathrm{id}_A \otimes m) \circ (\Delta \otimes \mathrm{id}_A) = \Delta \otimes m = (m \otimes \mathrm{id}_A) \circ (\mathrm{id}_A \otimes \Delta).$$

We will use the following graphical representation for the morphisms of a Frobenius algebra,

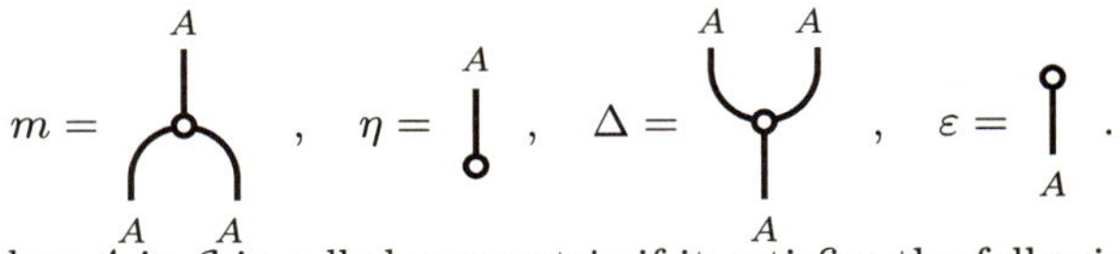

A Frobenius algebra A in $\mathcal{C}$ is called symmetric if it satisfies the following identity:

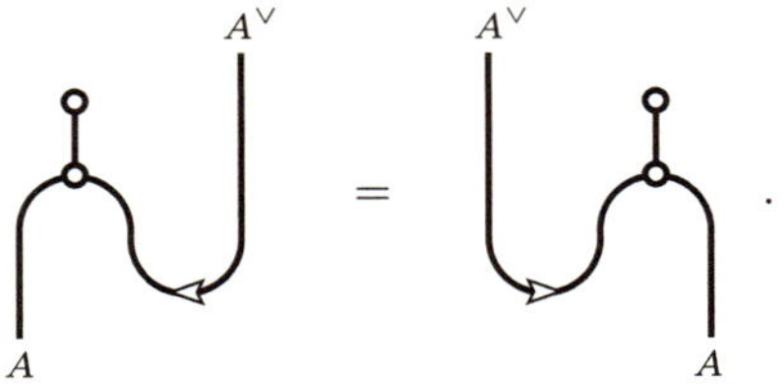

Let $\mathcal{C}$ be a modular tensor category with simple objects $U_i, i \in I$ where I is a finite set. If we replace the braiding and the twist in $\mathcal{C}$ by the antibraiding c^{-1} and the antitwist θ^{-1} respectively, we obtain another ribbon category structure on $\mathcal{C}$. In order to distinguish these two distinct structures, we denote $(\mathcal{C}, c, \theta)$ and $(\mathcal{C}, c^{-1}, \theta^{-1})$ by $\mathcal{C}_+$ and $\mathcal{C}_-$ respectively.

Let $\mathcal{D}$ be another modular tensor category. Let $W_j, j \in J$ be the representatives of simple objects in $\mathcal{D}$ where the index set J is finite. By $\mathcal{C} \boxtimes \mathcal{D}$ we mean the tensor product of $\mathbb{C}$-linear abelian categories [**BK**, def. 1.1.15], i.e. the category whose objects are direct sums of pairs $U \times W$ of objects $U \in \mathcal{C}$ and $W \in \mathcal{D}$ and whose morphism spaces are

$$\mathrm{Hom}_{\mathcal{C}\boxtimes\mathcal{D}}(U \times W, U' \times W') = \mathrm{Hom}_{\mathcal{C}}(U, U') \otimes_{\mathbb{C}} \mathrm{Hom}_{\mathcal{D}}(W, W')$$

for pairs, and direct sums of these if the objects are direct sums of pairs. The category $\mathcal{C} \boxtimes \mathcal{D}_-$ is also a modular tensor category which has simple objects $U_i \times W_j$ for $i \in I$ and $j \in J$. In the case of $\mathcal{C} = \mathcal{C}_{V^L}$ and $\mathcal{D} = \mathcal{C}_{V^R}$, we have $U_i \times W_j = U_i \otimes_{\mathbb{C}} W_j$, where $\otimes_{\mathbb{C}}$ is the usual vector space tensor product.

Theorem 3.11 ([**Ko1**]). *The category of genus-0 CFTs (or self-dual full field algebras) over $V^L \otimes V^R$ is isomorphic to the category of commutative symmetric Frobenius algebras in $(\mathcal{C}_{V^L})_+ \boxtimes (\mathcal{C}_{V^R})_-$.*

Remark 3.12. The commutativity and the associativity of the Frobenius algebra in Theorem 3.11 is precisely the categorical formulation of the commutativity and the associativity of full field algebra discussed in Section 3.5.

On a physical level of rigor, Sonoda [**So**] showed that higher genus theories only provide one additional condition: the modular invariance condition of 1-point genus-1 correlation functions [**HK3**, Thm 3.8]. For CFTs over $V^L \otimes V^R$, the categorical formulation of this *modular invariance condition* [**Ko3**] is given in equation (3.7). Therefore, it is natural to have the following conjecture.

Conjecture 3.13 ([**Ko3**]). *The category of CFTs over $V^L \otimes V^R$ is isomorphic to the category of commutative symmetric Frobenius algebras A_{cl} in $(\mathcal{C}_{V^L})_+ \boxtimes (\mathcal{C}_{V^R})_-$ satisfying the following modular invariance condition:*

(3.7) $\dfrac{\dim U_i \dim W_j}{\sqrt{\mathrm{Dim}\,\mathcal{C}_{V^L}\,\mathrm{Dim}\,\mathcal{C}_{V^R}}}$

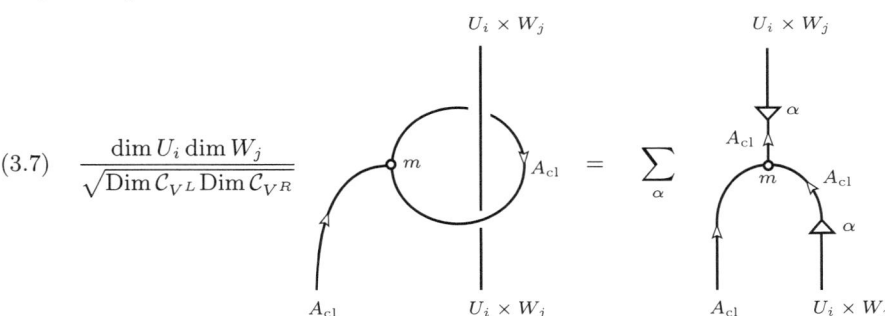

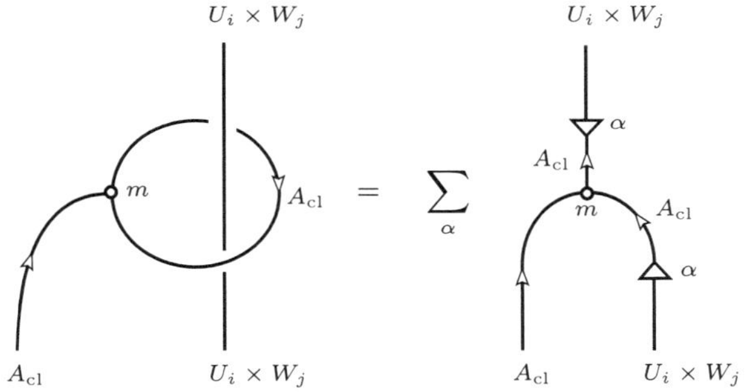

holds for all $i \in I$.

Remark 3.14. Above categorical formulation of modular invariance condition is based on many earlier influential works on this subject, in particular Zhu's proof of the modular transformation properties of q-traces on a rational VOA and its modules [**Z**], Huang's proof of modular transformation properties of general genus-one correlation functions involving intertwining operators [**H9**], Huang's proof of Verlinde conjecture [**H9**][**H10**] and Huang's construction of modular tensor categories [**H11**].

The modular invariance condition is hard to check. In a special case it can be replaced by an easy-to-check condition. A $\mathcal{C}$-algebra A is called haploid if $\dim \mathrm{Hom}(\mathbf{1}, A) = 1$.

Theorem 3.15 ([**KR2**]). *A haploid commutative symmetric Frobenius algebra A_{cl} in $\mathcal{C}_+ \boxtimes \mathcal{C}_-$ is modular invariant if and only if* $\dim A_{\mathrm{cl}} = \mathrm{Dim}(\mathcal{C})$.

When $V^L = V^R = V$, examples of modular invariant commutative symmetric Frobenius algebras in $(\mathcal{C}_V)_+ \boxtimes (\mathcal{C}_V)_-$ can be constructed explicitly. For example, let $U_i, i \in I$ be inequivalent irreducible V-modules. Then the object

$$A_{\mathrm{cl}} = \oplus_{i\in I} U_i^\vee \times U_i \quad \in (\mathcal{C}_V)_+ \boxtimes (\mathcal{C}_V)_-,$$

where $U_i^\vee \times U_i \cong U_i^\vee \otimes_{\mathbb{C}} U_i$ as vector spaces, has a structure of modular invariant commutative symmetric Frobenius $\mathcal{C}_{V\otimes V}$-algebra [**Ko3**][24]. This construction is called the "charge conjugation construction" or the "Cardy case". More constructions will be given in Section 4.6. In general, V^L and V^R can be different. For example, $V^R = V^L \otimes_{\mathbb{C}} V^\natural$ where $V^\natural$ is the Monster-Moonshine VOA.

4. Open-closed CFTs

Open-closed (or boundary) CFT was first developed in physics by Cardy [**C1, C2, C3, CL**]. It has important applications in the study of certain critical phenomena on surfaces with boundaries in condensed matter physics. It is also a powerful tool in the study of D-branes in string theory. Mathematically, it is easy to extend Segal's definition of a closed CFT to that of an open-closed CFT [**H8**][**HKr**]. As we discussed in Section 3, such a definition is not convenient if we want to apply the theory of VOA to the study of CFT. We will again take Vafa-Huang's approach to open-closed CFTs.

4.1. Basic definitions. The category $\mathcal{RS}^p_{\mathrm{cl-op}}$: First, consider Riemann surfaces with unparametrized boundaries. A puncture living in the interior of the surface is called an interior puncture; that living on a boundary component is called a boundary puncture. The parametrization of an interior puncture is defined the same as before. The parametrization of a boundary puncture p is the germ of a conformal map from neighborhoods of p to neighborhoods of zero in the upper half-plane such that it is real analytic on the boundary, together with an orientation $\epsilon_p = \pm$. We introduce a unified notation for both interior punctures and boundary punctures: $(p, \sigma_p, f_p, \epsilon_p)$ where σ_p takes the value of "interior" or "boundary", or equivalently "closed string" and "open string". A Riemann surface (not necessarily connected) with k ordered positively oriented punctures (with the splitting $k = k^{\mathrm{cl}} + k^{\mathrm{op}}$ according to their σ-values) and l ordered negatively oriented punctures (with $l = l^{\mathrm{cl}} + l^{\mathrm{op}}$) will be called a $(k^{\mathrm{cl}}, k^{\mathrm{op}}; l^{\mathrm{cl}}, l^{\mathrm{op}})$-surface and will be denoted by

$$\{\Sigma|(p_1, \sigma_{p_1}, f_{p_1}, +), \dots, (p_k, \sigma_{p_k}, f_{p_k}, +); (q_1, \sigma_{q_1}, f_{q_1}, -), \dots, (q_l, \sigma_{q_l}, f_{q_l}, -)\}$$

or Σ for simplicity. Two $(k^{\mathrm{cl}}, k^{\mathrm{op}}; l^{\mathrm{cl}}, l^{\mathrm{op}})$-surfaces are conformally equivalent if there is a biholomorphic map between them that preserves the orders, the orientations and the local coordinates of the punctures. We denote the conformal equivalence classes of Σ by $[\Sigma]$ and the moduli space of all $(k^{\mathrm{cl}}, k^{\mathrm{op}}; l^{\mathrm{cl}}, l^{\mathrm{op}})$-surfaces by $\mathbb{E}_{(k^{\mathrm{cl}}, k^{\mathrm{op}}; l^{\mathrm{cl}}, l^{\mathrm{op}})}$. We set $\mathbb{E} := \cup_{k^{\mathrm{cl}}, k^{\mathrm{op}}; l^{\mathrm{cl}}, l^{\mathrm{op}}} \mathbb{E}_{(k^{\mathrm{cl}}, k^{\mathrm{op}}; l^{\mathrm{cl}}, l^{\mathrm{op}})}$. On the one hand, the subspace of $\mathbb{E}$ containing only surfaces without boundary is nothing but $\mathcal{E}$. On the other hand, one can double a surface in $\mathbb{E}$ along its boundary components to obtain a closed surface with only interior punctures. This doubling defines a natural embedding of the moduli spaces:

$$\delta : \mathbb{E}_{(k^{\mathrm{cl}}, k^{\mathrm{op}}; l^{\mathrm{cl}}, l^{\mathrm{op}})} \hookrightarrow \mathcal{E}_{(2k^{\mathrm{cl}} + k^{\mathrm{op}}, 2l^{\mathrm{cl}} + l^{\mathrm{op}})}.$$

From now on, we identity $\mathbb{E}_{(k^{\mathrm{cl}}, k^{\mathrm{op}}; l^{\mathrm{cl}}, l^{\mathrm{op}})}$ with its image in $\mathcal{E}_{(2k^{\mathrm{cl}} + k^{\mathrm{op}}, 2l^{\mathrm{cl}} + l^{\mathrm{op}})}$. Notice that $\mathbb{E}$ inherits a smooth structure from that of $\mathcal{E}$. We also define

$$\mathbb{E}^g_{(k^{\mathrm{cl}}, k^{\mathrm{op}}; l^{\mathrm{cl}}, l^{\mathrm{op}})} := \mathbb{E}_{(k^{\mathrm{cl}}, k^{\mathrm{op}}; l^{\mathrm{cl}}, l^{\mathrm{op}})} \cap \mathcal{E}^g_{(2k^{\mathrm{cl}} + k^{\mathrm{op}}, 2l^{\mathrm{cl}} + l^{\mathrm{op}})}.$$

We define $\mathbb{E}_{(k^{\mathrm{cl}}, k^{\mathrm{op}}; l^{\mathrm{cl}}, l^{\mathrm{op}})} := \cup_g \mathbb{E}^g_{(k^{\mathrm{cl}}, k^{\mathrm{op}}; l^{\mathrm{cl}}, l^{\mathrm{op}})}$, $\mathbb{E}^g := \cup_{k^{\mathrm{cl}}, k^{\mathrm{op}}; l^{\mathrm{cl}}, l^{\mathrm{op}}} \mathbb{E}^g_{(k^{\mathrm{cl}}, k^{\mathrm{op}}; l^{\mathrm{cl}}, l^{\mathrm{op}})}$ and $\mathbb{E} := \cup_g \mathbb{E}^g$.

The sewing operation defined on $\mathcal{E}$ is closed on the subspace $\mathbb{E}$ as long as we only allow sewing between oppositely oriented punctures with the same σ-values. Therefore, sewing operations are well-defined on $\mathbb{E}$. Now we are ready to give the following definition.

---

[24]This result (but in a different framework) first appeared in [**Fe**] and later proved by Fjelstad, Fuchs, Runkel and Schweigert [**FRS1, FjFRS1**] (see Remark 4.24). The equivalence of two frameworks is known [**KR3**].

Definition 4.1. The partial category $\mathcal{RS}^p_{\text{cl}-\text{op}}$ consists of

(1) objects: finite ordered sets S with elements decorated by two colors: "closed string" and "open string", i.e. there is a splitting $S = S^{\text{cl}} \cup S^{\text{op}}$ such that $S^{\text{cl}} \cap S^{\text{op}} = \emptyset$.
(2) morphisms: $\text{Hom}_{\mathcal{RS}^p_{\text{cl}-\text{op}}}(S_1, S_2) := \mathbb{E}_{(|S_1^{\text{cl}}|,|S_1^{\text{op}}|;|S_2^{\text{cl}}|,|S_2^{\text{op}}|)}$.

The composition maps are defined by the sewing operations of Riemann surfaces.

Definition 4.2. An open-closed CFT is a real-analytic projective symmetric monoidal functor $\mathcal{F} : \mathcal{RS}^p_{\text{cl}-\text{op}} \to \mathcal{GV}$.

4.2. Operads. The category $\mathcal{RS}^p_{\text{cl}-\text{op}}$ is generated by two one-point sets: {a closed string}, {an open string}. We denote their $\mathcal{F}$-images by V_{cl}, V_{op} respectively. We will be interested in a few substructures of $\mathcal{RS}^p_{\text{cl}-\text{op}}$:

(1) the set $D := \cup_{n=0}^{\infty} \mathbb{E}^0_{0,n;0,1}$ gives a partial operad,
(2) the set $\mathbb{D} := \cup_{m=0,n=0}^{\infty} \mathbb{E}^0_{0,n;0,m}$ gives a partial dioperad,
(3) the set $S := \cup_{m=0,n=0}^{\infty} \mathbb{E}^0_{m,n;0,1}$ gives the so-called Swiss-cheese partial operad,
(4) the set $\mathbb{S} := \cup_{k=0,l=0,m=0,n=0} \mathbb{E}^0_{k,l;m,n}$ gives the so-called Swiss-cheese partial dioperad.

The structure $(V_{\text{op}}, \{\mathcal{F}([\Sigma])\}_{[\Sigma]\in D})$[25] is equivalent to a projective D-algebra structure on V_{op}, and that of $(V_{\text{op}}, \{\mathcal{F}([\Sigma])\}_{[\Sigma]\in \mathbb{D}})$ is equivalent to a projective $\mathbb{D}$-algebra. We will be interested in how to construct such algebras. The next two theorems tell us how to construct these algebras from two yet-to-be-introduced notions which will be discussed in the later subsections.

Theorem 4.3 ([**HK1**]). *An open-string VOA A canonically gives a projective D-algebra. If A is further equipped with a non-degenerate invariant bilinear form, then A gives a projective $\mathbb{D}$-algebra.*

Remark 4.4. The original theorem given in [**HK1**] says that the category of open-string VOAs of central charge c is isomorphic to the category of holomorphic algebras (not projective!) over a certain partial-operad extension of D by the $\frac{c}{2}$-th power of the determinant line bundle over D and satisfying additional natural properties. We only give the result in above form to avoid technicalities.

Theorem 4.5 ([**Ko2**]). *An open-closed field algebra A canonically gives a projective S-algebra. If it is self-dual, then A gives a projective $\mathbb{S}$-algebra.*

4.3. Open-string vertex operator algebras. The notion of open-string VOA was introduced in [**HK1**]. We will not give a precise definition of it. Instead, we will use the structure of a projective D-algebra $(V_{\text{op}}, \{\mathcal{F}([\Sigma])\}_{[\Sigma]\in D})$ to illustrate the basic ingredients and the properties of an open-string VOA. Note that there is a natural embedding $D \hookrightarrow K$ by doubling the disks in D so that all the punctures are located on the equator of the resulting sphere. Therefore, we can adapt the same notation as (3.1) for elements in D.

(1) An open-string VOA V_{op} is an $\mathbb{R}$-graded vector space such that $\dim(V_{\text{op}})_{(n)} < \infty$ and the grading is truncated from below, i.e. $(V_{\text{op}})_{(n)} = 0$ for $n << 0$.
(2) Similar to VOA, we have $[\hat{\mathbb{C}}, (\infty, f_\infty, -))] \in D \subset K$ where $f_\infty : w \mapsto \frac{-1}{w}$. This gives rise to a distinguished element $\mathcal{F}([\hat{\mathbb{C}}, (\infty, f_\infty, -))])$ in $\overline{V}_{\text{op}}$, denoted by $\mathbf{1}_{\text{op}}$. In an open-string VOA, $\mathbf{1}_{\text{op}} \in (V_{\text{op}})_{(0)}$.
(3) The $\mathcal{F}$ image of $[\hat{\mathbb{C}}, (0, f_0, +), (\infty, f_\infty, -))]$, where $f_0 : w \mapsto w$ and $f_\infty : w \to \frac{-1}{w}$, is $\text{id}_{V_{\text{op}}}$.

[25]Only compositions associated to sewing of surfaces resulting in surfaces in D are allowed.

(4) Consider the following element,

$$P(r) := [(\hat{\mathbb{C}}, (r, f_r, +), (0, f_0, +), (\infty, f_\infty, -))] \in D \subset K$$

where $r > 0$, $f_r : w \to w - r$, $f_0 : w \to w$, $f_\infty : w \to \frac{-1}{w}$. The linear map $\mathcal{F}(P(z))$ gives a vertex operator $Y_{\rm op}(\cdot, r)\cdot := \mathcal{F}(P(r)) : V_{\rm op} \otimes V_{\rm op} \to \overline{V}_{\rm op}$ for $r > 0$. For $u, v \in V_{\rm op}$, we have (recall equation (3.2))

$$u \otimes v \mapsto Y_{\rm op}(u, r)v = \sum_{n\in\mathbb{R}} u_n v\, r^{-n-1},$$

where $u_n : V_{\rm op} \to \overline{V}$ and the nontrivial summands are those n lying in $Q + \mathbb{Z}$ where Q is a finite set in $\mathbb{R}$. In an open-string VOA, $u_n \in \mathrm{End}(V)$. Moreover, if $u \in (V_{\rm op})_{(m)}$, then $u_n : (V_{\rm op})_{(k)} \to V_{(k+m-n-1)}$. Therefore, $u_n v = 0$ for $n >> 0$.

(5) By the axioms of CFT, the following sum:

$$\langle u', Y_{\rm op}(u_1, r_1)Y_{\rm op}(u_2, r_2)u_3\rangle := \sum_{n\in\mathbb{Z}} \langle u', Y_{\rm op}(u_1, r_1)P_n Y_{\rm op}(u_2, r_2)u_3\rangle$$

is absolutely convergent in $\overline{V}_{\rm op}$ when $r_1 > r_2 > 0$ for all $u_1, u_2, u_3 \in V_{\rm op}$ and $u' \in V'_{\rm op}$. The following sum:

$$\langle u', Y_{\rm op}(Y_{\rm op}(u_1, r_1 - r_2)u_2, r_2)u_3\rangle := \sum_{n\in\mathbb{Z}} \langle u', Y_{\rm op}(P_n Y_{\rm op}(u_1, r_1 - r_2)u_2, r_2)u_3\rangle$$

is absolutely convergent in $\overline{V}_{\rm op}$ when $r_2 > r_1 - r_2 > 0$ all $u_1, u_2, u_3 \in V_{\rm op}$ and $u' \in V'_{\rm op}$.

(6) We have the associativity of open-string VOA:

$$\langle u', Y_{\rm op}(u_1, r_1)Y_{\rm op}(u_2, r_2)u_3\rangle = \langle u', Y_{\rm op}(Y_{\rm op}(u_1, r_1 - r_2)u_2, r_2)u_3\rangle$$

when $r_1 > r_2 > r_1 - r_2 > 0$ for all $u_1, u_2, u_3 \in V$ and $u' \in V'$. This associativity is also called the OPE of boundary fields in physics.

(7) Consider the element $\Sigma_\omega := [(\hat{\mathbb{C}}, (\infty, f_\infty, -))] \in D$ with $f_\infty : w \to \frac{-1}{w-\varepsilon}$. It is clear that $\mathcal{F}(\Sigma_\omega)$ is an element in $\overline{V}_{\rm op}$. Then $\omega_{\rm op} := \frac{d}{d\varepsilon}|_{\varepsilon=0}\mathcal{F}(\Sigma_\omega)$ gives a distinguished element in $\overline{V}_{\rm op}$. Actually, $\omega_{\rm op} \in (V_{\rm op})_{(2)}$. One has

$$Y_{\rm op}(\omega_{\rm op}, r) = \sum_{n\in\mathbb{Z}} L(n)r^{-n-2} ,$$

where $L(n), n \in \mathbb{Z}$, can be shown to generate a Virasoro Lie algebra, i.e.

$$[L(m), L(n)] = (m-n)L(m+n) + \frac{c}{12}(m^3 - m)\delta_{m+n,0} , \qquad \forall m, n \in \mathbb{Z},$$

where $c \in \mathbb{C}$ is called the central charge. Moreover, $\omega = L(-2)\mathbf{1}$ and $L(0)$ gives the grading operator.

(8) We thus denote an open-string VOA by its basic ingredients: $(V_{\rm op}, Y_{\rm op}, \mathbf{1}_{\rm op}, \omega_{\rm op})$.

It is also straight forward to define the notion of a module over an open-string VOA. We will not do it here. Instead, we will explore a relation between the representation theory of VOAs and open-string VOAs. As we will show later that an open-string VOA is always a module over a VOA. This fact is extremely useful when the underlying VOA is rational.

We introduce a formal vertex operator $\mathcal{Y}_{\rm op}$ such that

$$\mathcal{Y}_{\rm op}(u, x)v = \sum_{n\in\mathbb{R}} u_n v x^{-n-1}.$$

for $u, v \in V_{\mathrm{op}}$. Namely, $\mathcal{Y}_{\mathrm{op}}(u,x)v|_{x=r} = Y_{\mathrm{op}}(u,r)v$. We define the so-called *meromorphic center* $C_0(V_{\mathrm{op}})$ of the open-string VOA V_{op} as follows:

$$C_0(V_{\mathrm{op}}) := \Big\{u \in \coprod_{n\in\mathbb{Z}} V_{(n)} | \mathcal{Y}_{\mathrm{op}}(u,x) \in (\mathrm{End}V)[[x, x^-1]], \quad \mathcal{Y}_{\mathrm{op}}(v,x)u = e^{xL(-1)}\mathcal{Y}_{\mathrm{op}}(u,-x)v, \forall v \in V_{\mathrm{op}}\Big\}$$

It is easy to show that $\mathbf{1}_{\mathrm{op}}, \omega_{\mathrm{op}} \in C_0(V_{\mathrm{op}})$ and the image of $C_0(V_{\mathrm{op}}) \otimes C_0(V_{\mathrm{op}})$ under $\mathcal{Y}_{\mathrm{op}}|_{C_0(V_{\mathrm{op}})}$ lies in $C_0(V_{\mathrm{op}})[[x, x^{-1}]]$. Moreover, we have the following result:

Proposition 4.6. *The quadruple $(C_0(V_{\mathrm{op}}), \mathcal{Y}_{\mathrm{op}}|_{C_0(V_{\mathrm{op}})}, \mathbf{1}_{\mathrm{op}}, \omega_{\mathrm{op}})$ is a VOA. V_{op} is a module over $C_0(V_{\mathrm{op}})$ and $\mathcal{Y}_{\mathrm{op}}$ is an intertwining operator of type $\binom{V_{\mathrm{op}}}{V_{\mathrm{op}}V_{\mathrm{op}}}$* [**FHL**] *with respect to $C_0(V_{\mathrm{op}})$.*

By the above results, an open-string VOA can be constructed and studied by the representation theory of VOAs. An open-string VOA containing a VOA V as a sub-VOA of its meromorphic center is called an open-string VOA over V. We denote the canonical embedding $V \hookrightarrow V_{\mathrm{op}}$ by ι_{op}. If V is a rational VOA, one can apply the tensor category theory to construct open-string VOAs.

An invariant bilinear form $(\cdot,\cdot)_{\mathrm{op}} : V_{\mathrm{op}} \otimes V_{\mathrm{op}} \to \mathbb{C}$ is called invariant if the following identity:

$$(w, Y_{\mathrm{op}}(u,z)v)_{\mathrm{op}} = (e^{-r^{-1}L(-1)}Y_{\mathrm{op}}(w, r^{-1})e^{-rL(1)}r^{-2L(0)}u, v)_{\mathrm{op}}$$

holds for $u, v, w \in V_{\mathrm{op}}$ and $r > 0$. An open-string VOA equipped with a non-degenerate symmetric invariant bilinear form is called self-dual. It was shown in [**Ko3**] that a self-dual open-string VOA canonically produces a projective algebra over the partial dioperad $\mathbb{D}$.

4.4. Open-closed field algebras. The notion of open-closed field algebra was introduced in [**Ko2**]. We will not give an explicit definition of it. Instead, by Theorem 4.5, we will illustrate the structures and the properties of an open-closed field algebra via the structures on S.

(1) The Swiss-cheese partial operad S contains the spherical partial operad K as a subset. Therefore, an open-closed field algebra contains a full field algebra V_{cl} as a substructure.
(2) The Swiss-cheese partial operad S also contains the partial operad D as a subset. Therefore, an open-closed field algebra contains an open-string VOA V_{op} as a substructure.
(3) Consider the following element, for $z \in \mathbb{H}$,

$$D(z) := [(\Sigma|(z, f_z, +), (0, f_0, +), (\infty, f_\infty, -))],$$

where $f_z : w \mapsto w - z$, $f_0 : w \mapsto w$, $f_\infty : w \mapsto \frac{-1}{w}$. It is clear that $\mathcal{F}(\Sigma) : V_{\mathrm{cl}} \otimes V_{\mathrm{op}} \to \overline{V}_{\mathrm{op}}$. We set $Y_{\mathrm{cl\text{-}op}}(\cdot; z, \bar{z})\cdot := \mathcal{F}(\Sigma)$.
(4) Since the entire Swiss-cheese partial operad S is generated by the elements in K, D and $D(z)$, an open-closed field algebra can be described by the data $V_{\mathrm{op}}, V_{\mathrm{cl}}, Y_{\mathrm{cl\text{-}op}}$. Therefore, we denote an open-closed field algebra by a triple $(V_{\mathrm{op}}|V_{\mathrm{cl}}, Y_{\mathrm{cl\text{-}op}})$.

The notion of open-closed field algebra introduced in [**Ko2**] is too general for the purpose of this paper. We will only consider the so-called *analytic open-closed field algebras* which satisfy the following analytic properties:

(1) (a) Y_{op} can be extended to a map $V_{\mathrm{op}} \otimes V_{\mathrm{op}} \times (\mathbb{C}^\times / -\mathbb{R}_+) \to \overline{V}_{\mathrm{op}}$:

$$Y_{\mathrm{op}}(v, r) = Y_{\mathrm{op}}(u, z)|_{z=r};$$

(b) $Y_{\text{cl-op}}$ can be extended to a map $V_{\text{cl}} \otimes V_{\text{op}} \times \mathbb{H} \times \overline{\mathbb{H}} \to \overline{V}_{\text{op}}$ such that for $z \in \mathbb{H}, \zeta \in \overline{\mathbb{H}}$

$$Y_{\text{cl-op}}(u; z, \bar{z}) = Y_{\text{cl-op}}(u; z, \zeta)|_{\zeta=\bar{z}}.$$

(2) For $n \in \mathbb{N}$, $v_1, \cdots, v_{n+1} \in V_{\text{op}}, v' \in (V_{\text{op}})'$ and $u_1, \cdots, u_n \in V_{\text{cl}}$ and $z_1, \cdots, z_n \in \mathbb{H}, \zeta_1, \cdots, \zeta_n \in \overline{\mathbb{H}}$, the series

$$\langle v', Y_{\text{cl-op}}(u_1; z_1, \zeta_1) Y_{\text{op}}(v_1, r_1) \cdots Y_{\text{cl-op}}(u_n; z_n, \zeta_n) Y_{\text{op}}(v_n, r_n) v_{n+1} \rangle$$

is absolutely convergent when $|z_1|, |\zeta_1| > r_1 > \cdots > |z_n|, |\zeta_n| > r_n > 0$ and can be extended to a (possibly multivalued) analytic function on:

$$\{(z_1, \zeta_1, r_1, \dots, z_n, \zeta_n, r_n) \in M_{\mathbb{C}}^{3n}\}$$

where $M_{\mathbb{C}}^n := \{(z_1, \dots, z_n) \in \mathbb{C}^n | z_i \neq z_j, \text{ for } i, j = 1, \dots, n \text{ and } i \neq j\}$.

(3) For $n \in \mathbb{N}$, $v', u_1, \dots, u_{n+1} \in V_{\text{cl}}$ and $z_1, \cdots, z_n, \zeta_1, \cdots, \zeta_n \in \mathbb{C}$ the series

$$\langle v', Y_{\text{cl}}(u_1; z_1, \zeta_1) \cdots Y_{\text{cl}}(u_n; z_n, \zeta_n) u_{n+1} \rangle$$

is absolutely convergent when $|z_1| > \cdots > |z_n| > 0$ and $|\zeta_1| > \cdots > |\zeta_n| > 0$ and can be extended to an analytic function on $M_{\mathbb{C}}^{2n}$.

(4) For $v' \in V_{\text{op}}', v_1, v_2 \in V_{\text{op}}, u \in V_{\text{cl}}$ and $z \in \mathbb{H}, \zeta \in \overline{\mathbb{H}}$, the series

$$\langle v', Y_{\text{op}}(Y_{\text{cl-op}}(u; z, \zeta) v_1, r) v_2 \rangle$$

is absolutely convergent when $r > |z|, |\zeta| > 0$.

(5) For $v' \in V_{\text{op}}', v \in V_{\text{op}}, u_1, u_2 \in V_{\text{cl}}$ and $z \in \mathbb{H}, \zeta \in \overline{\mathbb{H}}$, the series

$$\langle v', Y_{\text{cl-op}}(Y_{\text{cl}}(u_1; z_1, \zeta_1) u_1; z_2, \zeta_2) v \rangle,$$

converges absolutely when $|z_2| > |z_1| > 0$, $|\zeta_2| > |\zeta_1| > 0$ and $|z_1| + |\zeta_1| < |z_2 - \zeta_2|$.

An analytic open-closed field algebra satisfies the following nice properties:

(1) Unit property: $Y_{\text{cl-op}}(\mathbf{1}_{\text{cl}}; z, \zeta) = \text{id}_{V_{\text{op}}}$.

(2) Associativity I: For $u \in V_{\text{cl}}, v_1, v_2 \in V_{\text{op}}, v' \in V_{\text{op}}'$ and $z \in \mathbb{H}, \zeta \in \overline{\mathbb{H}}$, we have

$$\langle v', Y_{\text{cl-op}}(u; z, \zeta) Y_{\text{op}}(v_1, r) v_2 \rangle = \langle v', Y_{\text{op}}(Y_{\text{cl-op}}(u; z - r, \zeta - r) v_1, r) v_2 \rangle$$

when $|z|, |\zeta| > r > 0$ and $r > |r - z|, |r - \zeta| > 0$.

(3) Associativity II: For $u_1, u_2 \in V_{\text{cl}}, v_1, v_2 \in V_{\text{op}}, v' \in V_{\text{op}}'$ and $z_1, z_2 \in \mathbb{H}, \zeta_1, \zeta_2 \in \overline{\mathbb{H}}$, we have

$$\begin{aligned} &\langle w', Y_{\text{cl-op}}(u_1; z_1, \zeta_1) Y_{\text{cl-op}}(u_2; z_2, \zeta_2) v_2 \rangle \\ &\quad = \langle v', Y_{\text{cl-op}}(Y_{\text{cl}}(u_1; z_1 - z_2, \zeta_1 - \zeta_2) u_2; z_2, \zeta_2) v_2 \rangle \end{aligned}$$

when $|z_1|, |\zeta_1| > |z_2|, |\zeta_2|$ and $|z_2| > |z_1 - z_2| > 0, |\zeta_2| > |\zeta_1 - \zeta_2| > 0$ and $|z_2 - \zeta_2| > |z_1 - z_2| + |\zeta_1 - \zeta_2|$.

(4) Commutativity I: The map $Y_{\text{cl-op}}$ can be uniquely extended to $V_{\text{cl}} \otimes V_{\text{op}} \times R$, where

$$R := \{(z, \zeta) \in \mathbb{C}^2 | z \in \mathbb{H} \cup \mathbb{R}_+, \zeta \in \overline{\mathbb{H}} \cup \mathbb{R}_+, z \neq \zeta\}.$$

For $u \in V_{\text{cl}}, v_1, v_2 \in V_{\text{op}}$ and $v' \in V_{\text{op}}'$,

$$\langle v', Y_{\text{cl-op}}(u; z, \zeta) Y_{\text{op}}(v_1, r) v_2 \rangle,$$

which is absolutely convergent when $z > \zeta > r > 0$, and

$$\langle v', Y_{\text{op}}(v_1, r) Y_{\text{cl-op}}(u; z, \zeta) v_2 \rangle,$$

which is absolutely convergent when $r > z > \zeta > 0$, are the analytic continuations of each other along the following path [26].

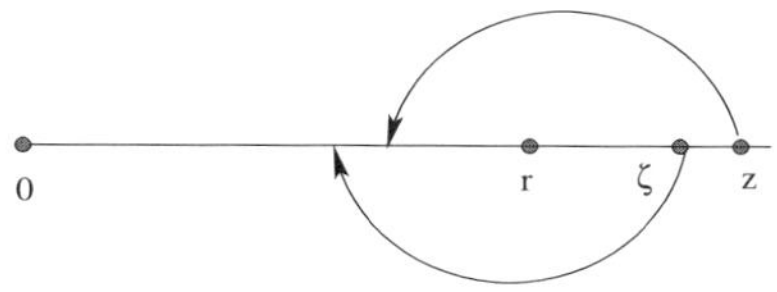

(5) Commutativity II: For $u_1, u_2 \in V_{\text{cl}}$, $v \in V_{\text{op}}$ and $v' \in V'_{\text{op}}$,

$$\langle v', Y_{\text{cl-op}}(u_1; z_1, \zeta_1) Y_{\text{cl-op}}(u_2; z_2, \zeta_2) v\rangle,$$

which is absolutely convergent when $z_1 > \zeta_1 > z_2 > \zeta_2 > 0$, and

$$\langle v', Y_{\text{cl-op}}(u_2; z_2, \zeta_2) Y_{\text{cl-op}}(u_1; z_1, \zeta_1) v\rangle,$$

which is absolutely convergent when $z_2 > \zeta_2 > z_1 > \zeta_1 > 0$, are analytic continuation of each other along the following paths.

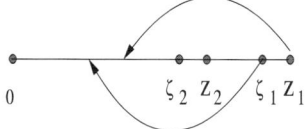

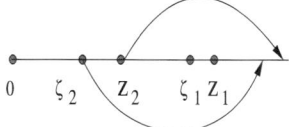

Remark 4.7. The commutativity II follows from the Associativity II and the commutativity of the full field algebra. This is an analogue of the statement that the action of a commutative algebra on its module is automatically commutative.

For a given open-closed field algebra $(V_{\text{op}}|V_{\text{cl}}, Y_{\text{cl-op}})$, the data $Y_{\text{cl-op}}$ can also be described by slightly simpler data. We define $\iota_{\text{cl-op}} : V_{\text{cl}} \times \mathbb{H} \times \overline{\mathbb{H}} \to \overline{V}_{\text{op}}$ as follows:

$$\iota_{\text{cl-op}}^{(z,\zeta)}(u) := Y_{\text{cl-op}}(u; z, \zeta)\mathbf{1}_{\text{op}}.$$

for $u \in V_{\text{cl}}$ and $v \in V_{\text{op}}$. Conversely, $Y_{\text{cl-op}}(u; z, \zeta)$ can be obtained from $\iota_{\text{cl-op}}$ by defining

$$Y_{\text{cl-op}}(u; z, \zeta) := Y_{\text{op}}(\iota_{\text{cl-op}}^{(z-r,\zeta-r)}(u), r)v$$

for $r > |z - r|, |\zeta - r| > 0$ and by the analytic continuations. Therefore, we can also denote the open-closed field algebra by a triple $(V_{\text{op}}|V_{\text{cl}}, \iota_{\text{cl-op}})$. Moreover, it is easy to show that this map $\iota_{\text{cl-op}}$ plays the role of an algebra homomorphism between V_{cl} and V_{op}. It becomes evident in the classification of open-closed rational CFTs (see Theorem 4.17).

Remark 4.8. In an open-closed field algebra $(V_{\text{op}}|V_{\text{cl}}, \iota_{\text{cl-op}})$, if we ignore the algebra structure on V_{op}, just viewing it as a module over one copy of the Virasoro algebra equipped with an action $Y_{\text{cl-op}}$ of V_{cl} satisfying the Associativity II and the unit property, we obtain some kind of a "V_{cl}-module" structure on V_{op} given by $Y_{\text{cl-op}}$. We will call it a chiral V_{cl}-module. Due to the fact that the map $\iota_{\text{cl-op}}$ is some kind of an algebraic homomorphism, a V_{op}-module is automatically a chiral V_{cl}-module.

From now on, we will be interested in open-closed field algebras over a VOA V, which consists of a full field algebra over $V \otimes V$ $(V_{\text{cl}}, V \otimes V \xrightarrow{\iota_{\text{cl}}} V_{\text{cl}})$ and an open-string VOA over V $(V_{\text{op}}, V \xrightarrow{\iota_{\text{op}}} V_{\text{op}})$. In this case, one can define two maps $h^L : V \to V_{\text{op}}$ and $h^R : V \to V_{\text{op}}$ as follows: for all $u, v \in V$,

$$h^L : u \mapsto \lim_{z\to 0} Y_{\text{cl-op}}(\iota_{\text{cl}}(u \otimes \mathbf{1}), z)\mathbf{1}_{\text{op}},$$

$$h^R : v \mapsto \lim_{\zeta\to 0} Y_{\text{cl-op}}(\iota_{\text{cl}}(\mathbf{1} \otimes v), \zeta)\mathbf{1}_{\text{op}}.$$

[26]The extended domain $R\backslash\{z = r, \text{or}, \zeta = r\}$ is simply connected for fixed $r > 0$, all possible paths of analytic continuation are homotopically equivalent.

Notice that h^L, h^R preserve the conformal weights. Namely

$$\begin{aligned} \mathrm{wt}\, h^L(u) &= \mathrm{wt}\,(u_{-1}\mathbf{1}_{\mathrm{op}}) = \mathrm{wt}^L u, \\ \mathrm{wt}\, h^R(v) &= \mathrm{wt}\,(v_{-1}\mathbf{1}_{\mathrm{op}}) = \mathrm{wt}^R v. \end{aligned}$$

Therefore, both h^L and h^R can be naturally extended to maps $\overline{V} \to \overline{V_{\mathrm{op}}}$. We still denote the extended maps by h^L and h^R respectively.

Definition 4.9. Let $(V, Y, \mathbf{1}, \omega)$ be a vertex operator algebra. An *open-closed field algebra over* V is an analytic open-closed field algebra

$$(V_{\mathrm{op}}|V_{\mathrm{cl}}, Y_{\mathrm{cl\text{-}op}}),$$

where $V_{\mathrm{cl}} = (V_{\mathrm{cl}}, Y_{\mathrm{cl}}, \mathbf{1}_{\mathrm{cl}}, \omega_{\mathrm{cl}})$ is a full field algebra over $V \otimes V$ and $(V_{\mathrm{op}}, Y_{\mathrm{op}}, \mathbf{1}_{\mathrm{op}}, \omega_{\mathrm{op}})$ is an open-string vertex operator algebra over V, satisfying the following conditions:

(1) *V-invariant boundary condition*: $h^L = h^R = \iota_{\mathrm{op}}$.

(2) *Chirality splitting property*: $\forall u \in V_{\mathrm{cl}}$, $u = u^L \otimes u^R \in W^L \otimes W^R \subset V_{\mathrm{cl}}$ for some V-modules W^L, W^R. There exist V-modules W_1, W_2 and intertwining operators $\mathcal{Y}^{(1)}, \mathcal{Y}^{(2)}, \mathcal{Y}^{(3)}, \mathcal{Y}^{(4)}$ of type $\binom{V_{\mathrm{op}}}{W^L W_1}$, $\binom{W_1}{W^R V_{\mathrm{op}}}$, $\binom{V_{\mathrm{op}}}{W^R W_2}$, $\binom{W_2}{W^L V_{\mathrm{op}}}$ respectively, such that

$$\langle w', Y_{\mathrm{cl\text{-}op}}(u; z, \zeta)w\rangle = \langle w', \mathcal{Y}^{(1)}(u^L, z)\mathcal{Y}^{(2)}(u^R, \zeta)w\rangle$$

when $|z| > |\zeta| > 0$, and

$$\langle w', Y_{\mathrm{cl\text{-}op}}(u; z, \zeta)w\rangle = \langle w', \mathcal{Y}^{(3)}(u^R, \zeta)\mathcal{Y}^{(4)}(u^L, z)w\rangle$$

when $|\zeta| > |z| > 0$ for all $u \in V_{\mathrm{cl}}, w \in V_{\mathrm{op}}, w' \in V_{\mathrm{op}}$.

In the case that V is generated by ω, i.e. $V = \langle\omega\rangle$, the $\langle\omega\rangle$-invariant boundary condition is also called the *conformally invariant boundary condition*. We also call an open-closed field algebra over $\langle\omega\rangle$ an *open-closed conformal field algebra*.

Remark 4.10. In order to obtain an open-closed CFT, the minimal requirement on the boundary is the conformally invariant boundary condition. By restricting ourselves to V-invariant boundary conditions and assuming V is rational, we will be able to obtain a nice categorical formulation of open-closed field algebra over V. The price we pay is that there are too few V-invariant D-branes to reveal an interesting geometry. To classify all conformally invariant D-branes for a given closed rational CFT is a central task in the program of SAG.

Definition 4.11. An open-closed field algebra $(V_{\mathrm{op}}|V_{\mathrm{cl}}, Y_{\mathrm{cl\text{-}op}})$ is called self-dual if there are non-degenerate symmetric invariant bilinear forms $(\cdot,\cdot)_{\mathrm{op}}$ and $(\cdot,\cdot)_{\mathrm{cl}}$ on V_{op} and V_{cl} respectively.

For a self-dual open-closed field algebra $(V_{\mathrm{op}}|V_{\mathrm{cl}}, \iota_{\mathrm{cl\text{-}op}})$, it is easy to obtain the Ishibashi states and the boundary states. Indeed, for $v \in V_{\mathrm{op}}$, we define the boundary state $\mathbf{B}(v) \in \overline{V_{\mathrm{cl}}}$ associated to v by

$$\mathbf{B}(v) = e^{L(-1)}(\bar{z}_0 - z_0)^{L(0)} \otimes e^{L(-1)}\overline{\bar{z}_0 - z_0}^{L(0)} \iota^*_{\mathrm{cl\text{-}op}}(z_0, \bar{z}_0)(v),$$

where $z_0 \in \mathbb{H}$ and $\iota^*_{\mathrm{cl\text{-}op}}$ is the adjoint of $\iota_{\mathrm{cl\text{-}op}}$, i.e. for $v \in V_{\mathrm{op}}$ and $u \in V_{\mathrm{cl}}$,

$$(v, \iota^{(z,\bar{z})}_{\mathrm{cl\text{-}op}}(u))_{\mathrm{op}} = ((\iota^{(z,\bar{z})}_{\mathrm{cl\text{-}op}})^*(v), u)_{\mathrm{cl}}.$$

Actually, $\mathbf{B}(v)$ is independent of z_0. This explains the notation.

Proposition 4.12 (Ishibashi States [**Ko3**]). *If $v \in V_{\mathrm{op}}$ is such that $L(-1)v = 0$, then, for $z_0 \in \mathbb{H}$, $\mathbf{B}(v)$ is an Ishibashi state, i.e.*

$$(L^L(n) - L^R(-n))\mathbf{B}(v) = 0, \qquad \forall n \in \mathbb{Z}.$$

An open-closed CFT containing only spheres and disks with arbitrary number of incoming and outgoing interior and boundary punctures is equivalent to a projective algebra over the Swiss-cheese partial dioperad [**Ko2**]. Since spheres and disks (under the doubling map) cover all genus-zero surfaces, a projective algebra over the Swiss-cheese partial dioperad is also called a genus-zero open-closed CFT.

Theorem 4.13 ([**Ko2**]). *A self-dual open-closed field algebra over a rational VOA V canonically gives a genus-zero open-closed CFT.*

Notice that all the surfaces in open-closed CFT can be obtained by gluing spheres and disks. Therefore, an open-closed field algebra over a rational V contains all the building blocks of an open-closed CFT. Higher genus surfaces (in the sense of the doubling map) only provide certain compatibility conditions. The modular invariance condition discussed in Section 3 is one of such compatibility conditions. It has long been conjectured that the only remaining compatibility condition is the so-called Cardy condition [**Lw**]. The Cardy condition comes from two ways of realizing the same world-sheet:

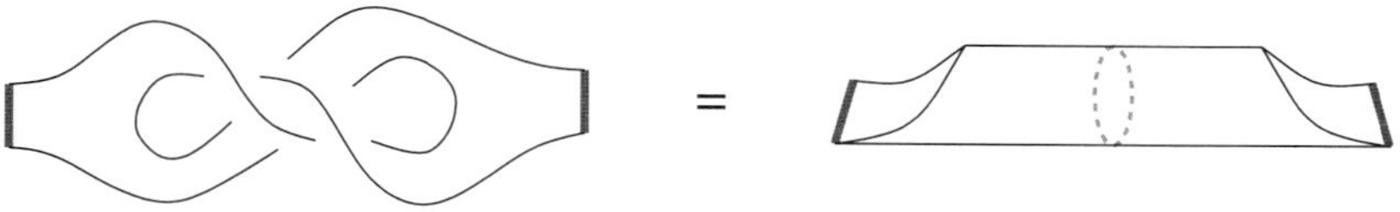

in which the two (red) line segments at two sides of the above surfaces represent incoming or outgoing open strings. A precise definition of the Cardy condition in terms of correlation functions has been worked out in [**Ko3**, Def 3.4, Thm 3.10]. We will not recall it here. We only need its categorical formulation [**Ko3**, eq. (5.31)][**KR2**, eq. (3.14)], which will be given in the next section.

Conjecture 4.14. *An open-closed field algebra over a rational VOA V satisfying the modular invariance condition and the Cardy condition gives a consistent open-closed CFT.*

4.5. Classification of open-closed CFTs over V. Now we assume that V is rational. Namely, the category $\mathcal{C}_V$ of V-modules is a modular tensor category. In this case, $\mathcal{C}_V \boxtimes (\mathcal{C}_V)_-$ is canonically equivalent to the monoidal center $Z(\mathcal{C}_V)$ of $\mathcal{C}_V$ as modular tensor categories. For simplicity, we set $\mathcal{C}_V^{(2)} := \mathcal{C}_V \boxtimes (\mathcal{C}_V)_-$. We denote the tensor product functor $A \times B \to A \otimes B$ from $\mathcal{C}_V^{(2)}$ to $\mathcal{C}_V$ by T. We denote the right adjoint of T by R. Actually, R is also a left adjoint of T. T is a monoidal functor. It was shown in [**KR2**] that R is both a lax and a colax functor. Moreover, it is a Frobenius functor such that it maps a Frobenius algebra in $\mathcal{C}_V$ to a Frobenius algebra in $\mathcal{C}_V \boxtimes (\mathcal{C}_V)_-$.

We need the notion of center of an algebra A in $\mathcal{C}_V$ in order to state our classification result. We first define what a left center of an algebra in $\mathcal{C}_V^{(2)}$. We denote the braiding of $\mathcal{C}_V^{(2)}$ by $c_{X,Y} : X \otimes Y \to Y \otimes X$. Then the left center of an algebra (B, m_B, ι_B) in $\mathcal{C}_V^{(2)}$ is the maximal subobject $e : Z \hookrightarrow B$ such that $m_B \circ (e \otimes \mathrm{id}_B) = m_B \circ c_{B,B} \circ (e \otimes \mathrm{id}_B)$.

Definition 4.15 ([**FjFRS2**]). The center of A in $\mathcal{C}_V$ is the left center $C_l(R(A))$ of $R(A)$ in $\mathcal{C}_V^{(2)}$.

Remark 4.16. The above definition of center of an algebra is not obviously natural. We will briefly describe its naturalness in this remark via its universal property. The monoidal center $Z(\mathcal{C})$ of a tensor category $\mathcal{C}$ is the category of pairs (Z, z) where $Z \in \mathcal{C}$ and $z : Z \otimes - \to - \otimes Z$ is the half braiding (satisfying certain properties). We denote the forgetful functor $Z(\mathcal{C}) \to \mathcal{C}$ by F. When $\mathcal{C}$ is a modular tensor category, there is a canonical equivalence of ribbon categories $\phi : \mathcal{C}_V^{(2)} \xrightarrow{\cong} Z(\mathcal{C})$. Moreover, we have $T \cong F \circ \phi$. Therefore, one can equivalently define the center of A in $\mathcal{C}$ as an object in $Z(\mathcal{C})$. More precisely, Davydov [**Da**] showed that the center of A can be equivalently defined by a pair $((Z,z), Z \xrightarrow{e_Z} A)$, where (Z,z) is an object in $Z(\mathcal{C})$ and e_Z is a morphism in $\mathcal{C}$, such that it is terminal among all such pairs $((X,x), e_X)$ satisfying the following commutative diagram:

$$\begin{array}{ccccc} X \otimes A & \xrightarrow{e_X \otimes \mathrm{id}_A} & A \otimes A & & \\ \downarrow{\scriptstyle x} & & & \searrow^{m_A} & \\ A \otimes X & \xrightarrow{\mathrm{id}_A \otimes e_X} & A \otimes A & \xrightarrow{m_A} & A\,. \end{array}$$

Similar to the case of algebras over vector spaces, the above definition of center is equivalent to a certain internal endomorphism of the identity functor on the category A-modules [**Da**].

Given a morphism $f : A \to B$ between two Frobenius algebras $(A, m_A, \eta_A, \Delta_A, \epsilon_A)$ and $(B, m_B, \eta_B, \Delta_B, \epsilon_B)$, we define the right adjoint of f to be

$$f^* := ((\epsilon_B \circ m_B) \otimes \mathrm{id}_A) \circ (\mathrm{id}_B \otimes f \otimes \mathrm{id}_A) \circ (\mathrm{id}_B \otimes (\Delta_A \circ \eta_A)).$$

Now we are ready to give a categorical formulation of open-closed field algebra over a rational V satisfying additional properties. This result was first obtained in [**Ko3**, Thm. 5.15]. The following version is taken from [**KR2**].

Theorem 4.17 ([**Ko3**][**KR2**]). *An open-closed field algebra over V satisfying the modular invariance condition and the Cardy condition is equivalent to a triple*

$$(A_{\mathrm{op}} | A_{\mathrm{cl}}, \iota_{\text{cl-op}})$$

where

(1) *A_{op} is a symmetric Frobenius algebra in $\mathcal{C}_V$;*
(2) *A_{cl} is a modular invariant commutative symmetric Frobenius algebra in $\mathcal{C}_V^{(2)}$;*
(3) *$\iota_{\text{cl-op}} : A_{\mathrm{cl}} \to R(A_{\mathrm{op}})$ is an algebra homomorphism factoring through $e : Z(A_{\mathrm{op}}) \hookrightarrow R(A_{\mathrm{op}})$*

satisfying the Cardy condition:

(4.1) $$\iota_{\text{cl-op}} \circ \iota^*_{\text{cl-op}} \;=$$ 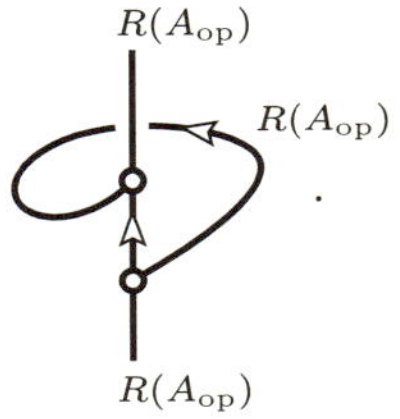

.

where ι^* is the right adjoint of $\iota_{\text{cl-op}}$.

Following [**Ko3**][**KR2**], we will call the triple defined in Theorem 4.17 a Cardy $\mathcal{C}_V|\mathcal{C}_V^{(2)}$-algebra.

Remark 4.18. If Conjecture 4.14 is correct, Theorem 4.17 gives a classification of open-closed rational CFTs satisfying the V-invariant boundary condition. A similar classification result has been obtained in another approach towards rational CFTs via the theory of conformal nets [**LR1, LR2, R**]. In this approach, no higher genus surface is involved. The modular invariance condition and the Cardy condition is somehow captured by a so-called Haag dual condition (see [**KR4**] for a review).

Definition 4.19. For a given full field algebra A_{cl} over $V \otimes V$, a *V-invariant D-brane* is a symmetric Frobenius algebra A in $\mathcal{C}_V$ together with an algebra homomorphism $\iota_{\text{cl-op}} : A_{\mathrm{cl}} \to R(A)$ such that the triple $(A|A_{\mathrm{cl}}, \iota_{\text{cl-op}})$ gives a Cardy $\mathcal{C}_V|\mathcal{C}_V^{(2)}$-algebra.

It is easy to see that all V-invariant D-branes naturally form a category instead of a set. In general, such V-invariant D-branes are too few to give an interesting geometry. For later purposes, one should consider symmetry-broken D-branes, i.e. U-invariant D-branes where U is a sub-VOA of V. In particular, when $U = \langle\omega\rangle$, they are also called *conformally invariant D-branes*. Therefore, to have a good picture of SAG, we need to classify all the conformally invariant D-branes associated to a closed CFT. In particular, we need to understand how V and V-modules decompose into modules over $\langle\omega\rangle$ the Virasoro sub-VOA of V. But in general, $\langle\omega\rangle$ is very small and irrational. Therefore, in order to understand SAG associated to a closed rational CFT, it is still unavoidable to study irrational theory.

4.6. Constructions. An algebra A in a modular tensor category $(\mathcal{C}, \otimes, \mathbf{1})$ is called haploid if $\dim \mathrm{Hom}_{\mathcal{C}}(\mathbf{1}, A) = 1$.

Definition 4.20. A Frobenius algebra $(A, m_A, \eta_A, \Delta_A, \epsilon_A)$ is called separable if

$$m_A \circ \Delta_A = \beta_A \mathrm{id}_A \quad \text{and} \quad \epsilon_A \circ \eta_A = \beta'_A \mathrm{id}_{\mathbf{1}} \quad \text{and} \quad \beta_A, \beta'_A \in \mathbb{C}^\times.$$

Proposition 4.21 ([**KR2**]). *Let A_{cl} be a haploid commutative symmetric Frobenius algebra in $\mathcal{C}_V^{(2)}$. If A is modular invariant, then A_{cl} is also separable.*

Proposition 4.22 ([**KR2**]). *Let $(A_{\mathrm{op}}|A_{\mathrm{cl}}, \iota_{\mathrm{cl\text{-}op}})$ be a Cardy $\mathcal{C}|\mathcal{C}^{(2)}$-algebra. If A_{cl} is simple and* $\dim A_{\mathrm{op}} \neq 0$, *then A_{op} is simple and separable.*

Now we give the reconstruction theorem for the Cardy $\mathcal{C}|\mathcal{C}^{(2)}$-algebra.

Theorem 4.23 ([**KR2**]). *Let A be a separable symmetric Frobenius $\mathcal{C}$-algebra. The triple $(A|Z(A), e)$, where $e : Z(A) \hookrightarrow R(A)$ is the canonical embedding, is a Cardy $\mathcal{C}|\mathcal{C}^{(2)}$-algebra.*

Remark 4.24. That an open-closed rational CFT can be constructed from a separable symmetric Frobenius algebra in $\mathcal{C}_V$ was first obtained by Fuchs, Runkel and Schweigert in a series of papers [**FRS1**][**FRS3**] [**FRS4**][**FjFRS1**], in which the rational CFT is studied as a holographic boundary of a 3-dimensional topological field theory. In particular, they start from a separable symmetric Frobenius algebra in a modular tensor category $\mathcal{C}$ and use 3-dimensional TFT techniques to construct a so-called *solution of sewing constraints* [**FjFRS1**], the notion of which can be proved to be equivalent to the notion of Cardy $\mathcal{C}|\mathcal{C}^{(2)}$-algebra [**KR3**]. I will refer to their approach to rational CFT as the FRS-framework.

Examples: $X \otimes X^\vee$ for any $X \in \mathcal{C}$ are separable symmetric Frobenius algebras in $\mathcal{C}$.

(1) Let $A = \mathbf{1}$, then $Z(\mathbf{1}) \cong \oplus_{i\in I} U_i^\vee \times U_i$.

(2) Let $A = X \otimes X^\vee$, then $Z(A) \cong Z(\mathbf{1})$ as Frobenius algebras.

$A_{\mathrm{cl}} = Z(\mathbf{1})$ case is also called the Cardy case. We will show later that all V-invariant D-branes associated to $Z(\mathbf{1})$ are given by $X \otimes X^\vee$. This is an indication of Morita equivalence which will be discussed later.

Remark 4.25. In Cardy case ($A_{\mathrm{cl}} = Z(\mathbf{1})$), all V-invariant D-branes are parametrized by objects X in $\mathcal{C}$ and obtained by constructing the internal hom $[X, X] = X \otimes X^\vee$. The category of V-invariant D-branes has objects X in $\mathcal{C}$ as objects and $[X, Y] = Y \otimes X^\vee$ as morphisms. Moreover, the functor $Y \mapsto [X, Y]$ is an equivalence between $\mathcal{C}$ and the category of $[X, X]$-modules [**O**]. Actually, one can show that the category of $Z(\mathbf{1})$-modules (internal to $\mathcal{C}_V^{(2)}$) is monoidally equivalent to $\mathcal{C}$ [**ENO2**]. Therefore, in this case, all D-branes are also parametrized by the modules over the closed CFT. This is not true for cases other than Cardy. But to some extent, other cases in rational CFT can all be viewed as certain twists of the Cardy case by some internal symmetries [**O**].

On the other hand, we can also examine the chiral modules over $Z(\mathbf{1})$ (recall Remark 4.8). A chiral $Z(\mathbf{1})$-module that respects the chiral symmetry given by a rational VOA V is nothing but a module in $\mathcal{C}$ over the algebra $T(Z(\mathbf{1})) = \oplus_{i\in I} U_i^\vee \otimes U_i$. It is easy to show that the category of chiral modules over $Z(\mathbf{1})$ is equivalent to $\mathcal{C}^{\oplus |I|}$. Therefore, the category of boundary conditions is given by an indecomposable component of the category of chiral modules over A_{cl}.

A generalization of this result to cases other than Cardy is given in Theorem 5.3. For irrational CFTs, one can still use the notion of chiral module because it is defined at the level of vertex operators without using tensor category language. The existence of the OPE of vertex operators predicts that a certain weak version of tensor product should exist even in the irrational CFTs. I believe that V_{op} should be obtained from M again by a certain internal-hom type of construction because the internal hom is nothing but the right adjoint functor of the tensor product. This should be the correct picture for irrational theories.

5. Holographic principle and defects

In Section 3, we have shown that a closed CFT is stringy commutative. By our philosophy, it provides a new geometry SAG. In this section, we will use rational CFTs to see some basic properties of SAG.

5.1. Holographic Principle. The Holographic Principle says that the information of the universe is contained in a part of it. In particular, such parts can have codimension higher than 1 or even be a point. In SAG, a D-brane plays the role of a generalized point. So it is natural to ask if a single V-invariant D-brane determines the bulk theory uniquely. Indeed, we have the following uniqueness theorem:

Theorem 5.1 ([**FjFRS2**][**KR2**]). *Let $(A|A_{\mathrm{cl}}, \iota_{\mathrm{cl\text{-}op}})$ be a Cardy $\mathcal{C}_V|\mathcal{C}_V^{(2)}$-algebra such that $\dim A \neq 0$ and A_{cl} is simple. Then A is separable and $(A|A_{\mathrm{cl}}, \iota_{\mathrm{cl\text{-}op}}) \cong (A|Z(A), e)$ as Cardy algebras.*

Remark 5.2. The above uniqueness result was first obtained by Fjelstad, Fuchs, Runkel and Schweigert [**FjFRS2**] in the FRS-framework. In the context of Cardy $\mathcal{C}_V|\mathcal{C}_V^{(2)}$-algebra, it was proved in [**KR2**]. The equivalence of the two approaches is explained in [**KR3**].

It is unclear if the above uniqueness result still holds for general conformally invariant D-branes. But we expect that it still holds at least for certain nice conformally invariant D-branes.

Conversely, for a closed CFT (or a bulk theory), we would like to know if there exists at least one D-brane and if it is unique. These questions are very important for us. The existence of D-branes means that the SAG associated to a given closed CFT is not empty, and the non-uniqueness means that the SAG contains more than one points. In the rational theories, we have the following result of the existence of V-invariant D-branes.

Theorem 5.3 ([**KR2**]). *If A_{cl} is a simple modular invariant commutative symmetric Frobenius algebra in $\mathcal{C}_V^{(2)}$, then there exist a simple separable symmetric Frobenius algebra A in $\mathcal{C}$ and a morphism $\iota_{\mathrm{cl\text{-}op}} : A_{\mathrm{cl}} \to R(A)$ such that*

(1) *$A_{\mathrm{cl}} \cong Z(A)$ as Frobenius algebras;*
(2) *$(A|A_{\mathrm{cl}}, \iota_{\mathrm{cl\text{-}op}})$ is a Cardy $\mathcal{C}_V|\mathcal{C}_V^{(2)}$-algebra;*
(3) *$T(A_{\mathrm{cl}}) \cong \oplus_{\kappa \in J} M_\kappa^\vee \otimes_A M_\kappa$ as algebras, where $\{M_\kappa\}_{\kappa \in J}$ is a set of representatives of the isomorphism classes of simple A-left modules.*

Remark 5.4. Notice that V-invariant D-branes can be parametrized by $\mathcal{C}_A$, the category of A-modules. On the other hand, $\mathcal{C}_{T(A_{\mathrm{cl}})} \cong \mathcal{C}_A^{\oplus |J|}$. Therefore, the category of boundary conditions is again an indecomposable component of the category of chiral modules over A_{cl}.

In the Cardy case (Section 4.6), we see that D-branes are not unique. The ambiguity is controlled by the so-called Morita equivalence.

Definition 5.5. *Two algebras A and B in a tensor category $\mathcal{C}$ are called Morita equivalent if there are an A-B-bimodule P and a B-A-bimodule Q such that $P \otimes_B Q \cong A$ and $Q \otimes_A P \cong B$ as bimodules.*

In the Cardy case, all V-invariant D-branes $X^\vee \otimes X$ are Morita equivalent to $\mathbf{1}$. Another example of Morita equivalence is given in Theorem 5.3, where all algebras $M_\kappa^\vee \otimes_A M_\kappa$ for $\kappa \in \mathcal{J}$ are Morita equivalent to A. Another equivalent way to define Morita equivalence is by the equivalence of the categories of modules over these two algebras. For an A-module X, $[X, X]$ is an algebra Morita equivalent to A and the functor $Y \mapsto [Y, X], \forall Y \in \mathcal{C}_A$ is an equivalence between these two categories [**O**]. In general, we have the following result.

Theorem 5.6 ([**KR2**]). *If $(A_{\mathrm{op}}^{(i)} | A_{\mathrm{cl}}^{(i)}, \iota_{\mathrm{cl\text{-}op}}^{(i)}), i = 1, 2$ are two Cardy $\mathcal{C}_V|\mathcal{C}_V^{(2)}$-algebras such that $A_{\mathrm{cl}}^{(i)}$ is simple and $\dim A_{\mathrm{op}}^{(i)} \neq 0$ for $i = 1, 2$, then $A_{\mathrm{cl}}^{(1)} \cong A_{\mathrm{cl}}^{(2)}$ as algebras if and only if $A_{\mathrm{op}}^{(1)}$ and $A_{\mathrm{op}}^{(2)}$ are Morita equivalent.*

Remark 5.7. The above theorem is equivalent to the statement that two simple separable symmetric Frobenius algebras A and B in $\mathcal{C}_V$ are Morita equivalent if and only if $Z(A) \cong Z(B)$ as Frobenius algebras. In other words, a single V-invariant D-brane determines uniquely the bulk theory; and the bulk theory determines uniquely the Morita class of the boundary theories.

Since a V-invariant D-brane is very symmetric and thus very scarce, we cannot see a continuum geometry here. I believe that by breaking the chiral symmetry down to conformal symmetry, one should be able to recover a rich and continuum geometry. For general conformally invariant D-branes, we need deal with irrational theories. The above strong result (Theorem 5.3) is perhaps not true for all conformally invariant D-branes. To study the general cases, it is useful to call two open-string VOAs *quasi-Morita equivalent* if they are both conformally invariant D-branes of the same closed CFT. It is important to understand this quasi-Morita equivalence in SAG. In some sense, such an equivalence can be viewed as the inverse statement of Holographic Principle which provides the foundation of SAG.

5.2. Beauty and the Beast. We will briefly discuss a result of CFTs with defects that might be relevant to SAG. For more details of this subject, readers can consult [**FrFRS**][**RS**] and another contribution to this book [**DKR2**].

As we discussed in the introduction, we expect that spacetime are emergent from the structures of a CFT. For example, it is known to physicists that a space can emerge as the moduli space of D0-branes [**As**], which are certain 0-dimensional objects in the category of D-branes, or emerge in a certain limit of a family of CFTs [**KS**][**Soi**]. What about time? Alain Connes observed that there is a God given embedding of the real number $\mathbb{R}$ in the group of outer automorphisms of a von Neumann algebra factor of type III [**Con**]. He believes that it should be understood as a time evolution. It appeared later as Connes-Rovelli's Thermo Time Hypothesis [**CR**]. This deep observation seems to suggest that the emergence of time in SAG might also be related to the automorphism group (or duality group[27]) of a CFT.

It has been known in many QFTs that defects play mysterious roles in various dualities [**Sav**]. Their relation has been clarified in the framework of rational CFTs by Jürg Fröhlich, Jürgen Fuchs, Ingo Runkel and Christoph Schweigert [**FrFRS**]. In particular, they show explicitly how an invertible defect can produce a duality of the bulk CFT[28]. A V-invariant

[27]In physics, duality usually can be weaker than an automorphism. But in this work, by duality we mean a true automorphism.

[28]They also studied more general defects (called group-like) defects and their relation to non-invertible dualities.

defect in a CFT over a rational VOA V is a defect line separating two bulk phases and respecting the chiral algebra V [**FrFRS**]. If two bulk phases are determined by their boundary conditions, i.e. two simple separable symmetric Frobenius algebras A and B in $\mathcal{C}_V$, then a V-invariant defect is given by an A-B-bimodule M. Such a bimodule is invertible if there is another A-B-bimodule N such that $M \otimes_B N \cong A$ and $N \otimes_A M \cong B$ as bimodules. The equivalence classes of invertible A-A-defects form a group called the Picard group denoted by $\mathrm{Pic}(A)$. A V-invariant automorphism of a bulk CFT is defined to be an automorphism of the bulk algebra $A_{\mathrm{cl}} = Z(A)$ such that it is the identity map on $V \otimes V$. We denote such V-invariant automorphism group by $\mathrm{Aut}(A_{\mathrm{cl}})$. Then we have the following exact correspondence between Beauty (the automorphism group) and the Beast (the invertible defects).

Theorem 5.8 ([**DKR1**]). *For a simple separable symmetric Frobenius algebra A in a modular tensor category $\mathcal{C}$, we have* $\mathrm{Aut}(Z(A)) \cong \mathrm{Pic}(A)$.

Remark 5.9. The correspondence between duality and defects is not an isolated phenomenon. For example, a categorified version of Theorem 5.8 is proved in [**ENOM**][29]. More precisely, for a finite fusion category $\mathcal{C}$ we also have $\mathrm{Aut}(Z(\mathcal{C})) \cong \mathrm{Pic}(\mathcal{C})$, where $Z(\mathcal{C})$ is the monoidal center of $\mathcal{C}$. This can be viewed as a precise correspondence between duality and defects in Turaev-Viro TQFT (or Levin-Wen models) [**KK**]. The relation between duality and defects is known in many other QFTs [**KW**][**KT**][**DGG**]. I believe that under certain finite conditions, such as Hopkins-Lurie's fully dualizable condition [**Lu2**], an exact correspondence between dualities and invertible defects is true in many QFTs.

On the practical side, the invertible defects are much easier to compute than the dualities of a bulk theory. On the philosophical side, it suggests that the automorphism group of a QFT can emerge from the moduli of certain invertible defects. I believe that we can recover continuum automorphism groups if we study conformally invariant defects. It suggests that the emergence of time might be related to invertible defects. Notice that we have only discussed the automorphisms of the closed CFT. How about the automorphism groups of an open CFT? It is very possible that each D-brane carries its own time evolution. Moreover, it was proposed by Connes and Rovelli [**CR**] in their Thermo Time Hypothesis that a time involution depends on the choice of a thermo-state. So it is desirable to define thermo-states or KMS states in the context of CFT. We will return to these questions in our future studies.

Because the Holographic Principle is so important in our program, it is worthwhile to explore its properties further. Since a boundary theory often determines the bulk theory by taking the center, we would like to ask if this process is functorial. By considering the groupoid of those algebras A described in the Theorem 5.8 and invertible bimodules, and the groupoid of commutative algebras in $Z(\mathcal{C})$ and isomorphisms of algebras, we obtain a groupoid version of Theorem 5.8 [**DKR1**]. Namely, it is an equivalence of groupoid. It turns out that this equivalence of groupoid can be extended to a richly structured functor which take an algebra to its center. Namely, the notion of the center is functorial. This suggests that the Holographic Principle is also functorial. More interestingly, this functoriality demands defects of all codimensions to appear. As a purely algebraic statement, this functoriality properly encodes all the information of defects without attributing it to a geometric bordism category. This seems to suggest a purely algebraic way to understand extended TQFT [**Lu2**] (see also [**Ber**] in this book). See [**DKR2**] in this book for more details on the functoriality of the center and its relation to TFTs with defects.

[29]This result was obtained independently by Kitaev and myself [**KK**] and was announced in May 2009 in a conference on TQFT held at Northwestern University. In [**KK**] we identify the elements in $\mathrm{Pic}(\mathcal{C})$ as the physical invertible defects of a lattice model and the simple objects in $Z(\mathcal{C})$ as anyonic excitations in the bulk phase.

6. Conclusions and outlook

In this work, we have outlined a dream of a new algebraic geometry (SAG) based on 2-dimensional conformal field theories. We also reviewed some recent progress on the mathematical foundations of rational open-closed CFTs. Our understanding on V-invariant D-branes in a rational CFT is quite satisfying. But the real challenge lies in how to understand conformally invariant D-branes in concrete examples in order to see a rich geometry. In this paper, we did not touch upon open-closed superconformal field theory which is the real hero behind many miracles in geometry. But its mathematical foundation is still lacking. We will return to these issues in the future publications.

Actually, the philosophy of SAG itself can have useful applications in other fields. Many structures that have appeared in CFT also appeared in many different contexts. For example, the Holographic Principle appears in the context of E_n-algebra or E_n-category as the so-called generalized Deligne conjecture [**De**]; defects reincarnate in algebraic geometry as Fourier-Mukai transformations, etc. Therefore, questions asked here can also be transported to other fields, and vice versa. Let us discuss two concrete cases below.

- String topology was invented by Chas and Sullivan [**ChSu**]. It can be viewed as an open-closed homological CFT [**Go**]. Let M be a simply connected and closed manifold. The closed string topology is an algebraic structure on $H_*(LM)$, where LM denotes the loop space. This is the closed algebra in this case. By our philosophy of SAG, we would like to ask what lies in the spectrum. Namely, what are the compatible open algebras? Can they recover all the "points" in M? Let N be a submanifold of M. The chain space $C_*(\mathcal{P}_{N,N})$ of the path space between N and N is a differential graded algebra. On the level of homology, this algebra structure is the string topology product introduced by Sullivan [**Su**]. The dg-algebra $C_*(\mathcal{P}_{N,N})$ is an open algebra in this case. The notion of center in this case is given by Hochschild cohomology. Then our question on the spectrum of the closed string topology is related to the following question asked by Blumberg, Cohen and Teleman [**BCT**]: For what submanifolds N the relation:
$$HH^*(C_*(\mathcal{P}_{N,N}), C_*(\mathcal{P}_{N,N})) \cong H_*(LM)$$
holds? As pointed out in [**BCT**], the answer is affirmative for a point in M and $N = M$ by works in the 1980s. They went on to give more answers to their question including the inclusion $N \hookrightarrow M$ being null homotopic or being the inclusion of the fiber of a fibration $M \to B$. One can also ask many other obvious questions. For example, can one give a geometric meaning (as some kind of a boundary condition) of a certain module over the closed algebra $H_*(LM)$ so that the open algebra $C_*(\mathcal{P}_{N,N})$ arises as the internal hom of this module?

- A crucial result in the local geometric Langlands correspondence (GLC) says that the center of the vertex algebra $V_{-h^\vee}(\mathfrak{g})$ associated to a simple Lie algebra $\mathfrak{g}$ at the critical level $k = -h^\vee$ is isomorphic to the function algebra over the space of ${}^L\mathfrak{g}$-opers on the formal disk [**Fr**, Theorem 9]. Actually, the critical level is exactly the case of non-CFT because the Sugawara construction of Virasoro algebra fails exactly in this case. The vertex operator algebra $V_k(\mathfrak{g})$ for $k \neq -h^\vee$ can be viewed as a D-brane in the Cardy case. Its center gives the closed CFT V_{cl}. Moreover, by SAG, the spectrum of this closed CFT is just the category of D-branes. Therefore, we expect what appears on the one side of GLC for noncritical levels is the category of D-branes of the closed CFT V_{cl} associated to $\mathfrak{g}$. It is very possible that the other side of GLC is also given by the category of D-branes of a closed CFT associated to ${}^L\mathfrak{g}$. This is supported by the intuition from physics. The physics origin of GLC is a conjectured duality, which is derived from Montonen-Olive duality by topological twisting, between two 4-dimensional topological gauge theories with gauge group G and ${}^LG$ respectively [**KW**]. One consequence of this duality is a conjectured equivalence between two 3-categories of boundary conditions of these two topological gauge theories [**Kap**].

Although the speculation of a new geometry SAG discussed in this paper is still too naive and premature, it indeed motivated some of my own works on CFTs. I hope that this naive picture can inspire more serious works in this direction. Another reason for me to write about it is to respond to some opinions I heard in various occasions that 2-dimensional CFTs are well understood, and there are not many interesting things left to do. I hope that this paper can convince some of my readers that there are still a lot of interesting questions in 2-dimensional CFT waiting to be studied.

References

[As] P. S. Aspinwall, *A point's point of view of stringy geometry*, J. High Energy Phys. (2003) no. 1, 002.

[At] M. Atiyah, *Topological quantum field theories*, Inst. Hautes Etudes Sci. Publ. Math. **68** (1989) 175-186.

[BK] B. Bakalov and A.A. Kirillov, *Lectures on Tensor Categories and Modular Functors*, AMS, Providence, 2001.

[Bek] D. Bekenstein, *Information in the holographic universe*, Scientific American, **289** (2003) 58-65.

[BFN] D. Ben-Zvi, J. Francis, and D. Nadler, *Integral transforms and Drinfeld centers in derived algebraic geometry*, J. Amer. Math. Soc. **23** (2010) no. 4, 909-966.

[Ber] J. E. Bergner, *Models for (∞, n)-categories and the cobordism hypothesis*, Hisham Sati, Urs Schreiber (eds.), Mathematical Foundations of Quantum Field and Perturbative String Theory, Proceedings of Symposia in Pure Mathematics, AMS.

[BCT] A. J. Blumberg, R. L. Cohen, C. Teleman, *Open-closed field theories, string topology, and Hochschild homology*, Alpine perspectives on algebraic topology, 53-76, Contemp. Math., 504, Amer. Math. Soc., Providence, RI, 2009.

[BDLR] I. Brunner, M. Douglas, A. Lawrence, C. Romelsberger, *D-branes on the Quintic*, J. High Energy Phys no. 8 (2000) 15.

[BO] A. I. Bondal, D. O. Orlov, *Reconstruction of a variety from the derived category and groups of autoequivalences*, Compos. Math. 125 (2001), 327-344.

[Bor] R.E. Borcherds, *Vertex algebras, Kac-Moody algebras, and the Monster*, Proc. Natl. Acad. Sci. USA **83** (1986), 3068-3071.

[Bou] R. Bousso, *The holographic principle*, Rev. Mod. Phys. 74 (2002) 825-874.

[BPZ] A. A. Belavin, A. M. Polyakov and A. B. Zamolodchikov, *Infinite conformal symmetries in two-dimensional quantum field theory*, Nucl. Phys. B **241** (1984) 333-380.

[C1] J. Cardy, *Conformal invariance and surface critical behavior*, Nucl. Phys. B **240**, 514-532 (1984);

[C2] J. Cardy, *Operator content of two-dimensional conformal invariant theories*, Nucl. Phys. B **270** (1986) 186-204.

[C3] J. Cardy, *Boundary conditions, fusion rules and the Verlinde formula*, Nucl. Phys. B **324** (1989) 581-596.

[CL] J.L. Cardy, D.C. Lewellen, *Bulk and boundary operators in conformal field theory*, Phys. Lett. B **259** (1991) 274-278.

[CHKL] S. Carpi, R. Hillier, Y. Kawahigashi, R. Longo, *Spectral triples and the super-Virasoro algebra*, Commun. Math. Phys. **295** (2010) 71-97.

[CHKLX] S. Carpi, R. Hillier, Y. Kawahigashi, R. Longo, F. Xu, in preparation.

[ChSu] M, Chas, D. Sullivan, *String Topology*, [arXiv:math.GT/9911159].

[Con] A. Connes *Noncommutative Geometry*, Academic Press, Inc., San Diego, CA, 1994.

[CR] A. Connes, C. Rovelli, *Von Neumann Algebra Automorphisms and Time-Thermodyna- mics Relation in Generally Covariant Quantum Theories*, Class. Quantum Grav. **11**, 2899-2917 (1994)

[CS] F. Conrady, C. Schweigert, *Topologizations of chiral representations*, Commun. Math. Phys. **245** (2004), no. 3, 429-448.

[Cos] K. J. Costello, *Topological conformal field theories and Calabi-Yau categories*, Adv. Math. 210 (2007), no. 1, 165-214.

[Da] A. Davydov, *Centre of an algebra*, Adv. Math. **225** (2010) 319-348.

[De] See n-lab for the references of Deligne Conjecture, http://ncatlab.org/nlab/show/Deligne+conjecture

[D] M. Douglas, *Topics in D-geometry*, Class.Quant.Grav. **17** (2000) 1057-1070.

[DAG] See related pages in n-lab for references in derived algebraic geometry: http://ncatlab.org/nlab/show/derived+algebraic+geometry

[DGG] N. Drukker, D. Gaiotto, J. Gomis, *The virtue of defects in 4D gauge theories and 2D CFTs*, [arXiv:1003.1112]

[DKR1] A. Davydov, L. Kong, I. Runkel, *Invertible defects and isomorphisms of rational CFTs*, [arXiv:1004.4725]

[DKR2] A. Davydov, L. Kong, I. Runkel, *Field theories with defects and the centre functor*, Hisham Sati, Urs Schreiber (eds.), Mathematical Foundations of Quantum Field and Perturbative String Theory, Proceedings of Symposia in Pure Mathematics, AMS. [arXiv:1107.0495]

[DMZ] C.-Y. Dong, G. Mason, Y.-C. Zhu, *Discrete series of the Virasoro algebra and the moonshine module*, Algebraic Groups and Their Generalizations: Quantum and infinite-dimensional Methods, Proc. 1991 American Math. Soc. Summer Research Institute, ed. by W. J. Haboush and B. J. Parshall, Proc. Symp. Pure Math. **56**, Part 2, Amer. Math. Soc., Providence, 1994, 295-316.

[ENO1] P.I. Etingof, D. Nikshych, V. Ostrik, *On fusion categories*, Ann. Math. **162** (2005) 581-642.

[ENO2] P.I. Etingof, D. Nikshych, V. Ostrik, *Weakly group-theoretical and solvable fusion categories*, Adv. Math. **226** (2011), no. 1, 176-205.

[ENOM] P.I. Etingof, D. Nikshych, V. Ostrik, with an appendix by E. Meir, *Fusion categories and homotopy theory*, Quantum Topol. 1 (2010), no. 3, 209-273.

[Fr] E. Frenkel, *Lectures on the Langlands program and conformal field theory*, Frontiers in number theory, physics, and geometry. II, 387-533, Springer, Berlin, 2007.

[F] I.B. Frenkel, a talk given at the Institute for Advanced Study, 1988.

[Fe] G. Felder, J. Fröhlich, J. Fuchs, C. Schweigert, *Correlation functions and boundary conditions in rational conformal field theory and three-dimensional topology*, Compositio Math. **131** (2002) 189-237.

[FG] J. Fröhlich, K. Gawedzki, *Conformal Field Theory and geometry of strings*, Mathematical quantum theory. I. Field theory and many-body theory (Vancouver, BC, 1993), 57-97, CRM Proc. Lecture Notes, 7, Amer. Math. Soc., Providence, RI, 1994.

[FHL] I.B. Frenkel, Y.-Z. Huang, J. Lepowsky, *On axiomatic approaches to vertex operator algebras and modules*, Mem. Amer. Math. Soc. **104** (1993), no. 494.

[FLM1] I. B. Frenkel, J. Lepowsky, and A. Meurman, *A natural representation of the Fischer-Griess monster with the modular function J as character*, Proc. Natl. Acad. Sci. USA **81** (1984) 3256-3260.

[FLM2] I. B. Frenkel, J. Lepowsky, and A. Meurman, *Vertex operator algebras and the Monster*, Pure and Appl. Math., **134**, Academic Press, New York, 1988.

[FFFS] G. Felder, J. Fröhlich, J. Fuchs, C. Schweigert, *The geometry of WZW branes*, J. Geom. Phys. **34** (2000) 162-190.

[FjFRS1] J. Fjelstad, J. Fuchs, I. Runkel and C. Schweigert, *TFT construction of RCFT correlators V: Proof of modular invariance and factorisation*, Theor. Appl. Categor. **16** (2006) 342-433.

[FjFRS2] J. Fjelstad, J. Fuchs, I. Runkel and C. Schweigert, Uniqueness of open/closed rational CFT with given algebra of open states, Adv. Theor. Math. Phys. **12** (2008) 1283-1375.

[FriS] D. Friedan and S. Shenker, *The analytic geometry of two-dimensional conformal field theory*, Nucl. Phys. **B281** (1987) 509-545.

[FMS] P. Francesco, P. Mathieu, D. Senechal, *Conformal Field Theory*, Springer-Verlag, New York, 1997.

[FrFRS] J. Fröhlich, J. Fuchs, I. Runkel and C. Schweigert, *Duality and defects in rational conformal field theory*, Nucl. Phys. B **763** (2007) 354-430.

[FRS1] J. Fuchs, I. Runkel and C. Schweigert, *TFT construction of RCFT correlators. I: Partition functions*, Nucl. Phys. B **646** (2002) 353-497.

[FRS2] J. Fuchs, I. Runkel and C. Schweigert, *TFT construction of RCFT correlators II: Unoriented world sheets*, Nucl.Phys. B **678** (2004) 511-637.

[FRS3] J. Fuchs, I. Runkel and C. Schweigert, *TFT construction of RCFT correlators III: simple currents*, Nucl. Phys. B **694** (2004) 277-353.

[FRS4] J. Fuchs, I. Runkel and C. Schweigert, *TFT construction of rcft correlators IV: structure constants and correlation functions*, Nucl. Phys. B **715**(3) (2005) 539-638.

[FS] J. Fuchs and C. Schweigert, *Category theory for conformal boundary conditions*, Fields Institute Commun. **39** (2003) 25-71.

[Ga] W. L. Gan, *Koszul duality for dioperads*, Math. Res. Lett. **10**(1) (2003) 109-124.

[Go] V. Godin, *Higher string topology operations*, [arXiv:0711.4859]

[GS] T. Gomez E. R. Sharpe, *D-branes and Scheme Theory*, [arXiv:hep-th/0008150]

[Hu] B.-L. Hu, *Emergent/Quantum Gravity: Macro/Micro Structures of Spacetime*, J. Phys. Conf. Ser.174 (2009) 012015.

[H1] Y.-Z. Huang, *On the geometric interpretation of vertex operator algebras*, Ph.D thesis, Rutgers University, 1990.

[H2] Y.-Z. Huang, *Geometric interpretation of vertex operator algebras*, Proc. Natl. Acad. Sci. USA **88** (1991), 9964-9968.

[H3] Y.-Z. Huang, *A theory of tensor products for module categories for a vertex operator algebra, IV*, J. Pure Appl. Alg. **100** (1995) 173-216.

[H4] Y.-Z. Huang, *Two-dimensional conformal geometry and vertex operator algebras*, Progress in Mathematics, Vol. 148, Birkhäuser, Boston, 1997.

[H5] Y.-Z. Huang, *Generalized Rationality and a "Jacobi Identity" for Intertwining Operator Algebras*, [arXiv:q-alg/9704008]

[H6] Y.-Z. Huang, *A functional-analytic theory of vertex (operator) algebras, I*, Commun. Math. Phys. **204** (1999), 61-84.

[H7] Y.-Z. Huang, *A functional-analytic theory of vertex (operator) algebras, II*, Commun. Math. Phys. **242** (2003), 425-444.

[H8] Y.-Z. Huang, *Riemann surfaces with boundaries and the theory of vertex operator algebras*, Vertex operator algebras in mathematics and physics (Toronto, ON, 2000), 109-125, *Fields Inst. Commun.*, 39, Amer. Math. Soc., Providence, RI, 2003.

[H9] Y.-Z. Huang, *Differential equations, duality and modular invariance*, Commun. Contemp. Math. **7** (2005) 649-706.

[H10] Huang, Y.-Z.: Vertex operator algebras and the verlinde conjecture. *Commun. Contemp. Math.* 10(1) (2008) 103-154.

[H11] Y.-Z. Huang, *Rigidity and modularity of vertex tensor categories*, Commun. Contemp. Math. **10** (2008) 871-911

[HK1] Y.-Z. Huang and L. Kong, *Open-string vertex algebra, category and operads*, Commun. Math. Phys. **250** (2004) 433-471.

[HK2] Y.-Z. Huang and L. Kong, *Full field algebras*, Commun. Math. Phys. **272** (2007) 345-396.

[HK3] Y.-Z. Huang and L. Kong, *Modular invariance for conformal full field algebras*, Trans. Amer. Math. Soc. **362** (2010), no. 6, 3027-3067.

[HKr] P. Hu, I. Kriz, *Closed and open conformal field theories and their anomalies*, Commun. Math. Phys. **254** (2005), no. 1, 221-253.

[HL1] Y.-Z. Huang, J. Lepowsky, *A theory of tensor products for module categories for a vertex operator algebra, I*, Selecta Math. (N.S.) 1 (1995) 699-756.

[HL2] Y.-Z. Huang, J. Lepowsky, *A theory of tensor products for module categories for a vertex operator algebra, II*, Selecta Math. (N.S.) 1 (1995) 757-786.

[HL3] Y.-Z. Huang, J. Lepowsky, *A theory of tensor products for module categories for a vertex operator algebra, III*, J. Pure Appl. Alg. **100** (1995) 141-171.

[HL4] Y.-Z. Huang, J. Lepowsky, *A theory of tensor products for module categories for a vertex operator algebra, V*, unpublished.

[Ka1] M. Kapranov, *Noncommutative geometry and path integrals*, Algebra, arithmetic, and geometry: in honor of Yu. I. Manin, Vol. II, 49-87, Progr. Math., 270, Birkhuser Boston, Inc., Boston, MA, 2009.

[Ka2] M. Kapranov, *Free Lie algebroids and the space of paths*, Selecta Math. (N.S.) 13 (2007), no. 2, 277-319.

[KV1] M. Kapranov, E. Vasserot, *Vertex algebras and the formal loop space*, *Publ. Math. Inst. Hautes tudes Sci.* No. 100 (2004), 209-269.

[KV2] M. Kapranov, E. Vasserot, *Supersymmetry and the formal loop space*, [arXiv:1005.4466]

[Kap] A. Kapustin, *Topological field theory, higher categories, and their applications*, Proceedings of ICM 2010, [arXiv:1004.2307]

[KT] A.N. Kapustin, M. Tikhonov, *Abelian duality, walls and boundary conditions in diverse dimensions*, J. High Energy Phys. 0911 (2009) 006.

[KW] A.N. Kapustin, E. Witten, *Electric-magnetic duality and the geometric Langlands program*, Commun. Number Theory Phys. **1** (2007) 1-236.

[KK] A. Kitaev, L. Kong, *Models for gapped boundaries and domain walls*, [arXiv:1104.5047]

[Ko1] L. Kong, *Full field algebras, operads and tensor categories*, Adv. Math. **213** (2007) 271-340.

[Ko2] L. Kong, *Open-closed field algebras*, Commun. Math. Phys. **280** (2008) 207-261.

[Ko3] L. Kong, *Cardy condition for open-closed field algebras*, Commun. Math. Phys. **283** (2008) 25-92.

[KO] A.A. Kirillov, V. Ostrik, *On q-analog of McKay correspondence and ADE classification of $\widehat{sl}(2)$ conformal field theories*, Adv. Math. **171** (2002) 183-227.

[KR1] L. Kong, I. Runkel, *Morita classes of algebras in modular tensor categories*, Adv. Math. **219** (2008) 1548-1576.

[KR2] L. Kong, I. Runkel, *Cardy algebra and sewing constraints, I*, Commun. Math. Phys. **292** (2009) 871-912.

[KR3] L. Kong, I. Runkel, *Cardy algebra and sewing constraints, II*, in preparation.

[KR4] L. Kong, I. Runkel, *Algebraic Structures in Euclidean and Minkowskian Two-Dimensional Conformal Field Theory*, Noncommutative structures in mathematics and physics, 217-238, K. Vlaam. Acad. Belgie Wet. Kunsten (KVAB), Brussels, 2010.

[KS] M. Kontsevich, Y. Soibelman, *Homological mirror symmetry and torus fibrations*, Symplectic geometry and mirror symmetry (Seoul, 2000), 203-263, World Sci. Publ., River Edge, NJ, 2001.

[Lb] Wikipedia: `http://en.wikipedia.org/wiki/Philosophy_of_space_and_time`

[Lw] D.C. Lewellen, *Sewing constraints for conformal field theories on surfaces with boundaries*, Nucl. Phys. B372 (1992) 654-682.

[LL] J. Lepowsky, H.-S. Li, *Introduction to vertex operator algebras and their representations*, Progress in Mathematics, 227. Birkhuser Boston, Inc., Boston, MA, 2004.

[LR1] R. Longo and K.H. Rehren, *Nets of subfactors*, Rev. Math. Phys. 7 (1995) 567-597.

[LR2] R. Longo and K.H. Rehren, *Local fields in boundary conformal QFT*, Rev. Math. Phys. **16** (2004) 909-960.

[LY1] C.-H. Liu, S. T. Yau, *Azumaya-type noncommutative spaces and morphism therefrom: Polchinski's D-branes in string theory from Grothendieck's viewpoint*, [arXiv:0709.1515]

[LY2] C.-H. Liu, S. T. Yau, *D-branes and Azumaya noncommutative geometry: From Polchinski to Grothendieck*, [arXiv:1003.1178]

[Lu1] J. Lurie, *Derived algebraic geometry, II, III, IV, V, VI*, [arXiv:math/0702299, 0703204, 0709.3091, 0905.0459, 0911.0018]

[Lu2] J. Lurie, *On the Classification of Topological Field Theories*, Current developments in mathematics, 2008, 129-280, Int. Press, Somerville, MA, 2009.

[MSV] A. Malikov, V. Schechtman, A. Vaintrob, *Chiral de Rham complex*, Commun. math. Phys. **204** (1999), 439-473.

[Ma] J. M. Maldacena, *The large N limit of superconformal field theories and super-gravity*, Adv. Theor. Math. Phys. **2** (1998), no. 2, 231-252.

[MMS] J. M. Maldacena, G. Moore, N. Seiberg, *Geometrical interpretation of D-branes in gauged WZW models*, J. High Energy Phys. 2001, no. 7, 46-108.

[Mo1] G. Moore, *Some comments on branes, G-flux, and K-theory*, Int. J. Mod. Phys. **A16** (2001) 936-944.

[Mo2] G. Moore, *What is a brane*, Notices of the AMS, Vol 52, No. 2, (2005)

[MSeg] G. Moore and G. Segal, D-branes and K-theory in 2D topological field theory, [hep-th/0609042].

[MSei] G. Moore, N. Seiberg, *Classical and quantum conformal field theory*, Commun. Math. Phys. **123** (1989) 177-254.

[O] V. Ostrik, *Module categories, weak Hopf algebras and modular invariants*, Transform. Group 8 (2003), no. 2, 177-206.

[MW] S. Morrison, K. Walker, *The blob complex*, [arXiv:1009.5025]

[Pi] D. Pickrell, *$P(\phi)_2$ Quantum field theories and Segal's Axioms*, Commun. Math. Phys. 280 (2008), no. 2, 403-425.

[R] K.H. Rehren *Canonical tensor product subfactors*, Commun. Math. Phys. 211 (2000) 395-406.

[RS] I. Runkel, R. Suszek, *Gerbe-holonomy for surfaces with defect networks*, Adv. Theor. Math. Phys. **13** (2009) 1137-1219.

[RT] N. Reshetikhin and V.G. Turaev, *Invariants of 3-manifolds via link polynomials and quantum groups*, Inv. Math. **103** (1991) 547-597.

[RW1] D. Roggenkamp, K. Wendland, *Limits and degenerations of unitary conformal field theories*, Commun. Math. Phys. 251 (2004), no. 3, 589-643.

[RW2] D. Roggenkamp, K. Wendland, *Decoding the geometry of conformal field theories*, [arXiv:0803.0657]

[Sak] A. D. Sakharov, *Vacuum quantum fluctuations in curved space and the theory of gravitation*, Sov. Phys. Dokl. (1968) 12:1040-1041.

[Sav] R. Savit, *Duality in field theory and statistical systems*, Rev. Mod. Phys. 52 (1980) 453

[Sh] E. R. Sharpe, *D-Branes, Derived Categories, and Grothendieck Groups*, Nucl. Phys. B **561** (1999) 433-450.

[Se] G. Segal, *The definition of conformal field theory*, preprint 1988; U. Tillmann (ed.), Topology, geometry and quantum field theory, London Math. Soc. Lect. Note Ser. **308** (2002) 421-577.

[Soi] Y. Soibelman, *Collapsing conformal field theories, spaces with non-negative Ricci curvature and non-commutative geometry*, Hisham Sati, Urs Schreiber (eds.), Mathematical Foundations of Quantum Field and Perturbative String Theory, Proceedings of Symposia in Pure Mathematics AMS.

[So] H. Sonoda, *Sewing conformal field theories, I, II*, Nucl. Phys. B **311** (1988) 401-416, 417-432.

[Su] D. Sullivan, *Open and closed string field theory interpreted in classical algebraic topology*, Topology, geometry and quantum eld theory, London Math. Soc. Lecture Notes, vol. 308, Cambridge Univ. Press, Cambridge, 2004, 344-357.

[Sus] L. Susskind, *The world as a hologram*, J. Math. Phys. **36** (1995) 6377-6396.

[tH] G. 't Hooft, *Dimensional reduction in quantum gravity*, [arXiv:gr-qc/9310026].

[T] V.G. Turaev, *Quantum Invariants of Knots and 3-Manifolds*, de Gruyter, New York, 1994.

[Va] C. Vafa, *Conformal theories and punctured surfaces*, Phys. Lett. B **199** (1987) 195-202.

[Ve] E. Verlinde, *On the origin of gravity and the laws of Newton*, J. High Energy Phys. 1104:029 (2011).

[We] I. Weiss, *From operads to dendroidal sets*, Hisham Sati, Urs Schreiber (eds.), Mathematical Foundations of Quantum Field and Perturbative String Theory, Proceedings of Symposia in Pure Mathematics, AMS, [arXiv:arXiv:1012.4315]

[Wi] E. Witten, *Quantum field theory and the Jones polynomial*, Commun. Math. Phys. **121** (1989), no. 3, 351-399.

[Z] Y.-C. Zhu, *Modular invariance of vertex operator algebras*, J. Amer. Math. Soc. **9** (1996) 237-302.

Institute for Advanced Study, Tsinghua University, Beijing 100084 China
E-mail address: kong.fan.liang@gmail.com

Proceedings of Symposia in Pure Mathematics
Volume **83**, 2011

COLLAPSING CONFORMAL FIELD THEORIES, SPACES WITH NON-NEGATIVE RICCI CURVATURE AND NON-COMMUTATIVE GEOMETRY

YAN SOIBELMAN

CONTENTS

1. Introduction

1.1. **Motivations.** This paper is a slightly revised and shorten version of my notes "Collapsing Conformal Field Theories and quantum spaces with non-negative Ricci curvature" started in 2003, which can be found on my home page. I will refer below to that draft as "Notes". The main motivation for the "Notes" was the analogy between the moduli spaces of Conformal Field Theories (CFTs) and the space of isomorphism classes of compact metric-measure spaces equipped with the measured Gromov-Hausdorff topology.

The algebro-geometric approach to the concept of moduli space as a space representing the functor of "isomorphism classes of families" is not very useful for Riemannian manifolds (the same can be said about many other "functorial" concepts). On the other hand, the "moduli space" (i.e. the set of isometry classes) of compact Riemannian manifolds carries some natural Hausdorff topologies (e.g. Gromov-Hausdorff). Therefore one can compactify it in a larger set consisting of isometry classes of compact metric spaces. It is well-known that some differential-geometric structures of Riemannian manifolds admit generalizations to the points of compactified moduli space. As an example we mention the notion of sectional curvature extended to Alexandrov spaces or, more recent, the property to have non-negative Ricci curvature extended to compact metric-measure spaces (see [LV], [St]).

It is natural to ask whether this philosophy can be applied in the case of (properly defined) non-commutative Riemannian manifolds. To my knowledge, this program is still in its infancy.

We are motivated by Segal's axioms of unitary Conformal Field Theory (see [Seg]) as well as by the approach to non-commutative Riemannian geometry developed by A. Connes in [Co1]. Many structures of Connes's approach (e.g. spectral triples, see [Co1], [CoMar]) are closely related to the structures considered in present paper.

We should warn the reader that the paper does not contain results which can be called "new" by mathematical or physical standards. It is a review and discussion of various existing concepts united by the author's wish to see their analogs in the widely understood framework of non-commutative ("quantum") geometry. In which sense they are really non-commutative is a different question.

1.2. **Collapsing CFTs from metric point of view.** One of our goals is to define a quantum Riemannian manifold (or, more generally, "Riemannian space") by a set of axioms similar to Segal's axioms of a unitary CFT. The set of "isometry classes" (i.e. the moduli space) should be treated similarly to the set of isometry classes of compact Riemannian manifolds with restrictions on the diameter and Ricci curvature The central charge plays a role of the dimension of the space, and the "spectral gap" for the Virasoro operator $L_0 + \overline{L}_0$ plays a role of the (square root of the inverse of) diameter. In the spirit of Gromov, Cheeger, Colding, Fukaya and others we would like to compactify the "moduli space" of such objects by their "Gromov-Hausdorff degenerations". Notice that we treat the moduli space of CFTs (or, more generally, QFTs) as an object of metric geometry rather than one of algebraic geometry. We used this philosophy in [KoSo1], where the concept of collapsing family of unitary CFTs was introduced with the aim to explain Mirror Symmetry. The main idea of [KoSo1] is that the "moduli space of CFTs with bounded central charge and spectral gap (i.e. the minimal positive eigenvalue of the Virasoro operator $L_0 + \overline{L}_0$) bounded from below by a non-negative constant", is precompact in some natural topology. The topology itself was not specified in [KoSo1]. Since the notion of collapse depends on the spectral properties of the Virasoro operator $L_0+\overline{L}_0$, one should use the topology which gives continuity of the spectral data, e.g. measured Gromov-Hausdorff topology. It was argued in loc. cit. that if the spectral gap approaches zero, then the collapsing family of unitary CFTs gives rise to a topological space, which contains a dense open Riemannian manifold. It was clear to the authors (but probably not stated very explcitly in loc.cit) that the Riemannian metric has non-negative Ricci curvature. The restriction on the Ricci curvature follows from the unitarity of the theory. We discuss more details below in relation to Bakry's "curvature-dimension inequalities". The above discussion can be summarized into a slogan:

Collapsing unitary two-dimensional CFTs= Riemannian manifolds (possibly singular) with non-negative Ricci curvature.

From this point of view the geometry of (possibly singular) Riemannian manifolds with non-negative Ricci curvature should be thought of as a limit of the (quantum) geometry of certain Quantum Field Theories (namely, two-dimensional unitary CFTs). Riemannian manifolds themselves appear as target spaces for non-linear sigma models. Hence non-linear sigma-models provide a partial compactification of the moduli space of two-dimensional unitary CFTs.

1.3. **Quantum Riemannian spaces via Segal's axioms.** From the point of view of Segal's axioms, the geometry which underlies a collapsing family of CFTs is the geometry of 2-dimensional compact oriented Riemannian manifolds degenerating into metrized graphs. The algebra of the CFT is encoded in the operator product expansion (OPE). Its collapse gives rise to a commutative algebra A. The (rescaled) operator $L_0 + \overline{L}_0$ collapses into a second order differential operator on A. The rest of the conformal group does not survive. Hence the family of CFTs collapses into a QFT. The latter, according to [KoSo1], should be thought of as a Gromov-Hausdorff limit of the former.

The above considerations suggest the following working definitions in the framework of quantum geometry. A quantum compact Riemannian space (more precisely, 2-space, if we want to stress that surfaces are 2-dimensional) is defined by the following data: a separable complex Hilbert space H, an operator $S(\Sigma) : H^{\otimes n} \to H^{\otimes m}$ called the amplitude of Σ which is given for each compact Riemannian 2-dimensional oriented manifold Σ with n marked "input" circles and m marked "output" circles. The kernel of $S(\Sigma)$ is the tensor $K_\Sigma \in H^{\otimes m} \otimes (H^{\otimes n})^* = Hom((H^{\otimes m})^* \otimes H^{\otimes n}, \mathbf{C})$ called the correlator. In the same vein we define quantum compact Riemannian 1-spaces. Having in mind degenrations we should allow "singular" spaces which are given by similar data assigned to *metrized graphs* with marked input and output vertices. Natural gluing axioms should be satisfied in both cases, as well as continuity of the data with respect to some natural topology. In particular we should allow degenerations of Riemannian 2-spaces into singular Riemannian 1-spaces. At the level of geometry this means that metrized graphs are Gromov-Hausdorff limits of compact oriented Riemannian 2-dimensional manifolds with boundary. At the level of algebra, all "algebraic data" (e.g. spaces of states, operators) associated with graphs are limits of the corresponding data for the surfaces. The notion of limit should have "geometric" and "algebraic" counterparts. Geometrically it can be the measured Gromov-Hausdorff limit, while for Hilbert spaces it can be any notion of limit which respects continuity of the spectral data of the associated positive self-adjoint operators. We are going to review several possibilities in the main body of the paper. We should point out that we consider singular Riemannian 1-spaces, but do not consider a similar notion for an arbitrary $d > 1$. Hence are picture of collapse is restricted to the case $d = 2 \to d = 1$. More general story should probably include QFTs with the space-time being an arbitrary metric-measure space.

Also, we can (and should) relax the condition that H is a Hilbert space, since the limit of Hilbert spaces can be a locally convex vector space of more general type (e.g. a nuclear space). In order to obtain a "commutative" Riemannian geometry we require that the amplitude operator (or correlator) associated with a surface or a graph is invariant with respect to the (separate) permutations of inputs and outputs.

If we accept the above working definitions of quantum Riemannian spaces, then many natural questions arise, in particular:

a) What is the dimension of a quantum Riemannian space?

b) What is the diameter?

c) Which quantum Riemannian spaces should be called manifolds?

d) What are various curvature tensors, e.g. Ricci curvature?

When we speak about quantum Riemannian 1-spaces, Riemannian 2-spaces or, more generally, Riemannian d-spaces, the number d corresponds to the dimension of the *world-sheet*, not the *space-time*. In particular, one can associate a Riemannian 1-space with a compact Riemannian manifold of any dimension. Having in mind possible relationship with CFT we should allow the number d to be non-integer ("central charge").

By analogy with the commutative Riemannian geometry one can ask about the structure of the "moduli space" of quantum Riemannian manifolds rigidified by some geometric data. In particular, we can ask about the "space of isometry classes" of quantum compact Riemannian d-spaces equipped with a non-commutative version of the Gromov-Hausdorff topology. Then one can ask about analogs of classical precompactness and compactness theorems, e.g. those which claim precompactness of the "moduli space" of Riemannian manifolds having fixed dimension, diameter bounded from above and the Ricci curvature bounded from below (see e.g. [Gro1]). Introducing a non-commutative analog of measure, one can ask about non-commutative analogs of theorems due to Cheeger, Colding, Fukaya and others for the class of *metric-measure spaces*, i.e. metric spaces equipped with a Borel probability measure. To compare with the case of unitary CFTs we remark that the space of states of a unitary CFT plays the role of L_2-space (of the loop space of a manifold), with the vacuum expectation value playing the role of the measure. The unitarity condition ("reflection positivity" in the language of Euclidean Quantum Field Theory) turns out to be an analog of the non-negativeness of the Ricci curvature. Normalized boundary states should correspond to probability measures.

The above discussion motivates the idea to treat unitary CFTs and their degenerations as "quantum metric-measure spaces with bounded diameter and non-negative Ricci curvature". One expects that this moduli space is precompact and complete in the natural topology.[1] One hopes for a similar picture for QFTs which

[1]More precisely, non-commutativity arises from CFTs with boundary conditions. Our point of view differs from [RW] where a unitary CFT already gives rise to a non-commutative space.

live on a space-time of a more general type than just a manifold. Leaving aside possible physical applications, one can ask (motivated by metric geometry) "what is a QFT with the space-time, which is a compact metric-measure space ?" There are vague ideas about a possible answer (see e.g. Appendix in the preliminary version of current paper on the author's home page).

Defined in this way, quantum Riemannian spaces enjoy some functorial properties, well-known at the level of CFTs (e.g. one can take a tensor product of quantum Riemannian spaces).

1.4. **Spectral triples, Bakry calculus and Wasserstein spaces.** The idea that QFTs should be studied by methods of non-commutative geometry was suggested by Alain Connes (see [Co1]). The idea to use Connes's approach for the description of CFTs and their degenerations goes back to [FG]. It was further developed in [RW] in an attempt to interpet the earlier approach of [KoSo1] from the point of view of Connes's spectral triples. The measure was not included in the list of data neither in [FG] nor in [RW], since in the framework of spectral triples the measure can be recovered from the rest of the data. Ricci curvature was not defined in the framework of spectral triples. In particular, it was not clear how to define a spectral triple with non-negative Ricci curvature. The present paper can be thought of as a step in this direction.

Recall (see e.g. [Co1]) that a spectral triple is given by a unital C^*-algebra of bounded operators in a Hilbert space and a 1-parameter semigroup continuously acting on the space. It is assumed that the semigroup has an infinitesimal generator D which is a positive unbounded self-adjoint operator with compact resolvent, and the commutator $[D, f]$ with any algebra element f is bounded. Similar structures appear in the theory of random walks and Markov semigroups on singular spaces (see e.g. [Ba], [BaEm], [LV], [Led], [St]). In that case one also has a probability measure which is invariant with respect to the semigroup. This similarity makes plausible the idea that the "abstract calculus" of Markov semigroups developed by Bakry and Emery (see [BaEm], [Ba]) can be used for the description of the topological space obtained from a collapsing family of unitary CFTs. This idea was proposed by Kontsevich in a series of talks in 2003. In those talks Kontsevich introduced the notion of "singular Calabi-Yau manifold" defined in terms of what he called Graph Field Theory (and what we call commutative Riemannian 1-geometry below). One hopes that the "moduli space of singular Calabi-Yau manifolds" with bounded dimension and fixed diameter, being equipped with a (version of) Gromov-Hausdorff (or measured Gromov-Hausdorff) topology, is compact. More generally one can expect a similar result for singular quantum Riemannian 1-spaces which have bounded dimension, diameter bounded from above and Ricci curvature bounded from below.[2] To our knowledge there is no precompactness theorem for the space of "abstract Bakry-Emery data".[3] In a similar vein we mention

We prefer to axiomatize the structure arising from the full space of states rather than from the subspace of invariants with respect to a W-algebra.

[2]In the preprint [En] the precompactness of the moduli space of commutative measured Riemannian 1-spaces and the usual bounds on the diameter and Ricci curvature was proved. Methods of [En] are based on explicit estimates of the heat kernel as well as classical results by Cheeger and Colding [ChC3]. We do not see how to extend them to the non-commutative case.

[3]Bakry-Emery data naturally lead to the metric on the space of states of a C^*-algebra coinciding with the metric introduced by Connes and generalized later by Rieffel (see [Rie]) in his notion of "quantum metric space". Although the precompactness theorem was formulated and proved

precompactness theorems for a class of metric-measure spaces which generalizes the class of Riemannian manifolds with non-negative Ricci curvature (see [LV], [St]). The authors introduced in [LV], [St] the notion of N-Ricci curvature, $N \in [1, \infty]$. The notion of N-Ricci curvature is defined in terms of geodesics in the space of probability measures equipped with the Wasserstein L_2-metric (see Section 6). We hope that there is a non-commutative generalization of the notion of N-Ricci curvature as well as of the Wasserstein metric, so that the class of quantum metric-measure spaces with non-negative N-Ricci curvature and bounded diameter is compact with respect to the non-commutative generalization of the measured Gromov-Hausdorff topology or non-commutative generalization of the metric introduced in [St].

Let us make few additional remarks about the relationship of our approach with the one of Connes (see [Co1]). In the notion of spectral triple he axiomatized the triple (A, H, Δ) where A is the algebra of smooth functions on a compact closed Riemannian manifold M (considered as a complex algebra with an anti-linear involution), $H = L_2(M, vol_M)$ is the Hilbert space of functions, which are square-integrable with respect to the volume form associated with the Riemannian metric, and Δ is the Laplace operator associated with the metric.[4] From a slightly different perspective, the data are: involutive algebra A, a positive linear functional $\tau(f) = \int_M f\, vol_M$ which defines the completion H of A with respect to the scalar product $\tau(fg^*)$, the $*$-representation $A \to End(H)$, and the 1-parameter semigroup $exp(-t\Delta), t \geq 0$ acting on H by means of trace-class operators. The generator of the semigroup is a non-negative self-adjoint unbounded operator Δ with discrete spectrum, and the algebra A being naturally embedded to H belongs to the domain of Δ. Thus A encodes the topology of M, while τ encodes the measure, and Δ encodes the Riemannian structure. Let $B_1(f, g)$ be a bilinear form $A \otimes A \to A$ given by $2B_1(f, g) = \Delta(fg) - f\Delta(g) - \Delta(f)g$. The formula $d(\phi, \psi) = sup_{B_1(f,f) \leq 1} |\varphi(f) - \psi(f)|$ defines the distance function on the space of states of the C^*-completion of A in terms of the spectral triple data (the C^*-completion can be spelled out intrinsically in terms of the operator norm derived from the $*$-representation $A \to End(H)$). Every point $x \in M$ gives rise to a state (delta-function δ_x). One sees that the above formula recovers the Riemannian distance function on M without use of the language of points, so it can be generalized to the case of non-commutative algebra A. There are many non-trivial examples of spectral triples which do not correspond to commutative Riemannian manifolds (see e.g. [Co1], [CoMar]). Let us observe that in the case of Riemannian manifolds the 1-parameter semigroup $exp(-t\Delta), t \geq 0$ assigns a trace-class operator $exp(-l\Delta)$ to every segment $[0, l]$, which we can view as a very simple metrized graph with one input and one output. Moreover, the multiplication $m_A : A \otimes A \to A$ gives rise to the family of operators $S_{l_1,l_2,l_3}\ H^{\otimes 2} \to H$ such that

$$x_1 \otimes x_2 \mapsto exp(-l_3\Delta)(m_A(exp(-l_1\Delta)(x_1) \otimes exp(-l_2\Delta)(x_2))),$$

for any $l_1, l_2, l_3 > 0$.

In a bit more symmetric way, one has a family of trilinear forms $H^{\otimes 3} \to \mathbf{C}$ such that $(x_1, x_2, x_3) \mapsto \tau(m_A(m_A \otimes id)(exp(-l_1\Delta)(x_1) \otimes exp(-l_2\Delta)(x_2) \otimes$

by Rieffel for compact quantum metric spaces, it is not clear how to extend his approach to the case of quantum Riemannian 1-spaces discussed in this paper.

[4]In fact Connes considered the case of spin manifolds, so he used the Dirac operator D instead of $\Delta = D^2$, and H was the space of square-integrable sections of the spinor bundle.

$exp(-l_3\Delta)(x_3)))$. Hence, starting with a commutative spectral triple, we can produce trace-class operators associated with two types of metrized graphs:

a) to a segment $I_l := [0, l]$ we associate an operator $exp(-l\Delta) := S_{I_l} : H \to H$, assuming that for $l = 0$ we have the identity operator;

b) to the Y-shape graph Γ_{l_1,l_2,l_3} with different positive lengths of the three edges we associate an operator $S_{l_1,l_2,l_3} := S(\Gamma_{l_1,l_2,l_3}) : H^{\otimes 2} \to H$. Notice that $m_A(f_1 \otimes f_2) = lim_{l_1+l_2+l_3 \to 0} S(\Gamma_{l_1,l_2,l_3})(f_1 \otimes f_2), f_i \in A, i = 1, 2$, hence the multiplication on A can be recovered from operators associated to metrized graphs as long as we assume continuity of the operators with respect to the length of an edge of the tree.

From the point of view of Quantum Field Theory it is natural to consider more general graphs. This leads to the notion of singular quantum Riemannian 1-geometry (see below). It turns out that this language is suitable for spelling out various differential-geometric properties of Riemannian manifolds (non-commutative and singular in general), in particular, the property to have non-negative Ricci curvature.

1.5. **Relationship to physics.** One can speculate about possible applications of the ideas of this paper. The Gromov-Hausdorff (or measured Gromov-Hausdorff) topology is coarser than topologies typically used in physics. The question is: are these "Gromov-Hausdorff type" topologies "physical enough" to derive interesting properties of the moduli spaces? For example, it is interesting whether measured Gromov-Hausdorff topology can be useful in the study of the so-called "string landscape" and the problem of finiteness of the volume of the corresponding moduli spaces of QFTs (see e.g. [Dou 1], [Dou 2], [Dou L], [Va], [OVa]). For example, the problem of statistics of the string vacua leads to the counting of the number of critical points of a certain function (prepotential) on the moduli space of certain CFTs (see e.g. [Dou 1], [Z]). Finiteness of the volume of the moduli space of CFTs (the volume gives the first term of the asymptotic expansion of the number of critical points) is crucial. According to the previous discussion, the "Gromov-Hausdorff type" moduli space of unitary CFTs with bounded central charge and spectral gap bounded from below is expected to be compact. Probably the requirement that the spectrum is discrete leads to the finiteness of the volume of the moduli space with respect to the measure derived from the Zamolodchikov metric (cf. [Va]). This would give a bound for the number of string vacua. I should also point out an interesting paper [Dou 3] which contains a discussion of the moduli of QFTs from a perspective close to "Notes" and [KoSo1].

1.6. *Acknowledgements.* I thank Jean-Michel Bismut, Kevin Costello, Alain Connes, Michael Douglas, Boris Feigin, Misha Gromov, Kentaro Hori, David Kazhdan, Andrey Losev, John Lott, Yuri Manin, Matilde Marcolli, Nikolay Reshetikhin, Dmitry Shklyarov, Katrin Wendland for useful conversations and correspondence. I am especially grateful to Maxim Kontsevich for numerous discussions about CFTs and Calabi-Yau manifolds, which influenced this work very much. I thank referee for the careful reading and pointing out many inaccuracies. I thank to IHES for hospitality and excellent research and living conditions. This work was partially supported by an NSF grant.

2. Reminder on degenerating Conformal Field Theories

This Section contains the material borrowed from [KoSo1].

Unitary Conformal Field Theory is well-defined mathematically thanks to Segal's axiomatic approach (see [Seg]). We are going to recall Segal's axioms later. In the case of the complex line $\mathbf{C}$ the data defining a unitary CFT can be summarized as follows:

1) A real number $c \geq 0$ called central charge.

2) A bi-graded pre-Hilbert *space of states* $H = \oplus_{p,q \in \mathbf{R}_{\geq 0}} H^{p,q}, p - q \in \mathbf{Z}$ such that $dim(\oplus_{p+q \leq E} H^{p,q})$ is finite for every $E \in \mathbf{R}_{\geq 0}$. Equivalently, there is an action of the Lie group $\mathbf{C}^*$ on H, so that $z \in \mathbf{C}^*$ acts on $H^{p,q}$ as $z^p \bar{z}^q := (z\bar{z})^p \bar{z}^{q-p}$.

3) An action of the product of Virasoro and anti-Virasoro Lie algebras $Vir \times \overline{Vir}$ (with the same central charge c) on H, so that the space $H^{p,q}$ is an eigenspace for the generator L_0 (resp. $\overline{L}_0$) with the eigenvalue p (resp. q).

4) The space H carries some additional structures derived from the operator product expansion (OPE). The OPE is described by a linear map $H \otimes H \to H \widehat{\otimes} \mathbf{C}\{z, \bar{z}\}$. Here $\mathbf{C}\{z, \bar{z}\}$ is the topological ring of formal power series $f = \sum_{p,q} c_{p,q} z^p \bar{z}^q$ where $c_{p,q} \in \mathbf{C}$, $p, q \to +\infty$, $p, q \in \mathbf{R}$, $p - q \in \mathbf{Z}$. The OPE satisfies some axioms which I do not recall here (see e.g. [Gaw]). One of the axioms is a sort of associativity of the OPE.

Let $\phi \in H^{p\,q}$. Then the number $p + q$ is called the *conformal dimension* of ϕ (or the *energy*), and $p - q$ is called the *spin* of ϕ. Notice that, since the spin of ϕ is an integer number, the condition $p + q < 1$ implies $p = q$.

The central charge c can be described intrinsically by the formula $dim(\oplus_{p+q \leq E} H^{p,q}) = exp(\sqrt{4/3\pi^2 c E}(1 + o(1))$ as $E \to +\infty$. It is expected based on examples (but not rigorously justified) that all possible central charges form a well-ordered subset of $\mathbf{Q}_{\geq 0} \subset \mathbf{R}_{\geq 0}$. If $H^{0,0}$ is a one-dimensional vector space, the corresponding CFT is called irreducible. A general CFT is a sum of irreducible ones. The *trivial* CFT has $H = H^{0,0} = \mathbf{C}$ and it is the unique irreducible unitary CFT with $c = 0$.

Remark 2.0.1. *There is a version of the above data and axioms for Superconformal Field Theory (SCFT). In that case each $H^{p,q}$ is a hermitian super vector space. There is an action of a super extension of the product of Virasoro and anti-Virasoro algebra on H. In the discussion of the moduli spaces below we will speak about CFTs, not SCFTs. Segal's axiomatics for SCFT is not available from the published literature.*

2.1. **Moduli space of Conformal Field Theories.** For a given CFT one can consider its group of symmetries (i.e. automorphisms of the space $H = \oplus_{p,q} H^{p,q}$ preserving all the structures).

Conjecture 2.1.1. *The group of symmetries is a compact Lie group of dimension less or equal than $dim\, H^{1,0}$.*

Of course discrete symmetries can be present, but they do not affect the dimension of the group.

Let us fix $c_0 \geq 0$ and $E_{min} > 0$, and consider the moduli space $\mathcal{M}_{c \leq c_0}^{E_{min}}$ of all irreducible CFTs with the central charge $c \leq c_0$ and

$$min\{p + q > 0 | H^{p,q} \neq 0\} \geq E_{min}.$$

Conjecture 2.1.2. *$\mathcal{M}_{c \leq c_0}^{E_{min}}$ is a compact real analytic stack of finite local dimension. The dimension of the base of the minimal versal deformation of a given CFT is less or equal to $dim\, H^{1,1}$.*

We define $\mathcal{M}_{c\leq c_0} = \cup_{E_{min}>0}\mathcal{M}^{E_{min}}_{c\leq c_0}$. We would like to compactify this stack by adding boundary components corresponding to certain degenerations of the theories as $E_{min} \to 0$. The compactified space is expected to be a compact stack $\overline{\mathcal{M}}_{c\leq c_0}$. In what follows we will loosely use the word "moduli space" instead of the word "stack".

Remark 2.1.3. *There are basically two classes of rigorously defined CFTs: the rational theories (RCFT) and the lattice CFTs. They are defined algebraically (e.g. in terms of braided monoidal categories or vertex algebras). Physicists often consider so-called sigma models defined in terms of maps of two-dimensional Riemannian surfaces (world-sheets) to a Riemannian manifold (world-volume, or target space). Such a theory depends on a choice of the Lagrangian, which is a functional on the space of such maps. Descriptions of sigma models as path integrals do not have precise mathematical meaning. Segal's axioms arose from an attempt to treat the path integral categorically. As we will explain below, there is an alternative way to speak about sigma models in terms of degenerations of CFTs. Roughly speaking, sigma models "live" near the boundary of the compactified moduli space* $\overline{\mathcal{M}}_{c\leq c_0}$.

2.2. **Physical picture of a simple collapse.** In order to compactify $\mathcal{M}_{c\leq c_0}$ we consider degenerations of CFTs as $E_{min} \to 0$. A degeneration is given by a one-parameter (discrete or continuous) family $H_\varepsilon, \varepsilon \to 0$ of bi-graded spaces as above, where $(p,q) = (p(\varepsilon), q(\varepsilon))$, equipped with OPEs, and such that $E_{min} \to 0$. The subspace of fields with conformal dimensions vanishing as $\varepsilon \to 0$ with the highest rate gives rise to a commutative algebra $H^{small} = \oplus_{p(\varepsilon)\ll 1} H_\varepsilon^{p(\varepsilon),p(\varepsilon)}$ (the algebra structure is given by the terms in OPEs which vanish with the slowest rate as $\varepsilon \to 0$). This is general phenomenon of keeping the lowest modes which is well-known e.g. in the "large volume limit" (see e.g. [FG]). In such a limit geometry emerges and one recovers the target space from the OPE. The spectrum X of H^{small} is expected to be a compact space ("manifold with singularities") such that $dim\, X \leq c_0$. It follows from the conformal invariance and the OPE, that the grading of H^{small} (rescaled as $\varepsilon \to 0$) is given by the eigenvalues of a second order differential operator defined on the smooth part of X (see discussion below in Section 6.6). The operator has positive eigenvalues and is determined up to multiplication by a scalar. It is natural to treat it as a Laplace-Beltrami operator of a Riemannian metric. Thus we can say that the smooth part of X carries a metric g_X, which is defined up to multiplication by a scalar. Other terms in OPEs give rise to additional differential-geometric structures on X.

Thus, as a first approximation to the "true" picture, we assume the following description of a "simple collapse" of a family of CFTs. The degeneration of the family is described by the point of the boundary of $\overline{\mathcal{M}}_{c\leq c_0}$ which is a triple $(X, \mathbf{R}_+^* \cdot g_X, \phi_X)$, where the metric g_X is defined up to a positive scalar factor, and $\phi_X : X \to \mathcal{M}_{c\leq c_0 - dim\, X}$ is a map. One can have some extra conditions on the data. For example, the metric g_X can satisfy the Einstein equation.

Although the scalar factor for the metric is arbitrary, one should imagine that the curvature of g_X is "small", and the injectivity radius of g_X is "large". The map ϕ_X appears naturally from the point of view of the simple collapse of CFTs described above. Indeed, in the limit $\varepsilon \to 0$, the space H_ε becomes an H^{small}-module. It can be thought of as a space of sections of an infinite-dimensional vector bundle $W \to X$. One can argue that fibers of W generically are spaces of states of CFTs with central charges less or equal than $c_0 - dim\, X$. This is encoded in the map ϕ_X.

In the case when CFTs from $\phi_X(X)$ have non-trivial continuous symmetry groups, one expects a kind of a gauge theory on X as well.

Purely bosonic sigma-models correspond to the case when $c_0 = c(\varepsilon) = dim\, X$ and the residual theories (CFTs in the image of ϕ_X) are all trivial. The target space X in this case should carry a Ricci flat metric. In the $N = 2$ supersymmetric case the target space X is a Calabi-Yau manifold, and the residual bundle of CFTs is a bundle of free fermion theories.

Remark 2.2.1. *It was conjectured in [KoSo1] that all compact Ricci flat manifolds (with the metric defined up to a constant scalar factor) appear as target spaces of degenerating CFTs. Thus, the construction of the compactification of the moduli space of CFTs should include a compactification of the moduli space of Einstein manifolds. As we already mentioned in the Introduction, there is a deep relationship between the compactification of the moduli space of CFTs and Gromov's compactification. Moreover, as we will discuss below, all target spaces appearing as limits of CFTs have in some sense non-negative Ricci curvature. More precisely, the limit of the rescaled Virasoro operator $L_0 + \overline{L}_0$ satisfies the Bakry curvature-dimension inequality $CD(0, \infty)$. In the case of compact Riemannian manifolds the latter is equivalent to non-negativeness of Ricci curvature.*

2.3. **Multiple collapse and the structure of the boundary.** In terms of the Virasoro operator $L_0 + \overline{L}_0$ the degeneration of CFT ("collapse") is described by a subset (cluster) S_1 in the set of eigenvalues of $L_0 + \overline{L}_0$ which approaches to zero "with the same speed" provided $E_{min} \to 0$. The next level of the collapse is described by another subset S_2 of eigenvalues of $L_0 + \overline{L}_0$. Elements of S_2 approach zero "modulo the first collapse" (i.e. at the same speed, but "much slower" than elements of S_1). One can continue to build a tower of degenerations. It leads to a hierarchy of boundary strata. Namely, if there are further degenerations of CFTs parametrized by X, one gets a fiber bundle over the space of triples $(X, \mathbf{R}^*_+ \cdot g_X, \phi_X)$ with the fiber which is the space of triples of similar sort. Finally, we obtain the following conjectural qualitative geometric picture of the boundary $\partial\overline{\mathcal{M}}_{c \leq c_0}$.

A boundary point is given by the following data:

1) A finite tower of maps of compact topological spaces $p_i : \overline{X}_i \to \overline{X}_{i-1}, 0 \leq i \leq k$, $\overline{X}_0 = \{pt\}$.

2) A sequence of smooth manifolds $(X_i, g_{X_i}), 0 \leq i \leq k$, such that X_i is a dense subspace of $\overline{X}_i$, and $dim\, X_i > dim\, X_{i-1}$, and p_i defines a fiber bundle $p_i : X_i \to X_{i-1}$.

3) Riemannian metrics on the fibers of the restrictions of p_i to X_i, such that the diameter of each fiber is finite. In particular the diameter of X_1 is finite, because it is the only fiber of the map $p_1 : X_1 \to \{pt\}$.

4) A map $X_k \to \mathcal{M}_{c \leq c_0 - dim\, X_k}$.

The data above are considered up to the natural action of the group $(\mathbf{R}^*_+)^k$ (it rescales the metrics on fibers).

There are some additional data, like non-linear connections on the bundles $p_i : X_i \to X_{i-1}$. The set of data should satisfy some conditions, like differential equations on the metrics. It is an open problem to describe these conditions in the general case. In the case of $N = 2$ SCFTs corresponding to sigma models with Calabi-Yau target spaces these geometric conditions were formulated as a conjecture in [KoSo1].

2.4. Example: Toroidal models. A non-supersymmetric toroidal model is described by the so-called Narain lattice, endowed with some additional data. More precisely, let us fix the central charge $c = n$ which is a positive integer number. What physicists call the Narain lattice $\Gamma^{n,n}$ is the unique even unimodular lattice of rank $2n$ and signature (n,n). It can be described as $\mathbf{Z}^{2n}$ equipped with the quadratic form $Q(x_1,...,x_n,y_1,...,y_n) = \sum_i x_i y_i$. The moduli space of toroidal CFTs is given by

$$\mathcal{M}^{tor}_{c=n} = O(n,n,\mathbf{Z})\backslash O(n,n,\mathbf{R})/O(n,\mathbf{R}) \times O(n,\mathbf{R}).$$

Equivalently, it is a quotient of the open part of the Grassmannian $\{V_+ \subset \mathbf{R}^{n,n} |\, dim\, V_+ = n, Q_{|V_+} > 0\}$ by the action of $O(n,n,\mathbf{Z}) = Aut(\Gamma^{n,n}, Q)$. Let V_- be the orthogonal complement to V_+. Then every vector of $\Gamma^{n,n}$ can be uniquely written as $\gamma = \gamma_+ + \gamma_-$, where $\gamma_\pm \in V_\pm$. For the corresponding CFT one has

$$\sum_{p,q} dim(H^{p,q}) z^p \bar{z}^q = \Big|\prod_{k\geq 1}(1-z^k)\Big|^{-2n} \sum_{\gamma\in\Gamma^{n,n}} z^{Q(\gamma_+)/2} \bar{z}^{-Q(\gamma_-)/2}$$

Let us describe the (partial) compactification of the moduli space $\mathcal{M}^{tor}_{c=n}$ by collapsing toroidal CFTs. Suppose that we have a one-parameter family of toroidal theories such that $E_{min}(\varepsilon)$ approaches zero. Then for the corresponding vectors in H_ε one gets $p(\varepsilon) = q(\varepsilon) \to 0$. It implies that $Q(\gamma(\varepsilon)) = 0, Q(\gamma_+(\varepsilon)) \ll 1$. It is easy to see that the vectors $\gamma(\varepsilon)$ form a semigroup with respect to addition. Thus one obtains a (part of the) lattice of rank less than or equal to n. In the case of "maximal" simple collapse the rank will be equal to n. One can see that the corresponding points of the boundary give rise to the following data: $(X, \mathbf{R}^*_+ \cdot g_X, \phi^{triv}_X; B)$, where (X, g_X) is a flat n-dimensional torus, $B \in H^2(X, i\mathbf{R}/\mathbf{Z})$ and ϕ^{triv}_X is the constant map from X to the trivial theory point in the moduli space of CFTs. These data in turn give rise to a toroidal CFT, which can be realized as a sigma model with the target space (X, g_X) and given B-field B. The residual bundle of CFTs on X is trivial.

Let us consider a 1-parameter family of CFTs given by $(X, \lambda g_X, \phi^{triv}_X; B = 0)$, where $\lambda \in (0, +\infty)$. There are two degenerations of this family, which define two points of the boundary $\partial\overline{\mathcal{M}}^{tor}_{c=n}$. As $\lambda \to +\infty$, we get a toroidal CFT defined by $(X, \mathbf{R}^*_+ \cdot g_X, \phi^{triv}_X; B = 0)$. As $\lambda \to 0$ we get $(X^\vee, \mathbf{R}^*_+ \cdot g_{X^\vee}, \phi^{triv}_X; B = 0)$, where $(X^\vee, g_{X^\vee})$ is the dual flat torus.

There might be further degenerations of the lattice. Thus one obtains a stratification of the compactified moduli space of lattices (and hence CFTs). Points of the compactification are described by flags of vector spaces $0 = V_0 \subset V_1 \subset V_2 \subset ... \subset V_k \subset \mathbf{R}^n$. In addition one has a lattice $\Gamma_{i+1} \subset V_{i+1}/V_i$, considered up to a scalar factor. These data give rise to a tower of torus bundles $X_k \to X_{k-1} \to ... \to X_1 \to \{pt\}$ over tori with fibers $(V_{i+1}/V_i)/\Gamma_{i+1}$. If $V_k \simeq \mathbf{R}^{n-l}, l \geq 1$, then one also has a map from the total space X_k of the last torus bundle to the point $[H_k]$ in the moduli space of toroidal theories of smaller central charge: $\phi_n : X_k \to \mathcal{M}^{tor}_{c=l}$, $\phi_k(X_k) = [H_k]$.

2.5. Example: WZW model for $SU(2)$. In this case we have a discrete family with $c = \frac{3k}{k+2}$, where $k \geq 1$ is an integer number called *level*. In the limit $k \to +\infty$ one gets $X = SU(2) = S^3$ equipped with the standard metric. The corresponding bundle is the trivial bundle of trivial CFTs (with $c = 0$ and $H = H^{0,0} = \mathbf{C}$).

Analogous picture holds for an arbitrary compact simply connected simple group G.

2.6. **Example: minimal models.** This example has been worked out in [RW], Section 4. Notice that they considered the asymptotics as $\varepsilon \to 0$ of the 3-point function of primary fields. This is equivalent to our picture of collapse with the only reservation that the set of primary fields can be infinite. One has a sequence of unitary CFTs H_m with the central charge $c_m = 1 - \frac{6}{m(m+1)}, m \to +\infty$. In this case $c_m \to 1$, and the limiting space is the interval $[0, \pi]$. The metric is given by $g(x) = \frac{4}{\pi^2} sin^4 x, x \in [0, \pi]$. The corresponding volume form is $vol_g = \sqrt{g(x)}dx = \frac{2}{\pi} sin^2 x dx$. In all above examples the volume form on the limiting space is the one associated with the Riemannian metric.

2.7. **A-model and B-model of $N = 2$ SCFT as boundary strata.** In the case of $N = 2$ Superconformal Field Theory one should modify the above considerations in a natural way. As a result one arrives at the following conjectural picture of the simple collapse.

The boundary of the compactified moduli space $\overline{\mathcal{M}}^{N=2}$ of $N = 2$ SCFTs with a given central charge contains an open stratum given by sigma models with Calabi-Yau targets. Each stratum is parametrized by the equivalence classes of quadruples $(X, J_X, \mathbf{R}_+^* \cdot g_X, B)$ where X is a compact real manifold, J_X a complex structure, g_X is a Calabi-Yau metric, and $B \in H^2(X, i\mathbf{R}/\mathbf{Z})$ is a B-field. The residual bundle of CFTs is a bundle of free fermion theories.

As a consequence of supersymmetry, the moduli space $\mathcal{M}^{N=2}$ of superconformal field theories is a complex manifold which is locally isomorphic to the product of two complex manifolds.[5] Some physicists believe that this decomposition (up to certain corrections) is global. Also, there are two types of sigma models with Calabi-Yau targets: A-models and B-models. Hence, the traditional picture of the compactified moduli space looks as follows:

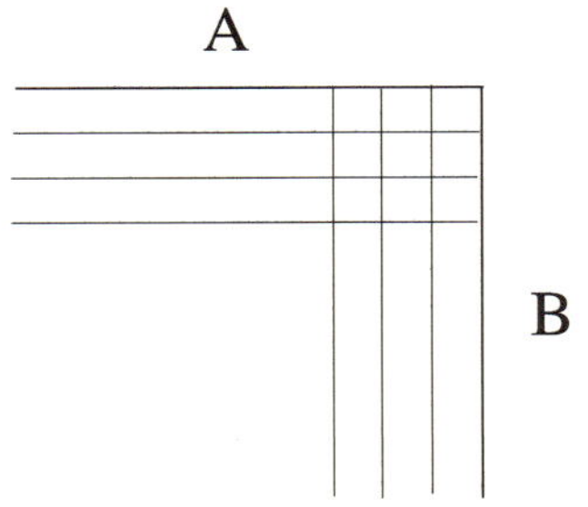

Here the boundary consists of two open strata (A-stratum and B-stratum) and a mysterious meeting point. This point corresponds, in general, to a submanifold of codimension one in the closure of A-stratum and of B-stratum.

As we explained in [KoSo1] this picture should be modified. Namely, there is another open stratum of $\partial\overline{\mathcal{M}}^{N=2}$ (called T-stratum in [KoSo1]). It consists of toroidal models (i.e. CFTs associated with Narain lattices), parametrized by a

[5]Strictly speaking, one should exclude models with chiral fields of conformal dimension $(2, 0)$, e.g. sigma models on hyperkähler manifolds. This can be explained from the deformation theory point of view.

manifold Y with a Riemannian metric defined up to a scalar factor. This T-stratum meets both A and B strata along the codimension one stratum corresponding to the double collapse. Therefore the "true" picture is obtained from the traditional one by the real blow-up at the corner:

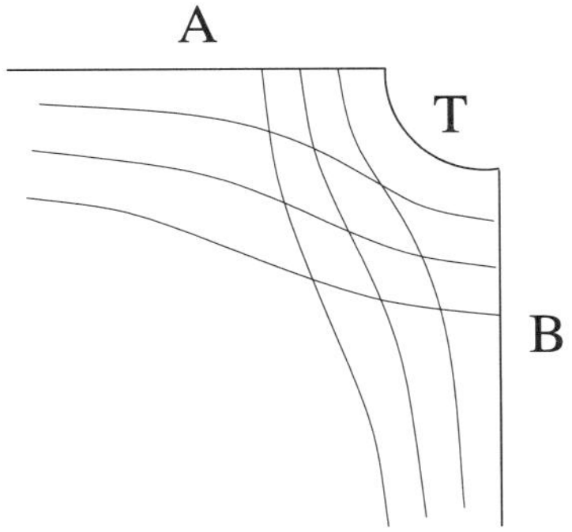

2.8. **Mirror symmetry and the collapse.** Mirror symmetry is related to the existence of two different strata of the boundary $\partial\overline{\mathcal{M}}^{N=2}$ which we called A-stratum and B-stratum. As a corollary, the same quantities admit different geometric descriptions near different strata. In the traditional picture, one can introduce natural coordinates in a small neighborhood of a boundary point corresponding to $(X, J_X, \mathbf{R}_+^* \cdot g_X, B)$. Skipping X from the notation, one can say that the coordinates are (J, g, B) (complex structure, Calabi-Yau metric and the B-field). Geometrically, the pairs (g, B) belong to the preimage of the Kähler cone under the natural map $Re : H^2(X, \mathbf{C}) \to H^2(X, \mathbf{R})$ (more precisely, one should consider B as an element of $H^2(X, i\mathbf{R}/\mathbf{Z})$). It is usually said, that one considers an open domain in the complexified Kähler cone with the property that it contains together with the class of metric $[g]$ also the ray $t[g], t \gg 1$. Mirror symmetry gives rise to an identification of neighborhoods of $(X, J_X, \mathbf{R}_+^* \cdot g_X, B_X)$ and $(X^\vee, J_{X^\vee}, \mathbf{R}_+^* \cdot g_{X^\vee}, B_{X^\vee})$ such that J_X is interchanged with $[g_{X^\vee}] + iB_{X^\vee})$ and vice versa.

We can describe this picture in a different way. Using the identification of complex and Kähler moduli, one can choose $([g_X], B_X, [g_{X^\vee}], B_{X^\vee})$ as local coordinates near the meeting point of the A-stratum and the B-stratum. There is an action of the additive semigroup $\mathbf{R}_{\geq 0} \times \mathbf{R}_{\geq 0}$ in this neighborhood. It is given explicitly by the formula $([g_X], B_X, [g_{X^\vee}], B_{X^\vee}) \mapsto (e^{t_1}[g_X], B_X, e^{t_2}[g_{X^\vee}], B_{X^\vee})$ where $(t_1, t_2) \in \mathbf{R}_{\geq 0} \times \mathbf{R}_{\geq 0}$. As $t_1 \to +\infty$, a point of the moduli space approaches the B-stratum, where the metric is defined up to a positive scalar only. The action of the second semigroup $\mathbf{R}_{\geq 0}$ extends by continuity to the non-trivial action on the B-stratum. Similarly, in the limit $t_2 \to +\infty$ the flow retracts the point to the A-stratum.

This picture should be modified, if one makes a real blow-up at the corner, as we discussed before. Again, the action of the semigroup $\mathbf{R}_{\geq 0} \times \mathbf{R}_{\geq 0}$ extends continuously to the boundary. Contractions to the A-stratum and B-stratum carry non-trivial actions of the corresponding semigroups isomorphic to $\mathbf{R}_{\geq 0}$. Now, let us choose a point in, say, the A-stratum. Then the semigroup flow takes it along the boundary to the new stratum, corresponding to the double collapse. The semigroup $\mathbf{R}_{\geq 0} \times \mathbf{R}_{\geq 0}$ acts trivially on this stratum. A point of the double collapse is also a limiting point of a 1-dimensional orbit of $\mathbf{R}_{\geq 0} \times \mathbf{R}_{\geq 0}$ acting on the T-stratum.

Explicitly, the element (t_1, t_2) changes the size of the tori defined by the Narain lattices, rescaling them with the coefficient $e^{t_1-t_2}$. This flow carries the point of the T-stratum to another point of the double collapse, which can then be moved inside of the B-stratum. The whole path, which is the intersection of $\partial\overline{\mathcal{M}}^{N=2}$ and the $\mathbf{R}_{\geq 0} \times \mathbf{R}_{\geq 0}$-orbit, connects an A-model with the corresponding B-model through the stratum of toroidal models. We can depict it as follows:

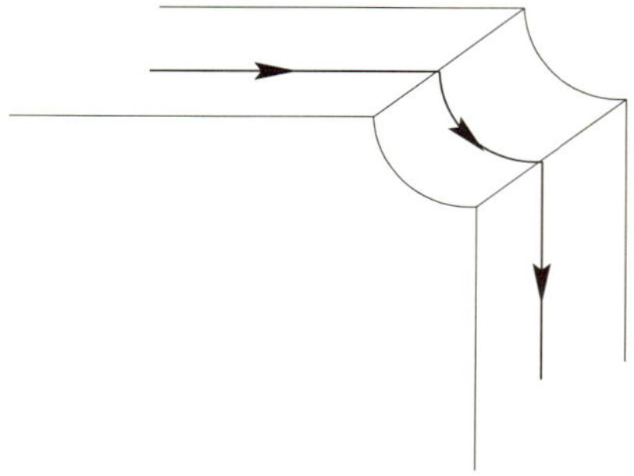

The conclusion of the above discussion is the following: in order to explain the mirror symmetry phenomenon it is not necessary to build full SCFTs. It is sufficient to work with simple toroidal models on the boundary of the compactified moduli space $\overline{\mathcal{M}}^{N=2}$.

3. Segal's axioms and collapse

3.1. **Segal's axioms.** Let us recall Segal's axioms of a 2-dimensional unitary CFT (see e.g. [FFRS], [Gaw], [Ru], [Seg]). Similarly to [FFRS] and [Ru] we use surfaces with a metric rather than surfaces with complex structure. To simplify the description we use oriented surfaces only. The data of a unitary CFT are described as follows:

1) A real number c called *central charge*.

2) 2-dimensional *world-sheets*. The latter are defined by:

2a) an oriented 2-dimensional manifold Σ, possibly disconnected, with finitely many boundary components $(C_i)_{i\in I}$ labeled by elements of the disjoint union of finite sets: $I = I_- \sqcup I_+$. Components labeled by the elements of I_- (resp. I_+) are called *incoming* (resp. *outcoming*). The orientation of Σ induces orientations of all C_i as boundary components. This orientation is called canonical;

2b) a Riemannian metric g_Σ on Σ;

2c) for some $\varepsilon > 0$ (which depends on $i \in I$) a real-analytic conformal embedding to (Σ, g_Σ) of either the flat annulus $A_\varepsilon^- = \{z \in \mathbf{C} | 1 - \varepsilon < |z| \leq 1\}$ (for $i \in I_-$) or the flat annulus $A_\varepsilon^+ = \{z \in \mathbf{C} | 1 \leq |z| < 1 + \varepsilon\}$ (for $i \in I_+$) (the annuli are equipped with the standard flat metric $\frac{dz\, d\overline{z}}{z\overline{z}}$) such that: $f_i^{\pm}(|z| = 1) = C_i$ and f_i^- (resp f_i^+) is orientation preserving (resp. reversing) with respect to the canonical orientation. The map f_i^+ (resp.f_i^-) is called a *parametrization* of C_i.

3) A complex separable Hilbert space H[6] equipped with an antilinear involution $x \mapsto \sigma(x)$.

[6]This condition can be relaxed so one can assume that H is a locally compact vector space, e.g. a nuclear vector space.

4) A trace-class operator (*amplitude*) $S(\Sigma, (f_i)_{i\in I}, g_\Sigma) : \otimes_{i\in I_-} H \to \otimes_{i\in I_+} H$ (by convention the empty tensor product is equal to $\mathbf{C}$). We will sometimes denote by H_i the tensor factor corresponding to $i \in I$.

These data are required to satisfy the following axioms:

CFT 1) If $(\Sigma, (f_i), g_\Sigma) = \sqcup_\alpha(\Sigma_\alpha, (f_i^\alpha), g_{\Sigma_\alpha})$ then $S(\Sigma, (f_i), g_\Sigma) = \otimes_\alpha S(\Sigma_\alpha, (f_i^\alpha), g_{\Sigma_\alpha})$.

CFT 2) Let $i_0 \in I$, and $\overline{f}_{i_0}(z, \overline{z}) = f_{i_0}(\frac{1}{z}, \frac{1}{\overline{z}})$ (i.e. $\overline{f}_{i_0}$ induces the opposite orientation on C_{i_0}). For the new world-sheet we require that if i_0 was in I_- (resp. I_+) then it is now in I_+ (resp. I_-). The condition says: $\langle S(\Sigma, \overline{f}_{i_0}, (f_i)_{i\in I\setminus i_0}, g_\Sigma) x_{i_0} \otimes x, y\rangle = \langle S(\Sigma, (f_i)_{i\in I}, g_\Sigma)(\sigma(x_{i_0}) \otimes x, y\rangle$, where $\langle,\rangle$ denotes the hermitian scalar product on H.

CFT 3) Let $i_0 \in I_-, j_0 \in I_+$, and f_{i_0} and f_{j_0} be the corresponding parametrizations. Let us change (by a real-analytic change of coordinates in the annulus A_ε^+) the parametrization f_{j_0} in such a way that the pull-backs of the metric g_Σ under f_{i_0} and f_{j_0} coincide at the corresponding points of the circle $|z| = 1 \subset A_\varepsilon^\pm$. Let us keep the same notation f_{j_0} for the new parametrization. Identifying points $f_{i_0}(z, \overline{z})$ and $f_{j_0}(z, \overline{z})$, for $|z| = 1$ we obtain a new 2-dimensional oriented surface Σ_{i_0,j_0} such that $C_{i_0}, i_0 \in I_-$ and $C_{j_0}, j_0 \in I_+$ are isometrically identified. By construction, the surface Σ_{i_0,j_0} carries a smooth metric induced by g_Σ. In order to formulate the next condition we need to introduce the notion of partial trace. Let $A : V\otimes H_{i_0} \to W\otimes H_{j_0}$ be a linear map. Using the antilinear involution σ let us identify $H_{j_0} = H$ with the dual space $H_{i_0}^* = H^*$ as follows: $y \mapsto l_y = \langle \bullet, \sigma(y)\rangle$. Then we define $Tr_{i_0,j_0}(A) : V \to W$ by the formula $Tr_{i_0,j_0}(A)(v\otimes x_{i_0}) = \sum_m l_{e_m}(A(v\otimes x_{i_0}))$, where the sum is taken over elements e_m of an orthonormal basis of H_{j_0}. The condition says: $S(\Sigma_{i_0,j_0}, (f_i)_{i\in I\setminus\{i_0,j_0\}}, g_{\Sigma_{i_0,j_0}}) = Tr_{i_0,j_0} S(\Sigma, (f_i)_{i\in I}, g_\Sigma)$.

CFT 4) Let $\overline{\Sigma}$ denote the same 2-dimensional manifold Σ, but with the orientation changed to the opposite one, and with I_- being interchanged with I_+, but all parametrizations remaining the same. Using the involution σ we identify each dual space H_i^* with $H_i, i \in I$ as above. The condition says that the operator corresponding to $\overline{\Sigma}$ coincides with the dual operator $S(\Sigma, (f_i), g_\Sigma)^* : \otimes_{i\in I_+} H \to \otimes_{i\in I_-} H$.

CFT 5) If the metric g_Σ is replaced by $e^h g_\Sigma$ where h is a real-valued smooth function, then

$$S(\Sigma, (f_i), e^h g_\Sigma) = exp(\frac{c}{96\pi} D(h)) S(\Sigma, (f_i), g_\Sigma),$$

where

$$D(h) = \int_\Sigma (|\nabla h|^2 + 4Rh) d\mu_\Sigma,$$

R is the scalar curvature of the metric g_Σ, $d\mu_\Sigma$ is the measure corresponding to g_Σ, and ∇h is the gradient of h with respect to g_Σ.

CFT 6) The operator $S(\Sigma, (f_i), g_\Sigma)$ is invariant with respect to isometries of world-sheets which respect labelings and parametrizations of the boundary components.

3.2. **Collapse of CFTs as a double-scaling limit.** Collapse of a family of unitary CFTs admits a description in the language of Segal's axioms. Let us consider the set $\mathcal{W}$ of isomorphism classes of worldsheets defined in the previous subsection. An isomorphism of worldsheets is an isometry of two-dimensional manifolds with boundary, which respects separately labelings of incoming and outcoming circles as well as parametrizations.

For a fixed non-negative integer number $g \geq 0$ let us consider a subset $\mathcal{P}_g \subset \mathcal{W}$ consisting of worldsheets Σ which are surfaces of genus $g \geq 0$ glued from a collection of spheres with three holes (a.k.a. pants) joined by flat cylinders. The number of cylinders depend on the genus g only. Every worldsheet is conformally equivalent to one like this.

Without giving a definition of the topology on the moduli space of unitary CFTs we would like to describe "a path to infinity" in the moduli space.

Namely, let us consider the amplitudes for a family of unitary CFTs with the same central charge c, depending on the parameter $\varepsilon \to 0$, which satisfy the following properties:

1) The minimal eigenvalue $\lambda_{min}(\varepsilon)$ of the non-negative operator $L_0 + \overline{L}_0$ corresponding to the standard cylinder $S^1_{R=1} \otimes [0,1]$ satisfies the property $\lambda_{min}(\varepsilon) = \lambda_0 \varepsilon \to 0 + o(\varepsilon)$.

2) Consider a family of worldsheets $\Sigma = \Sigma(l_1, ..., l_m, R_1, ..., R_m)$ of genus g such that $min_i\, l_i \to +\infty, max_i R_i/l_i \to 0$ and which are glued from 2-dimensional spheres with three holes each, connected by m "long tubes", which are flat cylinders each of which is isometric to $S^1_{R_i} \times [0, l_i]$ (see Figure below), where $S^1_{R_i}$ is the circle of radius R_i, $1 \leq i \leq m$. Then the following holds

$$l_i \lambda_{min}(\varepsilon) \to l_i^0 < \infty$$

as $l_i \to \infty, \varepsilon \to 0$ uniformly for $1 \leq i \leq m$.

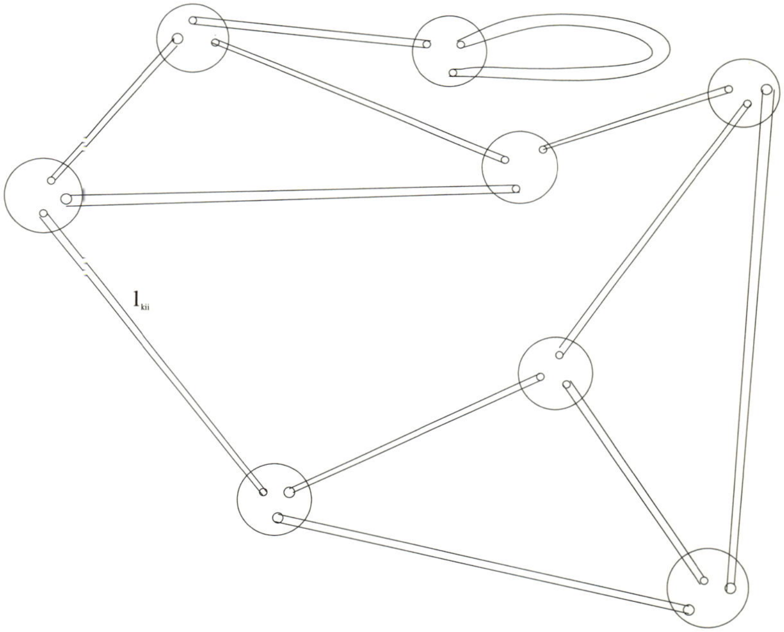

Let us normalize the partition function Z of the standard sphere S^2 in such a way that $Z(S^2) = 1$.

Conjecture 3.2.1. *For the family of closed surfaces Σ as above we have*

$$Z(\Sigma) exp(-\frac{c}{6} \cdot l_i \lambda_{min}) \leq const,$$

where the constant does not depend on Σ.

In order to formulate the next conjecture we need to use the terminology of subsequent sections.

Conjecture 3.2.2. *After rescaling* $L_0 + \overline{L}_0$ *by the factor* λ_{min}^{-1}, *there is a limit in the sense of Section 4.2 of the unitary CFTs, which is a quantum Riemannian 1-space in the sense of Section 5.2, such that the limiting for* $(L_0+\overline{L}_0)/\lambda_{min}$ *operator* L *satisfies the curvature-dimension inequality* $CD(0,\infty)$ *from Section 6.1.*

In this sense CFTs collapse to QFTs with the space-time having non-negative Ricci curvature.

4. Quantum Riemannian d-geometry

4.1. **2-dimensional case and CFT.** Here we follow [Ru]. For any $\varepsilon > 0$ we denote by A_ε the annulus $\{z \in \mathbf{C} | 1-\varepsilon < |z| < 1+\varepsilon\} = A_\varepsilon^- \cup A_\varepsilon^+$. Let us consider the (non-unital) category $Riem_2$ whose objects are $(k+1)$-tuples $X = (\varepsilon; f_1, ..., f_k), k \geq 0$, where $\varepsilon > 0$ and $f_i : A_\varepsilon \to \mathbf{R}_{>0}, 1 \leq i \leq k$ is a smooth function. Notice that each function f_i defines a metric $f_i(x,y)(dx^2 + dy^2)$ on A_ε. Let $Y = (\eta; g_1, ..., g_l)$ be another object of $Riem_2$. A morphism $\phi : X \to Y$ is by definition a triple (M, g_M, j_-, j_+), where (M, g_M) is a 2-dimensional compact oriented Riemannian manifold with boundary, and $j_- : \sqcup_{1\leq i\leq k} A_\varepsilon^- \to M$ and $j_+ : \sqcup_{1\leq i\leq l} A_\eta^- \to M$ orientation preserving (reversing for j_+) isometric embeddings such that $Im(j_-) \cap Im(j_+) = \emptyset$, and ∂M is the union of the images of $S^1 = \{z \in A_\varepsilon | |z| = 1\}$ under $j_\pm$. The embeddings $j_\pm$ are called parametrizations. Boundary components of M parametrized by j_- (resp. j_+) are called *incoming* (resp. *outcoming*). Composition of morphisms $\phi \circ \psi$ is defined in the natural way: one identifies the point $j_+(z), z \in S^1$ on the outcoming boundary component of the surface defined by ψ with the point $j_-(z), z \in S^1$ on the corresponding incoming boundary component of the surface defined by ϕ. One makes $Riem_2$ into a rigid symmetric monoidal category via disjoint union operation on objects: $(\varepsilon_1, f_1) \otimes (\varepsilon_2, f_2) = (\varepsilon, f_1, f_2)$, where $\varepsilon = min\{\varepsilon_1, \varepsilon_2\}$. The duality functor is given by $(\varepsilon, f(z))^* = (\varepsilon, \overline{f(\overline{z})})$, where the bar means complex conjugation. On morphisms the tensor product is given by the disjoint union of surfaces together with taking the minimal ε for all incoming and outcoming boundary components. Finally, for a morphism $\phi : X \otimes Y \to Y \otimes Z$ one has a trace map $tr_Y(\phi) : X \to Z$ obtained by identification of points $j_+(z), z \in S^1$ and $j_-(z), z \in S^1$ of outcoming and incoming boundary components of Y.

Definition 4.1.1. *A (commutative) Riemannian 2-space is a monoidal functor* $F : Riem_2 \to Hilb_{\mathbf{C}}$ *to the rigid monoidal category of separable complex Hilbert spaces.*

As usual, the tensor product of Hilbert spaces is defined as the completed algebraic tensor product. We will denote it simply by $\otimes$, changing sometimes the notation to $\widehat{\otimes}$ in order to avoid a confusion.

A commutative Riemannian 2-space defines a unitary CFT with central charge $c \in \mathbf{R}$ if F is a functor to the category of complex Hilbert spaces satisfying the following properties:

a) F depends on the isometry class of a surface defined by a morphism, not by the surface itself. It assigns the same Hilbert space H to all objects $(\varepsilon; f_1, ..., f_k)$;

b) F preserves trace maps;

c) if $\phi : X \to Y$ is a morphism and $\overline{\phi} : Y \to X$ is obtained by reversing the orientation of the corresponding surface then $F(\overline{\phi})$ is the Hermitian conjugate to $F(\phi)$;

d) change of the metric $g_M \mapsto e^h g_M$ on the surface (recall that h is a smooth real function) defined by a morphism $\phi = (M, g_M, j_-, j_+)$ leads to a new morphism ϕ_h (between the same objects according to a)) such that $F(\phi_h) = exp(\frac{c}{96\pi}D(h))F(\phi)$, and $D(h)$ is the Liouville action described in CFT 5) in the previous Section.

In order to define a *quantum Riemannian 2-space*, we start with the category $Riem_2^{NC}$ with the objects which are $(k+1)$-tuples $(\varepsilon; f_1, ..., f_k)$, as before. A morphism is defined as a quadruple (M, g_M, j_+, j_-, p) where (M, g_M, j_+, j_-) is a Riemannian 2-dimensional surface with parametrized neighborhood of the boundary, as before, and p is a *marking* of the incoming (resp. outcoming) boundary components of ∂M by finite sets $\{1, ..., k\}$ (resp. $\{1, ..., l\}$). On the set of objects we now introduce a rigid *monoidal* structure, which is not symmetric. The tensor product $(\varepsilon_1, f_1) \otimes (\varepsilon_2, f_2)$ is defined as $(min\{\varepsilon_1, \varepsilon_2\}, f_1, f_2)$, as before. But now there is no commutativity isomorphism $X_1 \otimes X_2 \to X_2 \otimes X_1$ if f_1 is not equal to f_2, since isomorphisms in $Riem_2^{NC}$ must respect markings. A quantum Riemannian 2-space is defined by a monoidal functor $F : Riem_2^{NC} \to Hilb_{\mathbf{C}}$. In general F does not commute with the natural action of the product of symmetric groups $S_k \times S_l$ on the markings by the sets $\{1, ..., k\}$ (incoming circles) and $\{1, ..., l\}$ (outcoming circles).

4.2. **General case and spaces with measure.** Let us fix a non-negative integer d. A *quantum Riemannian d-space* is defined as a monoidal functor $F : Riem_d^{NC} \to Hilb_{\mathbf{C}}$, where $Riem_d^{NC}$ is a rigid monoidal category of marked Riemannian manifolds described below (cf. [Seg], comments to Section 4):

a) An object of $Riem_d^{NC}$ is an isometry class of a neighborhood of a compact oriented Riemannian $(d-1)$-manifold in an oriented d-manifold (called *Riemannian d-germ*) together with a bijection between the set of its connected components and a finite set $\{1, ..., n\}$ for some $n \geq 1$.

b) Morphisms are oriented compact Riemannian d-dimensional bordisms. The Riemannian metric is trivialized near the boundary as the product of the given metric on the boundary and the standard flat metric on the interval (this is what we called parametrization in the two-dimensional case). Composition of morphisms is defined via pasting and gluing operation which respects the metric, similarly to the case $d = 2$ considered above.

c) The tensor product $X_1 \otimes X_2$ is defined as the disjoint union of Riemannian d-germs $X_1 \sqcup X_2$ equipped with the marking such that connected components of X_1 are marked first. In this way $X_1 \otimes X_2$ is not necessarily isomorphic to $X_2 \otimes X_1$. Duality corresponds to the reversing of the orientation.

Similarly to the two-dimensional case we have an operation $tr_Y(\phi) : X \to Z$ associated with a morphism $\phi : X \otimes Y \to Y \otimes Z$.

The Hilbert space $H_X := F(X)$ is called the *space of states* assigned to X, for a morphism $f : X \to Y$ the linear map $S(X, Y) := F(f)$ is called the *amplitude* assigned to f.

Next, we would like to introduce a topology on the space of quantum Riemannian d-spaces specified by the definition of convergence. The latter depends on the notion of convergence of Hilbert vector spaces which carry self-adjoint positive operators (each operator is a generator of the semigroup corresponding to the intervals $[0, t]$).

Let us discuss this notion of convergence. Suppose that we have a sequence of Hilbert spaces H_α parametrized by $\alpha \in \mathbf{Z}_+ \cup \{\infty\}$, where $H_\infty := H$. Assume that every H_α carries a self-adjoint non-negative (in general unbounded) operator L_α where $L_\infty := L$. We would like to define a topology on the union of H_α. We will do that following [KS].

Definition 4.2.1. *1) We say that a sequence H_α converges to H as $\alpha \to \infty$, if there exist an open dense subspace $D \subset H$, and for any α there is a continuous linear map $R_\alpha : D \to H_\alpha$, such that $lim_{\alpha\to\infty}||R_\alpha(x)||_{H_\alpha} = ||x||_H$ for any $x \in D$.*

2) Suppose that H_α converges to H in the above sense. We say that a sequence $x_\alpha \in H_\alpha$ strongly converges to $x \in H$ if there is a sequence $y_\beta \in D$ converging to x such that

$$lim_{\beta\to\infty} lim_{\alpha\to\infty}||R_\alpha(y_\beta) - x_\alpha||_{H_\alpha}.$$

3) We say that a sequence (H_α, L_α) strongly converges to (H, L) if D contains the domain of L, the sequence H_α converges to H is the sense of 1), and for any sequence $x_\alpha \in H_\alpha$ which strongly converges to $x \in H$ and any continuous function $\varphi : [0,\infty) \to \mathbf{R}$ with compact support the sequence $\varphi(L_\alpha)(x_\alpha)$ strongly converges to $\varphi(L)(x)$.

4) We say that a sequence of quantum Riemannian d-spaces $\{M_\alpha\}_{\alpha\in\mathbf{Z}_+}$ converges to a quantum Riemannian d-space M as $\alpha \to \infty$, if for any morphism $f : X \to Y$ in $Riem_d^{NC}$ the corresponding sequence $(H^*_{X,\alpha} \otimes H_{Y,\alpha}, S(X, Y, \alpha))$ strongly converges to $(H^*_X \otimes H_Y, S(X, Y))$.

Remark 4.2.2. *One can define the notion of quantum locally convex d-geometry over a complete normed field K by considering monoidal functors to the monoidal category $Vect_K^{lc}$ of locally convex topological spaces over K. Probably the case of locally compact K (e.g. the field of p-adic numbers) is of some interest.*

A natural generalization of the above definitions gives rise to the notion of *quantum Riemannian d-space with a measure.* It is an analog of the notion of Riemannian manifold M equipped with a probability measure, which is absolutely continuous with respect to the measure $dvol_M/vol(M)$ defined by the Riemannian metric. Namely, we assume that an additional datum is given: a continuous linear functional $\tau : H \to \mathbf{C}$ such that $\tau(e^{-tL}x) = \tau(x)$ for any $x \in H, t \geq 0$, and which satisfies the "trace" property described below. Let $\Gamma^{in_1,in_2,out}_{l_1,l_2,l}$ be a Y-shape graph with two inputs (marked by in_1, in_2), one output (marked by out) and three edges with the lengths l_i (outcoming from $in_i, i = 1, 2$) and l (incoming to out). Then the trace property says:

$$\tau \circ S(\Gamma^{in_1,in_2,out}_{l_1,l_2,l}) = \tau \circ S(\Gamma^{in_2,in_1,out}_{l_2,l_1,l}).$$

This is an analog of the property: $\tau(fg) = \tau(gf)$ in case when H is obtained by the Gelfand-Naimark-Segal construction from a C^*-algebra and a tracial state τ on it.

We define the category $Riem_d^{NC,mes}$ of *measured $(d-1)$-dimensional Riemannian germs* with the objects which are $(d-1)$-Riemannian germs as before, equipped with probability measures which are absolutely continuous with respect to the volume measure associated with the germ of the Riemannian metric. Morphisms between measured $(d-1)$-Riemannian germs are measured compact Riemannian d-dimensional manifolds, such that in the neighborhood of a boundary both metric and measure are trivialized as products of the given metric and measure on the boundary and the standard metric and measure on the interval.

Then we modify the above Property 4) by the following requirement:

if $x_\alpha \to x, x_\alpha \in H_\alpha, x \in H$ (strong convergence), then $\tau_\alpha(x_\alpha) \to \tau(x)$ for the corresponding linear functionals. In this way we obtain a topology on quantum measured Riemannian d-spaces.

In the above discussion of convergence we spoke about convergence of quantum Riemannian spaces, not underlying germs of Riemannian manifolds. In other words, having a sequence of quantum Riemannian d-spaces given by functors $F_\alpha : Riem_d^{NC} \to Hilb_{\mathbf{C}}, \alpha \in \mathbf{Z}_+$ we say that the sequence converges to a quantum Riemannian d-space given by a functor $F : Riem_d^{NC} \to Hilb_{\mathbf{C}}$ if for any morphisms $f : X \to Y$ in $Riem_d^{NC}$ we have $F_\alpha(f) \to F(f)$ as $\alpha \to \infty$. The latter convergence is defined in the Definition 4.2.1. On the other hand, we have the Gromov-Hausdorff (or measured Gromov-Hausdorff) topology on the objects of $Riem_d^{NC}$ (we leave it as an exercise to the reader to work out the modification of either topology which takes into account labelings of the boundary components). Moreover we have those topologies on the spaces of morphisms as well. This can be used to define the notion of *continuous* functor.

Definition 4.2.3. *We say that a functor $F : Riem_d^{NC} \to Hilb_{\mathbf{C}}$ is continuous if for any sequence of morphisms $f_\alpha : X_\alpha \to Y_\alpha$ in $Riem_d^{NC}$ which converges in the Gromov-Hausdorff sense to $f : X \to Y$ we have: $F(f_\alpha)$ converges to $F(f)$ in the sense of Definition 4.2.1.*

In particular, if F is continuous then $X_\alpha \to X, \alpha \to \infty$ implies $F(X_\alpha) \to F(X), \alpha \to \infty$.

Continuity allows to contract e.g. loops in the $d = 1$ case. A different property which we will use instead is the *non-collapsing continuity*. It is the same as the one given in the above definition with the restriction that for the limit $f_\alpha \to f$ we require $vol(f_\alpha)(B(0,r)) \geq const > 0$ for any ball $B(0,r) \subset f_\alpha$ of the Riemannian d-bordism f_α. In particular, for $d = 1$ non-collapsing continuity prohibits contraction not only of loops of a graph but of edges as well.

There is a natural modification of the above definitions to the case of quantum measured Riemannian d-spaces, which we leave as an exercise to the reader. In this case we have that measured Gromov-Hausdorff convergence of objects in $Riem_d^{NC,mes}$ implies the above-mentioned convergence of Hilbert spaces and linear functionals.

5. Graphs and singular quantum Riemannian 1-geometry

The content of this section is heavily influenced by conversations with Maxim Kontsevich. The notion of commutative Riemannian 1-space is basically what he called Graph Field Theory.

5.1. **Singular quantum Riemannian 1-spaces.** We introduced quantum Riemannian d-spaces in Section 4.2. The definition mimicks Segal's definition of a unitary CFT. We also considered two kinds of limits of quantum Riemannian d-spaces: the one which does not involve a limit of underlying Riemannian germs and the one which does. The former notion covers the case of the limit of CTFs which is again a CFT. The latter notion deals with functors which are continuous in non-collapsing sense. Therefore the above axiomatics does not include the case of collapsing CFTs when 2-dimensional worldsheets degenerate into metrized graphs (see Section 3.2). The latter degeneration does not preserve conformal symmetry, hence the limit cannot be a CFT. On the other hand such limits give a (partial) compactification of the "moduli of quantum Riemannian 2-spaces". Their study is natural from the point of view of Gromov's precompactness theory. We are not going to develop a general axiomatics of the corresponding singular quantum Riemannian d-manifolds. We will concentrate on the case $d = 1$. In the case $d = 1$ we want to allow singular worldsheets, i.e. metrized graphs. This leads to the following definition.

A *singular* quantum Riemannian 1-space is defined by the following data:

1) A class G of *metrizable labeled graphs* (Γ, I_-, I_+, l, p) described in 1a), 1b) below:

1a) $\Gamma \in G$ is a finite graph with external vertices having the valency one and labeled by elements of the disjoint union of finite sets: $I = I_- \sqcup I_+$ (the letter p above denotes the labeling). Vertices parametrized by the elements of I_- (resp. I_+) are called *incoming* (resp. *outcoming*), or, simply, *in* (resp. *out*) vertices. We denote by $V_{in}(\Gamma)$ the set of inner vertices of Γ, by $E(\Gamma)$ the set of edges of Γ, etc.

1b) A length function $l : E(\Gamma) \to \mathbf{R}_{>0}$, where $E(\Gamma)$ is the set of edges of Γ.

2) A separable real Hilbert space H (or a complex Hilbert space with real structure).[7]

3) To each (Γ, l, I_-, I_+) a trace-class operator $S(\Gamma, l, I_-, I_+) : \otimes_{I_-} H \to \otimes_{I_+} H$, which we will often denote simply by $S(\Gamma)$.

These data are required to satisfy the following axioms:

QFT 1) If Γ is obtained by gluing Γ_1 and Γ_2 (with an obvious definition of the sets $I_\pm$ and the length function) then $S(\Gamma) = S(\Gamma_1) \circ S(\Gamma_2)$ (composition of operators).

QFT 2) Operators $S(\Gamma)$ are invariant with respect to isometries.

QFT 3) If $\Gamma^\vee$ is obtained from Γ by relabeling so that all incoming vertices are declared outcoming and vice versa then $S(\Gamma^\vee) = S(\Gamma)^*$ (conjugate operator). Here we use the scalar product in order to identify H and H^*.

We will add two more axioms below. They will allow us to compactify various "moduli spaces" of quantum Riemannian 1-spaces.

QFT 4) The operators $S(\Gamma)$ enjoy the non-collapsing continuity property. In other words, a small deformation of a metrized graph Γ leads to a small change of $S(\Gamma)$ in the *normed* operator topology. In addition, the operator $S(\Gamma)$ is a continuous function (in the *strong* operator topology) with respect to the length of an edge which is not a loop. Moreover if $F \subset \Gamma$ is a subforest (i.e. a collection of internal edges without loops), and $(\Gamma/F, I_-, I_+, l)$ is the metrized graph obtained by contracting all of the edges from F, then $S(\Gamma/F)$ coincides with the limiting operator $lim_{l(F)\to 0} S(\Gamma)$, which is the limit of $S(\Gamma)$ as lengths of all edges belonging to F simultaneously approach 0.

QFT 5) If Γ_ε^0 is a graph, which is a segment $[0, \varepsilon] \subset \mathbf{R}$ such that $I_- = \{0\}$ and $I_+ = \{\varepsilon\}$ then $lim_{\varepsilon\to 0} S(\Gamma_\varepsilon^0) = id_H$, where the limit is taken in the strong operator topology. Similarly, if Γ_ε^1 is a graph, which is the same segment but with $I_- = \{0, \varepsilon\}$ and $I_+ = \{\emptyset\}$, then $lim_{\varepsilon\to 0} S(\Gamma_\varepsilon^1) : H \otimes H \to \mathbf{R}$ is the scalar product on H.

[7]As in in the case of CFT we can relax this condition assuming that H is a locally compact, in particular, a nuclear space. In case if H is a complex vector space the formulas below have to be modified in order to include complex conjugation.

Example 5.1.1. *Let M be a compact Riemannian manifold with metric g_M. We can normalize the Riemannian measure $d\mu = \sqrt{\det g_M}\, dx$ so that $\int_M d\mu(M) = 1$. Let $H = L^2(M, d\mu)$. Then with each $t > 0$ we associate a trace-class operator $P_t = exp(-t\Delta)$, where Δ is the Laplace operator on M. One has an integral representation $(P_t f)(x) = \int_M G_t(x,y) f(y) d\mu(y)$, where $G_t(x,y)$ is the heat kernel. To a graph $(\Gamma, l) \in G$ we assign the following function on $M^{I_- \cup I_+}$:*

$$K(\Gamma, l)((x_i)_{i \in I_-}, (y_j)_{j \in I_+}) = \int_{M^{V_{in}(\Gamma)}} d\mu^{V_{in}(\Gamma)} \prod_{e \in E(\Gamma)} G_{l(e)}(x, y),$$

where $x = (x_i)_{i \in I_-}, y = (y_j)_{j \in I_+}$ and $d\mu^{V_{in}(\Gamma)} = \prod_{v \in V_{in}(\Gamma)} d\mu$ is the product measure on $M^{V_{in}(\Gamma)}$, associated with $d\mu$. In other words, we assign a measure $d\mu$ to every internal vertex, the kernel $G_{l(e)}(x,y)$ (propagator) to every edge and then integrate over all internal vertices.

We obtain a function whose variables are parametrized by input and output vertices. Finally, we define $S(\Gamma, l) : \otimes_{I_-} H \to \otimes_{I_+} H$ as the integral operator with the kernel $K(\Gamma, l)$.

We see that every Riemannian compact manifold defines a singular quantum Riemannian 1-space (which can be non-singuilar as in the above example). In fact it is a commutative Riemannian 1-space, since the operators $S(\Gamma, l)$ are invariant with respect to the action of the product of symmetric groups $S_{I_-} \times S_{I_+}$ on the set $I_- \times I_+$ (and hence the product on the algebra of smooth functions $A \subset H$ defined by the Y-shape graph is commutative).

Let us restate the above example in terms more suitable for non-commutative generalization. In order to define a collection of operators $S(\Gamma, l) : \otimes_{I_-} H \to \otimes_{I_+} H$ it suffices to have tensors $K(\Gamma, l) \in (\otimes_{I_-} H^*) \otimes (\otimes_{I_+} H)$ satisfying the composition property: $K(\Gamma_1 \circ \Gamma_2, l_1 \circ l_2) = ev^{\otimes_{I_+} H}(K(\Gamma_1, l_1) \otimes K(\Gamma_1, l_2))$, where $ev : H^* \otimes H \to \mathbf{C}$ is the natural pairing, $I_\pm$ denote the sets of input and output vertices for the composed graph $\Gamma_1 \circ \Gamma_2$ (see QFT 1)), and $l_1 \circ l_2$ is the length function obtained by the natural extension of l_1 and l_2 to $\Gamma_1 \circ \Gamma_2$.

Suppose that to each vertex $v \in V(\Gamma)$ we assigned a tensor $T_v \in \otimes_{i \in Star(v)} H$, where $Star(v)$ is the set of adjacent vertices (i.e. vertices $w \in V(\Gamma)$ such that (w, v) is an edge), and to every edge $e \in E(\Gamma)$ with the endpoints v_1, v_2 we assigned a linear functional $\tau_e : H \otimes H \to \mathbf{C}$. We define $T(\Gamma) := \otimes_{v \in V(\Gamma)} T_v$. Let τ_{I_+} be the tensor product $\otimes_e \tau_e$ taken over all edges which are not of the form $e = (v, w)$, where $w \in I_+$ (i.e. e does not have an endpoint which is an *out* vertex). Then the element $\tau_{I_+}(T(\Gamma))$ belongs to $\otimes_{i \in I_+} H$.

In the above example we thought of $G_{l(e)}(x,y)$ as of an element of the space $H \otimes H$, where tensor factors are assigned to the endpoints of e. To such an edge we assigned the pairing $\tau_e(f, g) = \int d\mu_x d\mu_y G_{l(e)}(x,y) f(x) g(y)$, where $d\mu_x = d\mu_y = d\mu$. We remark, that in order to define $T(\Gamma)$ it suffices to have vectors $z_i \in H, i \in I_- \cup I_+$ and tensors $T_e \in H \otimes H$ for every edge $e = (v_1, v_2) \in E(\Gamma)$.

5.2. **Spectral triples and quantum Riemannian 1-geometry.** As we explained in the Introduction, a spectral triple in the sense of Connes gives rise to a quantum Riemannian 1-geometry. Let us make this point more precise. We will use a slightly different definition than in [Co1], since we would like to use an abstract version of the Laplace operator rather than the Dirac operator. More precisely, let us consider a triple (A, H, L) which consists of an unital complex $*$-algebra A, a separable Hilbert left A-module H (i.e. the $*$-algebra A acts on H by bounded operators), a self-adjoint non-negative unbounded operator L on H, such that the operator $P_t = exp(-tL) : H \to H$ has finite trace for all $t > 0$. The latter implies that that the spectrum of L consists of eigenvalues only, and the operator $(1+L)^{-1}$ is compact. To formulate the second assumption, for any $a \in A$ let us consider the function $\phi_a(t) = e^{ta}Le^{-ta}, t \geq 0$. It takes values in the unbounded operators in H, and the domains of all $\phi_a(t)$ belong to the domain of L for all $a \in A, t \geq 0$. We say that $\phi_a(t)$ has k-th derivative at $t = 0$ if it can be represented as a sum $\phi_a(t) = C_0(a) + C_1(a)t + \frac{C_2(a)}{2!}t^2 + ... + \frac{C_k(t,a)}{k!}t^k$ such that $C_i(a)$ are some (possibly unbounded, but with a non-empty common domain) operators in H and $C_k(0,a) := C_k(a)$ is a bounded operator in H. We will denote it by $ad_a^k(L)$. Now the second assumption says that $ad_a^2(L)$ exists for all $a \in A$. Basically this means that the double commutator $[[L,a],a]$ exists and bounded for all $a \in A$.

Let us assume that the spectral triple has finite metric dimension n (a.k.a. n-summable, see [Co1], [CoMar]) and satisfies the regularity conditions (see loc cit.) such that the volume functional

$$\tau(f) = Tr_\omega(fL^{-n/2})$$

is finite for all $f \in exp(-tL)(H)$ (here Tr_ω denotes the Dixmier trace, see loc. cit.). For simplicity we will also assume that H is obtained from A by a GNS completion with respect to the scalar product $\tau(fg^*)$ (this assumption can be relaxed). Then it looks plausible that one can define a quantum Riemannian 1-geometry similarly to the last paragraph of the previous subsection.

6. Ricci curvature, diameter and dimension: probabilistic and spectral approaches to precompactness

In this Section we review some results presented in [BBG], [Ba], [Led], [LV], [St], [KS], [KMS], [KaKu1-2] (see also [U]). Recall that there are basically three approaches to precompactness of metric (and metric-measure) spaces with the diameter bounded from above and the Ricci curvature bounded from below. "Geometric" approach (which goes back to Gromov, see e.g. [Gro1], and which was developed in a deep and non-trivial way by Cheeger, Colding, Fukaya and many others, see e.g. [Fu], [ChC1-3]) deals with Gromov-Hausdorff (or measured Gromov-Hausdorff) topology, and (very roughly speaking) embeds a compact metric space (X,d) into the Banach space $C(X)$ of continuous functions via $x \mapsto d(x,\bullet)$. Then precompactness follows from a version of the Arzela-Ascoli theorem, since the space X is approximated by a finite metric space. By the nature of this approach one needs the notion of Ricci curvature to be defined "locally", in terms of points of X.

In the "spectral" approach (see e.g. [BBG], [KaKu1-2]) one embeds the metric-measure space $(X,d,d\mu)$ into $L_2(X,d\mu)$ via $x \mapsto K_t(x,\bullet)$ where $K_t(x,y)$ is the "heat kernel" (which needs to be defined if (X,d) is not a Riemannian manifold). Then precompactness follows from a version of Rellich's theorem, since the assumptions on the diameter and curvature imply that the image of X belongs to a Sobolev space, which is compactly embedded in $L_2(X,d\mu)$. Precompactness relies on the estimates for eigenvectors and eigenvalues of the "generalized Laplacian". The former are still local while the latter depend on the global geometry of X. The spectral topology in general does not coincide with the measured Gromov-Hausdorff topology (restrictions on the diameter and Ricci curvature can make these topologies equivalent, see e.g. [KS], Remark 5.1).

In the "probabilistic" approach (see a good review in [L], or original proofs in [LV], [St]) one uses the ideas of optimal transport (different point of view is presented in [AGS]). It is a mixture of the previous two approaches, since one studies the "heat flow" on the space $P(X)$ of probabilistic Borel measures on X, but proves the precompactness theorem in the measured Gromov-Hausdorff topology via a kind of Arzela-Ascoli argument. The point is that the heat flow on measures can be interpreted either as a gradient line of a functional (entropy) or as a geodesic for some metric on $P(X)$. More precisely the space $P(X)$ carries a family $W_p, p \geq 1$ of the so-called Wasserstein metrics. In the case $p = 1$ such a metric (called also Monge-Kantorovich metric) being restricted to delta-functions $\delta_x, x \in X$ reproduces the original distance on X. The distance $W_1(\phi, \psi)$ coincides with the "non-commutative distance" on the space of states on $C(X)$ discussed in the Introduction, thus making a connection with Connes's approach to non-commutative Riemannian geometry. In order to use the above ideas of optimal transport one needs a non-commutative analog of the Wasserstein metric W_2. This is an interesting open problem[8]

6.1. **Semigroups and curvature-dimension inequalities.** We closely follow [Ba], [Led]. Let $(X, d\mu)$ be a space with a measure $d\mu$ (which is assumed to be a Borel probability measure), and $P_t, t \geq 0$ a semigroup of bounded operators acting continuously in the operator norm topology on the Hilbert space of real-valued functions $L_2(X, d\mu)$, and such that

$$(P_t f)(x) = \int_X G_t(x, y) f(y) d\mu,$$

where the kernels $G_t(x, y)d\mu$ are non-negative for all $t \geq 0$. It is also assumed that $P_t(1) = 1$, which is true for semigroups arising from Markov processes (main application of loc.cit).

Example 6.1.1. *For the Brownian motion in* $\mathbf{R}^n$ *starting from the origin one has*

$$G_t(x, y)d\mu = \frac{1}{(2\pi t)^{n/2}} e^{-|x-y|^2/2t} dy.$$

One defines the generator of the semigroup P_t as

$$Lf = lim_{t \to 0} \frac{(P_t f - f)}{t}.$$

Then on the domain of the non-negative operator L one has $\frac{\partial}{\partial t} P_t(f) = LP_t(f)$. For the Brownian motion the operator L is just the standard Laplace operator on $\mathbf{R}^n$. We will assume that L is symmetric on its domain (this corresponds to the so-called time reversible measures). This also implies that the finite measure $d\mu$ is invariant with respect to the semigroup $P_t, t \geq 0$. We will also assume that the domain of L contains a dense unital subalgebra A (typically, the algebra of

[8] I thank Dima Shlyakhtenko who pointed the paper [BiVo] out to me, where the non-commutative analogs of the Wasserstein metrics W_p were introduced in the framework of the free probability theory. Since we define the tensor product of quantum spaces by means of the tensor product of algebras, rather than their free product, it is not clear how to use that definition for the purposes of quantum Riemannian 1-geometry.

real-valued smooth functions on a manifold, or the algebra of real-valued Lipschitz functions on a metric space). Following Bakry we introduce a sequence of bilinear forms $A \otimes A \to A$:

0) $B_0(f,g) = fg$,

1) $2B_1(f,g) = LB_0(f,g) - fL(g) - L(f)g = L(fg) - fL(g) - L(f)g$,

2) $2B_n(f,g) = LB_{n-1}(f,g) - B_{n-1}(f,Lg) - B_{n-1}(Lf,g), n \geq 2$.

In the case when $L = \Delta$ is the Laplace operator in $\mathbf{R}^n$ (Brownian motion case) one has $B_2(f,f)(x) = |Hess(f)|^2(x) := \sum_{1 \leq i,j \leq n} (\partial^2 f / \partial x_i \partial x_j)^2(x)$. If L is the Laplace operator on the Riemannian manifold M then

$$B_1(f,f) = g_M(\nabla f, \nabla f),$$

$$B_2(f,f) = Ric(\nabla f, \nabla f) + |Hess(f)|^2,$$

where Ric denotes the Ricci tensor. In the case of a general Riemannian manifold the Hessian matrix $Hess(f) := \nabla^2(f)$ can be defined as a second derivative of f in the Riemannian structure, so the above equality for $B_2(f,f)$ still holds. All that can be axiomatized as follows.

Suppose that we are given a triple (A, H, L) such that

1) H is a separable real Hilbert space;

2) L is a (possibly unbounded) operator on H, which is symmetric on its domain;

3) $A \subset H$ is a unital real algebra, dense in the domain of L, such that $L(1) = 0$.

Definition 6.1.2. *a) An operator L satisfies a curvature-dimension condition $CD(R,N)$, where $R \in \mathbf{R}, N \geq 1$ if for all $f \in A$ one has*

$$B_2(f,f) \geq RB_1(f,f) + \frac{(Lf)^2}{N}.$$

b) An operator L has the Ricci curvature greater than or equal to R if it satisfies $CD(R,\infty)$, i.e.

$$B_2(f,f) \geq R\,B_1(f,f),$$

for any $f \in A$.

Remark 6.1.3. *The above considerations can be generalized to the case of complex Hilbert spaces with the real structure defined by an involution $x \mapsto x^*, x \in H$, which is compatible with the involution on $A \subset H$. In that case we define a sequence of bilinear forms $B_n(f, g^*)$. In what follows for simplicity we will discuss real Hilbert spaces.*

Suppose that a triple (A, H, L) satisfies the above conditions 1)-3).

Definition 6.1.4. *We say that (A, H, L) has Ricci curvature greater than or equal to R if the operator L satisfies $CD(R,\infty)$.*

Example 6.1.5. *a) If $d\mu = e^h dvol_M$, where $g = g_M$ is a Riemannian metric on the n-dimensional manifold M, $dvol_M$ is the corresponding volume form, and h is a smooth real function then $L = \Delta + \nabla(h)\nabla$, where Δ is the Laplace operator associated with the metric g. It was shown by Bakry and Emery (see [BaEm]) that $CD(R,N), N \geq n$ for L is equivalent to the following inequality of symmetric tensors:*

$$Ric \geq Hess(h) + Rg_M + \frac{dh \otimes dh}{(N-n)}.$$

In particular, $CD(R,n)$ implies $h = const$, and hence $Ric \geq Rg_M$ everywhere.

b) If $L = (d/dx)^2 - q(x)d/dx$ then $CD(R,N)$ is equivalent to

$$q' \geq R + \frac{q^2}{N-1}.$$

Definition 6.1.6. *A spectral triple (A, L, H) is called measured spectral triple if we are also given Gelfand-Naimark-Segal state γ on A which is invariant with respect to the semigroup $exp(-tL), t > 0$.*

6.2. **Wasserstein metric and N-curvature tensor.** Here we recall definitions and results of [LV], [St].

Let $f : \mathbf{R}_{\geq 0} \to \mathbf{R}$ be a continuous convex function, such that $f(0) = 0$. If $x^N f(x^{-N})$ is convex on $(0,\infty)$, we will say that the function f is N-convex, where $1 \leq N < \infty$. If $e^x f(e^{-x})$ is convex on $(-\infty,\infty)$ we will say that the function f is ∞-convex.

Example 6.2.1. *a) The function $f_N(x) = Nx(1 - x^{-1/N})$ is N-convex for $1 < N < \infty$.*

b) The function $f_\infty(x) = x\log x$ is ∞-convex.

Let $(X, d\mu)$ be a compact Hausdorff space equipped with a finite probability measure, and f be an arbirtary continuous convex function as above. For any probability measure $d\nu = \rho\, d\mu$, which is absolutely continuous with respect to $d\mu$ we define the f-relative entropy of $d\nu$ with respect to $d\mu$ by the formula

$$E^f_{d\mu}(d\nu) = \int_X f(\rho(x))d\mu$$

(one can slightly modify this formula in order to include measures with a non-trivial singular part in the Lebesgue decomposition).

Now let (X, d) be a compact metric space. We define the L_2-Wasserstein metric (or, simply, the Wasserstein metric, since we will not consider other Wasserstein metrics) on the space $P(X)$ of all Borel probability measures on X by the formula

$$W_2(d\nu_1, d\nu_2)^2 = inf\{\int_{X\times X} d^2(x_1,x_2)d\xi\},$$

where the infimum is taken over all probability measures $d\xi \in P(X \times X)$ such that $(\pi_i)_* d\xi = d\nu_i, i = 1, 2$, where π_i are the natural projections of $X \times X$ to the factors. Then $(P(X), W_2)$ becomes a compact metric space (in fact a length space, if X is a length space), and the corresponding metric topology coincides with the weak $*$-topology on measures.

Definition 6.2.2. *([LV], [St])*

a) We say that the compact metric-measure length space $(X, d, d\mu)$ has a non-negative N-Ricci curvature, $1 \leq N < \infty$ if for all $d\nu_0, d\nu_1 \in P(X)$ which have supports belonging to $supp(d\mu)$ there is a Wasserstein geodesic $d\nu_t, 0 \leq t \leq 1$ joining $d\nu_0$ and $d\nu_1$ such that for all N-convex functions f and all $0 \leq t \leq 1$ one has

$$E^f_{d\mu}(d\nu_t) \leq tE^f_{d\mu}(d\nu_1) + (1-t)E^f_{d\mu}(d\nu_0).$$

In other words, the f-relative entropy $E^f_{d\mu}$ is convex along a geodesic joining $d\nu_0$ and $d\nu_1$.

b) Given $R \in \mathbf{R}$ we say that $(X, d, d\mu)$ has ∞-Ricci curvature bounded below by R if for all $d\nu_0, d\nu_1$ as in part a) there is a Wasserstein geodesic $d\nu_t$ joining $d\nu_0$ and $d\nu_1$ such that for all ∞-convex functions f and all $t \in [0,1]$ one has:

$$E^f_{d\mu}(d\nu_t) \leq tE^f_{d\mu}(d\nu_1) + (1-t)E^f_{d\mu}(d\nu_0) - \frac{1}{2}\lambda(f)t(1-t)W_2(d\nu_0, d\nu_1)^2,$$

and $\lambda = \lambda_R$ is a certain map from ∞-convex functions to $\mathbf{R} \cup \{-\infty\}$ defined in [LV], Section 5. For $f = x\log x$ one can take $\lambda(f) = R$.

Now let $N \in [1, \infty]$, and $d\mu = e^h\, dvol_g$ be a probability measure on a smooth compact connected Riemannian manifold $(M, g), dim\, M = n$, associated with an arbitrary smooth real function h. One defines the Ricci N-curvature Ric_N as follows (see [LV]):

a) $Ric_N = Ric - Hess(h)$, if $N = \infty$,

b) $Ric_N = Ric - Hess(h) - \frac{dh \otimes dh}{N-n}$, if $n < N < \infty$,

c) $Ric_N = Ric - Hess(h) - \infty \cdot (dh \otimes dh)$, if $N = n$. Here by convention $0 \cdot \infty = 0$.

d) $Ric_N = -\infty$, if $N < n$.

The following theorem was proved in [LV], [St].

Theorem 6.2.3. *1) For $N \in [1, \infty)$ the measured length space $(M, g, d\mu)$ has non-negative N-Ricci curvature iff $Ric_N \geq 0$ as a symmetric tensor.*

2) It has ∞-Ricci curvature bounded below by R iff $Ric_\infty \geq Rg$.

Notice that in the above assumptions the condition $Ric_N \geq Rg$ is equivalent to Bakry's $CD(R, N)$ condition for the operator $\Delta + \nabla(h)\nabla$.

One defines the measured Gromov-Hausdorff topology on the set of metric-measure spaces in the usual way: a sequence $(X_i, d_i, d\mu_i)$ converges to $(X, d, d\mu)$ if there are ε_i-approximations $f_i : (X_i, d_i) \to (X, d)$ such that $\varepsilon_i \to 0$ as $i \to \infty$, which satisfy the condition that the direct images $(f_i)_* d\mu_i$ converge (in the weak topology) to $d\mu$.

Theorem 6.2.4. *([LV])*

a) For all $1 \leq N < \infty$ the set of length metric-measure spaces with non-negative N-Ricci curvature is precompact and complete in the measured Gromov-Hausdorff topology.

b) For $N = \infty$ the same is true for the set of length metric-measure spaces with the ∞-Ricci curvature greater than or equal to fixed R.

A similar result was proved in [St], where the precompactness theorem was established with respect to the following distance function on metric-measure spaces:

$$\mathbf{D}^2((X_1, d_1, d\mu_1), (X_2, d_2, d\mu_2)) = inf\{\int_{X_1 \times X_2} d^2(x_1, x_2)d\chi(x_1, x_2)\},$$

where the infimum is taken over all $d\chi \in P(X_1 \times X_2)$ which project onto $d\mu_1$ and $d\mu_2$ respectively under the natural projections, and all metrics d on $X_1 \sqcup X_2$ which coincide with the given metrics $d_i, i = 1, 2$ on $X_i, i = 1, 2$.

6.3. **Remark about the Laplacian.** As we have seen, the notion of Laplacian (maybe a generalized one) plays an important role in the description of the collapsing CFTs. In the case of metric-measure spaces one can use the following approach to the notion of Laplacian (see [Kok]). Let $(X, d, d\mu)$ be a metric-measure space.

For any point $x \in X$ and the open ball $B(x,r)$ with center at x we define the operator "mean value" $f \mapsto \langle f\rangle_{B(x,r)}$, where

$$\langle f\rangle_{B(x,r)} = \frac{1}{\mu(B(x,r))}\int_{B(x,r)} f(y)d\mu(y).$$

Let us assume that the measure of every ball is positive. Then one defines the Laplacian $\Delta_\mu(f)$ by the formula

$$\Delta_\mu(f)(x) := lim_{r\to 0} sup_{r>0}\frac{2}{r^2}\langle (f-f(x))\rangle_{B(x,r)} =$$

$$lim_{r\to 0} inf_{r>0}\frac{2}{r^2}\langle (f-f(x))\rangle_{B(x,r)},$$

provided the last two limits exist and coincide. It was observed in [Kok] that this *measured* Laplacian coincides with $\frac{1}{n+2}\Delta_g$ on a Riemannian n-dimensional manifold (M,g). Moreover, for a large class of metric-measure spaces it is a symmetric non-negative operator on certain classes of functions built from Lipschitz functions on $(X,d,d\mu)$. In a different framework of spaces with diffusion and codiffusion the notion of the heat operator and the Laplacian was introduced in [Gro2], Sect.3.3. The Laplacian introduced in [Gro2] is in fact a vector field, which is the gradient of the energy function. Probably in all cases when the Laplacian or the heat operator can be defined as a symmetric non-negative operator, there is an interesting quantum Riemannian 1-geometry of spaces with non-negative Ricci curvature.

6.4. **Spectral metrics.** Here we follow [BBG].

Let (M,g_M) be a compact closed Riemannian manifold, $d\mu = dvol_M$ denotes the associated Riemannian measure. Let us choose an orthonormal basis $\psi_j, j\geq 0$ of eigenvectors of the Laplacian $\Delta = \Delta_{g_M}$, which is an unbounded non-negative self-adjoint operator on the Hilbert space $H = L_2(M,d\mu)$. Let λ_j be the eigenvalue corresponding to ψ_j. Then for every $t>0$ one defines a map $\Phi_t : M \to l_2(\mathbf{Z}_+)$ by the formula

$$\Phi_t(x) = \sqrt{Vol(M)}(e^{-\lambda_j t/2}\psi_j(x))_{j\geq 0}.$$

In this way one obtains an embedding of M into $l_2(\mathbf{Z}_+)$. For any two Riemannian manifolds (M_1,g_{M_1}) and (M_2,g_{M_2}) as above, and any two choices of the orthonormal bases one can compute the Hausdorff distance d_H between the compact sets $\Phi_t(M_1,g_{M_1})$ and $\Phi_t(M_2,g_{M_2})$ inside the metric space $l_2(\mathbf{Z}_+)$. This number depends on the choices of orthonormal bases for the Laplace operators on M_1 and M_2, but one can remedy the problem by taking *sup inf* d_H over all possible pairs of choices. This gives the distance $d_t((M_1,g_{M_1}),(M_2,g_{M_2}))$ defined in [BBG]. It was proved there that the distance (for a fixed $t>0$) is equal to zero if and only if the Riemannian manifolds are isometric. A more invariant way to spell out this embedding is via the heat kernel. Namely, one considers the map $M \to L_2(M,d\mu)$ such that $x \mapsto K_M(t/2,x,\bullet)$, where $K_M(t,x,y) = \sum_{j\geq 0} e^{-\lambda_j t}\psi_j(x)\psi_j(y)$ is the heat kernel. Subsequently one can identify $L_2(M,d\mu)$ with $l_2(\mathbf{Z}_+)$ since any two separable Hilbert spaces are isometric. Thus we obtain the above embedding.

Let $\mathcal{M}(n,R,D)$ be the set of compact closed Riemannian manifolds such that $dim\, M = n, Ric(M)\geq R, diam(M)\leq D$. It was proved in [BBG] that the Φ_t-image of $\mathcal{M}(n,R,D)$ belongs to a bounded subset of the Sobolev space $h^1(\mathbf{Z}_+)$ (the latter consists of sequences $(a_0,a_1,...)\in l_2(\mathbf{Z}_+)$ such that $\sum_{j\geq 0}(1+j^{2/n})a_j^2<\infty$). By Rellich's theorem the embedding $h^1(\mathbf{Z}_+)\to l_2(\mathbf{Z}_+)$ is a compact operator. This implies that the image of the embedding of $\mathcal{M}(n, R, D)$ via Φ_t is precompact in $l_2(\mathbf{Z}_+)$. Moreover, the eigenvalues of the Laplacian are continuous with respect to the spectral distance d_t. Since only smooth manifolds were considered in [BBG] the measure $d\mu$ was always the one associated with the Riemannian metric. The approach of [BBG] was further developed and generalized in [KaKu1-2]. In the loc. cit the authors discussed the compactification of $\mathcal{M}(n, R, D)$ with respect to their version of the spectral distance, which is different from the one in [BBG].

6.5. **Spectral structures and measured Gromov-Hausdorff topology.** Here we briefly recall the approach suggested in [KS].

As we have seen above, the Laplacian and the heat kernel can be defined for more general spaces than just Riemannian manifolds (see e.g. [KMS], [S] for the case of Alexandrov spaces). The notion of spectral structure was introduced in [KS] with the purpose to study the behavior of eigenvalues of the Laplacian with respect to perturbations of the metric and topology of not necessarily compact Riemannian manifolds. A *spectral structure* is a tuple of compatible data: $(L, Q, E, U_t, R(z), H)$ which consists of a separable Hilbert space H (complex or real), a self-adjoint non-negative linear operator $L : H \to H$, a densely defined quadratic form Q generated by $\sqrt{L}$, a spectral measure $E = E_\lambda(L)$, a pointwise continuous contraction semigroup U_t with the infinitesimal generator L, pointwise continuous resolvent $(z - L)^{-1}$. One can associate a spectral structure with a a pointed locally compact metric space equipped with a Radon measure (or with a not necessarily pointed compact metric-measure space). Two types of topologies on the set of spectral triples were introduced in [KS]: strong topology and compact topology. For both topologies the natural forgetful map from "geometric" spectral structures to the corresponding metric-measure spaces is continuous. The main results of [KS] concern convergence of spectral structures under the condition that the underlying metric-measure spaces converge. In particular, one has such a convergence for the class of Riemannian complete (possibly non-compact) pointed n-dimensional manifolds with the Ricci curvature bounded from below ([KS], Theorem 1.3). The results of [KS] can be considered as a generalization of the results of [Fu], [ChC1] about continuity of eigenvalues of the Laplacian with respect to measured Gromov-Hausdorff topology. Every commutative Riemannian 1-space associated with a complete pointed Riemannian manifold (or compact closed non-pointed Riemannian manifold) gives rise to a spectral structure. Therefore, applying results of [KS] one can deduce precompactness of the moduli space of such Riemannian 1-spaces. Unfortunately, [KS] does not contain any precompactness results about the moduli space of "abstract" spectral structures, i.e. those which are not associated with meatric-measure spaces. The same is true for [BBG].

6.6. **Non-negative Ricci curvature for quantum 1-geometry.** Suppose that we are given a singular quantum Riemannian 1-space which satisfies the following property: there is a dense pre-Hilbert subspace $A \subset H$, such that for any graph $\Gamma \in G$ the tensor product $A^{\otimes I_-}$ belongs to the domain of the operator $S(\Gamma)$. Then, taking a Y-shape graph, as in the Introduction, we recover from the axioms QFT 1)-QFT 4) an associative product on A. Following Kontsevich we impose the following three conditions:

QFT 6) (spectral gap, or boundness of the diameter) Let L be a generator of the semigroup $S(\Gamma_\varepsilon^0)$ (see QFT 5)). Then the spectrum of L belongs to the set $\{0\} \cup \{a\} \cup [b, +\infty)$, where $b > a > 0$ are some numbers.

QFT 7) (7-term relation) For a graph $\Gamma(l_1, l_2, l_3, l_4)$ with one internal vertex and four attached edges of length $l_i, 1 \leq i \leq 4$ one has the following identity:

$$\sum_{1\leq i\leq 4} \frac{\partial}{\partial l_i} S_{\Gamma(l_1,l_2,l_3,l_4)} = \sum_{1\leq j\leq 3} \frac{\partial}{\partial \varepsilon_j}|_{\varepsilon_j=0} S_{\Gamma(l_1,l_2,l_3,l_4,\varepsilon_j)},$$

where $\Gamma(l_1, l_2, l_3, l_4, \varepsilon_j)$ is a graph obtained from $\Gamma(l_1, l_2, l_3, l_4)$ by inserting one internal edge of the length ε_j (the three summands on the RHS correspond to the three different ways of pairing of four external vertices of the graph $\Gamma(l_1, l_2, l_3, l_4)$, see the Figure below).

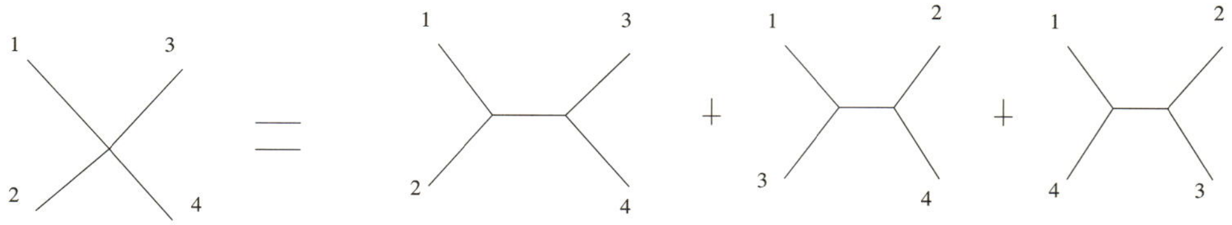

QFT 8) (non-negativeness of Ricci curvature) If $\Gamma(l_1, l_1, l_3, l_3, \varepsilon)$ is the graph $\Gamma(l_1, l_2, l_3, l_4, l)$ from QFT 7) with $l_2 = l_1, l_4 = l_3, \varepsilon = l$ then the following identity holds:

$$(\partial/\partial l_1 - \partial/\partial l)^2 S_{\Gamma(l_1,l_1,l_3,l_3,l)} \geq 0.$$

This condition is equivalent to the following one:

$$(\frac{d}{dt})^2|_{t=0}(e^{tL}(e^{-tL}(f) \cdot e^{-tL}(f))) \geq 0,$$

for all $f \in A$. The LHS of this inequality is equal to $B_2(f, f)$. Another way to state the above inequality is to say that the amplitude $S_{\Gamma(l_1+t,l_1+t,l_3,l_3,l-t)}$ is convex at $t = 0$. The continuity of the amplitudes $S_{\Gamma,l}$ (see QFT 4)) implies that the condition $B_2(f, f) \geq 0$ is preserved if we contract a subforset of Γ (i.e. we allow to contract a disjoint union of trees, while the contraction of loops is prohibited).

Remark 6.6.1. *a) The condition QFT 6) is motivated by the property that for any $t > 0$ we have $Tr(e^{-tL}) < \infty$. This implies that there is an interval $(0, \lambda), \lambda > 0$ which contains finitely many points of the spectrum of L. If $L = \Delta$, the Laplace operator on a compact Riemannian manifold M, then the spectral gap λ is of the magnitude $1/(diam M)^2$.*

b) The condition QFT 7) is motivated by the case $L = \Delta$. In this case

$$\Delta(f_1f_2f_3) - \Delta(f_1f_2)f_3 + ... + \Delta(f_1)f_2f_3 - \Delta(1)f_1f_2f_3 = 0$$

(7-term relation for the second order differential operator Δ).

c) The condition QFT 8) is motivated by the $CD(0, \infty)$ inequalities of Bakry.

It was shown by Kontsevich that Segal's axioms of unitary CFT imply that a collapsing sequence of unitary CFTs gives rise to a commutative Riemannian 1-space, which satisfies axioms QFT1)-QFT 8). The underlying Riemannian manifold M is the smooth part of $X = Spec(H^{small})$ (see Section 2.2) with L derived from *the limit of the rescaled Virasoro operators* $(L_0 + \overline{L}_0)/\varepsilon$, *where* ε *is the minimal eigenvalue of* $(L_0 + \overline{L}_0)$, $\varepsilon \to 0$.

Recall that for a non-negative self-adjoint operator L acting in a Hilbert space H a *spectral gap* $\lambda_1(L)$ is the smallest positive eigenvalue of L. Let us consider singular quantum Riemannian 1-spaces with measure, for which the Hilbert space H is obtained by the Gelfand-Naimark-Segal construction from a unital involutive algebra A and a state $\tau : A \to \mathbf{C}$. To every such quantum space one can associate a spectral triple (A, H, L).

Conjecture 6.6.2. *The space of isomorphism classes of singular quantum Riemannian* 1*-spaces with measure, which have non-negative Ricci curvature, spectral gap bounded below by a given number* C, *and such that the corresponding spectral triples have dimensional spectrum belonging to a given interval* $[a, b]$ *(see [CoMar] for the definition of the dimensional spectrum) is precompact in the topology defined in Section 4.2.*

Finally, we are going to discuss an example in which a version of the above conjecture was verified. Let $M_j, j \geq 1$ be a sequence of compact Riemannian manifolds of the same dimension n with diameter equal to 1 (recall that rescaling of the metric does not change the Ricci curvature). Then, following Example 5.1.1 we can associate with this sequence a sequence of non-commutative Riemannian 1-spaces $V(M_j), j \geq 1$. Suppose that N is a measured Gromov-Hausdorff limit of M_j as $j \to \infty$. Then it follows from [ChC3] that N carries a generalized Laplacian, and a measure, hence the same formulas as in Example 5.1.1 allows us to associate with N a quantum Riemannian 1-space $V(N)$. Let us say that $V(M_j)$ *weakly converges* to $V(N)$ if for any sequence of Lipschitz functions $f_j : M_j \to \mathbf{R}, f : N \to \mathbf{R}$ such that $|f_j \circ \psi_j - f|_{L_\infty(N)}$ as $j \to \infty$, and for any metrized graph Γ, we have: $S_\Gamma(f_j) \to S_\Gamma(f)$ as $j \to \infty$. Here $\psi_j : N \to M_j$ is any sequence of ε_j-approximations such that $\varepsilon_j \to 0$ as $j \to \infty$. Weak convergence gives rise to the topology on the space of equivalence classes of quantum Riemannian 1-spaces of *geometric origin* (i.e. those which correspond to metric-measured spaces). Then the following result holds (see [En] for the proof).

Theorem 6.6.3. *The subspace of the space of the above quantum Riemannian* 1*-spaces corresponding to manifolds with non-negative Ricci curvature is precompact in the weak topology.*

Having in mind results of [LV] and [St] we expect that it is in fact compact.

References

[AGS] L. Ambrosio, N. Gigli, G. Savare, Gradient Flows: In Metric Spaces and in the Space of Probability Measures, Birkhauser, 2005.

[BBG] P. Berard, G. Besson, S. Gallot, Embedding Riemannian manifolds by their heat kernel, Geom. Funct. Anal., 4:4, 1994, 373-398.

[Ba] D. Bakry, Functional inequalities for Markov semigroups, preprint, available at: http://www.lsp.ups-tlse.fr/Bakry/

[BaEm] D. Bakry, M. Emery, Diffusions hypercontractives, Lect. Notes in Math. no. 1123, 1985, 177-206.

[BiVo] P. Biane, D. Voiculescu, A Free Probability Analogue of the Wasserstein Metric on the Trace-State Space, math.OA/0006044.

[ChC1] Cheeger, T.H. Colding, On the structure of spaces with Ricci curvature bounded below I, J. Diff. Geom., 46, 1997, 37-74.

[ChC2] Cheeger, T.H. Colding, On the structure of spaces with Ricci curvature bounded below I, J. Diff. Geom., 54:1, 2000, 13-35.

[ChC3]J. Cheeger, T.H. Colding, On the structure of spaces with Ricci curvature bounded below III, J. Diff. Geom., 54:1, 2000, 37-74.

[Co1] A. Connes, Non-commutative geometry, Academic Press, 1994.

[CoKr] A. Connes, D. Kreimer, Renormalization in quantum field theory and the Riemann-Hilbert problem, hep-th/9909126.

[CoMar] A Connes, M. Marcolli, A walk in the non-commutative garden, math.QA/0601054.

[Dou 1] M. Douglas, The statistics of string/M theory vacua, hep-th/0303194.

[Dou 2] M. Douglas, Talk at the String-2005 Conference, http://www.fields.utoronto.ca/audio/05-06/strings/douglas.

[Dou3] M. Douglas, Spaces of Quantum Field Theories, arXiv:1005.2779.

[Dou L] M. Douglas, Z. Lu, Finiteness of volume of moduli spaces, hep-th/0509224.

[En] A. Engoulatov, Heat kernel and applications to the convergence of Graph Field Theories, preprint, 2006.

[FFRS] J. Fjelstad, J. Fuchs, I. Runkel, C. Schweigert, Topological and conformal field theory as Frobenius algebras, math.CT/0512076.

[FG] J. Frölich, K. Gawedzki, Conformal Field Theory and geometry of strings, hep-th/9310187.

[FM] W. Fulton, R. Macpherson, A compactification of configuration spaces, Annals Math., 139(1994), 183-225.

[Fu1] K. Fukaya, Collapsing of Riemannian manifolds and eigenvalues of Laplace operator, Invent. Math., 87, 1987, 517-547.

[Gaw] K. Gawedzki, Lectures on Conformal Field Theory, in: Quantum Fields and Strings: a course for mathematicians, AMS,1999, vol. 2, 727-805.

[Gi1] V. Ginzburg, Lectures on Noncommutative Geometry, math.AG/0506603.

[GiKa] V. Ginzburg, M. Kapranov, Koszul duality for operads, Duke Math. J. 76 (1994), 203-272.

[Gro1] M. Gromov, Metric structures for Riemannian and non-Riemannian spaces, Birkhäuser, 1999.

[Gro2] M. Gromov, Random walks in random groups, Geom. Funct. Anal. 13:1, 2003, 73-146.

[Kaw 1] Y. Kawahigashi, Classification of operator algebraic conformal field theories in dimensions one and two, math-ph/0308029.

[Kaw 2] Y. Kawahigashi, Classification of operator algebraic conformal field theories, math.OA/0211141.

[Kaw-Lo] Y. Kawahigashi, R. Longo, Noncommutative Spectral Invariants and Black Hole Entropy, math-ph/0405037.

[KS] K. Kuwae, T. Shioya, Convergence of spectral structures: a functional analytic theory and its applications to spectral geometry, Comm. Anal. Geom., 11:4, 2003, 599-673.

[KMS] K. Kuwae, Y. Machigashira, T. Shioya, Sobolev spaces, Laplacian and heat kernel on Alexandrov spaces, 1998.

[KaKu1] A. Kasue, H.Kumura, Spectral convergence of Riemannian manifolds, Tohoku Math. J., 46, 1994, 147-179.

[KaKu2] A. Kasue, H.Kumura, Spectral convergence of Riemannian manifolds, II Tohoku Math. J., 48.1996, 71-120.

[Kok] S. Kokkendorff, A Laplacian on metric measure spaces. Preprint of Technical University of Denmark, March 2006.

[KoSo1] M. Kontsevich, Y. Soibelman, Homological Mirror Symmetry and torus fibrations, math.SG/0011041.

[KoSo2] M. Kontsevich, Y. Soibelman, Deformations of algebras over operads and Deligne conjecture, math.QA/0001151, published in Lett. Math. Phys. (2000).

[KoSo3] M. Kontsevich, Y. Soibelman, Deformation theory, (book in preparation).

[Li] H. Li, C*-algebraic quantum Gromov-Hausdorff distance, math.OA/0312003.

[L] J. Lott, Optimal transport and Ricci curvature for metric-measure spaces, math.DG/06101542.

[LV] J. Lott, C. Villani, Ricci curvature for metric-measure spaces via optimal transport, math.DG/0412127.

[Led] M. Ledoux, The geometry of Markov diffusion generators, preprint, available at: http://www.lsp.ups-tlse.fr/Ledoux/

[OVa] H. Ooguri, C. Vafa, On the geometry of the string landscape and the swampland, hep-th/0605264.

[Rie] M. Rieffel, Gromov-Hausdorff Distance for Quantum Metric Spaces, math.OA/0011063.

[RW] D. Roggenkamp, K. Wendland, Limits and degenerations of unitary Conformal Field Theories, hep-th/0308143.

[Ru] I. Runkel, Algebra in braided tensor categories and conformal field theory, preprint.

[S] T. Shioya, Convergence of Alexandrov spaces and spectrum of Laplacian, 1998.

[Seg] G.Segal, The definition of Conformal Field Theory, in: Topology, Geometry and Quantum Field Theory, Cambridge Univ. Press, 2004, 421-577.

[Si] L. Silberman, Addendum to "Random walks on random groups" by M. Gromov, Geom. Funct. Anal., 13:1, 2003.

[St] K-T. Sturm, On the geometry of metric measure spaces, preprint 203, Bonn University, 2004.

[ST] S. Stolz, P. Teichner, Supersymmetric field theories and integral modular functions, in preparation.

[T] D. Tamarkin, Formality of chain operad of small squares, math.QA/9809164.

[U] H. Urakawa, Convergence rates to equilibrium of the heat kernels on compact Riemannian manifolds, preprint.

[V] C. Villani, Optimal transport, old and new, book in preparation.

[Va] C. Vafa, The string landscape and the swampland, hep-th/0509212.

[W] Wei Wu, Quantized Gromov-Hausdorff distance, math.OA/0503344.

[Z] S. Zelditch, Counting string/M vacua, math-ph/0603066.

address: Yan Soibelman, Department of Mathematics, Kansas State University, Manhattan, KS 66506, USA
email: soibel@math.ksu.edu

Proceedings of Symposia in Pure Mathematics
Volume **83**, 2011

Supersymmetric field theories and generalized cohomology

Stephan Stolz and Peter Teichner

CONTENTS

1. Results and conjectures

This paper is a survey of our mathematical notions of Euclidean field theories as models for (the cocycles in) a cohomology theory. This subject was pioneered by Graeme Segal [**Se1**] who suggested more than two decades ago that a cohomology theory known as *elliptic cohomology* can be described in terms of 2-dimensional (conformal) field theories. Generally what we are looking for are isomorphisms of the form

$$(1.1) \qquad \{\text{supersymmetric field theories of degree } n \text{ over } X\}/_{\text{concordance}} \cong h^n(X)$$

where a field theory over a manifold X can be thought of as a family of field theories parametrized by X, and the abelian groups $h^n(X), n \in \mathbb{Z}$, form some (generalized) cohomology theory. Such an isomorphism would give geometric cocycles for these cohomology groups in terms of objects from physics, and it would allow us to use the computational power of algebraic topology to determine families of field theories up to concordance.

To motivate our interest in isomorphisms of type (1.1), we recall the well-known isomorphism

$$(1.2) \qquad \{\text{Fredholm bundles over } X\}/_{\text{concordance}} \cong K^0(X),$$

Supported by grants from the National Science Foundation, the Max Planck Society and the Deutsche Forschungsgemeinschaft.
[2010] Primary 55N20 Secondary 11F23 18D10 55N34 57R56 81T60.

where $K^0(X)$ is complex K-theory. We showed in [**HoST**] that the space of Euclidean field theories of dimension $1|1$ has the homotopy type of the space of Fredholm operators, making the connection to (1.1).

The isomorphism (1.2) is one of the pillars of index theory in the following sense. Let $\pi\colon M \to X$ be a fiber bundle whose fibers are $2k$-manifolds. Let us assume that the tangent bundle along the fibers admits a spin^c-structure. This assumption guarantees that we can construct the Dirac operator on each fiber and that these fit together to give a bundle of Fredholm operators over the base space X. Up to concordance, this family is determined by the element in $K^0(X)$ it corresponds to via the isomorphism (1.2). The *Family Index Theorem* describes this element as the image of the unit $1 \in K^0(M)$ under the (topological) *push-forward map*

$$\pi_*\colon K^0(M) \longrightarrow K^{-2k}(X) \cong K^0(X). \tag{1.3}$$

The construction of the map π_* does not involve any analysis – it is described in homotopy theoretic terms.

There is also a physics interpretation of the push-forward map (1.3).The Dirac operator on a Riemannian spin^c manifold N determines a Euclidean field theory of dimension $1|1$ (physicists would call it "supersymmetric quantum mechanics" on N), which should be thought of as the quantization of the classical system consisting of a superparticle moving in N. We can think of the bundle of Dirac operators associated to a fiber bundle of spin^c manifolds $N \to M \to X$ as a $1|1$-dimensional field theory over X. This allows the construction of a (physical) push-forward map π_*^q as the fiberwise *quantization* of π. The Feynman-Kac formula implies that π_*^q equals the analytic push-forward and hence the equality $\pi_* = \pi_*^q$ is equivalent to the family index theorem.

Our proposed model for 'elliptic cohomology' will be given by $2|1$-dimensional Euclidean field theories. More precisely, we conjecture that these describe the (periodic version of the) universal theory of topological modular forms TMF* introduced by Hopkins and Miller [**Ho**]. There is a topological push-forward map $\pi_* : \mathrm{TMF}^0(M) \to \mathrm{TMF}^{-k}(X)$ if $\pi : M \to X$ is a fiber bundle of k-dimensional string manifolds. The above discussion then has a conjectural analogue which would lead to a family index theorem on loop spaces. The fiberwise quantization π^q requires the existence of certain integrals over mapping spaces of surfaces to M.

In the next section, we provide a very rough definition of our notion of field theory and describe our main results and conjectures. A detailed description of the rest of the paper can be found at the end of that section. This paper only contains ideas and outlines of proofs, details will appear elsewhere.

Acknowledgements: It is a pleasure to thank Dan Freed, Dmitri Pavlov, Ingo Runkel, Chris Schommer-Pries, Urs Schreiber and Ed Witten for valuable discussions. We also thank the National Science Foundation, the Deutsche Forschungsgemeinschaft and the Max Planck Institute for Mathematics in Bonn for their generous support. Most of the material of this paper was presented at the Max Planck Institute as a IMPRS-GRK summer school in 2009.

1.1. Field theories. More than two decades ago, Atiyah, Kontsevich and Segal proposed a definition of a field theory as a functor from a suitable bordism category to the category of topological vector spaces. Our notion of field theory is a refinement of theirs for which the definition is necessarily quite intricate because we have to add the precise notion of "supersymmetry" and "degree" (another refinement, namely "locality" still needs to be implemented). While about half of this paper is devoted to explaining the definition, this is not a complete account. Fortunately, for the description of our results and conjectures, only a cartoon picture of what we mean by a field theory is needed.

Roughly speaking, a d-dimensional (topological) field theory E assigns to any closed smooth $(d-1)$-manifold Y a topological vector space $E(Y)$, and to a d-dimensional bordism Σ from Y_0 to Y_1 a continuous linear map $E(\Sigma)\colon E(Y_0) \to E(Y_1)$. There are four requirements:

(1) If Σ, Σ' are bordisms from Y_0 to Y_1 which are diffeomorphic relative boundary, then $E(\Sigma) = E(\Sigma')$.

(2) E is compatible with composition; i.e., if Σ_1 is a bordism from Y_0 to Y_1 and Σ_2 is a bordism from Y_1 to Y_2, and $\Sigma_2 \cup_{Y_1} \Sigma_1$ is the bordism from Y_0 to Y_2 obtained by gluing Σ_1 and Σ_2 along Y_1, then

$$E(\Sigma_2 \cup_{Y_1} \Sigma_1) = E(\Sigma_2) \circ E(\Sigma_1).$$

(3) E sends disjoint unions to tensor products, i.e., $E(Y \amalg Y') = E(Y) \otimes E(Y')$ for the disjoint union $Y \amalg Y'$ of closed $(d-1)$-manifolds Y, Y', and

$$E(\Sigma \amalg \Sigma') = E(\Sigma) \otimes E(\Sigma')\colon E(Y_0) \otimes E(Y_0') \to E(Y_1) \otimes E(Y_1')$$

if Σ is a bordism from Y_0 to Y_1 and Σ' is a bordism from Y_0' to Y_1'.

(4) The vector space $E(Y)$ should depend smoothly on Y, and the linear map $E(\Sigma)$ should depend smoothly on Σ.

The first two requirements can be rephrased by saying that E is a functor from a suitable bordism category d-Bord to the category TV of topological vector spaces. The objects of d-Bord are closed $(d-1)$-manifolds; morphisms are bordisms up to diffeomorphism relative boundary. The third condition amounts to saying that the functor $E\colon d\text{-}\mathsf{Bord} \to \mathsf{TV}$ is a *symmetric monoidal functor* (with monoidal structure given by disjoint union on d-Bord, and (projective) tensor product on TV). Making the last requirement precise is more involved; roughly speaking it means that we need to replace the domain and range categories by their *family versions* whose objects are *families* of closed $(d-1)$-manifolds (respectively topological vector spaces) parametrized by smooth manifolds (and similarly for morphisms). In technical terms, d-Bord and TV are refined to become fibered categories over the Grothendieck site of smooth manifolds.

REMARK 1.4. The empty set is the monoidal unit with respect to disjoint union, and hence requirement (3) implies that $E(\emptyset)$ is a 1-dimensional vector space (here we think of $\emptyset$ as a closed manifold of dimension $d-1$). If Σ is a closed d-manifold, we can consider Σ as a bordism from $\emptyset$ to itself, and hence

$$E(\Sigma) \in \mathrm{Hom}(E(\emptyset), E(\emptyset)) = \mathbb{C}.$$

More generally, if Σ is a family of closed d-manifolds parametrized by some manifold S (i.e., Σ is a fiber bundle over S), then the requirement (4) implies that $E(\Sigma)$ is a smooth function on S.

There are many possible variations of the above definition of field theory by equipping the closed manifolds Y and the bordisms Σ with additional structure. For example if the additional structure is a conformal structure, E is called a *conformal* field theory; if the additional structure is a Riemannian metric, we will refer to E

as a *Riemannian* field theory. If no additional structure is involved, it is customary to call E a *topological* field theory, although it would seem better to call it a smooth field theory (and reserve the term 'topological' for functors on the bordism category of topological manifolds).

Our main interest in this paper is in *Euclidean field theories*, where the additional structure is a Euclidean structure, i.e., a flat Riemannian metric. In physics, the word 'Euclidean' is typically used to indicate a Riemannian metric as opposed to a Lorentzian metric without our flatness assumption on that metric, and so our terminology might be misleading (we will stick with it since 'Euclidean structure' is a mathematical notion commonly used in rigid geometry, and the alternative terminology 'flat Riemannian field theory' has little appeal).

An important invariant of a field theory E is its *partition function* Z_E, obtained by evaluating E on all closed d-manifolds Σ with the appropriate geometric structure, and thinking of $\Sigma \mapsto E(\Sigma)$ as a function on the moduli stack of closed d-manifolds equipped with this structure, see Definition 4.13. We will only be interested in the partition function of conformal or Euclidean 2-dimensional field theories restricted to surfaces of genus 1. Hence the following low-brow definition:

DEFINITION 1.5. Let E be a conformal or Euclidean field theory of dimension 2. Then the *partition function of* E is the function

$$Z_E\colon \mathfrak{h} \longrightarrow \mathbb{C} \qquad \tau \mapsto E(T_\tau)$$

where $\mathfrak{h}$ is the upper half-plane $\{\tau \in \mathbb{C} \mid \mathrm{im}(\tau) > 0\}$, and T_τ is the torus $T_\tau = \mathbb{C}/\mathbb{Z}\tau + \mathbb{Z}$ with the flat metric induced by the standard metric on the complex plane.

We note that for $A = \left(\begin{smallmatrix} a & b \\ c & d \end{smallmatrix}\right) \in SL_2(\mathbb{Z})$ the torus T_τ is conformally equivalent (but generally not isometric) to $T_{\tau'}$ for $\tau' = \frac{a\tau+b}{c\tau+d}$. In particular, the partition function of a 2-dimensional conformal field theory is invariant under the $SL_2(\mathbb{Z})$-action (but generally not of a Euclidean field theory).

1.2. Field theories over a manifold. Another possible additional structure on bordisms Σ is to equip them with smooth maps to a fixed manifold X. The resulting field theories are called *field theories over* X. We think of a field theory over X as a family of field theories parametrized by X: If E is a field theory over X, and x is a point of X, we obtain a field theory E_x by defining $E_x(Y) := E(Y \xrightarrow{c_x} X)$, where c_x is the constant map that sends every point of Y to x, and similarly for bordisms Σ.

REMARK 1.6. Let Σ be a closed d-manifold, $\mathrm{map}(\Sigma, X)$ the space of smooth maps and

$$\mathrm{ev}\colon \mathrm{map}(\Sigma, X) \times \Sigma \to X$$

the evaluation map. We think of the trivial fiber bundle $\mathrm{map}(\Sigma, X) \times \Sigma \to \mathrm{map}(\Sigma, X)$ and the evaluation map as a smooth family of d-manifolds with maps to X, parametrized by the mapping space. In particular, if E is a d-dimensional field theory over X, we can evaluate E on $(\mathrm{map}(\Sigma, X) \times \Sigma, \mathrm{ev})$ to obtain a smooth function $Z_{E,\Sigma} \in C^\infty(\mathrm{map}(\Sigma, X))$ (this mapping space is not a finite dimensional manifold, but that is not a problem, since one can work with presheaves of manifolds). We note that $Z_{E,\Sigma}$ can be interpreted as part of the partition function of

E. It follows from requirement (1) that

$$Z_{E,\Sigma} \quad \text{belongs to} \quad C^\infty(\text{map}(\Sigma, X))^{\text{Aut}(\Sigma)}, \tag{1.7}$$

the functions invariant under the action of the automorphisms of Σ. Here $\text{Aut}(\Sigma)$ is the diffeomorphism group of Σ if E is a topological field theory, and the group of structure preserving diffeomorphisms if E is a field theory corresponding to some additional geometric structure.

We observe that d-dimensional topological field theories over X are extremely familiar objects for $d = 0, 1$: If E is a 0-dimensional field theory over X, we obtain the function $Z_{E,\text{pt}} \in C^\infty(\text{map}(\text{pt}, X)) = C^\infty(X)$ associated to the 0-manifold pt consisting of a single point. This construction gives an isomorphism between 0-dimensional topological field theories over X and $C^\infty(X)$. A slightly stronger statement holds: the groupoid of 0-TFT's over X is equivalent to the discrete groupoid with object set $C^\infty(X)$.

To a vector bundle with a connection $V \to X$ we can associate a 1-dimensional field theory E_V over X as follows. We think of a point $x \in X$ as a map $x\colon \text{pt} \to X$ and decree that E_V should associate to x the fiber V_x; any object in the bordism category is isomorphic to a disjoint union of these, and hence the functor E_V is determined on objects. If $\gamma\colon [a, b] \to X$ is a path in X from $x = \gamma(a)$ to $y = \gamma(b)$, we think of γ as a morphism from x to y in the bordism category and define $E_V(\gamma)\colon E_V(x) = V_x \to E_V(y) = V_y$ to be the parallel translation along the path γ. The proof of the following result [**DST**] is surprisingly subtle. Note that smooth paths give morphisms in our bordism category, but piecewise smooth paths do not. Therefore, one even needs to decide how to compose morphisms.

THEOREM 1.8 ([**DST**]). *The groupoid* 1-TFT(X) *of* 1*-dimensional topological field theories over a manifold* X *is equivalent to groupoid of finite-dimensional vector bundles over* X *with connections.*

1.3. Supersymmetric field theories. Another variant of field theories are *supersymmetric field theories of dimension* $d|\delta$, where δ is a non-negative integer. These are defined as above, but replacing d-dimensional manifolds (respectively $(d-1)$-manifolds) by supermanifolds of dimension $d|\delta$ (respectively $(d-1)|\delta$). The previous discussion is included since a supermanifold of dimension $d|0$ is just a manifold of dimension d. In order to formulate the right smoothness condition in the supersymmetric case, we need to work with families whose parameter spaces are allowed to be supermanifolds rather than just manifolds.

If E is a $0|1$-dimensional TFT over a manifold X we can consider the function $Z_{E,\Sigma}$ of Remark 1.6 for the $0|1$-supermanifold $\Sigma = \mathbb{R}^{0|1}$. It turns out that the algebra of smooth functions on the supermanifold $\text{map}(\mathbb{R}^{0|1}, X)$ can be identified with $\Omega^*(X)$, the differential forms on X [**HKST**], and that the subspace invariant under the $\text{Diff}(\mathbb{R}^{0|1})$-action is the space of closed 0-forms. This leads to the following result.

THEOREM 1.9 ([**HKST**]). *The groupoid* $0|1$-TFT(X) *of* $0|1$*-dimensional topological field theories over a manifold* X *is equivalent to the discrete groupoid with objects* $\Omega^0_{cl}(X)$.

The notion of Euclidean structures on manifolds can be generalized to supermanifolds, see 4.2. In particular, we can talk about (supersymmetric) Euclidean field theories of dimension $d|\delta$ ($\delta > 0$ means that we are talking about a supersymmetric theory, and hence the adjective 'supersymmetric' is redundant). If E is a $0|1$-EFT over X (Euclidean field theory of dimension $0|1$ over X), then as above we can consider the function $Z_{E,\mathbb{R}^{0|1}} \in C^\infty(\mathrm{map}(\mathbb{R}^{0|1}, X))$. Now this function is invariant only under the subgroup $\mathrm{Iso}(\mathbb{R}^{0|1}) \subset \mathrm{Diff}(\mathbb{R}^{0|1})$ consisting of the diffeomorphisms preserving the Euclidean structure on $\mathbb{R}^{0|1}$. We show in [**HKST**] that the invariant subspace

$$C^\infty(\mathrm{map}(\mathbb{R}^{0|1}, X)^{\mathrm{Iso}(\mathbb{R}^{0|1})} = \Omega^*(X)^{\mathrm{Iso}(\mathbb{R}^{0|1})}$$

is the space $\Omega^{ev}_{cl}(X)$ of closed, even-dimensional forms on X. This leads to

THEOREM 1.10 ([**HKST**]). *The groupoid* $0|1$-EFT(X) *of* $0|1$*-dimensional Euclidean field theories over a manifold* X *is equivalent to the discrete groupoid with objects* $\Omega^{ev}_{cl}(X)$.

We are intrigued by the fact that field theories over X are quite versatile objects; depending on the dimension and the type of field theory we get such diverse objects as smooth functions (0-TFT's), vector bundles with connections (1-TFT's) or closed even-dimensional differential forms ($0|1$-EFT's). Higher dimensional field theories over X typically don't have interpretation as classical objects. For example, Dumitrescu has shown that a $\mathbb{Z}/2$-graded vector bundle $V \to X$ equipped with a Quillen superconnection $\mathbb{A}$ leads to a $1|1$-EFT $E_{V,\mathbb{A}}$ over X [**Du**]. However, not every $1|1$-EFT over X is isomorphic to one of these.

There are *dimensional reduction* constructions that relate field theories of different dimensions. For example, there is a *dimensional reduction functor*

$$\mathrm{red}\colon 1|1\text{-EFT}(X) \longrightarrow 0|1\text{-EFT}(X) \tag{1.11}$$

In his thesis [**Ha**], Fei Han has shown that if $V \to X$ is a vector bundle with a connection ∇, then the image of the Dumitrescu field theory $E_{V,\nabla} \in 1|1$-EFT(X) under this functor, interpreted as a closed differential form via Theorem 1.10, is the Chern character form of (V, ∇).

1.4. Concordance classes of field theories. Next we want to look at field theories over X from a topological perspective. A smooth map $f\colon X \to Y$ induces a functor $d|\delta$-EBord$(X) \to d|\delta$-EBord(Y) and by pre-composition this gives a functor

$$f^*\colon d|\delta\text{-EFT}(Y) \longrightarrow d|\delta\text{-EFT}(X).$$

In other words, field theories over manifolds are *contravariant* objects, like smooth functions, differential forms or vector bundles with connections.

DEFINITION 1.12. Let $E_\pm$ be two field theories over X of the same type and dimension. Then E_+ is *concordant* to E_- if there is a field theory E over $X \times \mathbb{R}$ such that the two bundles $i_\pm^* E$ are isomorphic to $\pi_\pm^* E_\pm$. Here $i_\pm\colon X \times (\pm 1, \pm\infty) \hookrightarrow X \times \mathbb{R}$ are the two inclusion maps and $\pi_\pm\colon X \times (\pm 1, \pm\infty) \to X$ are the projections. We will write $d|\delta$-EFT$[X]$ for the set of *concordance classes* of $d|\delta$-dimensional EFT's over X.

We remark that passing to concordance classes forgets 'geometric information' while retaining 'homotopical information'. More precisely, the functors from manifolds to sets, $X \mapsto d|\delta$-EFT$[X]$, are homotopy functors. For example, two vector bundles with connections are concordant if and only if the vector bundles are isomorphic; two closed differential forms are concordant if and only if they represent the same de Rham cohomology class. In particular, Theorems 1.9 and 1.10 have the following consequence.

COROLLARY 1.13 ([**HKST**]). *$0|1$-TFT$[X]$ and $H^0_{dR}(X)$ are naturally isomorphic as rings. Similarly, there is an isomorphism between $0|1$-EFT$[X]$ and $H^{ev}_{dR}(X)$, the even-dimensional de Rham cohomology of X.*

1.5. Field theories of non-zero degree. Corollary 1.13 suggest the question whether there is a field theoretic description of $H^n_{dR}(X)$. The next theorem gives a positive answer to this question using the notion of 'degree n field theories' which will be discussed in Section 5.

THEOREM 1.14 ([**HKST**]). *Let X be a smooth manifold. Then there are equivalences of groupoids computing $0|1$-field theories of degree n as follows:*

$$0|1\text{-TFT}^n(X) \cong \Omega^n_{cl}(X) \qquad 0|1\text{-EFT}^n(X) \cong \begin{cases} \Omega^{ev}_{cl}(X) & n \text{ even} \\ \Omega^{odd}_{cl}(X) & n \text{ odd} \end{cases}$$

Moreover, there are isomorphisms of abelian groups

$$0|1\text{-TFT}^n[X] \cong H^n_{dR}(X)$$

$$0|1\text{-EFT}^n[X] \cong \begin{cases} H^{ev}_{dR}(X) & n \text{ even} \\ H^{odd}_{dR}(X) & n \text{ odd} \end{cases}$$

These isomorphisms are compatible with the multiplicative structure on both sides (given by the tensor product of field theories on the left, and the cup product of cohomology classes on the right).

We note that Fei Han's result mentioned at the end of section 1.3 implies the commutativity of the diagram, interpreting the Chern-character as dimensional reduction:

$$\begin{array}{ccc} K^0(X) & \xrightarrow{ch} & H^{ev}_{dR}(X) \\ \downarrow & & \downarrow \cong \\ 1|1\text{-EFT}[X] & \xrightarrow{\text{red}} & 0|1\text{-EFT}[X] \end{array}$$

Here ch is the Chern character, red is the dimensional reduction functor (1.11), and the left arrow is induced by mapping a vector bundle with connection to its Dumitrescu field theory. We believe that this arrow is an isomorphism because a very much related description of $K^0(X)$ via $1|1$-EFT's was given in [**HoST**].

1.6. $2|1$-EFT's and topological modular forms. The definition of a partition function for a 2-dimensional Euclidean field theory (see Definition 1.5) can be extended to Euclidean field theories of dimension $2|1$ (see Definition 4.13). For $E \in 2|1$-EFT, the restriction Z^{++}_E of its partition function to the non-bounding spin structure $++$ on the torus can be considered as a complex valued function on the upper half plane $\mathfrak{h}$, see Remark 4.15. As mentioned after Definition 1.5, the partition function of a *conformal* 2-dimensional field theory is smooth (but not necessarily holomorphic) and invariant under the usual $SL_2(\mathbb{Z})$-action, while no modularity properties would be expected for *Euclidean* field theories. It turns out that for a $2|1$-EFT E the supersymmetry forces Z^{++}_E to be holomorphic and $SL_2(\mathbb{Z})$-invariant:

THEOREM 1.15 ([**ST3**],[**ST6**]). *Let E be a Euclidean field theory of dimension $2|1$. Then the function $Z_E^{++}: \mathfrak{h} \to \mathbb{C}$ is a holomorphic modular function with integral Fourier coefficients. Moreover, every such function arises as Z_E^{++} from a $2|1$-Euclidean field theory E.*

We recall that a holomorphic modular function is a holomorphic function $f: \mathfrak{h} \to \mathbb{C}$ which is meromorphic at $\tau = i\infty$ and is $SL_2(\mathbb{Z})$-invariant. A modular function is *integral* if the coefficients a_k in its q-expansion $f(\tau) = \sum_{k=-N}^{\infty} a_k q^k$, $q = e^{2\pi i \tau}$ are integers. The restriction on partition functions coming from the above theorem is quite strong since any (integral) holomorphic modular function is an (integral) polynomial in the j-function.

In the following discussion, we use $2|0$-dimensional Euclidean field theories of degree n which will be explained in the following section. These use the bordism category of Euclidean *spin* manifolds.

THEOREM 1.16 ([**ST6**]). *There is a field theory $P \in 2|0\text{-EFT}^{-48}$ with partition function Z_P equal to the discriminant squared, which is a periodicity element in the sense that multiplication by P gives an equivalence of groupoids*

$$2|0\text{-EFT}^n(X) \xrightarrow{\cong} 2|0\text{-EFT}^{n-48}(X)$$

CONJECTURE 1.17. There is an isomorphism $2|1\text{-EFT}^n_{loc}[X] \cong \text{TMF}^n(X)$ compatible with the multiplicative structure. Here TMF* is the 24^2 periodic cohomology theory of topological modular forms mentioned above. The periodicity class has modular form the 24th power of the discriminant. We expect that our 48 periodicity will turn into a 24^2 periodicity after building in the locality.

$2|1\text{-EFT}^n_{loc}[X]$ are concordance classes of *local* (sometimes called *extended*) $2|1$-dimensional Euclidean field theories over X. These are more elaborate objects than the field theories discussed so far. For $n = 0$, they are 2-functors out of a bordism 2-category whose objects are $0|1$-manifolds, whose morphisms are $1|1$-dimensional bordisms, and whose 2-morphisms are $2|1$-dimensional manifolds with corners (all of them furnished with Euclidean structures and maps to X). The need for working with *local* field theories in order to obtain cohomology theories is explained in our earlier paper [**ST1**, Section 1.1]: for non-local theories we can't expect exactness of the Mayer-Vietoris sequence. Regarding terminology from that paper: an '(enriched) elliptic object over X' is now a '(local) 2-dimensional conformal field theory over X'.

While very general and beautiful results have been obtained in particular by Lurie on local versions of *topological* field theories [**Lu**], [**SP**], unfortunately even the *definition* of local Euclidean field theories has not been tied down.

1.7. Gauged field theories and equivariant cohomology. The definition of field theories over a manifold X in §1.2 can be generalized if X is equipped with a smooth action of a Lie group G. Let us equip bordisms Σ with additional structure that consists of a triple (P, f, ∇), where $P \to \Sigma$ is a principal G-bundle, $f: P \to X$ is a G-equivariant map, and ∇ is a connection on P, and similarly for Y's. We will call the corresponding field theories *G-gauged field theories over X*. If G is the trivial group, this is just a field theory over X in the previous sense. We think of a G-gauged field theory over X as a G-equivariant family of field theories parametrized by X. We write $d|1\text{-TFT}^n_G(X)$ for the groupoid of G-gauged

$d|1$-dimensional topological field theories of degree n over X, and $d|1\text{-TFT}^n_G[X]$ for the abelian group of their concordance classes.

THEOREM 1.18 ([**HSST**]). *The group* $0|1\text{-TFT}^n_G[X]$ *is isomorphic to the equivariant de Rham cohomology group* $H^n_{dR,G}(X)$.

Like in the non-equivariant case, the proof of this result is based on identifying the groupoid $0|1\text{-TFT}^n_G(X)$, which we show is equivalent to the discrete groupoid whose objects consist of the n-cocycles of the Weil model of equivariant de Rham cohomology [**GS**, (0.5),(0.7)]. We note that in the absence of a connection, the resulting TFT's give G-invariant closed differential forms and hence concordance classes lead to less interesting groups. For example, if G is compact and connected then one simply gets back ordinary de Rham cohomology $H^n_{dR}(X)$.

We believe that a different way of getting the Weil model would be to replace the target X by the stack $\widehat{X}_G$ of G-bundles with connection and G-map to X and then define $0|1\text{-TFT}^n(\widehat{X}_G)$ as for manifold targets (that definition extends from manifolds, aka. representable stacks, to all stacks). There is a geometric map

$$0|1\text{-TFT}^n_G(X) \longrightarrow 0|1\text{-TFT}^n(\widehat{X}_G)$$

which should be an isomorphism.

For $G = S^1$ we have also constructed an equivariant version of Euclidean field theories of dimension $0|1$ over X.

THEOREM 1.19 ([**HaST**]). *For an S^1-manifold X, the group* $0|1\text{-EFT}^n_{S^1}[X]$ *is isomorphic to the localized equivariant de Rham cohomology group* $H^{ev}_{dR,G}(X)[u^{-1}]$ *for even n, respectively* $H^{odd}_{dR,G}(X)[u^{-1}]$ *for odd n. Here* $u \in H^2_{dR}(BS^1)$ *is a generator.*

For an S^1-gauged *Euclidean* field theory, we require that the curvature of the principal S^1-bundle $P \to \Sigma$ is equal to a 2-form canonically associated to the Euclidean structure on Σ. The point of this condition is that the Euclidean structure on Σ and the connection on P give a canonical Euclidean structure on the $1|1$-manifold P. The result is a functor

$$1|1\text{-EFT}(X) \longrightarrow 0|1\text{-EFT}_{S^1}(LX)$$

which is a generalization of the 'dimensional reduction functor' (1.11). Passing to concordance classes we obtain a homomorphism

$$K^0(X) \longrightarrow 1|1\text{-EFT}[X] \longrightarrow 0|1\text{-EFT}_{S^1}[LX] \cong H^{ev}_{dR,S^1}(LX])[u^{-1}]$$

THEOREM 1.20 ([**HaST**]). *If V is a complex vector bundle over X, the image of* $[V] \in K^0(X)$ *under the above map is the Bismut-Chern character.*

This statement generalizes the field theoretic interpretation of the Chern character in terms of the reduction functor red (1.11) since the homomorphism

$$H^{ev}_{dR,S^1}(LX)[u^{-1}] \longrightarrow H^{ev}_{dR}(X)$$

induced by the inclusion map $X = LX^{S^1} \hookrightarrow LX$ maps the Bismut-Chern character of V to the Chern character of V.

1.8. Comparison with our 2004 survey. For the readers who are familiar with our 2004 survey [**ST1**], we briefly summarize some advances that are achieved by the current paper. The main new ingredient is a precise definition of *supersymmetric* field theories. This requires to define the right notion of geometric supermanifolds which make up the relevant bordism categories. We decided to use *rigid geometries* in the spirit of Felix Klein since these have a simple extension to odd direction. In supergeometry it is essential to work with families of objects that are parametrized by supermanifolds (the additional odd parameters). This forced us to work with the family versions (also known as fibered versions) of the bordism and vector space categories. In Segal's (and our) original notion of field theory, there was an unspecified requirement of 'continuity' of the symmetric monoidal functor. It turns out that our family versions implement this requirement in the following spirit A map is smooth if and only if it takes smooth functions to smooth functions.

In the 2-dimensional case, the rigid geometry we use corresponds to flat Riemannian structures on surfaces. Following Thurston and others, we call this a *Euclidean* structure. The flatness has the effect that only closed surfaces of genus one arise in the bordism category and hence our Euclidean field theories contain much less information then, say, a conformal field theory. Again we think of this as an advantage for several reasons. One is simply the fact that it becomes much easier to construct examples of field theories by a generator and relation method as discussed in Section 3.2. Another reason is the conjectured relation to topological modular forms, where also only genus one information is used. The last reason is our desire to express the Witten genus [**Wi**] of a closed Riemannian string manifold as the partition function of a field theory, the nonlinear $2|1$-dimensional Σ-model. It is well established in the physics community that such a field theory should exist and have modular partition function (a fact proven mathematically for the Witten genus by Don Zagier [**Za**]). However, it is also well known that this field theory can only be conformal if the Ricci curvature of the manifold vanishes. The question arises why a non-conformal field theory should have a modular (and in particular, holomorphic) partition function? One of the results in this paper is exactly this fact, proven precisely for our notion of $2|1$-dimensional Euclidean field theory. The holomorphicity is a consequence of the more intricate structure of the moduli stack of supertori.

In the conformal world, surfaces can be glued together along their boundaries by diffeomorphism, as we did in [**ST1**]. However, for most other geometries this is not possible any more and hence a precise notion of geometric bordism categories requires the introduction of collars. A precise way of doing this is one of the important, yet technical, contributions of this paper. We also decided to work with categories internal to a strict 2-category A as our model for 'weak 2-categories'. They are very flexible, allowing the introduction of fibered categories (needed for our family versions), symmetric monoidal structures (modeling Pauli's exclusion principle) and flips (related to the spin-statistics theorem) by just changing the ambient 2-category A. We also need the isometries (2-morphisms of an internal category) of bordisms to define the right notion of twisted field theories in Section 5. In order to fully model *local* 2-dimensional twisted field theories, we'll have to choose certain target 'weak 3-categories' in future papers. One possible model is introduced in the contribution [**DH**] in the current volume.

The 'adjunction transformations' in [**ST1**, Def. 2.1] are now completely replaced by allowing certain 'thin bordisms', for example L_0 in Section 3.2. Because of the existence of these geometric 1-morphisms, any functor must preserve the adjunctions automatically. In addition, we don't consider the (anti)-involutions on the categories any more, partially because we want to allow non-oriented field theories and partially because our $2|1$-dimensional bordism category does not have a real structure any more.

Finally, we enlarged the target category of a field theory to allow general topological vector spaces (locally convex, complete Hausdorff) because the smoothness requirement for a field theory sometimes does not hold for Hilbert spaces, see Remark 3.15. This has the additional advantage of being able to use the *projective* tensor product (leading to non-Hilbert spaces) for which inner products and evaluations are continuous operations (unlike for the Hilbert tensor product).

1.9. Summary of Contents. The next section leads up to the definition of field theories associated to rigid geometries in Definition 2.48. This includes Euclidean field theories of dimension d, which are obtained by specializing the geometry to be the Euclidean geometry of dimension d. Along the way we present the necessary categorical background (on internal categories in §2.2, categories with flip in §2.6 and fibered categories in §2.7) as well as geometric background (the construction of the Riemannian bordism category in §2.3 and the definition of families of rigid geometries in §2.5). Section 3 discusses 2-dimensional EFT's and their partition functions. The arguments presented in this section provide the first half of the outline of the proof of our modularity theorem 1.15 for the partition functions of Euclidean field theories of dimension $2|1$. The outline of the proof is continued in §4.4 after some preliminaries on supermanifolds in §4.1, on super Euclidean geometry in §4.2, and the definition of supersymmetric field theories associated to a supergeometry in §4.3; specializing to the super Euclidean geometry, these are supersymmetric Euclidean field theories. In section 5 we define *twisted field theories*. This notion is quite general and includes Segal's weakly conformal field theories (see §5.2) as well as Euclidean field theories of degree n (see §5.3). The last section contains an outline of the proof of the periodicity theorem 1.16.

2. Geometric field theories

2.1. Segal's definition of a conformal field theory. In this section we start with Graeme Segal's definition of a 2-dimensional conformal field theory and elaborate suitably to obtain the definition of a d-dimensional Euclidean field theory, and more generally, a field theory associated to every 'rigid geometry' (see Definitions 2.48 and 4.12). Segal has proposed an axiomatic description of 2-dimensional conformal field theories in a preprint that widely circulated for a decade and a half (despite the "do not copy" advice on the front) before it was published as [**Se2**]. In the published version, Segal added a foreword/postscript commenting on developments since the original manuscript was written in which he proposes the following definition of conformal field theories.

DEFINITION 2.1. **(Segal [Se2,** Postscript to section 4]**)** A *2-dimensional conformal field theory* (H, U) consists of the following two pieces of data:

(1) A functor $Y \mapsto H(Y)$ from the category of closed oriented smooth 1-manifolds to locally convex complete topological vector spaces, which takes disjoint unions to (projective) tensor products, and
(2) For each oriented cobordism Σ, with conformal structure, from Y_0 to Y_1 a linear trace-class map $U(\Sigma)\colon H(Y_0) \to H(Y_1)$, subject to
   (a) $U(\Sigma \circ \Sigma') = U(\Sigma) \circ U(\Sigma')$ when cobordisms are composed, and
   (b) $U(\Sigma \amalg \Sigma') = U(\Sigma) \otimes U(\Sigma')$.
   (c) If $f\colon \Sigma \to \Sigma'$ is a conformal equivalence between conformal bordisms, the diagram

$$\begin{array}{ccc} H(Y_0) & \xrightarrow{U(\Sigma)} & H(Y_1) \\ {\scriptstyle H(f_{|Y_0})}\downarrow & & \downarrow{\scriptstyle H(f_{|Y_0})} \\ H(Y_0') & \xrightarrow{U(\Sigma')} & H(Y_1') \end{array} \tag{2.2}$$

   is commutative.

Furthermore, $U(\Sigma)$ must depend smoothly on the conformal structure of Σ.

Condition (c) is not explicitly mentioned in Segal's postscript to section 4, but it corresponds to identifying conformal surfaces with parametrized boundary in his bordisms category if they are conformally equivalent relative boundary, which Segal does in the first paragraph of §4.

2.2. Internal categories. We note that the data (H, U) in Segal's definition of a conformal field theory (Definition 2.1) can be interpreted as a pair of symmetric monoidal functors. Here H is a functor from the groupoid of closed oriented smooth 1-manifolds to the groupoid of locally convex topological vector spaces. The domain of the functor U is the groupoid whose objects are conformal bordisms and whose morphisms are conformal equivalences between conformal bordisms. The range of U is the groupoid whose objects are trace-class operators (= nuclear operators) between complete locally convex topological vector spaces and whose morphisms are commutative squares like diagram (2.2). The monoidal structure on the domain groupoids of H and U is given by the disjoint union, on the range groupoids it is given by the tensor product.

Better yet, the two domain groupoids involved fit together to form an *internal category* in the category of symmetric monoidal groupoids. The same holds for the two range groupoids, and the pair (H, U) is a functor between these internal categories. It turns out that internal categories provide a convenient language not only for field theories a la Segal; rather, *all refinements that we'll incorporate in the following sections fit into this framework.* What changes is the *ambient* strict 2-category, which now is the 2-category SymCat of symmetric monoidal categories, or equivalently, the 2-category Sym(Cat) of symmetric monoidal objects in Cat, the 2-category of categories. Later we will replace Cat by Cat/Man (respectively Cat/csM) whose objects are categories fibered over the category Man of smooth manifolds (respectively csM of supermanifolds).

Internal categories are described e.g., in section XII.1 of the second edition of Mac Lane's book [**McL**], but his version of internal categories is too strict to define the internal bordism category we need as domain. A weakened version of internal categories and functors is defined for example by Martins-Ferreira in [**M**] who calls them *pseudo categories*. This is a good reference which describes completely explicitly internal categories (also known as pseudo categories, §1), functors between them (also known as pseudo functors, §2), natural transformations (also known as pseudo-natural transformations, §3), and modifications between natural transformations (also known as pseudo-modifications, §4).

REMARK 2.3. There is a slight difference between the definition of natural transformation in [**M**] and ours (see Definition 2.19) in that we won't insist on invertibility of a 2-cell which is part of the data of a natural transformation. It should be emphasized that our field theories of degree n will be defined as natural transformations of functors between internal categories (see Definitions 5.2 and 5.10), and here it is crucial that we don't insist on invertibility (see Remark 5.8). The main result of [**M**, Theorem 3] implies that for fixed internal categories C, D the functors from C to D are the objects of a bicategory whose morphisms are natural transformations (its 2-morphisms are called 'modifications'). That result continues to hold since the construction of various compositions does not involve taking inverses.

For much of the material on internal categories covered in this subsection, we could simply refer to [**M**] for definitions. For the convenience of the reader and since internal categories are the categorical backbone of our description of field theories, we will describe them in some detail. We start with the definition of an internal category in an ambient category A. Then we explain why this is too strict to define our internal bordism category and go on to show how this notion can be suitably weakened if the ambient category A is a strict 2-category.

DEFINITION 2.4. **(Internal Category)** Let A be a category with pull-backs (here A stands for 'ambient'). An *internal category* or *category object* in A consists of two objects $\mathsf{C}_0, \mathsf{C}_1 \in \mathsf{A}$ and four morphisms

$$s,t\colon \mathsf{C}_1 \longrightarrow \mathsf{C}_0 \qquad u\colon \mathsf{C}_0 \longrightarrow \mathsf{C}_1 \qquad c\colon \mathsf{C}_1 \times_{\mathsf{C}_0} \mathsf{C}_1 \longrightarrow \mathsf{C}_1$$

(source, target, unit morphism and composition), subject to the following four conditions expressing the usual axioms for a category:

$$s\,u = 1 = t\,u\colon \mathsf{C}_0 \longrightarrow \mathsf{C}_0, \tag{2.5}$$

(this specifies source and target of the identity map); the commutativity of the diagram

$$\begin{array}{ccccc} \mathsf{C}_1 & \xleftarrow{\pi_1} & \mathsf{C}_1 \times_{\mathsf{C}_0} \mathsf{C}_1 & \xrightarrow{\pi_2} & \mathsf{C}_1 \\ {\scriptstyle t}\downarrow & & \downarrow{\scriptstyle c} & & \downarrow{\scriptstyle s} \\ \mathsf{C}_0 & \xleftarrow{t} & \mathsf{C}_1 & \xrightarrow{s} & \mathsf{C}_0 \end{array} \tag{2.6}$$

(this specifies source and target of a composition); the commutativity of the diagram

$$\begin{array}{ccccc} \mathsf{C}_1 & \xrightarrow{u\,t\times 1} & \mathsf{C}_1 \times_{\mathsf{C}_0} \mathsf{C}_1 & \xleftarrow{1\times u\,s} & \mathsf{C}_1 \times_{\mathsf{C}_0} \mathsf{C}_0 \\ & {\scriptstyle 1}\searrow & \downarrow{\scriptstyle c} & \swarrow{\scriptstyle 1} & \\ & & \mathsf{C}_1 & & \end{array} \tag{2.7}$$

expressing the fact the u acts as the identity for composition, and the commutativity of the diagram

(2.8)
$$\begin{CD} C_1 \times_{C_0} C_1 \times_{C_0} C_1 @>{c\times 1}>> C_1 \times_{C_0} C_1 \\ @V{1\times c}VV @VV{c}V \\ C_1 \times_{C_0} C_1 @>{c}>> C_1 \end{CD}$$

expressing associativity of composition.

(Functors between internal categories) Following MacLane (§XII.1), a functor $f\colon C \to D$ between internal categories C, D in the same ambient category A is a pair of morphisms in A

$$f_0\colon C_0 \longrightarrow D_0 \qquad f_1\colon C_1 \longrightarrow D_1.$$

Thought of as describing the functor on "objects" respectively "morphisms", they are required to make the obvious diagrams commutative:

(2.9)
$$\begin{CD} C_1 @>{f_1}>> D_1 \\ @V{s}VV @VV{s}V \\ C_0 @>{f_0}>> D_0 \end{CD} \qquad \begin{CD} C_1 @>{f_1}>> D_1 \\ @V{t}VV @VV{t}V \\ C_0 @>{f_0}>> D_0 \end{CD}$$

(2.10)
$$\begin{CD} C_1 \times_{C_0} C_1 @>{f_1\times f_1}>> C_1 \\ @V{c_C}VV @VV{c_D}V \\ D_1 \times_{D_0} D_1 @>{f_1}>> D_1 \end{CD} \qquad \begin{CD} C_0 @>{f_0}>> D_0 \\ @V{u}VV @VV{u}V \\ C_1 @>{f_1}>> D_1 \end{CD}$$

(Natural transformations) If f, g are two internal functors $C \to D$, a *natural transformation* n from f to g is a morphism

$$n\colon C_0 \longrightarrow D_1$$

making the following diagrams commutative:

(2.11)
$$\begin{array}{ccccc} & & C_0 & & \\ & {}^{g_0}\swarrow & \downarrow{\scriptstyle n} & \searrow^{f_0} & \\ D_0 & \xleftarrow{\;t\;} & D_1 & \xrightarrow{\;s\;} & D_0 \end{array} \qquad \begin{CD} C_1 @>{g_1\times ns}>> D_1 \times_{D_0} D_1 \\ @V{nt\times f_1}VV @VV{c_D}V \\ D_1 \times_{D_0} D_1 @>{c_D}>> D_1 \end{CD}$$

We note that the commutativity of the first diagram is needed in order to obtain the arrows $gt \times f_1$, $g_1 \times ns$ in the second diagram. If the ambient category A is the category of sets, then n is a natural transformation from the functor f to the functor g; the first diagram expresses the fact that for an object $a \in C_0$ the associated morphism $n_a \in D_1$ has domain $f_0(a)$ and range $g_0(a)$. The second diagram expresses the fact that for every morphism $h\colon a \to b$ the diagram

$$\begin{CD} f_0(a) @>{n_a}>> g_0(a) \\ @V{f_1(h)}VV @VV{g_1(h)}V \\ f_0(b) @>{n_b}>> g_0(b) \end{CD}$$

is commutative.

As mentioned before, we would like to regard Segal's pair (H, U) as a functor between internal categories where the ambient category A is the category of symmetric monoidal groupoids. However, this is not quite correct due to the lack of associativity of the internal bordism category. In geometric terms, the problem is that if Σ_i is a bordism from Y_i to Y_{i+1} for $i = 1, 2, 3$, then $(\Sigma_3 \cup_{Y_3} \Sigma_2) \cup_{Y_2} \Sigma_1$ and $\Sigma_3 \cup_{Y_3} (\Sigma_2 \cup_{Y_2} \Sigma_1)$ are not strictly speaking *equal*, but only canonically conformally equivalent. In categorical terms, this means that the diagram (2.8) is not commutative; rather, the conformal equivalence between these bordisms is a morphism in the groupoid C_1 whose objects are conformal bordisms. This depends functorially on $(\Sigma_3, \Sigma_2, \Sigma_1) \in \mathsf{C}_1 \times_{\mathsf{C}_0} \mathsf{C}_1 \times_{\mathsf{C}_0} \mathsf{C}_1$ and hence it provides an invertible natural transformation α between the two functors of diagram (2.8)

(2.12)
$$\begin{array}{ccc} \mathsf{C}_1 \times_{\mathsf{C}_0} \mathsf{C}_1 \times_{\mathsf{C}_0} \mathsf{C}_1 & \xrightarrow{c \times 1} & \mathsf{C}_1 \times_{\mathsf{C}_0} \mathsf{C}_1 \\ {\scriptstyle 1 \times c} \downarrow & \stackrel{\alpha}{\cong} & \downarrow {\scriptstyle c} \\ \mathsf{C}_1 \times_{\mathsf{C}_0} \mathsf{C}_1 & \xrightarrow{c} & \mathsf{C}_1 \end{array}$$

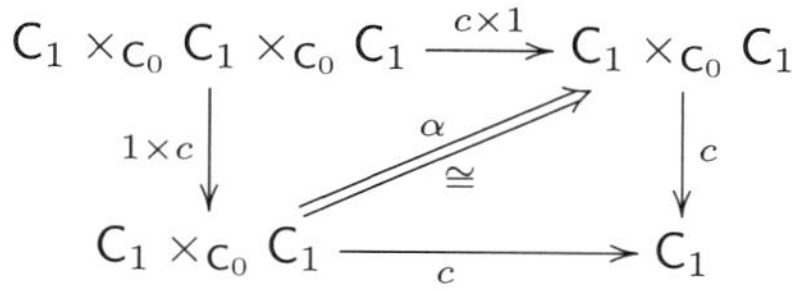

The moral is that we should relax the associativity axiom of an internal category by replacing the assumption that the diagram above is commutative by the weaker assumption that the there is an invertible 2-morphism α between the two compositions. This of course requires that the ambient category A can be refined to be a strict 2-category (which happens in our case, with objects, morphisms, respectively 2-morphisms being symmetric monoidal groupoids, symmetric monoidal functors, respectively symmetric monoidal natural transformations).

This motivates the following definition:

DEFINITION 2.13. An *internal category in a strict 2-category* A consists of the following data:

- objects C_0, C_1 of A;
- morphisms s, t, c of A as in definition 2.4;
- invertible 2-morphisms α from (2.12), and invertible 2-morphisms λ, ρ in the following diagram

(2.14)
The morphisms are required to satisfy conditions (2.7) and (2.6). The 2-morphisms are subject to two coherence diagrams (see [**M**, Diagrams (1.6) and (1.7)]); in particular, there is a pentagon shaped diagram involving the 'associator' α. In addition, it is required that composing the 2-morphisms α, λ, or ρ with the identity 2-morphism on s or t gives an identity 2-morphism.

For comparison with [**M**] it might be useful to note that in that paper the letter d (domain) is used instead of our s (source), c (codomain) instead of t (target), m (multiplication) instead of c (composition) and e instead of u (unit).

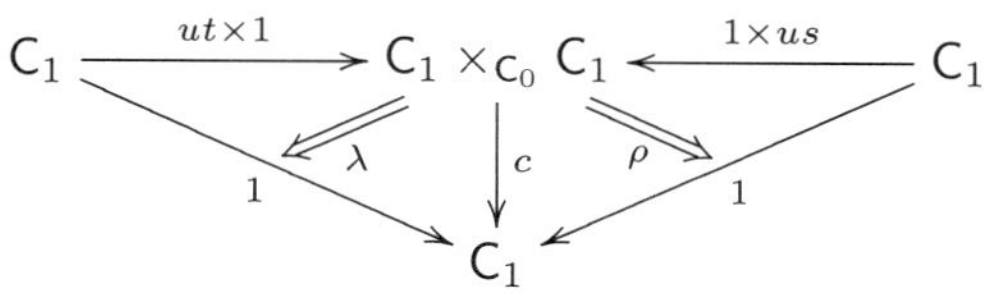

REMARK 2.15. In the special case of an internal category where C_0 is a terminal object of the strict 2-category A, this structure is called a *monoid in* A. The composition
$$c\colon C_1 \times C_1 \longrightarrow C_1,$$
is thought of as a multiplication on C_1 with unit $u\colon C_0 \to C_1$. For example, a monoid in $A = \mathsf{Cat}$ is a *monoidal category*.

Similarly, a *symmetric monoidal category* can be viewed as a *symmetric monoid* in Cat in the sense of the definition below. The point of this is that we will need to talk about symmetric monoids in other strict 2-categories, e.g., categories fibered over some fixed category S.

DEFINITION 2.16. A *symmetric monoid* in a strict 2-category A is a monoid $(C_1, c, u, \alpha, \lambda, \rho)$ in A together with an invertible 2-morphism σ called *braiding isomorphism*:

(2.17)
$$\begin{array}{c}
C_1 \times C_1 \xrightarrow{m} C_1 \\
\tau \downarrow \quad \sigma \nearrow_m
\\
C_1 \times C_1
\end{array}$$

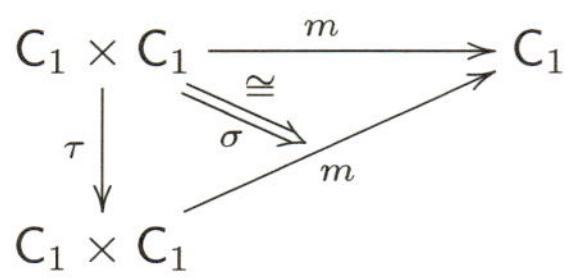

Here τ is the morphism in A that switches the two copies of C_1. The 2-morphisms σ, α, λ and ρ are subject to coherence conditions well-known in the case $A = \mathsf{Cat}$ [**McL**, Ch. XI, §1].

Next we define functors between categories internal to a 2-category by weakening Definition 2.4.

DEFINITION 2.18. Let C, D be internal categories in a strict 2-category A. Then a *functor* $f\colon C \to D$ is a quadruple $f = (f_0, f_1, \mu, \epsilon)$, where $f_0\colon C_0 \to D_0$, $f_1\colon C_1 \to D_1$ are morphisms, and μ, ϵ are invertible 2-morphisms

$$\begin{array}{cc}
C_1 \times_{C_0} C_1 \xrightarrow{c_C} C_1 & \quad C_0 \xrightarrow{f_0} D_0 \\
f_1 \times f_1 \downarrow \; \mu \; \downarrow f_1 & \quad u \downarrow \; \epsilon \; \downarrow u \\
D_1 \times_{D_0} D_1 \xrightarrow{c_D} D_1 & \quad C_1 \xrightarrow{f_1} D_1
\end{array}$$

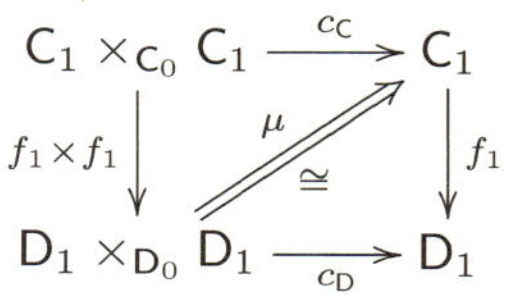

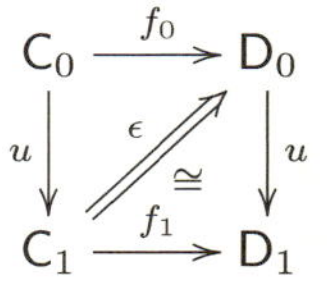

It is required that the diagrams (2.9) commute. The 2-morphisms μ, ϵ are subject to three coherence conditions (see [**M**, Diagrams (2.5) and (2.6)]) as well as the usual condition that horizontal composition with the identity 2-morphisms 1_s or 1_t results in an identity 2-morphism.

DEFINITION 2.19. Let $f, g\colon C \to D$ be internal functors between internal categories in a strict 2-category A. A *natural transformation* from $f = (f_0, f_1, \mu^f, \epsilon^f)$ to $g = (g_0, g_1, \mu^g, \epsilon^g)$ is a pair $n = (n, \nu)$, where $n\colon C_0 \to D_1$ is a morphism, and ν is a 2-morphism:

(2.20)
$$\begin{array}{c}
C_1 \xrightarrow{n\, t \times f_1} D_1 \times_{D_0} D_1 \\
g_1 \times n\, s \downarrow \; \nu \; \downarrow c_D \\
D_1 \times_{D_0} D_1 \xrightarrow{c_D} D_1
\end{array}$$

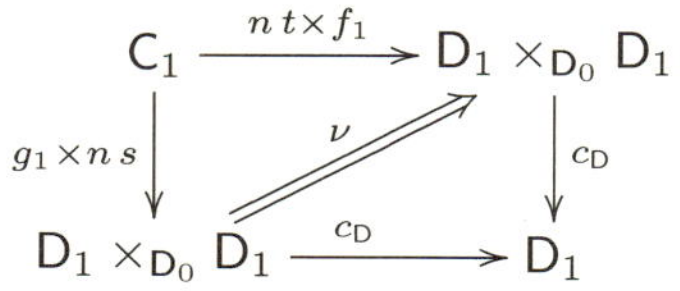

It is required that the first diagram of (2.11) is commutative. There are two coherence conditions for the 2-morphism ν (see [**M**, Diagrams (3.5) and (3.6)]), and the usual requirement concerning horizontal composition with 1_s and 1_t (these are the identity 2-morphisms of the morphisms s respectively t).

Note that the 2-morphism ν is not required to be invertible. Some authors add the word 'lax' in this more general case. See Remark 5.8 for the reason why this more general notion is important for twisted field theories.

2.3. The internal Riemannian bordism category. Now we are ready to define d-RBord, the category of d-dimensional Riemannian bordisms. This is a category internal to $\mathsf{A} = \mathsf{SymGrp}$, the strict 2-category of symmetric monoidal groupoids.

We should mention that in our previous paper [**HoST**] we defined the Riemannian bordism category d-RB. This is just a category, rather than an internal category like d-RBord. In this paper we are forced to deal with this more intricate categorical structure, since we wish to consider *twisted* field theories, in particular field theories of *non-trivial degree.*

Before giving the formal definition of d-RBord, let us make some remarks that hopefully will motivate the definition below. Roughly speaking, $d\text{-RBord}_0$ is the symmetric monoidal groupoid of closed Riemannian manifolds of dimension $d-1$, and $d\text{-RBord}_1$ is the symmetric monoidal groupoid of Riemannian bordisms of dimension d. The problem is to define the composition functor

$$c\colon d\text{-RBord}_1 \times_{d\text{-RBord}_0} d\text{-RBord}_1 \longrightarrow d\text{-RBord}_1.$$

If Σ_1 is a Riemannian bordism from Y_0 to Y_1 and Σ_2 is a Riemannian bordism from Y_1 to Y_2, c should map $(\Sigma_2, \Sigma_1) \in d\text{-RBord}_1 \times_{d\text{-RBord}_0} d\text{-RBord}_1$ to the Riemannian manifold $\Sigma := \Sigma_2 \cup_{Y_1} \Sigma_1$ obtained by gluing Σ_2 and Σ_1 along their common boundary component Y_1. The problem is that the Riemannian metrics on Σ_1 and Σ_2 might not fit together to give a Riemannian metric on Σ. A necessary, but not sufficient condition for this is that the second fundamental form of Y_1 as a boundary of Σ_1 matches with the second fundamental form of Y_1 as a boundary of Σ_2.

In the usual gluing process of d-dimensional bordisms, the two glued bordisms intersect in a closed $(d-1)$-dimensional manifold Y, the object of the bordism category which is the source (respectively target) of the bordisms to be glued. For producing a Riemannian structure on the glued bordism (actually, even for producing a smooth structure on it), it is better if the intersection is an *open* d-manifold on which the Riemannian structures are required to match. This suggests to refine an *object* of the bordism category $d\text{-RBord}_0$ to be a pair (Y, Y^c), where Y is an open Riemannian d-manifold, and $Y^c \subset Y$ (the *core* of Y) is a closed $(d-1)$-dimensional submanifold of Y. We think of Y as *Riemannian collar* of the $(d-1)$-dimensional core manifold Y^c (this core manifold is the only datum usually considered). In order to distinguish the domain and range of a bordism, we will in addition require a decomposition $Y \setminus Y^c = Y^+ \amalg Y^-$ of the complement into disjoint open subsets, both of which contain Y^c in its closure. Domain and range of a bordism is customarily controlled by comparing the given orientation of the closed manifold Y^c with the orientation induced by thinking of it as a part of the boundary of an oriented bordism Σ. Our notion makes it unnecessary to furnish our manifolds with orientations.

Our main goal here is to define the d-dimensional Euclidean bordism category d-EBord We first define the Riemannian bordism category d-RBord and then d-EBord as the variation where we insist that all Riemannian metrics are *flat* and that the cores $Y^c \subset Y$ are totally geodesic. We want to provide pictures and it's harder to draw interesting pictures of flat surfaces (e.g., the flat torus doesn't embed in $\mathbb{R}^3$).

DEFINITION 2.21. The *d-dimensional Riemannian bordism category* d-RBord is the category internal to the strict 2-category $A = \mathsf{SymGrp}$ of symmetric monoidal groupoids defined as follows. Note that all Riemannian manifolds that arise are without boundary.

The object groupoid d-RBord$_0$. The *objects* of the groupoid d-RBord$_0$ are quadruples $(Y, Y^c, Y^\pm)$, where Y is a Riemannian d-manifold (without boundary and usually non-compact) and $Y^c \subset Y$ is a *compact* codimension 1 submanifold which we call the *core* of Y. Moreover, we require that $Y \setminus Y^c = Y^+ \amalg Y^-$, where $Y^\pm \subset Y$ are disjoint open subsets whose closures contain Y^c. Often we will suppress the data $Y^c, Y^\pm$ in the notation and write Y for an object of d-RBord.

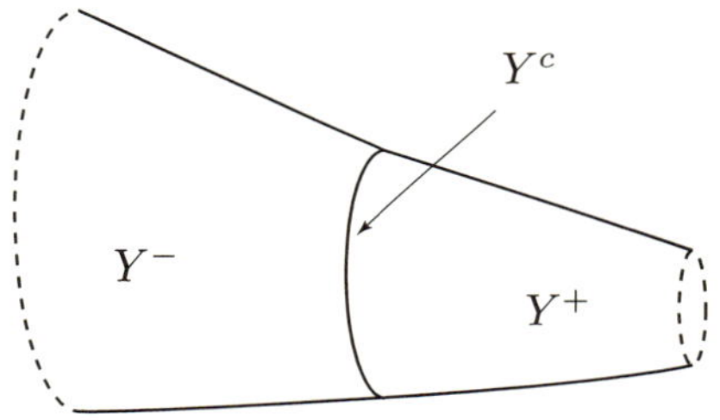

FIGURE 1. An object $(Y, Y^c, Y^\pm)$ of 2-RBord$_0$

An *isomorphism* in d-RBord from Y_0 to Y_1 is the germ of an invertible isometry $f \colon W_0 \to W_1$. Here $W_j \subset Y_j$ are open neighborhoods of Y_j^c and f is required to send Y_0^c to Y_1^c and $W_0^\pm$ to $W_1^\pm$ where $W_j^\pm := W_j \cap Y_j^\pm$. As usual for germs, two such isometries represent the *same* isomorphism if they agree on some smaller open neighborhood of Y_0^c in Y_0.

We remark that if $(Y, Y^c, Y^\pm)$ is an object of d-RBord, and $W \subset Y$ is an open neighborhood of Y^c, then $(Y, Y^c, Y^\pm)$ is isomorphic to $(W, Y^c, Y^\pm \cap W)$. In particular, we can always assume that Y is diffeomorphic to $Y^c \times (-1, +1)$, since by the tubular neighborhood theorem, there is always a neighborhood W of Y^c such that the pair (W, Y^c) is diffeomorphic to $(Y^c \times (-1, +1), Y^c \times \{0\})$.

The morphism groupoid d-RBord$_1$ is defined as follows. An object of d-RBord$_1$ consists of a pair $Y_0 = (Y_0, Y_0^c)$, $Y_1 = (Y_1, Y_1^c)$ of objects of d-RBord$_0$ (the source respectively target) and a *Riemannian bordism* from Y_0 to Y_1, which is a triple (Σ, i_0, i_1) consisting of a Riemannian d-manifold Σ and smooth maps $i_j \colon W_j \to \Sigma$. Here $W_j \subset Y_j$ are open neighborhoods of the cores Y_j^c. Letting $i_j^\pm : W_j^\pm \to \Sigma$ be the restrictions of i_j to $W_j^\pm := W_j \cap Y_j^\pm$, we require that

(+) i_j^+ are isometric embeddings into $\Sigma \setminus i_1(W_1^- \cup Y_1^c)$ and

(c) the *core* $\Sigma^c := \Sigma \setminus \big(i_0(W_0^+) \cup i_1(W_1^-)\big)$ is compact.

Particular bordisms are given by isometries $f\colon W_0 \to W_1$ as above, namely by using $\Sigma = W_1$, $i_1 = \mathrm{id}_{W_1}$ and $i_0 = f$. Note that in this case the images of i_0^+ and i_1^+ are *not disjoint* but we didn't require this condition.

Below is a picture of a Riemannian bordism; we usually draw the domain of the bordism to the right of its range, since we want to read compositions of bordisms, like compositions of maps, from right to left. Roughly speaking, a bordism between

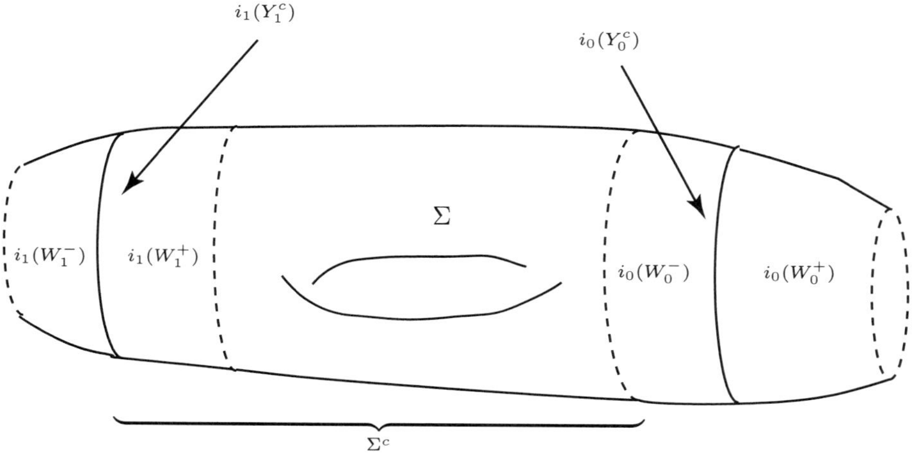

FIGURE 2. A Riemannian bordism (object of 2-RBord_1)

objects Y_0 and Y_1 of d-RBord_0 is just an ordinary bordism Σ^c from Y_0^c to Y_1^c equipped with a Riemannian metric, thickened up a little bit at its boundary to make gluing possible..

A morphism from a bordism Σ to a bordism Σ' is a germ of a triple of isometries

$$F\colon X \to X' \qquad f_0\colon V_0 \to V_0' \qquad f_1\colon V_1 \to V_1'.$$

Here X (respectively V_0, V_1) is an open neighborhood of $\Sigma^c \subset \Sigma$ (respectively $Y_0^c \subset W_0 \cap i_0^{-1}(X), Y_1^c \subset W_1 \cap i_1^{-1}(X)$) and similarly for X', V_0', V_1'. We require the conditions for f_j to be a morphism from Y_j to Y_j' in d-RBord_0, namely $f_j(Y_j^c) = (Y_j')^c$ and $f_j(V_j^\pm) = (V_j')^\pm$. In addition, we require that these isometries are compatible in the sense that the diagram

$$\begin{array}{ccccc} V_1 & \xrightarrow{i_1} & X & \xleftarrow{i_0} & V_0 \\ \downarrow{\scriptstyle f_1} & & \downarrow{\scriptstyle F} & & \downarrow{\scriptstyle f_0} \\ V_1' & \xrightarrow[i_1']{} & X' & \xleftarrow[i_0']{} & V_0' \end{array}$$

is commutative. Two such triples (F, f_0, f_1) and (G, g_0, g_1) represent the same germ if there there are smaller open neighborhoods X'' of $\Sigma^c \subset X$ and V_j'' of $Y_j \subset V_j \cap i_j^{-1}(X'')$ such that F and G agree on X'', and f_j and g_j agree on V_j'' for $j = 0, 1$.

Source, target, unit and composition functors. The functors

$$s,t\colon d\text{-}\mathsf{RBord}_1 \longrightarrow d\text{-}\mathsf{RBord}_0$$

send a bordism Σ from Y_0 to Y_1 to Y_0 respectively Y_1. There is also the functor

$$u\colon d\text{-}\mathsf{RBord}_0 \longrightarrow d\text{-}\mathsf{RBord}_1$$

that sends (Y,Y^c) to the Riemannian bordism given by the identity isometry $\mathrm{id}\colon Y \to Y$. These functors are compatible with taking disjoint unions and hence they are symmetric monoidal functors, i.e., morphisms in SymGrp.

There is also a *composition functor*

$$c\colon d\text{-}\mathsf{RBord}_1 \times_{d\text{-}\mathsf{RBord}_0} d\text{-}\mathsf{RBord}_1 \longrightarrow d\text{-}\mathsf{RBord}_1$$

given by gluing bordisms. Let us describe this carefully, since there is a subtlety involved here due to the need to adjust the size of the Riemannian neighborhood along which we glue. Let Y_0, Y_1, Y_2 be objects of d-RBord_0, and let Σ, Σ' be bordisms from Y_0 to Y_1 respectively from Y_1 to Y_2. These data involve in particular smooth maps

$$i_1\colon W_1 \longrightarrow \Sigma \qquad i_1'\colon W_1' \longrightarrow \Sigma',$$

where W_1, W_1' are open neighborhoods of $Y_1^c \subset Y_1$. We set $W_1'' \stackrel{\text{def}}{=} W_1 \cap W_1'$ and note that our conditions guarantee that i_1 (respectively i_1') restricts to an isometric embedding of $(W_1'')^+ \stackrel{\text{def}}{=} W_1'' \cap Y_1^+$ to Σ (respectively Σ'). We use these isometries to glue Σ and Σ' along $(W_1'')^+$ to obtain Σ'' defined as follows:

$$\Sigma'' \stackrel{\text{def}}{=} \left(\Sigma' \smallsetminus i_1'((W_1')^+ \smallsetminus (W_1'')^+)\right) \cup_{(W_1'')^+} \left(\Sigma \smallsetminus i_1(W_1^- \cup Y_1^c)\right)$$

The maps $i_0\colon W_0 \to \Sigma$ and $i_2\colon W_2 \to \Sigma'$ can be restricted to maps (on smaller open neighborhoods) into Σ'' satisfying our conditions. This makes Σ'' a bordism from Y_0 to Y_2.

As explained above (see Equation (2.12)), the composition functor c is not strictly associative, but there is a natural transformation α as in diagram (2.12) which satisfies the pentagon identity.

REMARK 2.22. We point out that conditions (+) and (c) in the above definition of a Riemannian bordism also make sure that the composed bordism is again a Hausdorff space. In other words, gluing two topological spaces along open subsets preserves conditions like 'locally homeomorphic to $\mathbb{R}^n$' and structures like Riemannian metrics. However, it can happen that the glued up space is not Hausdorff, for example if one glues two copies of $\mathbb{R}$ along the interval $(0,1)$. The reader is invited to check that our claim follows from the following easy lemma.

LEMMA 2.23. *Let X, X' be manifolds and let U be an open subset of X and X'. Then $X \cup_U X'$ is a manifold if and only if the natural map $U \to X \times X'$ sends U to a* closed *set.*

2.4. The internal category of vector spaces.

DEFINITION 2.24. The category TV of (complete locally convex) topological vector spaces internal to SymGrp (the strict 2-category category of symmetric monoidal groupoids) is defined by the object respectively morphism groupoids as follows.

TV_0 is the groupoid whose objects are complete locally convex topological vector spaces over $\mathbb{C}$ and whose morphisms are invertible continuous linear maps. The completed projective tensor product gives TV_0 the structure of a symmetric monoidal groupoid.

TV_1 is the symmetric monoidal groupoid whose objects are continuous linear maps $f\colon V_0 \to V_1$. The morphisms from $f\colon V_0 \to V_1$ to $f'\colon V_0' \to V_1'$ are a pair of isomorphisms (g_0, g_1) making the diagram

$$\begin{array}{ccc} V_0 & \xrightarrow[\cong]{g_0} & V_0' \\ {\scriptstyle f}\downarrow & & \downarrow{\scriptstyle f'} \\ V_1 & \xrightarrow[\cong]{g_1} & V_1' \end{array}$$

commutative. It is a symmetric monoidal groupoid via the projective tensor product.

There are obvious source, target, unit and composition functors

$$s, t\colon \mathsf{TV}_1 \longrightarrow \mathsf{TV}_0 \qquad u\colon \mathsf{TV}_0 \longrightarrow \mathsf{TV}_1 \qquad c\colon \mathsf{TV}_1 \times_{\mathsf{TV}_0} \mathsf{TV}_1 \longrightarrow \mathsf{TV}_1$$

which make TV an internal category in SymGrp. This is a *strict* internal category in the sense that associativity holds on the nose (and not just up to natural transformations).

Now we are ready for a preliminary definition of a d-dimensional Euclidean field theory, which will be modified by adding a smoothness condition in the next section.

DEFINITION 2.25. **(Preliminary!)** A d-dimensional *Riemannian field theory* over a smooth manifold X is a functor

$$E\colon d\text{-}\mathsf{RBord}(X) \longrightarrow \mathsf{TV}$$

of categories internal to SymGrp, the strict 2-category of symmetric monoidal groupoids. A functor $E\colon d\text{-}\mathsf{RBord}(\mathrm{pt}) = d\text{-}\mathsf{RBord} \to \mathsf{TV}$ is a d-dimensional Riemannian field theory.

Similarly, a d-dimensional *Euclidean field theory* over a smooth manifold X is a functor

$$E\colon d\text{-}\mathsf{EBord}(X) \longrightarrow \mathsf{TV}$$

of categories internal to SymGrp, where the Euclidean bordism category d-EBord is defined completely analogously to d-RBord by using Euclidean structures (= flat Riemannian metrics) instead of Riemannian metrics, and by requiring that for any object $(Y, Y^c, Y^{\pm})$ the core Y^c is a totally geodesic submanifold of Y.

The feature missing from the above definition is the requirement that E should be *smooth*. Heuristically, this means that the vector space $E(Y)$ associated to an object Y of the bordism category as well as the operator $E(\Sigma)$ associated to a bordism Σ should depend smoothly on Y respectively Σ. To make this precise, we replace the categories d-EBord, TV by their *family versions* whose objects and morphisms are *smooth families* of objects/morphisms of the original category parametrized by some manifold.

2.5. Families of rigid geometries. In this subsection we define families of Euclidean d-manifolds, or more generally a families of manifolds equipped with a rigid geometry. This leads to the definition of the bordism category $(G,\mathbb{M})$-Bord of manifolds with rigid geometry $(G,\mathbb{M})$ (see Definition 2.46) and to the notion of a field theory based on a rigid geometry (see Definition 2.48).

Let G be a Lie group acting on a manifold $\mathbb{M}$. We want to think of $\mathbb{M}$ as the local model for *rigid geometries* with isometry group G. This idea is very well explained in [**Th**] and goes back to Felix Klein.

DEFINITION 2.26. A $(G,\mathbb{M})$*-structure* on a manifold Y is a maximal atlas consisting of *charts* which are diffeomorphisms

$$Y \supseteq U_i \xrightarrow[\cong]{\varphi_i} V_i \subseteq \mathbb{M}$$

between open subsets of Y and open subsets of $\mathbb{M}$ such that the U_i's cover Y, and a collection of elements $g_{ij} \in G$ which determine the transition functions in the sense that for every i,j the diagram

$$\begin{array}{ccc} & U_i \cap U_j & \\ {\scriptstyle\varphi_j}\swarrow & & \searrow{\scriptstyle\varphi_i} \\ \mathbb{M} & \xrightarrow{\quad g_{ij} \quad} & \mathbb{M} \end{array}$$

is commutative (here we interpret $g_{ij} \in G$ as an automorphism of $\mathbb{M}$ via the action map $G \times \mathbb{M} \to \mathbb{M}$). The g_{ij}'s are required to satisfy the cocycle condition

$$g_{ij} \cdot g_{jk} = g_{ik} \tag{2.27}$$

We note that if G acts effectively on $\mathbb{M}$, then the element $g_{ij} \in G$ is determined by φ_i and φ_j. In particular, the cocycle condition is automatic, and the definition can be phrased in a simpler way as in our previous paper [**HoST**, Definition 6.13]. In this paper we also wish to consider cases where G does not act effectively on $\mathbb{M}$ as in the second example below.

For our definition of the bordism category $(G,\mathbb{M})$-Bord we will also need the notion of a $(G,\mathbb{M})$-structure on a pair (Y,Y^c) consisting of a manifold Y and a codimension one submanifold Y^c.

DEFINITION 2.28. Let $(G,\mathbb{M})$ be a geometry and let $\mathbb{M}^c \subset \mathbb{M}$ be a codimension one submanifold of $\mathbb{M}$ (there is no condition relating $\mathbb{M}^c$ and the G-action). From now on, a 'geometry' will refer to such a triple $(G,\mathbb{M},\mathbb{M}^c)$, but we will suppress $\mathbb{M}^c$ in our notation. A $(G,\mathbb{M})$*-structure on a pair* (Y,Y^c) is a maximal atlas consisting of charts $\{(U_i,\varphi_i)\}$ for Y as in Definition 2.26 above with the additional requirement that φ_i maps $U_i \cap Y^c$ to $\mathbb{M}^c \subset \mathbb{M}$.

EXAMPLE 2.29. Let $\mathbb{M} := \mathbb{E}^d := \mathbb{R}^d$ be the d-dimensional Euclidean space, given by the manifold $\mathbb{R}^d$ equipped with its standard Riemannian metric. Let $G := \mathrm{Iso}(\mathbb{E}^d)$ be the isometry group of $\mathbb{E}^d$. More explicitly, $\mathrm{Iso}(\mathbb{E}^d)$ is the semi-direct product $\mathbb{E}^d \rtimes O(d)$ of $\mathbb{E}^d$ (acting on itself by translations) and the orthogonal group $O(d)$. Let $\mathbb{M}^c := \mathbb{R}^{d-1} \times \{0\} \subset \mathbb{E}^d$.

A *Euclidean structure* on a smooth d-manifold Y is an $(\mathrm{Iso}(\mathbb{E}^d),\mathbb{E}^d)$-structure in the above sense. It is clear that such an atlas determines a flat Riemannian metric on Y by transporting the metric on $\mathbb{E}^d$ to U_i via the diffeomorphism φ_i.

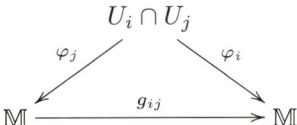

Conversely, a flat Riemannian metric on a manifold Y can be used to construct such an atlas. An $(\mathrm{Iso}(\mathbb{E}^d), \mathbb{E}^d)$-pair amounts to:

- a flat Riemannian manifold Y;
- a totally geodesic codimension one submanifold $Y^c \subset Y$;

EXAMPLE 2.30. A *Euclidean spin structure* on a d-manifold Y is exactly an $(\mathrm{Iso}(\mathbb{E}^{d|0}), \mathbb{E}^{d|0})$-structure on Y. Here $\mathbb{E}^{d|0} = \mathbb{E}^d$, and $\mathrm{Iso}(\mathbb{E}^{d|0}) = \mathbb{E}^d \rtimes Spin(d)$ is a double covering of $\mathbb{E}^d \rtimes SO(d) \subset \mathbb{E}^d \rtimes O(d)$ and acts on $\mathbb{E}^d$ via the double covering map. The peculiar notation $(\mathrm{Iso}(\mathbb{E}^{d|0}), \mathbb{E}^{d|0})$ is motivated by later generalizations: in section 4.2 we will define the super Euclidean space $\mathbb{E}^{d|\delta}$ (a supermanifold of dimension $d|\delta$) and the super Euclidean group $\mathrm{Iso}(\mathbb{E}^{d|\delta})$ (a super Lie group which acts on $\mathbb{E}^{d|\delta}$).

We note that the action of $G = \mathrm{Iso}(\mathbb{E}^{d|0})$ on $\mathbb{E}^d$ is *not effective*, since $-1 \in Spin(d)$ acts trivially. However, it lifts to an effective action on the principal $Spin(d)$-bundle $\mathbb{E}^d \times Spin(d) \to \mathbb{E}^d$ where the translation subgroup $\mathbb{E}^d$ acts trivially on $Spin(d)$, and $Spin(d) \subset G$ acts by left-multiplication. This implies that the transition elements $g_{ij} \in G$ determine a *spin-structure* on Y, i.e., a principal $Spin(d)$-bundle $Spin(Y) \to Y$ which is a double covering of the oriented frame bundle $SO(Y)$. Hence a $(\mathrm{Iso}(\mathbb{E}^{d|0}), \mathbb{E}^{d|0})$-structure on a manifold Y determines a flat Riemannian metric and a spin structure. Conversely, a spin structure on a flat Riemannian manifold Y determines a $(\mathrm{Iso}(\mathbb{E}^{d|0}), \mathbb{E}^{d|0})$-structure.

The last example shows that the definition of morphisms between manifolds equipped with $(G, \mathbb{M})$-structures requires some care: if Y, Y' are manifolds with $(\mathrm{Iso}(\mathbb{E}^{d|0}), \mathbb{E}^{d|0})$-structures, a morphisms $Y \to Y'$ should not just be an isometry $f\colon Y \to Y'$ compatible with the spin structures, but it should include an *additional datum*, namely a map $Spin(Y) \to Spin(Y')$ of the principal $Spin(d)$-bundles. In particular, each $(\mathrm{Iso}(\mathbb{E}^{d|0}), \mathbb{E}^{d|0})$-manifold Y should have an involution which acts trivially on Y, but is multiplication by $-1 \in Spin(d)$ on the principal bundle $Spin(Y) \to Y$.

It may be informative to compare this rigid geometry to other version of geometric structures on manifolds. If a Lie group H acts on a finite dimensional vector space V, an *H-structure* on a smooth manifold X is an H-principal bundle P together with an isomorphism $P \times_H V \cong TX$. An example of such an H-structure is the *flat H-structure* on the vector space V itself. It is given by $P := V \times H \to V$ and the isomorphism is given by $P \times_H V = (V \times H) \times_H V \cong V \times V \cong TV$ via translation in V.

DEFINITION 2.31. An H-structure on X is *integrable* if it is locally flat.

Examples of this notion include

(1) An integrable $GL_n(\mathbb{C})$-structure (on $V = \mathbb{C}^n$) is a complex structure: There are complex charts as discussed below.
(2) For $H = Sp(2n)$ and $V = \mathbb{R}^{2n}$ integrable structures are symplectic structures by Darboux's theorem.
(3) For $H = O(n)$ and $V = \mathbb{R}^n$ integrable structures are flat Riemannian metrics.
(4) For $H = U(n)$ and $V = \mathbb{C}^n$ integrable structures are flat Kähler structures.

The total space of the cotangent bundle T^*X carries a canonical (exact) symplectic structure. Moreover, a Riemannian metric on X induces one on T^*X. This metric is integrable if and only if the original metric is flat. Nevertheless, it turns out that T^*X is always Kähler.

There is a chart version of integrable H-structures. Choose a covering collection of charts on the manifold X with codomain being an open subset of a vector space V. We can now say that an integrable H-structure is a lift of the derivatives of transition functions φ_{ij} along the map $\rho\colon H \to GL(V)$ that satisfies the usual cocycle conditions.

An H-structure on X is *rigid* (and in particular integrable) if each transition function φ_{ij} is the (locally constant) restriction of the action on V by the semi-direct product $G = (V, +) \rtimes H$ of 'translations and rotations', see Definition 2.26. For $H = O(n)$ and $V = \mathbb{R}^n$ we get the notion of a rigid Euclidean manifold used in this paper (which is equivalent to an integrable $O(n)$-structure). For a general rigid geometry one would generalize the model space from a vector space to an arbitrary (homogenous) H-space.

We now turn to the morphisms between manifolds with rigid geometries.

DEFINITION 2.32. We denote by $(G, \mathbb{M})$-Man the category whose objects are $(G, \mathbb{M})$-manifolds. If Y, Y' are $(G, \mathbb{M})$-manifolds, a morphism from Y to Y' consists of a smooth map $f\colon Y \to Y'$ and elements $f_{i'i} \in G$ for each pair of charts (U_i, φ_i), $(U'_{i'}, \varphi'_{i'})$ with $f(U_i) \subset U'_{i'}$ such that the diagram

$$\begin{array}{ccc} U_i & \xrightarrow{\ f\ } & U_{i'} \\ {\scriptstyle\varphi_i}\downarrow & & \downarrow{\scriptstyle\varphi'_{i'}} \\ \mathbb{M} & \xrightarrow{\ f_{i'i}\ } & \mathbb{M} \end{array}$$

commutes. These elements of G are required to satisfy the coherence condition

$$f_{j'j} \cdot g_{ji} = g'_{j'i'} \cdot f_{i'i}$$

If the G-action on $\mathbb{M}$ is effective, then the elements $f_{i'i} \in G$ are determined by f and the charts φ_i, $\varphi_{i'}$, and so an isometry is simply a map $f\colon Y \to Y'$ satisfying a condition (namely, the existence of the $f_{i'i}$'s satisfying the requirements above). We note that these conditions do not imply that f is surjective or injective; e.g., a $(G, \mathbb{M})$-structure on Y induces a $(G, \mathbb{M})$-structure on any open subset $Y' \subset Y$, or covering space $Y' \to Y$ and these maps $Y' \to Y$ are morphisms of $(G, \mathbb{M})$-manifolds in a natural way.

Next, we define a 'family version' or 'parametrized version' of the category $(G, \mathbb{M})$-Man.

DEFINITION 2.33. A *family of* $(G, \mathbb{M})$*-manifolds* is a smooth map $p\colon Y \to S$ together with a maximal atlas consisting of charts which are diffeomorphisms φ_i between open subsets of Y and open subsets of $S \times \mathbb{M}$ making the following diagram commutative:

$$\begin{array}{ccc} Y \supseteq U_i & \xrightarrow[\cong]{\ \varphi_i\ } & V_i \subseteq S \times \mathbb{M} \\ & {\scriptstyle p}\searrow \quad \swarrow{\scriptstyle p_1} & \\ & S & \end{array}$$

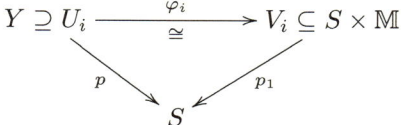

In addition, there are *transition data* which are smooth maps $g_{ij}\colon p(U_i \cap U_j) \to G$ which make the diagrams

(2.34)

$$\begin{array}{ccccc} & & U_i\cap U_j & & \\ & \swarrow^{\varphi_j} & & \searrow^{\varphi_i} & \\ p(U_i\cap U_j)\times\mathbb{M} & \xrightarrow{\mathrm{id}\times g_{ij}\times\mathrm{id}} & p(U_i\cap U_j)\times G\times\mathbb{M} & \xrightarrow{\mathrm{id}\times\mu} & p(U_i\cap U_j)\times\mathbb{M}\end{array}$$

and

$$\begin{array}{ccc} p(U_i\cap U_j\cap U_k) & & \\ {\scriptstyle g_{ij}\times g_{jk}}\downarrow & \searrow^{g_{ik}} & \\ G\times G & \xrightarrow{\mu} & G\end{array}$$

commutative. Here μ is the multiplication map of the Lie group G. We note that the conditions imply in particular that p is a *submersion* and $p(U_i \cap U_j) \subseteq S$ is open.

A *family of* $(G, \mathbb{M})$*-pairs* is a smooth map $p\colon Y \to S$, a codimension one submanifold $Y^c \subset Y$, and a maximal atlas consisting of charts $\{(U_i, \varphi_i)\}$ for Y as above, with the additional requirement that φ_i maps $U_i \cap Y^c$ to $S \times \mathbb{M}^c \subset S \times \mathbb{M}$.

If $Y \to S$ and $Y' \to S'$ are two families of $(G, \mathbb{M})$-manifolds, a *morphism* from Y to Y' consists of the following data:

- a pair of maps $(f, \hat{f})$ making the following diagram commutative:

$$\begin{array}{ccc} Y & \xrightarrow{\hat{f}} & Y' \\ \downarrow & & \downarrow \\ S & \xrightarrow{f} & S' \end{array}$$

- a smooth map $f_{i'i}\colon p(U_i) \to G$ for each pair of charts (U_i, φ_i) of Y respectively $(U'_{i'}, \varphi_{i'})$ of Y' with $\widehat{f}(U_i) \subset U'_{i'}$ making the diagrams

$$\begin{array}{ccccc} U_i & & \xrightarrow{\widehat{f}} & & U'_{i'} \\ {\scriptstyle\varphi_i}\downarrow & & & & \downarrow{\scriptstyle p_2\circ\varphi_{i'}} \\ p(U_i)\times\mathbb{M} & \xrightarrow{f_{i'i}\times\mathrm{id}} & G\times\mathbb{M} & \xrightarrow{\mu} & \mathbb{M}\end{array}$$

and

$$\begin{array}{ccccc} p(U_i\cap U_j) & & \xrightarrow{\mathrm{id}\times f_{j'j}} & & p(U_i\cap U_j)\times G \\ & & & & \downarrow{\scriptstyle g'_{i'j'}\times\mathrm{id}} \\ {\scriptstyle\mathrm{id}\times g_{ij}}\downarrow & & & & G\times G \\ & & & & \downarrow{\scriptstyle\mu} \\ p(U_i\cap U_j)\times G & \xrightarrow{f_{i'i}\times\mathrm{id}} & G\times G & \xrightarrow{\mu} & G\end{array}$$

commutative.

Abusing notation, we will usually write $(G,\mathbb{M})$-Man for this family category, but we use the notation $(G,\mathbb{M})$-$\mathsf{Man}/\mathsf{Man}$ if we want to emphasize that we talk about the family version.

2.6. Categories with flips. Let $(G,\mathbb{M})$ be a geometry in the sense discussed in the previous subsections. We note that an element g in the center of G determines an automorphism $\theta_Y\colon Y\to Y$ for any $(G,\mathbb{M})$-manifold Y. This automorphism is induced by multiplication by g on our model space $\mathbb{M}$ (more precisely, in terms of Definition 2.32 it is given by setting $f_{i'i}=g$ for every i,i'). For example, for $G=\mathrm{Iso}(\mathbb{E}^{d|0})=\mathbb{E}^d\rtimes Spin(d)$, the center of G is $\{\pm 1\}\subset Spin(d)$. If Y is a $(G,\mathbb{E}^d)$-manifold, i.e., a Euclidean manifold with spin structure (see Example 2.30), then θ_Y is multiplication by $-1\in Spin(d)$ on the principal $Spin(d)$-bundle $Spin(Y)\to Y$ (in particular, it is the identity on the underlying manifold). As in our previous paper [**HoST**] we will refer to θ_Y as 'spin-flip'.

Let E be a Euclidean spin field theory in the sense of the preliminary Definition 2.25, , i.e., a functor

$$E\colon d|0\text{-}\mathsf{EBord}\to\mathsf{TV}.$$

of categories internal to SymGrp, the strict 2-category of symmetric monoidal groupoids. Here $d|0$-EBord is the variant of the Euclidean bordism category d-EBord where 'Euclidean structures' are replaced by 'Euclidean spin structures' (see Example 2.30). Then E determines in particular a symmetric monoidal functor

$$E_0\colon d|0\text{-}\mathsf{EBord}_0\longrightarrow\mathsf{TV}_0 \tag{2.35}$$

which we can apply to an object Y of that category. Thought of as a morphism $\theta_Y\in d|0$-$\mathsf{EBord}_0(Y,Y)$, the spin-flip induces an involution on the vector space $E(Y)$; i.e., $E(Y)$ becomes a *super vector space* (for notational simplicity we drop the subscript E_0). If Y' is another object of the bordism category, then the symmetric monoidal functor E gives a commutative diagram

$$\begin{array}{ccc} EY\otimes EY' & \longrightarrow & E(Y\amalg Y')\\ {\scriptstyle\sigma_{EY,EY'}}\big\downarrow\cong & & \cong\big\downarrow{\scriptstyle E(\sigma_{Y,Y'})}\\ EY'\otimes EY & \longrightarrow & E(Y'\amalg Y)\end{array}$$

Here σ is the braiding isomorphism (in both categories), and the horizontal isomorphisms are part of the data of the symmetric monoidal functor E (see [**McL**, Ch. XI, §2]). Unfortunately, this is *not* what we want, and this reveals another shortcoming of the preliminary definition 2.25. We should emphasize that $\sigma_{EY,EY'}$ is the braiding isomorphism in the category of *ungraded* vector spaces given by

$$E(Y)\otimes E(Y')\longrightarrow E(Y')\otimes E(Y)\qquad v\otimes w\mapsto w\otimes v$$

What we want is the commutativity of the above diagram, but with $\sigma_{EY,EY'}$ being the braiding isomorphism in the category of super vector spaces given by

$$v\otimes w\mapsto(-1)^{|v||w|}w\otimes v$$

for homogeneous elements v,w of degree $|v|,|w|\in\mathbb{Z}/2$. That suggests to replace the symmetric monoidal category TV_0 by its 'super version' consisting of $\mathbb{Z}/2$-graded topological vector spaces, the projective tensor product, and the desired braiding isomorphism of super vector spaces. However, this doesn't solve the problem: $E(Y)$ now has two involutions: its grading involution θ_{EY} as super vector space and the involution $E(\theta_Y)$ induced by the spin-flip θ_Y. So we *require* that these two involutions agree, as in [**HoST**, Definition 6.44]. We will refer to θ_Y as a *flip* since the terminology 'twist' used in the context of balanced monoidal categories [**JS**, Def. 6.1] unfortunately conflicts with our use of 'twisted field theories'.

REMARK 2.36. The requirement that the grading involution on the quantum space space is induced by the spin flip of the world-sheet is motivated by the $1|1$-dimensional Σ-model with target a Riemannian spin manifold X. It turns out that the flip on the world-sheet $\mathbb{R}^{1|1}$ quantizes into the grading involution on the quantum state space $\Gamma(X;S)$, the spinors on X.

DEFINITION 2.37. A *flip* for a category C is a natural family of isomorphisms

$$\theta_Y : Y \xrightarrow{\cong} Y$$

for $Y \in \mathsf{C}$. If C, D are categories with flips, a functor $F\colon \mathsf{C} \to \mathsf{D}$ is *flip-preserving* if $F(\theta_Y) = \theta_{FY}$.

If $F, G\colon \mathsf{C} \to \mathsf{D}$ are two flip-preserving functors, and $N\colon F \to G$ is a natural transformation, we note that the commutativity of the diagram

$$\begin{array}{ccc} FY & \xrightarrow{NY} & GY \\ \downarrow{\scriptstyle \theta_{FY}} & & \downarrow{\scriptstyle \theta_{GY}} \\ FY & \xrightarrow{NY} & GY \end{array}$$

is automatic due to $\theta_{FY} = F(\theta_Y)$, $\theta_{GY} = G(\theta_Y)$. In other words, we don't need to impose any restrictions on natural transformations (other than being natural transformation between flip-preserving functors) to obtain a strict 2-category $\mathsf{Cat}^{\mathrm{f}\ell}$ whose objects are categories with flips, whose morphisms are flip preserving functors, and whose 2-morphisms are natural transformations.

Here are two basic examples of categories with flips.

EXAMPLE 2.38. Let SVect be the category of super vector spaces. An object of SVect is a vector space V equipped with a 'grading involution' $\theta_V\colon V \to V$ (which allows us to write $V = V^{ev} \oplus V^{odd}$, where V^{ev} (respectively V^{odd} is the $+1$-eigenspace (respectively -1-eigenspace of θ_V). Morphisms from V to W are linear maps $f\colon W \to V$ compatible with the grading involutions in the sense that $f \circ \theta_V = \theta_W \circ f$. In particular, θ_V is a morphism in SVect, and hence SVect is a *category with flip*.

In fact, this is a *symmetric monoid* in the strict 2-category $\mathsf{Cat}^{\mathrm{f}\ell}$ of categories with flip. For $V, W \in \mathsf{SVect}$, the tensor product $V \otimes W$ is the usual tensor product of vector spaces equipped with grading involution $\theta_{V\otimes W} = \theta_V \otimes \theta_W$, and the braiding isomorphism is described above. To check compatibility of the symmetric monoidal structure with the flip, we only need to check that the *functors* defining the symmetric monoid in Cat are compatible with the flip (as discussed above, there are no compatibility conditions for the natural transformations). These two functors are

- the tensor product

$$c\colon \mathsf{SVect} \times \mathsf{SVect} \longrightarrow \mathsf{SVect} \qquad (V, W) \mapsto V \otimes W$$

which is compatible with flips by construction of the grading involution,

- the unit functor u from the terminal object of Cat, the discrete category with one object, to SVect. The functor u maps the unique object to the monoidal unit $I \in$ SVect, which is the ground field with the trivial involution. In particular, $\theta_I = \mathrm{id}_I$, which is one way of saying that the functor u preserves the flip.

EXAMPLE 2.39. Let $(G, \mathbb{M})$ be a rigid geometry, and let g be an element of the center of G which acts trivially on the model space $\mathbb{M}$. Then as discussed at the beginning of this section, g determines an automorphism θ_Y for families of $(G, \mathbb{M})$-manifolds Y. In other words, g determines a flip θ for the category $(G, \mathbb{M})$-Man. The functor $p\colon (G, \mathbb{M})\text{-Man} \to \text{Man}$ is flip preserving if we equip Man with the trivial flip given by $\theta_M = \mathrm{id}_M$. This gives in particular a flip for the categories $d|0\text{-EBord}_i, i = 0, 1$, in our motivating example (2.35); using our new terminology we want to require that E_i are symmetric monoidal functors preserving the flip.

2.7. Fibered categories. Before defining the family versions of the bordism categories $(G, \mathbb{M})$-Bord associated to a geometry $(G, \mathbb{M})$ in the next section, we recall in this section the notion of a *Grothendieck fibration.* An excellent reference is [**Vi**], but we recall the definition for the convenience of the reader who is not familiar with this language. Before giving the formal definition, it might be useful to look at an example. Let Bun be the category whose objects are smooth fiber bundles $Y \to S$, and whose morphisms from $Y \to S$ to $Z \to T$ are smooth maps f, ϕ making the diagram

(2.40)
$$\begin{array}{ccc} Y & \xrightarrow{\phi} & Z \\ \downarrow & & \downarrow \\ S & \xrightarrow{f} & T \end{array}$$

commutative Let us consider the forgetful functor

(2.41)
$$p\colon \text{Bun} \longrightarrow \text{Man}$$

which sends a bundle to its base space. We note that if $Z \to T$ is a bundle, and $f\colon S \to T$ is a smooth map, then there is a pull-back bundle $f^*Z \to S$, and a tautological morphism of bundles $\phi\colon Y = f^*Z \to Z$ which maps to f via the functor p. The bundle morphism ϕ enjoys a universal property called *cartesian*, which more generally can be defined for any morphism $\phi\colon Y \to Z$ of a category B equipped with a functor $p\colon \mathsf{B} \to \mathsf{S}$ to another category S. In the following diagrams, an arrow going from an object $Y \in \mathsf{B}$ to an object $S \in \mathsf{S}$, written as $Y \mapsto S$, will mean $p(Y) = S$. Furthermore, the commutativity of the diagram

(2.42)
$$\begin{array}{ccc} Y & \xrightarrow{\phi} & Z \\ \downarrow & & \downarrow \\ S & \xrightarrow{f} & T \end{array}$$

will mean $p(\phi) = f$.

DEFINITION 2.43. Let $p\colon \mathsf{B} \to \mathsf{S}$ be a functor. An arrow $\phi\colon Y \to Z$ of B is *cartesian* if for any arrow $\psi\colon X \to Z$ in B and any arrow $g\colon p(X) \to p(Y)$ in S with

$p(\phi) \circ g = p(\psi)$, there exists a unique arrow $\theta\colon X \to Y$ with $p(\theta) = g$ and $\phi \circ \theta = \psi$, as in the commutative diagram

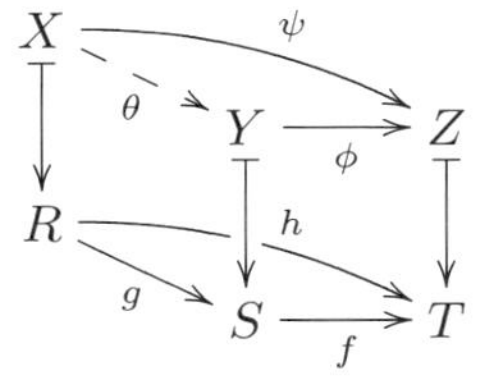

If $\phi\colon Y \to Z$ is cartesian, we say that the diagram (2.42) is a *cartesian square*.

In our example of the forgetful functor $p\colon \mathsf{Bun} \to \mathsf{Man}$, a bundle morphism (f, ϕ) as in diagram (2.40) is cartesian if and only if ϕ is a fiberwise diffeomorphism. In particular, the usual pullback of bundles provides us with many cartesian squares. This implies that the functor p is a *Grothendieck fibration* which is defined as follows.

DEFINITION 2.44. A functor $p\colon \mathsf{B} \to \mathsf{S}$ is a *(Grothendieck) fibration* if pull-backs exist: for every object $Z \in \mathsf{B}$ and every arrow $f\colon S \to T = p(Z)$ in S, there is a cartesian square

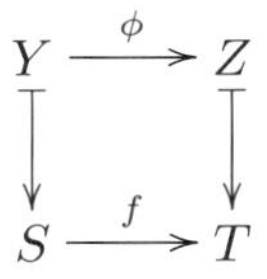

A *fibered category over* S is a category B together with a functor $p\colon \mathsf{B} \to \mathsf{S}$ which is a fibration. If $p_{\mathsf{B}}\colon \mathsf{B} \to \mathsf{S}$ and $p_{\mathsf{C}}\colon \mathsf{C} \to \mathsf{S}$ are fibered categories over S, then a *morphism of fibered categories* $F\colon \mathsf{B} \to \mathsf{C}$ is a base preserving functor ($p_{\mathsf{B}} \circ F = p_{\mathsf{C}}$) that sends cartesian arrows to cartesian arrows.

There is also a notion of base-preserving natural transformation between two morphisms from B to C. These form the 2-morphisms of a strict 2-category Cat/S whose objects are categories fibered over S and whose morphisms are morphisms of fibered categories.

REMARK 2.45. There is a close relationship between categories fibered over a category S and pseudo-functors $\mathsf{S}^{op} \to \mathsf{Cat}$ (see [**Vi**, Def. 3.10] for a definition of 'Pseudo-functor'; from an abstract point of view, this is a 2-functor, where we interpret S^{op} as a 2-category whose only 2-morphisms are identities). Any pseudo-functor determines a fibered category with an extra datum called 'cleavage', and conversely, a category C fibered over S with cleavage determines a pseudo-functor $\mathsf{S}^{op} \to \mathsf{Cat}$ (see [**Vi**, §3.1.2 and §3.1.3]). Since the 'space of cleavages' of a fibered category is in some sense contractible, it is mostly a matter of taste which language to use (see discussion in last paragraph of §3.1.3 in [**Vi**]). We prefer the language of fibered categories. An advantage of the Pseudo-functor approach is the definition of a 'stack' is a little easier, but also this can be done in terms of fibered categories (see [**Vi**, Ch. 4]).

In our example $\mathsf{Bun} \to \mathsf{Man}$ of a fibered category, there is a symmetric monoidal structure we haven't discussed yet: if $Y \to S$ and $Z \to S$ are fiber bundles over the same base manifold S, we can form the disjoint union $Y \amalg Z \to S$ to obtain a new bundle over S. This gives a morphism of fibered categories

$$c\colon \mathsf{Bun} \times_{\mathsf{Man}} \mathsf{Bun} \longrightarrow \mathsf{Bun}$$

which makes Bun a symmetric monoid in the strict 2-category $\mathsf{Cat}/\mathsf{Man}$ of categories fibered over Man (with the monoidal structure given by the categorical product). Of course, the other data u, α, λ, ρ of a symmetric monoid (see Def. 2.16) need to be specified as well – this is left to the reader.

2.8. Field theories associated to rigid geometries. The goal of this section is Definition 2.48 of a field theory associated to a 'rigid geometry' $(G, \mathbb{M})$ (see section 2.5). We begin by defining the (family) bordism category $(G, \mathbb{M})$-Bord. The definition will be modeled on Definition 2.21 of the Riemannian bordism category, but replacing Riemannian structures by $(G, \mathbb{M})$-structures. The bordism category $(G, \mathbb{M})$-Bord is a category internal to the strict 2-category $\mathsf{Sym}(\mathsf{Cat}^{\mathrm{f}\ell}/\mathsf{Man})$ of symmetric monoids in the 2-category of categories with flip fibered over Man. In fact, $(G, \mathbb{M})$-Bord_0 and $(G, \mathbb{M})$-Bord_1, the two categories over Man, have a useful additional property: they are *stacks* (see Remark 2.45). However, we won't discuss this property here, since it is not needed for the proofs of our results in this paper.

We recall that for any geometry $(G, \mathbb{M})$, there is a category $(G, \mathbb{M})$-Man of families of $(G, \mathbb{M})$-manifolds $Y \to S$. The obvious forgetful functor

$$(G, \mathbb{M})\text{-}\mathsf{Man} \to \mathsf{Man}$$

which sends a family to its parameter space is a Grothendieck fibration. Moreover, as discussed in Example 2.39, they are flip preserving. Hence it is an object in the strict 2-category $\mathsf{Cat}^{\mathrm{f}\ell}/\mathsf{Man}$ of categories with flip fibered over Man. In fact, like $\mathsf{Bun} \to \mathsf{Man}$, these categories over Man are symmetric monoids in $\mathsf{Cat}^{\mathrm{f}\ell}/\mathsf{Man}$. The monoidal structure is given by the disjoint union.

Now we can define the categories $(G, \mathbb{M})$-Bord_i for $i = 0, 1$, simply by repeating Definition 2.21 word for word, replacing 'Riemannian d-manifolds' by 'families of $(G, \mathbb{M})$-manifolds'.

DEFINITION 2.46. An object in $(G, \mathbb{M})$-Bord_0 is a quadruple $(Y, Y^c, Y^{\pm}) \to S$ consisting of a family of $(G, \mathbb{M})$-pairs (Y, Y^c) (see Definition 2.33) and a decomposition of $Y \setminus Y^c$ as the disjoint union of subspaces $Y^{\pm}$. It is required that $p \colon Y^c \to S$ is *proper*. This assumption is a family version of our previous assumption in Definition 2.21 that Y^c is compact, since it reduces to that assumption for $S = \mathrm{pt}$.

A morphism from $(Y_0, Y_0^c, Y_0^{\pm}) \to S_0$ to $(Y_1, Y_1^c, Y_1^{\pm}) \to S_1$ is a morphism from $W_0 \to S$ to $W_1 \to S$ in $(G, \mathbb{M})$-Man. Here $W_j \subset Y_j$ are open neighborhoods of Y_j^c, and it is required that this map sends $W_0 \cap Y_0^c$ to Y_1^c and $W_0 \cap Y_0^{\pm}$ to $Y_1^{\pm}$. More precisely, a morphism is a *germ* of such maps, i.e., two such maps represent the *same* morphism in $(G, \mathbb{M})$-Bord_0 if they agree on some smaller open neighborhood of Y_0^c in Y_0.

An object in $(G, \mathbb{M})$-Bord_1 consists of the following data

(1) a manifold S (the parameter space);
(2) a pair of objects $(Y_0, Y_0^c, Y_0^{\pm}) \to S$, $(Y_1, Y_1^c, Y_1^{\pm}) \to S$ of $(G, \mathbb{M})$-Bord_0 over the same parameter space S (the source respectively target);
(3) an object $\Sigma \in (G, \mathbb{M})$-$\mathsf{Man}_S$ (i.e., an S-family of $(G, \mathbb{M})$-manifolds);
(4) smooth maps $i_j \colon W_j \to \Sigma$ compatible with the projection to S. Here $W_j \subset Y_j$ are open neighborhoods of the cores Y_j^c.

Letting $i_j^{\pm} : W_j^{\pm} \to \Sigma$ be the restrictions of i_j to $W_j^{\pm} := W_j \cap Y_j^{\pm}$, we require that

(+) i_j^{+} are morphisms in $(G, \mathbb{M})$-Man_S from W_j^{+} to $\Sigma \smallsetminus i_1(W_1^{-} \cup Y_1^c)$ and (*c*) the restriction of p_Σ to the *core* $\Sigma^c := \Sigma \setminus \big(i_0(W_0^+) \cup i_1(W_1^-)\big)$ is proper.

Morphisms in the category $(G,\mathbb{M})$-Bord_1 are defined like the morphisms in d-RBord_1 in Definition 2.21, except that now the maps F, f_0, f_1 are morphisms in the category $(G,\mathbb{M})$-Man.

The categories $(G,\mathbb{M})$-Bord_0, $(G,\mathbb{M})$-Bord_1 are fibered over the category Man via the functor $p\colon (G,\mathbb{M})\text{-}\mathsf{Bord}_i \to \mathsf{Man}$ associating to an object in $(G,\mathbb{M})$-Man its parameter space $S \in \mathsf{Man}$. This functor preserves the flips (given on $(G,\mathbb{M})$-Bord_i by the choice of an element g in the center of G which acts trivially on $\mathbb{M}$; see Example 2.39) and the trivial flip on Man (i.e., $\theta_S = \mathrm{id}_S$ for all $S \in \mathsf{Man}$). This functor is also a Grothendieck fibration, and hence it is a symmetric monoid in the strict 2-category $\mathsf{Cat}^{\mathrm{f}\ell}/\mathsf{Man}$. The monoidal structure is given by the disjoint union. As explained in Definition 2.21 for the Riemannian bordism category d-RBord, the two objects $(G,\mathbb{M})$-Bord_i of the strict 2-category $\mathsf{Sym}(\mathsf{Cat}^{\mathrm{f}\ell}/\mathsf{Man})$ of symmetric monoids in $\mathsf{Cat}^{\mathrm{f}\ell}/\mathsf{Man}$ fit together to give an internal category, which we denote $(G,\mathbb{M})$-Bord (or $(G,\mathbb{M})$-$\mathsf{Bord}/\mathsf{Man}$ if we want to emphasize that we are thinking of families here). We call $(G,\mathbb{M})$-Bord the *bordism category of* $(G,\mathbb{M})$*-manifolds*. In the special case $(G,\mathbb{M}) = (\mathrm{Iso}(\mathbb{E}^d), \mathbb{E}^d)$, we call this internal category the *d-dimensional Euclidean bordism category*, and write d-EBord (or d-$\mathsf{EBord}/\mathsf{Man}$).

Next is the definition of the family version of the internal category TV. Abusing notation we will write TV for this family version as well, or TV/Man if we wish to distinguish it from its non-family version. Like the family bordism category $(G,\mathbb{M})$-Bord, the category TV is internal to the 2-category $\mathsf{Sym}(\mathsf{Cat}^{\mathrm{f}\ell}/\mathsf{Man})$. At first it seemed natural to us to think of a 'family of vector spaces' as a vector bundle over the parameter space. Later we noticed that we should let go of the local triviality assumption for reasons outlined in Remark 3.16.

DEFINITION 2.47. The internal category TV consists of categories TV_0, TV_1 with flip fibered over Man. The categories TV_0, TV_1 are defined as follows.

- TV_0 An object of TV_0 is a manifold S and a sheaf V over S of (complete, locally convex) $\mathbb{Z}/2$-graded topological modules over the structure sheaf $\mathcal{O}_S$ (of smooth functions on S). A morphism from a sheaf V over S to a sheaf W over T is a smooth map $f\colon S \to T$ together with a continuous $\mathcal{O}_T(U)$-linear map $V(f^{-1}(U)) \to W(U)$ for every open subset $U \subset T$ (here $\mathcal{O}_T(U)$ acts on $V(f^{-1})$ via the algebra homomorphism $f^*\colon \mathcal{O}_T(U) \to \mathcal{O}_S(f^{-1}(U))$. As in the category of super vector space SVect (see Example 2.38), the flip is given by the grading involution of the sheaf V.
- TV_1 An object of TV_1 consists of a manifold S, a pair of sheaves of topological $\mathcal{O}_S$-modules V_0, V_1 and an $\mathcal{O}_S$-linear map of sheaves $V_0 \to V_1$. We leave the definition of morphisms as an exercise to the reader.

Both of these fibered categories with flip are symmetric monoids in $\mathsf{Cat}^{\mathrm{f}\ell}/\mathsf{Man}$; the monoidal product is given by the projective tensor product over the structure sheaf.

DEFINITION 2.48. A $(G,\mathbb{M})$-field theory is a functor

$$(G,\mathbb{M})\text{-}\mathsf{Bord} \longrightarrow \mathsf{TV}$$

of categories internal to $\mathsf{Sym}(\mathsf{Cat}^{\mathrm{f}\ell}/\mathsf{Man})$, the strict 2-category of symmetric monoids in the 2-category of categories with flip fibered over Man. If X is a smooth manifold, a $(G,\mathbb{M})$-*field theory over* X is a functor

$$E\colon (G,\mathbb{M})\text{-}\mathsf{Bord}(X) \longrightarrow \mathsf{TV},$$

where $(G,\mathbb{M})$-$\mathsf{Bord}(X)$ is the generalization of $(G,\mathbb{M})$-Bord obtained by furnishing every $(G,\mathbb{M})$-manifold with the additional structure of a smooth map to X. Specializing the geometry $(G,\mathbb{M})$ to be the Euclidean geometry $(\mathrm{Iso}(\mathbb{E}^d),\mathbb{E}^d)$, where $\mathrm{Iso}(\mathbb{E}^d)$ is the group of isometries, we obtain the notion of a Euclidean field theory of dimension d.

3. Euclidean field theories and their partition function

3.1. Partition functions of 2-EFT's. In this subsection we will discuss Euclidean field theories of dimension 2 and $2|1$. The most basic invariant of a 2-dimensional EFT E is its *partition function*, which we defined in Definition 1.5 to be the function $Z_E\colon \mathfrak{h} \longrightarrow \mathbb{C}$, $\tau \mapsto E(T_\tau)$, where $T_\tau = \mathbb{C}/\mathbb{Z}\tau + \mathbb{Z}$. While this definition is good from the point of view that it makes contact with the definition of modular forms as functions on $\mathfrak{h}$, it is not good in the sense that a partition function should look at the value of E on every closed Euclidean 2-manifold, but not every closed Euclidean 2-manifold is isometric to one of the form T_τ. We observe that every closed oriented Euclidean 2-manifold is isometric to a torus

$$T_{\ell,\tau} := \ell(\mathbb{Z}\tau+\mathbb{Z})\backslash\mathbb{E}^2$$

obtained as the quotient of the subgroup $\ell(\mathbb{Z}\tau+\mathbb{Z}) \subset \mathbb{E}^2 \subset \mathrm{Iso}(\mathbb{E}^2)$ acting on $\mathbb{E}^2$ for some $\ell \in \mathbb{R}_+$, $\tau \in \mathfrak{h}$. Here we switch notation since we want to write objects in the bordism categories $(G,\mathbb{M})$-Bord systematically as quotients of the model space $\mathbb{M}$ by a left action of a subgroup of G acting freely on $\mathbb{M}$ (we insist on a *left* action since in the case of Euclidean structures on supermanifolds the group G is not commutative). Then we extend the domain of Z_E from $\mathfrak{h}$ to $\mathbb{R}_+ \times \mathfrak{h}$ by defining

$$Z_E\colon \mathbb{R}_+ \times \mathfrak{h} \longrightarrow \mathbb{C} \qquad \tau \mapsto E(T_{\ell,\tau}).$$

Abusing notation, we will again use the notation Z_E for this extension, and refer to it as 'partition function'.

A torus $T_{\ell,\tau}$ is isometric to $T_{\ell',\tau'}$ if and only if (ℓ,τ) and (ℓ',τ') are in the same orbit of the $SL_2(\mathbb{Z})$-action on $\mathbb{R}_+ \times \mathfrak{h}$ given by

$$\begin{pmatrix} a & b \\ c & d \end{pmatrix} (\ell,\tau) = \left(\ell|c\tau+d|, \frac{a\tau+b}{c\tau+d}\right) \tag{3.1}$$

If we forget about the first factor, the quotient stack $SL_2(\mathbb{Z})\backslash\!\!\backslash\mathfrak{h}$ has the well-known interpretation as the moduli stack of *conformal structures* on pointed tori. Similarly, the moduli stack of *Euclidean structures* on pointed tori can be identified with the quotient stack $SL_2(\mathbb{Z})\backslash\!\!\backslash(\mathbb{R}_+ \times \mathfrak{h})$. As in the conformal situation, the product $\mathbb{R}_+ \times \mathfrak{h}$ itself can be interpreted as the moduli space of Euclidean tori furnished with a basis for their integral first homology. Then the $SL_2(\mathbb{Z})$-action above corresponds to changing the basis.

What can we say about the partition function Z_E? First of all, it is a *smooth* function. To see this, we note that the tori $T_{\ell,\tau}$ fit together to a smooth bundle

$$p\colon \Sigma \to \mathbb{R}_+ \times \mathfrak{h} \tag{3.2}$$

with a fiberwise Euclidean structure, such that the fiber over $(\ell,\tau) \in \mathbb{R}_+ \times \mathfrak{h}$ is the Euclidean torus $T_{\ell,\tau}$. Applying E to this smooth family results in a smooth function on the parameter space $\mathbb{R}_+ \times \mathfrak{h}$. Compatibility of E with pullbacks guarantees that this is the function Z_E. Secondly, the partition function Z_E is *invariant under the $SL_2(\mathbb{Z})$-action*, since if (ℓ, τ), (ℓ', τ') are in the same orbit, then, as mentioned above, the tori $T_{\ell,\tau}$ and $T_{\ell',\tau'}$ are isometric, and hence $E(T_{\ell,\tau}) = E(T_{\ell',\tau'})$.

REMARK 3.3. If the function $Z_E(\ell, \tau)$ is independent of $\ell \in \mathbb{R}_+$, then the invariance of Z_E under the $SL_2(\mathbb{Z})$-action on $\mathbb{R}_+ \times \mathfrak{h}$ implies that $Z_E(1, \tau)$ has the transformation properties of a modular form of weight zero. This is the case e.g., if E is a *conformal* field theory, since the *conformal* class of the torus $T_{\ell,\tau}$ is independent of the scaling factor ℓ. If E is not conformal, there is no reason to expect $Z_E(\ell, \tau)$ to be independent of ℓ, and hence no reason for $Z_E(1, \tau)$ to be invariant under the $SL_2(\mathbb{Z})$-action. Similarly, even if E is a conformal theory, one shouldn't expect $Z_E(1, \tau)$ to be a *holomorphic* function, unless E is *holomorphic* in the sense that the operators associated to any bordism Σ depend holomorphically on the parameters determining the conformal structure on Σ. A precise definition of a holomorphic theory can be given in the terminology of this paper by working with families of conformal bordisms parametrized by complex instead of smooth manifolds.

However, as we will see in the proof of Theorem 1.15, if E has an extension to a supersymmetric Euclidean field theory of dimension $2|1$, then the function $Z_E(\ell, \tau)$ is independent of ℓ and holomorphic in τ.

As a step towards our proof of Theorem 1.15 we will prove the following result in the next section.

PROPOSITION 3.4. *Let $f\colon \mathfrak{h} \to \mathbb{C}$ be a $SL_2(\mathbb{Z})$-invariant holomorphic function, meromorphic at infinity with q-expansion $f(\tau) = \sum_{k=-N}^{\infty} a_k q^k$ with non-negative integral Fourier coefficients a_k. Then f is the partition function of a* 2-CEFT, *a conformal Euclidean field theory (see right below for a definition).*

3.2. Generators and relations of 2-EBord. A representation of a group G can be thought of as a functor $G \to \mathsf{Vect}$ from G, viewed as a groupoid with one object whose automorphism group is G, into the category of vector spaces. Hence a field theory, a functor from a bordism category to the category of (topological) vector spaces, can be thought of as a 'representation' of the bordism category. In the same way a presentation of G in terms of generators and relations is helpful when trying to construct a representation of G, a 'presentation' of the bordism category is helpful for the construction of field theories.

In this section we will construct a presentation of 2-CEBord, the *conformal (oriented) Euclidean bordism category*, a variant of 2-EBord based on the geometry

$$(\mathbb{E}^2 \rtimes (SO(2) \times \mathbb{R}_+), \mathbb{E}^2) \cong (\mathbb{C} \rtimes \mathbb{C}^\times, \mathbb{C})$$

where $\ell \in \mathbb{R}_+$ acts on $\mathbb{E}^2$ by multiplication by ℓ. The reason for our interest in 2-CEBord is that it is simpler to write down a presentation for 2-CEBord than for 2-EBord (see Proposition 3.14). Also, every conformal (oriented) Euclidean field theory gives an (oriented) Euclidean field theory by precomposing with the obvious functor between bordism categories. We will use 2-EFT's of this type to prove Proposition 3.4. We begin by describing particular objects of the categories 2-EBord_0 and 2-EBord_1.

The circle $K_\ell \in$ 2-EBord$_0$. The core of this object is the circle of length $\ell > 0$, which we prefer to think of as $\ell\mathbb{Z}\backslash\mathbb{E}^1$. The collar neighborhood is $Y = \ell\mathbb{Z}\backslash\mathbb{E}^2 \supset \ell\mathbb{Z}\backslash\mathbb{E}^1 = Y^c$; the complement $Y \setminus Y^c$ decomposes as the disjoint union of $Y^+ = \ell\mathbb{Z}\backslash\mathbb{E}^2_+$ and $Y^- = \ell\mathbb{Z}\backslash\mathbb{E}^2_-$, where $\mathbb{E}^2_\pm$ is the upper (respectively lower) open half plane. The group ℓZ acts on $\mathbb{E}^2$ via the embeddings $\ell\mathbb{Z} \subset \mathbb{R} \subset \mathbb{E}^2 \subset \mathrm{Iso}(\mathbb{E}^2)$. We note that there are simpler ways to describe this object, but it is this description that generalizes nicely to the case of Euclidean supermanifolds of dimension $2|1$. Below is a picture of the object S^1_ℓ.

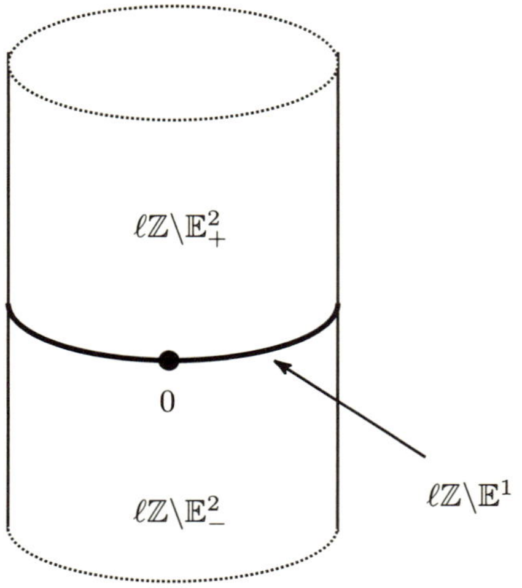

FIGURE 3. **The object K_ℓ of** 2-EBord$_0$

The cylinder $C_{\ell,\tau} \in$ 2-EBord$_1(K_\ell, K_\ell)$. We recall that an object of 2-EBord$_1$ is a pair Y_0, Y_1 of objects of 2-EBord$_0$ and bordism from Y_0 to Y_1, i.e., a triple

$$(W_1 \xrightarrow{i_1} \Sigma \xleftarrow{i_0} W_0)$$

where Σ is a Euclidean d-manifold, W_j is a neighborhood of $Y_j^c \subset Y_j$ for $j = 0, 1$, and i_0, i_1 are local isometries such that certain conditions are satisfied (see Definition 2.21 and figure 2). We make $C_{\ell,\tau}$ precise as an object of 2-EBord$_1$ by declaring it to be the following bordism from K_ℓ to itself:

$$C_{\ell,\tau} := \left(\ell\mathbb{Z}\backslash\mathbb{E}^2 \xrightarrow{\mathrm{id}} \ell\mathbb{Z}\backslash\mathbb{E}^2 \xleftarrow{\ell\tau} \ell\mathbb{Z}\backslash\mathbb{E}^2\right),$$

where $\ell\tau \in \mathfrak{h} \subset \mathbb{E}^2 \subset \mathrm{Iso}(\mathbb{E}^2)$ induces an isometry on the quotient $\ell\mathbb{Z}\backslash\mathbb{E}^2$ since it commutes with $\ell \in \mathrm{Iso}(\mathbb{E}^2)$, see figure 2.

The left cylinder $L_{\ell,\tau} \in$ 2-EBord$_1(K_\ell \amalg K_\ell, \emptyset)$:

$$L_{\ell,\tau} = \left(\emptyset \longrightarrow \ell\mathbb{Z}\backslash\mathbb{E}^2 \xleftarrow{\ell\tau \amalg -I} \ell\mathbb{Z}\backslash\mathbb{E}^2 \amalg \ell\mathbb{Z}\backslash\mathbb{E}^2\right),$$

where $-I = \left(\begin{smallmatrix} -1 & 0 \\ 0 & -1 \end{smallmatrix}\right) \in SO(2) \subset \mathrm{Iso}(\mathbb{E}^2)$. The terminology 'left' is motivated by reading bordisms from right to left, i.e., drawing the domain of the bordism on the right side, and its range on the left, as in figure 2.

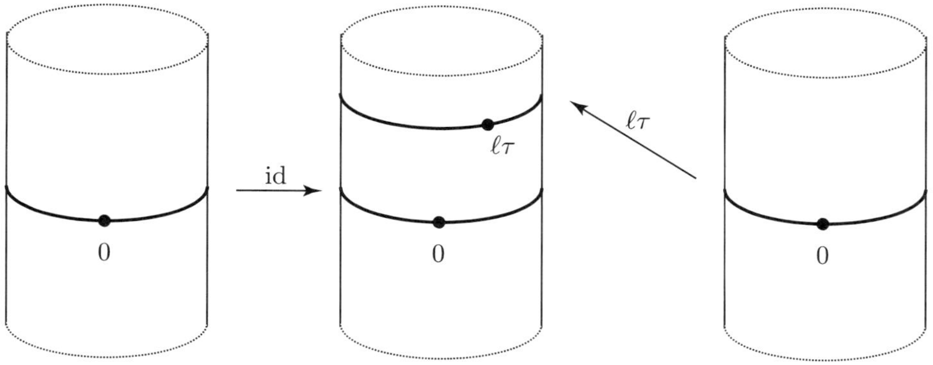

FIGURE 4. **The object** $C_{\ell,\tau} \in$ 2-$\mathsf{EBord}_1(K_\ell, K_\ell)$

The right cylinder $R_{\ell,\tau} \in$ 2-$\mathsf{EBord}_1(\emptyset, K_\ell \amalg K_\ell)$.

$$R_{\ell,\tau} = \left(\ell\mathbb{Z}\backslash\mathbb{E}^2 \amalg \ell\mathbb{Z}\backslash\mathbb{E}^2 \supset W_1 \xrightarrow{\ell\tau\circ(-I)\amalg I} \mathbb{R}^2/\ell\mathbb{Z} \longleftarrow \emptyset \right),$$

Here, using the terminology of Definition 2.21, W_1 is an open neighborhood of

$$Y_1^c = \ell\mathbb{Z}\backslash\mathbb{E}^1 \amalg \ell\mathbb{Z}\backslash\mathbb{E}^1 \quad \subset \quad \ell\mathbb{Z}\backslash\mathbb{E}^2 \amalg \ell\mathbb{Z}\backslash\mathbb{E}^2 = Y_1$$

which needs to be chosen carefully in order to satisfy conditions (+) and (c). The following choice works:

$$W_1 := \ell\mathbb{Z}\backslash(\mathbb{R} \times (-\infty, \epsilon)) \amalg \ell\mathbb{Z}\backslash(\mathbb{R} \times (-\infty, \epsilon)) \quad \subset \quad \ell\mathbb{Z}\backslash\mathbb{E}^2 \amalg \ell\mathbb{Z}\backslash\mathbb{E}^2 = Y_1,$$

where ϵ is any real number with $0 < \epsilon < \mathrm{im}(\tau)/2$. Note that in particular we need $\mathrm{im}(\tau) > 0$ for $R_{\ell,\tau}$ to exists, whereas $L_{\ell,\tau}, C_{\ell,\tau}$ make sense also for $\mathrm{im}(\tau) = 0$.

PROPOSITION 3.5. *The following isomorphisms hold in the category* 2-EBord_1*:*

$$L_{\ell,\tau} \circ \sigma \cong L_{\ell,\tau} \qquad \sigma \circ R_{\ell,\tau} \cong R_{\ell,\tau} \tag{3.6}$$

Here $\sigma\colon K_\ell \amalg K_\ell \to K_\ell \amalg K_\ell$ *is the isometry switching the two circles, i.e., the braiding isomorphism in the symmetric monoidal groupoid* 2-EBord_0*.*

$$T_{\ell,\tau} \cong T_{\ell',\tau'} \qquad \text{if } \ell' = \ell|c\tau + d|,\ \tau' = \frac{a\tau + b}{c\tau + d} \quad \text{for some } \left(\begin{smallmatrix} a & b \\ c & d \end{smallmatrix}\right) \in SL_2(\mathbb{Z}) \tag{3.7}$$

(3.8)
$$\left(\emptyset \xrightarrow{R_{\ell,\tau_1}\amalg R_{\ell,\tau_2}} K_\ell \amalg K_\ell \amalg K_\ell \amalg K_\ell \xrightarrow{\mathrm{id}\amalg L_{\ell,\tau_3}\amalg \mathrm{id}} K_\ell \amalg K_\ell\right) \cong R_{\ell,\tau_1+\tau_2+\tau_3}$$

(3.9)
$$\left(\emptyset \amalg K_\ell \xrightarrow{R_{\ell,\tau_1}\amalg \mathrm{id}} K_\ell \amalg K_\ell \amalg K_\ell \xrightarrow{\mathrm{id}\amalg L_{\ell,\tau_2}} K_\ell \amalg \emptyset = K_\ell\right) \cong C_{\ell,\tau_1+\tau_2}$$

$$\left(K_\ell \amalg K_\ell \xrightarrow{C_{\ell,\tau_1}\amalg \mathrm{id}} K_\ell \amalg K_\ell \xrightarrow{L_{\ell,\tau_2}} \emptyset\right) \cong L_{\ell,\tau_1+\tau_2} \tag{3.10}$$

$$\left(\emptyset \xrightarrow{R_{\ell,\tau_1}} K_\ell \amalg K_\ell \xrightarrow{L_{\ell,\tau_2}} \emptyset\right) \cong T_{\ell,\tau_1+\tau_2} \tag{3.11}$$

In the last 4 lines the parameter τ can be replaced by $\tau+1$, i.e., it really lies in $\mathbb{Z}\backslash\mathfrak{h}$.

These relations show in particular that $T_{\ell,\tau}$, $C_{\ell,\tau}$ and $L_{\ell,\tau}$ can be expressed in terms of $R_{\ell,\tau}$ and $L_{\ell,0}$. In fact *every* bordism (i.e., object of 2-EBord_1) can be expressed in terms of these. Still, a precise formulation of a 'presentation' of 2-EBord is tricky due to the dependence on the scale parameter. It becomes easier when we pass to the conformal Euclidean bordism category 2-CEBord, where objects become independent of the scale parameter ℓ (and hence we drop the subscript ℓ).

A 2-dimensional conformal Euclidean field theory $E\colon$ 2-$\mathsf{CEBord} \to \mathsf{TV}$ determines the following data:

(1) a topological vector space $V = E(K)$ with a smooth action of S^1;
(2) a smooth map $\rho\colon \mathbb{Z}\backslash\mathfrak{h} \to V \otimes V$, $\tau \mapsto E(R_\tau)$;
(3) a continuous linear map $\lambda = E(L_0)\colon V \otimes V \to \mathbb{C}$

These are subject to the following conditions

(a) λ and $\rho(-)$ are symmetric;
(b) the map
$$V\otimes V\otimes V\otimes V \xrightarrow{\mathrm{id}\otimes\lambda\otimes\mathrm{id}} V\otimes V\otimes V\otimes V$$
sends $\rho(\tau_1)\otimes\rho(\tau_2)$ to $\rho(\tau_1+\tau_2)$;
(c) The action map $\mathbb{Z}\backslash\mathbb{R}\times V \to V$ and the map $\mathbb{Z}\backslash\mathfrak{h}\times V \to V$, $(\tau, v)\mapsto (\gamma(\tau))(v)$ fit together to give a smooth map $\mathbb{Z}\backslash\bar{\mathfrak{h}}\times V \to V$. Here $\gamma(\tau)$ is the composition
$$V \cong \mathbb{C}\otimes V \xrightarrow{\rho(\tau)\otimes\mathrm{id}} V\otimes V\otimes V \xrightarrow{\mathrm{id}\otimes\lambda} V\otimes\mathbb{C}\cong V \tag{3.12}$$
(d) the function $\mathfrak{h}\to\mathbb{C}$, $\tau\mapsto\lambda(\rho(\tau))$ is $SL_2(\mathbb{Z})$-equivariant.

Concerning (1), we note that the automorphism group of $K \in$ 2-CEBord_1 is S^1 (which acts by rotation); by functoriality, $E(K)$ then inherits an S^1-action. The relations (a) are immediate consequences of the relations 3.6. Condition (b) is a consequence of relation 3.8, and condition (d) is a consequence of relation 3.7. Concerning (c) we note that relation 3.9 implies that $\gamma(\tau) = E(C_\tau)$ for $\tau\in\mathfrak{h}$. Unlike the right cylinders R_τ, the cylinders C_τ are defined for τ in the *closed* upper half plane $\bar{\mathfrak{h}}$. For $\tau\in\mathbb{R}$, the cylinder $C_\tau\in$ 2-CEBord_1 is the image of 'rotation by τ' under the canonical map from the endomorphisms of K in 2-CEBord_0 to the objects of 2-$\mathsf{CEBord}_1(K,K)$. It follows that $\gamma(\tau)$ for $\tau\in\mathbb{R}$ is given by the rotation action on V mentioned in (1). The C_τ's for $\tau \in \bar{\mathfrak{h}}$ fit together to form a smooth family of cylinders parametrized by $\bar{\mathfrak{h}}$. Applying the functor E, we see that

$$\gamma\colon \bar{\mathfrak{h}} \times V \longrightarrow V \tag{3.13}$$

is a smooth map.

The reader might wonder what we mean here by saying that γ is 'smooth' since $\bar{\mathfrak{h}}$ is a manifold with boundary. For this we require that for every smooth map $f\colon S \to \bar{\mathfrak{h}} \subset \mathbb{R}^2$ (where $S \in \mathsf{Man}$ is a manifold without boundary) the composition with f is smooth. In other words, we are thinking of manifolds with boundary as presheaves on Man. It is a formality to extend a category fibered over S to a category fibered over the category of presheaves on S. Similarly, a morphism of categories fibered over S extends canonically to a morphism of these larger categories fibered over the presheaves on S.

PROPOSITION 3.14 ([**ST5**]). *Given a triple (V, λ, ρ) as in (1)-(3) above, satisfying conditions (a)-(d), there is a unique conformal Euclidean field theory E of dimension 2 which realizes the triple in the sense that $E(K, L_0, R_\tau) = (V, \lambda, \rho(\tau))$.*

The proof – which we won't give in this paper – is based on two facts:

(1) every object $(Y, Y^c, Y^\pm) \in 2\text{-}\mathsf{CEBord}_0$ with connected core Y^c is equivalent to K.
(2) every object $(\Sigma, i_0, i_1) \in 2\text{-}\mathsf{CEBord}_1$ with connected core is isomorphic to C_τ, L_τ, R_τ or T_τ for some $\tau \in \mathbb{Z}\backslash\mathfrak{h}$.

The relations of Proposition 3.5 then imply that $E(C_\tau)$ and $E(L_\tau)$ can be reconstructed from the data λ and ρ. Using the monoidal structure, the functor E is then determined. This is basically the argument in the non-family version.

PROOF OF PROPOSITION 3.4. Let $f\colon \mathfrak{h} \to \mathbb{C}$ be a $SL_2(\mathbb{Z})$-invariant holomorphic function, meromorphic at infinity with q-expansion $f(\tau) = \sum_{k=-N}^{\infty} a_k q^k$. We will use Proposition 3.14 to construct $E \in 2\text{-EFT}$ whose partition function is f. We construct the data V, λ, ρ as follows:

- V is a completion, described below in Remark 3.15, of the algebraic direct sum
$$\bigoplus_{k=-N}^{\infty} V_k = \{(v_k)_{k=-N}^{\infty} \mid v_k \in V_k\}$$
Here $V_k = \mathbb{C}^{a_k}$, and the group S^1 acts on V_k by letting $q \in S^1$ act by scalar multiplication by q^k.
- $\lambda\colon V \otimes V \to \mathbb{C}$, $(v_k) \otimes (w_k) \mapsto \sum_{k=-N}^{\infty} \langle \bar{v}_k, w_k\rangle$, where $\langle \bar{v}_k, w_k\rangle$ is the Hermitian inner product of $\bar{v}_k$, the complex conjugate of v_k, and w_k (this depends $\mathbb{C}$-linearly on v_k and w_k).
- $\rho\colon \mathfrak{h} \to V \otimes V$, $\tau \mapsto \sum_{k=-N}^{\infty} \sum_{i=1}^{a_k} q^k e_i \otimes e_i$, where $\{e_i\}$ is the standard basis of $\mathbb{C}^{a_k}$.

It is straightforward to check that conditions (a)-(d) are satisfied, and hence there is a 2-dimensional conformal Euclidean field theory E that realizes these data. Let us calculate the partition function of E:

$$Z_E(\tau) = E(T_\tau) = E(L_0 \circ R_\tau) = E(L_0) \circ E(R_\tau) = \lambda(\rho) = \sum_{k=-N}^{\infty} a_k q^k = f(\tau)$$

REMARK 3.15. It is natural to complete the algebraic direct sum $\bigoplus_{k=-N}^{\infty} V_k$ to the Hilbert space direct sum

$$H = \{v = (v_k) \in \prod_k V_k \mid ||v|| := \sum_k |v_k|^2 < \infty\}$$

Unfortunately if we do this, the action map $S^1 \times H \to H$ is not smooth: its derivative at $\tau = 0$ is the operator $N\colon H \to H$ which is multiplication by $2\pi i k$ on V_k. In particular, the eigenvalues of N are unbounded (except if f is a Laurent polynomial which only happens if f is a constant modular function), and hence N is not continuous. At first glance it seems that this is a problem no matter which topology on V we pick: if the eigenvalues of N are unbounded, we expect that N is not continuous. However, this is not the case if we leave the world of Banach spaces behind: e.g., the operator $\frac{d}{d\theta}$ acting on $C^\infty(S^1)$ equipped with the Fréchet topology is continuous despite its eigenvalues being unbounded.

This example suggests to define

$$V := \left\{ v = (v_k) \in \prod_k V_k \mid ||v||_n := \sum_k |v_k|^2 k^{2n} < \infty \, \forall n \in \mathbb{N} \right\}$$

equipped with the Fréchet topology determined by the semi-norms $||\ \ ||_n$. It is clear that N is continuous in this topology and it can be shown that the action map $\mathfrak{h} \times V \to V$ is smooth using this topology on V.

□

REMARK 3.16. Unlike infinite dimensional separable Hilbert spaces, there are many non-isomorphic topological vector spaces constructed this way (for a complete isomorphism classification of these spaces see [**MV**, Prop. 29.1]). For a 2-EFT E the isomorphism class of the topological vector space $E(S^1_\ell)$ is an invariant of E, but we don't want it to be an invariant for the *concordance class of* E. This forces us to consider *sheaves* rather than locally trivial bundles of topological vector spaces as the appropriate notion of 'family of topological vector spaces' (see Definition 2.47).

3.3. Partition functions of spin 2-EFT's. The goal of this subsection is to show that every integral holomorphic modular function is the partition function of a *spin* 2-EFT. Our interest in spin EFT's comes from the fact that they are closely related to 2|1-EFT's: any 2|1-EFT E determines a reduced spin 2-EFT $\bar{E}$, and the partition function of E by definition is the partition function of $\bar{E}$.

PROPOSITION 3.17. *Every integral holomorphic modular function is the partition function of a* 2|0-CEFT.

PROOF. We've described the objects K_ℓ, $T_{\ell,\tau}$, $C_{\ell,\tau}$, etc as quotients of (subsets of) our model space $\mathsf{M} = \mathbb{E}^2$. These quotients are orbit spaces for a subgroup H of our symmetry group $G = \mathbb{R}^2 \rtimes SO(2)$ which acts freely on M. This subgroup H is contained in the translation group $\mathbb{R}^2 \subset G$ and is isomorphic to $\mathbb{Z}^2$ for $T_{\ell,\tau}$, and isomorphic to $\mathbb{Z}$ in the other cases.

We recall that in the case of Euclidean spin-structures, the model space is still $\mathbb{E}^2$, but the symmetry group now is $\mathbb{E}^2 \rtimes Spin(2)$, which acts on $\mathbb{E}^2$ via the double covering map $\mathbb{E}^2 \rtimes Spin(2) \to \mathbb{R}^2 \rtimes SO(2)$. In particular, the action of the symmetry group on the model space is no longer effective which makes understanding this structure more challenging. For the case at hand, it is helpful to think of the group $\mathbb{E}^2 \rtimes Spin(2)$ as acting not only on $\mathbb{E}^2$, but also compatibly on the spinor bundle of $\mathbb{E}^2$. This action is effective: the element $-1 \in Spin(2)$ in the kernel of $Spin(2) \to SO(2)$ acts trivially on $\mathbb{R}^2$, but by multiplication by -1 on its spinor bundle.

To specify the spin structure on $\mathbb{Z}\backslash\mathbb{E}^2$, we need to pick a lift of the generator of $\mathbb{Z} \subset \mathbb{E}^2 \subset \mathbb{E}^2 \rtimes SO(2)$ to $\mathbb{E}^2 \rtimes Spin(2)$; in other words, we need to pick an element of $\{\pm 1\} \subset Spin(2)$. The choice $+1$ is usually referred to as 'periodic spin structure', since sections of the spinor bundle of the quotient can be interpreted as $\mathbb{C}$-valued functions on $\mathbb{E}^2$ which are periodic (i.e., invariant) w.r.t. the $\mathbb{Z}$-action. The choice $\{-1\}$ is called 'anti-periodic spin structure' since spinors can be identified with functions which are anti-periodic, i.e., the generator acts by multiplication by -1. For a quotient of the form $\mathbb{Z}^2\backslash\mathbb{E}^2$, we need to specify an element in $\{\pm 1\}$ for *both* generators. We will use superscripts $s \in \{\pm\}$ to specify the spin structure, e.g., K_ℓ^s or $C_{\ell,\tau}^s$ (and in Section 5 we will use the notation $+,-$ for p,a). For the torus $T_{\ell,\tau} = \ell(\mathbb{Z}\tau + \mathbb{Z})\backslash\mathbb{E}^2$ we specify the spin structure by two superscripts: $T_{\ell,\tau}^{s_1 s_2}$ (with s_1 corresponding to the first generator $\ell\tau$; s_2 corresponds to the second generator ℓ).

Now the proof proceeds as in the previous case: a conformal Euclidean spin field theory determines the following algebraic data

- a $\mathbb{Z}/2$-graded topological vector space $V^s := E(K^s)$ for both spin structures $s = \pm$ with an action of $\mathbb{R}/\mathbb{Z}$ (for $s = +$) respectively $\mathbb{R}/2\mathbb{Z}$ (for $s = -$);
- a continuous linear map $\lambda^s = E(L_0^s)\colon V \otimes V \to \mathbb{C}$ for $s = \pm$;
- a smooth map $\rho^s \colon \mathfrak{h} \to V^s \otimes V^s$ given by $\tau \mapsto E(R_\tau^s)$ for $s = \pm$;

which are subject to compatibility conditions. Conversely, if these compatibility conditions are satisfied, then these data are realized by a conformal Euclidean spin field theory.

We recall that from the discussion of the non-spin case that the most interesting compatibility condition came from the transformation properties of the partition function $Z\colon \mathfrak{h} \to \mathbb{C}$, $\tau \mapsto E(T_\tau)$. This is the only condition that we will formulate and check in the spin case. The analog of the partition function in the spin case is the function

$$Z\colon \coprod_{s_1,s_2} \mathfrak{h}^{s_1 s_2} \longrightarrow \mathbb{C} \qquad \text{given by} \qquad \tau \mapsto E(T_\tau^{s_1 s_2}) \quad \text{for} \quad \tau \in \mathfrak{h}^{s_1 s_2}$$

Here the superscripts s_1, s_2 is just a way to distinguish the four copies of $\mathfrak{h}$ corresponding to the four spin structures.

In terms of the data above, Z is determined by

$$Z(\tau) = \begin{cases} \lambda^s((\alpha \otimes \mathrm{id})c(\rho_\tau^s)) & \tau \in \mathfrak{h}^{+s} \\ \lambda^s(c(\rho_\tau^s)) & \tau \in \mathfrak{h}^{-s} \end{cases}$$

where $c\colon V \otimes V \cong V \otimes V$ is the braiding isomorphism, and $\alpha\colon V \to V$ is the grading involution. It can be shown that the isometry $T_{\ell,\tau}^{s_1 s_2} \to T_{A(\ell,\tau)}^{s_1', s_2'}$ for $A \in SL_2(\mathbb{Z})$ is spin structure preserving if and only if

$$A \begin{pmatrix} s_1 \\ s_2 \end{pmatrix} = \begin{pmatrix} s_1' \\ s_2' \end{pmatrix}$$

Here we think of $\binom{s_1}{s_2}$ as a vector in $(\mathbb{Z}/2)^2$ (i.e., identifying $\mathbb{Z}/2$ and $\{\pm\}$). This implies that the function Z is $SL_2(\mathbb{Z})$-equivariant, where $SL_2(\mathbb{Z})$ acts on $\mathfrak{h}$ as in 3.1 and permutes the four copies of $\mathfrak{h}$ in the obvious way.

Now let $f\colon \mathfrak{h} \to \mathbb{C}$ be a holomorphic modular function with q-expansion $f(\tau) = \sum_{k=-N}^{\infty} a_k q^k$ with non-negative integral Fourier coefficients a_k. To find a spin 2-EFT with partition function f, we need to construct data V^s, λ^s and ρ^s_τ as above such that $Z(\tau) = f(\tau)$ for $\tau \in \mathfrak{h}^{++}$. Starting with f, we construct V, λ, ρ_τ as in the proof of Proposition 3.4 such that $\lambda(c(\rho_\tau)) = f(\tau)$. We note that

$$\lambda((\alpha \otimes \mathrm{id})c(\rho_\tau)) = \lambda(c(\rho_\tau)),$$

since V is an even vector space thanks to our assumption that the Fourier coefficients of f are non-negative. Hence defining $V^s := V$, $\lambda^s := \lambda$, $\rho^s_\tau := \rho_\tau$ for $s = \pm$, we deduce that $Z(\tau) = f(\tau)$ for $\tau \in \mathfrak{h}^{s_1 s_2}$ for *all* spin structures s_1, s_2. In particular, the function Z is $SL_2(\mathbb{Z})$-equivariant. Changing the $\mathbb{Z}/2$-grading on V^+, while not changing V^-, λ^s and ρ^s has the effect that $Z(\tau)$ changes sign for $\tau \in \mathfrak{h}^{++}$, while it doesn't change for $\tau \in \mathfrak{h}^{s_1 s_2}$, $s_1 s_2 \neq ++$. It follows that $-f$ can also be realized as partition function of a spin 2-EFT.

To finish the proof, we recall that the ring of integral weakly holomorphic forms is given by

$$MF^* = \mathbb{Z}[c_4, c_6, \Delta^{\pm 1}]/(c_4^3 - c_6^2 - 1728\Delta)$$

where

$$c_4 = 1 + 240 \sum_{k>0} \sigma_3(k) q^k \qquad c_6 = 1 - 504 \sum_{k>0} \sigma_5(k) q^k$$

are modular forms of weight 4 respectively 6. In particular, every element in MF^0 is an integral linear combination of positive powers of $j := c_4^{3n}/\Delta^n$. We note that $\Delta^{-1} = q^{-1} \prod_{m=1}^{\infty} \sum_i q^{mi}$ has non-negative Fourier coefficients and hence so does j. By our arguments above it follows that j and $-j$ are the partition functions of spin 2-EFT's. Forming sums and tensor products of field theories, we see that every linear combination of powers of j is realized as the partition function of a spin 2-EFT. □

3.4. Modularity of the partition function of 2|1-EFT's, part I. The proof of Theorem 1.15 (according to which the partition function of a 2|1-EFT is an weakly holomorphic integral modular form of weight zero) consists of two steps. This section provides the first of these steps, namely to show that if E is a spin 2-EFT (of degree zero) satisfying an additional condition, then the partition function of E is a weakly holomorphic integral modular form of weight zero (see Proposition 3.18). After defining 2|1-EFT's in section 4, we will define the partition function of a 2|1-EFT as the partition function of the spin 2-EFT E it determines. We will call E the *associated reduced Euclidean field theory.* If $E \in 2|0$-EFT is obtained this way, we say that E has a 'supersymmetric extension'. We will show in Proposition 4.16 that if E has a supersymmetric extension, then it satisfies the condition of Proposition 3.18. This is the second and final step in the proof of Theorem 1.15.

Let E be a spin 2-EFT, and for $s \in \{\pm\}$ let $V^s := E(K^s_\ell)$ be the topological vector space associated to K^s_ℓ (we recall from section 3.2 that K^s_ℓ is the object of 2|0-EBord_0 whose core $(K^s_\ell)^c$ is the circle of diameter ℓ with spin structure s; here $s = +$ corresponds to the periodic, and $s = -$ to the anti-periodic spin structure). Applying the functor E to the cylinder $C^s_{\ell,\tau}$, $\tau \in \bar{\mathfrak{h}}$, (an object in $2|0\text{-}\mathsf{EBord}_1(K^s_\ell, K^s_\ell)$; see section 3.2), we obtain an operator (i.e., a continuous linear map)

$$A^s_\tau = A^s(\tau) := E(C^s_{\ell,\tau})\colon V^s \longrightarrow V^s$$

We note that for fixed ℓ the map $\bar{\mathfrak{h}} \times V^s \longrightarrow V^s$, $(\tau, v) \mapsto A^s_\tau(v)$ is smooth as discussed above (see (3.13)).

PROPOSITION 3.18. *Let E be a spin EFT of dimension 2. We assume that for every $\tau \in \mathfrak{h}$ there is an odd operator $B_\tau\colon V \to V$ which commutes with A^+_τ and satisfies*

$$\frac{\partial A^+_\tau(v)}{\partial \bar{\tau}} = -B^2_\tau(v) \qquad \forall v \in V. \tag{3.19}$$

Then the partition function $Z^{++}_E\colon \mathfrak{h}^{++} \to \mathbb{C}$ is a holomorphic modular function with integral Fourier coefficients.

The proof of this proposition is based on a 'supersymmetry cancellation argument' (see proof of Lemma 3.27 below), an argument that seems to be fairly standard in the physics literature, at least on the Lie algebra level, i.e., for infinitesimal generators of the above families A_τ, B_τ.

The proof of this result consists of a number of steps that we formulate as lemmas.

LEMMA 3.20. *For any spin structure $s \in \{\pm\}$ and $\tau \in \mathfrak{h}$ the operator A^s_τ is* compact, *i.e., in the closure of the finite rank operators with the respect to the compact-open topology on the space of continuous linear maps $V^s \to V^s$.*

PROOF. We use a line of argument developed in our paper [**ST2**]. A central definition in that paper was the following.

DEFINITION 3.21. A morphism $f\colon X \to Y$ in a monoidal category C is *thick* if it can be factored in the form

$$X \cong I \otimes X \xrightarrow{t \otimes \mathrm{id}_X} Y \otimes Z \otimes X \xrightarrow{\mathrm{id}_Y \otimes b} Y \otimes I \cong Y \tag{3.22}$$

for morphisms $t\colon I \to Y \otimes Z$, $b\colon Z \otimes X \to I$.

We will consider thick morphisms in the fibered category $2|0\text{-}\overline{\mathsf{EBord}}$ over Man whose objects are the objects of $2|0\text{-}\mathsf{EBord}_0$. For $Y_0, Y_1 \in 2\text{-}\mathsf{EBord}_0$ we define the morphism set $2|0\text{-}\overline{\mathsf{EBord}}(Y_0, Y_1)$ to be the isomorphism classes of objects in $2|0\text{-}\mathsf{EBord}_1(Y_0, Y_1)$. We note that the functors s, t, u and c give $2|0\text{-}\overline{\mathsf{EBord}}$ the structure of a category. The monoidal structure on the groupoids $2|0\text{-}\mathsf{EBord}_i$, $i = 0, 1$, induces a monoidal structure on $2|0\text{-}\overline{\mathsf{EBord}}$. Moreover, the functor $E\colon 2|0\text{-}\mathsf{EBord} \to \mathsf{TV}$ (of categories internal to $\mathsf{Sym}(\mathsf{Cat}^{f\ell}/\mathsf{Man})$) induces a monoidal functor, fibered over Man,

$$\overline{E}\colon 2|0\text{-}\overline{\mathsf{EBord}} \longrightarrow \overline{\mathsf{TV}}$$

The isomorphism (3.8) of Proposition 3.5 holds as well in the bordism category $2|0\text{-}\mathsf{EBord}$ with a fixed spin structure s on the cylinders involved. This implies that the morphism $C^s_{\ell,\tau} \in 2|0\text{-}\overline{\mathsf{EBord}}(K^s_\ell, K^s_\ell)$ factors in the form

$$K^s_\ell = \emptyset \amalg K^s_\ell \xrightarrow{R^s_{\ell,\tau} \amalg \mathrm{id}} K^s_\ell \amalg K^s_\ell \amalg K^s_\ell \xrightarrow{\mathrm{id} \amalg L^s_{\ell,0}} K^s_\ell \amalg \emptyset = K^s_\ell\ .$$

In particular, $C^s_{\ell,\tau}$ is a thick morphism in the bordism category $2|0\text{-}\overline{\mathsf{EBord}}$ for $\tau \in \mathfrak{h}$ (note that unlike $C^s_{\ell,\tau}$ which is defined for τ in the *closed* upper half-plane, $R^s_{\ell,\tau}$ is only defined for $\tau \in \mathfrak{h}$; is is not hard to show that $C^s_{\ell,\tau}$ is not thick for $\tau \in \mathbb{R} \subset \bar{\mathfrak{h}}$).

It follows that $A^s_\tau = \overline{E}(C^s_{\ell,\tau})\colon V^s \to V^s$ is a thick morphism in the category $\overline{\mathsf{TV}}$. One of the main results of our paper [**ST2**, Theorem 4.27] gives a characterization of thick morphisms in the category TV as *nuclear operators* (known as *trace class operators* if domain and range are Hilbert spaces). We won't repeat here the definition of 'nuclear operator' (see e.g., [**ST2**, Def. 4.25, Def. 4.28 and Lemma 4.37]); it suffices here to know that any nuclear operator is compact. □

We observe that the relations (3.8) and (3.9) imply the isomorphism $C_{\ell,\tau_1} \circ C_{\ell,\tau_2} \cong C_{\ell,\tau_1+\tau_2}$ in $2\text{-}\mathsf{EBord}_1$ for $\tau_1, \tau_2 \in \mathfrak{h}$. The same relation holds in $2|0\text{-}\mathsf{EBord}$ if we consider cylinders $C^s_{\ell,\tau}$ for a fixed spin structure $s \in \{\pm\}$. This implies that for fixed ℓ, $A^s_\tau = E(C^s_{\ell,\tau})$ is a commutative semi-group of compact operators parametrized by the upper half-plane $\mathfrak{h}$. The following considerations are independent of the spin structure s and so we will suppress the superscript $s \in \{\pm\}$. By the spectral theorem we can decompose V into a sum of simultaneous (generalized) eigenspaces for these operators. The non-zero eigenvalues give smooth homomorphisms $\mathfrak{h} \to \mathbb{C}^*$ and hence can be written in the form

$$\mu(\tau) = e^{2\pi i(a\tau - b\bar{\tau})} = q^a \bar{q}^b \tag{3.23}$$

for some $a, b \in \mathbb{C}$. Let us denote by $V_{a,b} \subset V$ the generalized eigenspace corresponding to the eigenvalue function $\mu(\tau)$ given by equation (3.23). We note that the spaces $V_{a,b}$ are *finite dimensional*, since the operators A_τ are compact; in particular, any generalized eigenspace with non-zero eigenvalue is finite dimensional.

The element $\rho_\tau \in V \otimes V$ can now be studied on the $V_{a,b}$'s. The following result shows that the 'off-diagonal' entries vanish.

LEMMA 3.24. *Let $p_{a,b}\colon V \to V_{a,b}$ be the spectral projection onto $V_{a,b}$. Then for $(a,b) \neq (a',b')$*

$$p_{a,b} \otimes p_{a',b'}(\rho_\tau) = 0$$

PROOF. The isomorphisms (3.8) and (3.9) in Proposition 3.5 (or more precisely, their analog for $2|0\text{-}\mathsf{EBord}$) imply that the composition

$$\emptyset \xrightarrow{R^s_{\ell,\tau}} K^s_\ell \amalg K^s_\ell \xrightarrow{C^s_{\ell,\tau_1} \amalg C^s_{\ell,\tau_2}} K^s_\ell \amalg K^s_\ell$$

in $2|0\text{-}\overline{\mathsf{EBord}}$ is equal to $R^s_{\ell,\tau+\tau_1+\tau_2}$. Applying the functor $\overline{E}$, we obtain

$$(A_{\tau_1} \otimes A_{\tau_2})(\rho_\tau) = \rho_{\tau+\tau_1+\tau_2}$$

In particular, $(A_{\tau_1} \otimes \mathrm{id})(\rho_\tau) = (\mathrm{id} \otimes A_{\tau_1})(\rho_\tau)$. Restricted to $V_{a,b} \otimes V_{a',b'}$, the operator $A_{\tau_1} \otimes \mathrm{id}$ has eigenvalue $\mu_{a,b}(\tau_1)$, while $\mathrm{id} \otimes A_{\tau_1}$ has eigenvalue $\mu_{a',b'}(\tau_1)$, this shows that the projection of ρ_τ to $V_{a,b} \otimes V_{a',b'}$ must be zero. In fact, this argument only applies to the actual Eigenspaces, not the generalized Eigenspaces. However, using the Jordan normal form, we see that by an iterated application of our argument to the relevant filtration, we can conclude the same result. □

We recall that $Z^{++}(\tau) = \mu^+(\sigma(\rho^+_\tau))$ and we shall suppress the superscripts '+' from now on.

LEMMA 3.25. $\mu(\sigma(p_{a,b} \otimes p_{a,b})(\rho_\tau)) = \mathrm{str}((A_\tau)_{|V_{a,b}})$

PROOF. The restriction of the form $\lambda\colon V \otimes V \to \mathbb{C}$ to $V_{a,b}$ is non-degenerate (otherwise the restriction of A_τ, which can be expressed in terms of λ and ρ, would have a non-trivial kernel. Let $\{e_i\}$ be a basis of $V_{a,b}$ such that $\lambda(e_i \otimes e_j) = \delta_{ij}$. Writing $(p_{a,b} \otimes p_{a,b})(\rho_\tau)$ as a linear combination of the basis elements $e_i \otimes e_j$ for $V_{a,b} \otimes V_{a,b}$ and calculating both sides proves the lemma. $\square$

We conclude the last equality in the following equation. For the second equality more care is needed since V is not just the direct sum of the $V_{a,b}$'s. However, a careful estimate shows that for a fixed $a \in \mathbb{Z}$ the 'small' eigenvalues of A_τ don't contribute to the coefficient of q^a in the q-expansion of $Z(\tau)$, see [**ST3**]

$$Z(\tau) = \lambda(\sigma(\rho_\tau)) = \sum_{a,b} \operatorname{str}((A_\tau)_{|V_{a,b}}) = \sum_{a,b} \mu_{a,b}(\tau) \operatorname{sdim} V_{a,b} \tag{3.26}$$

LEMMA 3.27. $\operatorname{sdim} V_{a,b} = 0$ *for* $b \neq 0$.

PROOF. By assumption, the operator $B\colon V \to V$ commutes with the operators A_τ and hence it restricts to an operator $B_{|V_{a,b}}\colon V_{a,b} \to V_{a,b}$. Restricted to $V_{a,b}$, the operator A_τ has eigenvalue $\mu_{a,b}(\tau)$ and hence $\partial A_\tau/\partial\bar\tau$ has eigenvalue

$$\frac{\partial \mu_{a,b}(\tau)}{\partial \bar\tau} = \frac{\partial}{\partial\bar\tau} e^{2\pi i(a\tau - b\bar\tau)} = -2\pi i b \mu_{a,b}(\tau)$$

In particular, for $b \neq 0$, the operator $\partial A_\tau/\partial\bar\tau = -B^2$ is invertible, and hence $B\colon V_{a,b} \to V_{a,b}$ is an isomorphism. Since B is odd, it maps the even part $V_{a,b}^{ev}$ isomorphically to the odd part $V_{a,b}^{odd}$ and hence $\operatorname{sdim} V_{a,b} = 0$. $\square$

This then implies $Z_E(\tau) = \sum_{a \in \mathbb{Z}} \operatorname{sdim} V_{a,0}\, q^a$. Here a is an integer by the following argument. By construction, $C_{\ell,\tau} \cong C_{\ell,\tau'}$ if $\tau = \tau' \mod \mathbb{Z}$ and hence the operator $A_\tau = E(C_{\ell,\tau})$ depends only on $q = e^{2\pi i\tau}$. That forces $\mu_{a,b}(\tau) = q^a$ to be a function of q which forces a to be an integer.

This shows that $Z^{++}(\tau)$ is a *holomorphic* function with an integral q-expansion. Moreover, the fact that A_τ is a compact operator forces $V_{a,0} = 0$ for sufficiently negative integers a.

For the arguments so far, we fixed $\ell > 0$, and suppressed the ℓ-dependence in the notation. If we vary ℓ, the function $Z(\ell,\tau)$ and hence the coefficients of its q-expansion depend continuously on that parameter. Hence the *integrality* of the coefficients show that $Z(\ell,\tau)$ is in fact *independent of* ℓ. This implies by remark 3.3 that $Z(1,\tau)$ is a holomorphic modular function.

4. Supersymmetric Euclidean field theories

In this section we will define *supersymmetric Euclidean field theories* by replacing manifolds by supermanifolds and Euclidean structures by super Euclidean structures in the definitions of the previous two sections. In more detail, we will consider rigid geometries i.e., $(G,\mathbb{M})$-structures in the category of supermanifolds (Def. 4.4), and will introduce in particular the Euclidean geometry $(\operatorname{Iso}(\mathbb{E}^{d|\delta}), \mathbb{E}^{d|\delta})$ (see section 4.2). We will define super versions of our categories TV, $(G,\mathbb{M})$-Bord and d-EBord. From a categorical point of view, these categories will be internal to the strict 2-category $\mathsf{Sym}(\mathsf{Cat}^{fl}/\mathsf{csM})$ of symmetric monoids in the 2-category of categories with flip fibered over the category csM of supermanifolds. A $(G,\mathbb{M})$-field theory is then defined to be a functor $(G,\mathbb{M})$-$\mathsf{Bord} \to \mathsf{TV}$ of these internal categories

(Definition 4.12); a (supersymmetric) Euclidean field theory of dimension $d|\delta$ is the special case where $(G, \mathbb{M})$ is (super) Euclidean geometry of dimension $d|\delta$. We end this section with the definition of the partition function of a Euclidean field theory of dimension $2|1$ (Def. 4.13), and the proof that this partition function is an integral holomorphic modular function.

4.1. Supermanifolds. There are two notions of 'supermanifolds', based on commutative superalgebras over $\mathbb{R}$ and $\mathbb{C}$, respectively. Our interest in field theories of dimension $2|1$ in this paper forces us to work with the latter (see Remark 4.11). We recall that a *commutative superalgebra* is a commutative monoid in the symmetric monoidal category SVect of super vector spaces (see Example 2.38). More down to earth, it is a $\mathbb{Z}/2$-graded algebra such that for any homogeneous elements a, b we have $a \cdot b = (-1)^{|a||b|} b \cdot a$.

DEFINITION 4.1. A *supermanifold M of dimension $p|q$* is a pair $(M_{\mathrm{red}}, \mathcal{O}_M)$ consisting of a (Hausdorff, second countable) topological space M_{red} (called the *reduced manifold*) and a sheaf $\mathcal{O}_M$ (called the *structure sheaf*) of commutative superalgebras over $\mathbb{C}$ locally isomorphic to $(\mathbb{R}^p, \mathcal{C}^\infty(\mathbb{R}^p) \otimes \Lambda[\theta_1, \dots, \theta_q])$. Here $\mathcal{C}^\infty(\mathbb{R}^p)$ is the sheaf of smooth complex valued functions on $\mathbb{R}^p$, and $\Lambda[\theta_1, \dots, \theta_q])$ is the exterior algebra generated by elements $\theta_1, \dots, \theta_q$ (which is equipped with a $\mathbb{Z}/2$-grading by declaring the elements θ_i to be odd). It is more customary to require that $\mathcal{O}_M$ is a sheaf of real algebras; in this paper we will always be dealing with a *structure sheaf of complex algebras* (these are called *cs-manifolds* in [**DM**]). Abusing language, the global sections of $\mathcal{O}_M$ are called *functions* on M; we will write $C^\infty(M)$ for the algebra of functions on M.

As explained by Deligne-Morgan in [**DM**, §2.1], the quotient sheaf $\mathcal{O}_M/J$, where J is the ideal generated by odd elements, can be interpreted as a sheaf of smooth functions on M_{red}, giving it a smooth structure. Morphisms between supermanifolds are defined to be morphisms of ringed spaces.

EXAMPLE 4.2. Let N be a smooth p-manifold and $E \to N$ be a smooth complex vector bundle of dimension q. Then $\Pi E \overset{\mathrm{def}}{=} (N, \mathcal{C}^\infty(\Lambda E^\vee))$ is an example of a supermanifold of dimension $p|q$. Here $\mathcal{C}^\infty(\Lambda E^\vee))$ is the sheaf of sections of the exterior algebra bundle $\Lambda E^\vee = \bigoplus_{i=0}^q \Lambda^i(E^\vee)$ generated by $E^\vee$, the bundle dual to E; the Π in ΠE stands for *parity reversal.* We note that *every* supermanifold is isomorphic to a supermanifold constructed in this way (but not every morphism $\Pi E \to \Pi E'$ is induced by a vector bundle homomorphism $E \to E'$). In particular if $E = TN_{\mathbb{C}}$, the complexified tangent bundle of N, then the algebra of functions on the supermanifold $\Pi TN_{\mathbb{C}}$ is given by

$$C^\infty(\Pi TN_{\mathbb{C}}) = C^\infty(N, \Lambda TN_{\mathbb{C}}^\vee) = \Omega^*(N; \mathbb{C}), \tag{4.3}$$

where $\Omega^*(N; \mathbb{C})$ is the algebra of complex valued differential forms on N.

Now we discuss how the rigid geometries from section 2.5, their family versions, and the $(G, \mathbb{M})$-bordism groups (see Definition 2.46) can be generalized to supermanifolds. This is straightforward since we've been careful to express these definition in categorical terms, i.e., in terms of objects and morphisms of the category Man of smooth manifolds, which we now replace by the category csM of supermanifolds. When topological terms, like 'open subset of Y' appear in a definition, they have to be interpreted in term of the reduced manifold; e.g., an open subset of a supermanifold Y has to be interpreted as the supermanifold given by the structure sheaf $\mathcal{O}_Y$ restricted to some open subset U of the reduced manifolds Y_{red}.

DEFINITION 4.4. Let G be a super Lie group (i.e., a group object in the category csM), and suppose that G acts on some supermanifold $\mathbb{M}$ of dimension $d|\delta$. Then we can interpret the pair $(G, \mathbb{M})$ as a 'rigid geometry', and define as in Definition 2.26 the notion of a $(G, \mathbb{M})$-manifold. If in addition we specify a 1-codimensional submanifold $\mathbb{M}^c \subset \mathbb{M}$ (this is a manifold of dimension $d-1|\delta$) then we can define a $(G, \mathbb{M})$-structure on pairs as well.

Generalizing Definition 2.33 from manifolds to supermanifolds, we obtain a notion of families $Y \to S$ of $(G, \mathbb{M})$-manifolds respectively pairs (we note that the parameter space S of these families is in general a supermanifold). The construction of the bordism categories $(G, \mathbb{M})$-Bord_0, $(G, \mathbb{M})$-Bord_1 (Def. 2.46) generalizes as well; again, point topology assumptions, e.g., the properness assumption on the map $Y^c \to S$ in the definition of an object has to be interpreted as a statement about the reduced manifolds. The forgetful functor which sends a family of $(G, \mathbb{M})$-manifolds to its parameter space is now a functor from $(G, \mathbb{M})$-Bord_i to the category of supermanifolds csM, making them categories over csM.

We observe that the category csM comes equipped with a canonical flip θ: For every supermanifold M, we can define $\theta_M(f) = f$ respectively $\theta_M(f) = -f$ depending on whether $f \in C^\infty(M)$ is even respectively odd. If there is a central point $g : \mathrm{pt} \to G$ such that multiplication by g induces $\theta_{\mathbb{M}}$, we can define a flip θ for the categories $(G, \mathbb{M})$-Bord_i, making them objects in the strict 2-category $\mathsf{Sym}(\mathsf{Cat}^{f\ell}/\mathsf{csM})$. Here $\mathsf{Cat}^{f\ell}/\mathsf{csM}$ is the strict 2-category of categories with flip over csM, and $\mathsf{Sym}(\mathsf{Cat}^{f\ell}/\mathsf{csM})$ is the strict 2-category of symmetric monoids in that 2-category. So what we obtain in the end is a category $(G, \mathbb{M})$-Bord which is internal to $\mathsf{Sym}(\mathsf{Cat}^{f\ell}/\mathsf{csM})$. More generally, if X is a manifold, we have the bordism category $(G, \mathbb{M})$-$\mathsf{Bord}(X)$ of $(G, \mathbb{M})$-manifolds with maps to X.

4.2. Super Euclidean geometry. Next we define the super analogues of Euclidean manifolds, namely $(G, \mathbb{M})$-manifolds, where $\mathbb{M}$ is *super Euclidean space*, and G is the *super Euclidean group*. Our definitions are modeled on the definitions of *super Minkowski space* and *super Poincaré group* in [**DF**, §1.1], [**Fr**, Lecture 3].

To define super Euclidean space, we need the following data:

V	a real vector space with an inner product
Δ	a complex spinor representation of $Spin(V)$
$\Gamma\colon \Delta \otimes \Delta \to V_{\mathbb{C}}$	a $Spin(V)$-equivariant, non-degenerate symmetric pairing

Here $V_{\mathbb{C}}$ is the complexification of V. A complex representation of $Spin(V) \subset \mathbb{C}\ell(V)^{ev}$ is a *spinor representation* if it extends to a module over $\mathbb{C}\ell(V)^{ev}$, the even part of the complex Clifford algebra generated by V.

The supermanifold $V \times \Pi\Delta$ is the *super Euclidean space*. We note that this is the supermanifold associated to the trivial complex vector bundle $V \times \Delta \to V$, and hence the algebra of functions on this supermanifold is the exterior algebra (over $C^\infty(V)$) generated by the $\Delta^\vee$-valued functions on V, which we can interpret as *spinors on* V.

The pairing Γ allows us to define a multiplication on the supermanifold $V \times \Pi\Delta$ by

$$(V \times \Pi\Delta) \times (V \times \Pi\Delta) \longrightarrow V \times \Pi\Delta \tag{4.5}$$
$$(v_1, w_1), (v_2, w_2) \mapsto (v_1 + v_2 + \Gamma(w_1 \otimes w_2), w_1 + w_2), \tag{4.6}$$

which gives $V \times \Pi\Delta$ the structure of a super Lie group, i.e., a group object in the category of supermanifolds (see [**DM**, §2.10]). Here we describe the multiplication map in terms of the *functor of points approach* explained e.g., in [**DM**, SS2.8-2.9]: for any supermanifold S the set X_S of *S-points* in another supermanifold X consists of all morphisms $S \to X$. For example, an S-point of the supermanifold $V \times \Pi\Delta$ amounts to a pair (v, w) with $v \in C^\infty(S)^{ev} \otimes_{\mathbb{C}} V_{\mathbb{C}}$, $w \in C^\infty(S)^{odd} \otimes \Delta$ and $\bar{v}_{red} = v_{red}$ (where $v_{red} \in C^\infty(S_{red}) \otimes V_{\mathbb{C}}$ is the restriction of v to the reduced manifold, and $\bar{v}_{red}$ is its complex conjugate). A morphism of supermanifolds $X \to Y$ induces maps $X_S \to Y_S$ between the S-points of X and Y, which are functorial in S. Conversely, any collection of maps $X_S \to Y_S$ which is functorial in S comes from a morphism $X \to Y$ (by Yoneda's lemma).

We note that the spinor group $Spin(V)$ acts on the supermanifold $V \times \Pi\Delta$ by means of the double covering $Spin(V) \to SO(V)$ on V and the spinor representation on Δ. The assumption that the pairing Γ is $Spin(V)$-equivariant guarantees that this action is compatible with the (super) group structure we just defined. We define the *super Euclidean group* to be the semi-direct product $(V \times \Pi\Delta) \rtimes Spin(V)$. By construction, this supergroup acts on the supermanifold $V \times \Pi\Delta$ (the *translation* subgroup $V \times \Pi\Delta$ acts by group multiplication on itself, and $Spin(V)$ acts as explained above).

We will use the following notation and terminology

$$\mathbb{E}^{d|\delta} := V \times \Pi\Delta \qquad \text{super Euclidean space} \tag{4.7}$$
$$\mathrm{Iso}(\mathbb{E}^{d|\delta}) := (V \times \Pi\Delta) \rtimes Spin(V) \qquad \text{super Euclidean group,} \tag{4.8}$$

where $d = \dim_{\mathbb{R}} V$, $\delta = \dim_{\mathbb{C}} \Delta$.

REMARK 4.9. Up to isomorphism, there are two (respectively one) irreducible module(s) of dimension $2^{(d-2)/2}$ (respectively $2^{(d-1)/2}$) over $\mathbb{C}\ell(V)^{ev}$ if d is even (respectively odd). This implies that δ is a multiple of $2^{(d-2)/2}$ (respectively $2^{(d-1)/2}$). In particular if $\delta = 1$, the case we are interested in, this forces $d = 0, 1, 2$.

Similarly up to isomorphism, the inner product space V and hence the associated Euclidean group is determined by the dimension of V. By contrast, the isomorphism class of the data (V, Δ, Γ) is in general not determined by the superdimension $d|\delta = \dim_{\mathbb{R}} V | \dim_{\mathbb{C}} \Delta$. In particular, the (isomorphism class of the) pair $(\mathrm{Iso}(\mathbb{E}^{d|\delta}), \mathbb{E}^{d|\delta})$ might depend on (the isomorphism class of) (V, Δ, Γ), not just $d|\delta$, contrary to what the notation might suggest.

In this paper, we are only interested in the cases $d|\delta = 0|1, 1|1$ and $2|1$. We note that $\mathbb{C}\ell_0^{ev} = \mathbb{C}\ell_1^{ev} = \mathbb{C}$ and hence there is only one module Δ (up to isomorphism) of any given dimension δ. For $d = \dim V = 0$, the homomorphism Γ is necessarily trivial; for $d = 1$, $\delta = 1$, the homomorphism Γ is determined (up to isomorphism of the pair (Δ, Γ)) by the requirement that Γ is non-degenerate.

For $d = 2$, $\delta = 1$, there are two non-isomorphic modules Δ over $\mathbb{C}\ell_2^{ev} = \mathbb{C} \oplus \mathbb{C}$. To describe them explicitly as representations of $Spin(2)$, we identify the double covering map $Spin(2) \to SO(2) = S^1$ with $\mathbb{R}/2\mathbb{Z} \to \mathbb{R}/\mathbb{Z}$ by mapping

$\tau \in \mathbb{R}/\mathbb{Z}$ to $q = e^{2\pi i\tau} \in S^1$. The irreducible complex representations of $Spin(2)$ are parametrized by half integers $k \in \frac{1}{2}\mathbb{Z}$. For $k \in \frac{1}{2}\mathbb{Z}$ let us write $\mathbb{C}_k$ for the complex numbers equipped with a $Spin(2)$-action such that $\tau \in \mathbb{R}/2\mathbb{Z}$ acts by multiplication by $q^k = e^{2\pi i\tau k}$ (note that this is well-defined for $k \in \frac{1}{2}\mathbb{Z}$). Then up to isomorphism $\Delta = \mathbb{C}_k$ for $k = \pm 1/2$ and the S^1-equivariant homomorphism

$$\Delta \otimes \Delta = \mathbb{C}_{2k} \xrightarrow{\Gamma} V_{\mathbb{C}} = \mathbb{C}_1 \oplus \mathbb{C}_{-1}$$

is given by the inclusion map into the first summand (for $k = 1/2$) respectively second summand (for $k = -1/2$). For reasons that will become clear later, we fix our choice of (Δ, Γ) to be given by $k = -1/2$. Specializing the multiplication on the super Lie group $\mathbb{E}^{d|\delta} = V \times \Pi\Delta$ (see (4.5)) to the case at hand, we obtain

(4.10)

$$\mathbb{E}^{2|1} \times \mathbb{E}^{2|1} \longrightarrow \mathbb{E}^{2|1} \qquad (\tau_1, \bar\tau_1, \theta_1), (\tau_2, \bar\tau_2, \theta_2) \mapsto (\tau_1 + \tau_2, \bar\tau_1 + \bar\tau_2 + \theta_1\theta_2, \theta_1 + \theta_2),$$

where θ_i are odd functions on some parametrizing supermanifold S, and $\tau_i, \bar\tau_i$ are even functions whose restriction to S_{red} are complex conjugates of each other.

REMARK 4.11. If the module Δ over $\mathbb{C}\ell(V)^{ev}$ (the even part of the complex Clifford algebra generated by the real vector space V), is the complexification of a real module $\Delta_{\mathbb{R}}$ over $C\ell(V)^{ev}$ (the even part of the real Clifford algebra generated by V), then we can consider the *real supermanifold* $V \times \Pi\Delta_{\mathbb{R}}$, and work throughout in the category of real supermanifolds. This happens for $d = 0, 1$, $\delta = 1$ and prompted us to work with real supermanifolds in our papers [**HoST**], [**HKST**] which deal with these cases.

By contrast, for $d = \dim V = 2$, the algebra $C\ell(V)^{ev}$ is isomorphic to $\mathbb{C}$, and hence the smallest real module over this algebra has dimension two. This forces us to work with *supermanifolds* if we want to consider Euclidean structures or Euclidean field theories of dimension $2|1$.

4.3. Supersymmetric field theories. In Definition 2.47 we defined the categories TV_i, $i = 0, 1$, over Man resulting in categories with flip over Man, where the flip was given by the grading involution. These categories can be extended to categories with flip over $\mathsf{csM} \supset \mathsf{Man}$. An object of TV_0 is a supermanifold S and a sheaf V over S_{red} of $\mathbb{Z}/2$-graded topological $\mathcal{O}_S$-modules. We note that the flip $\theta_{S,V}$ of such an object is again the automorphism of (S, V) given by the grading involution. In contrast to the earlier situation where S was a manifold, this in now in general a non-trivial automorphism of S. In categorical terms, the forgetful functor $\mathsf{TV}_0 \to \mathsf{csM}$ preserves the flip defined on both categories by the grading involution. This construction then results in objects $\mathsf{TV}_0, \mathsf{TV}_1 \in \mathsf{Cat}^{\mathrm{f}\ell}/\mathsf{csM}$ (the strict 2-category of categories with flip fibered over csM). In fact, these are symmetric monoids in $\mathsf{Cat}^{\mathrm{f}\ell}/\mathsf{csM}$ and fit together to give a category TV internal to the strict 2-category $\mathsf{Sym}(\mathsf{Cat}^{\mathrm{f}\ell}/\mathsf{csM})$ of symmetric monoids of $\mathsf{Cat}^{\mathrm{f}\ell}/\mathsf{csM}$.

DEFINITION 4.12. Given a geometry $(G, \mathbb{M})$ as in Definition 4.4, a $(G, \mathbb{M})$-*field theory* is a functor

$$E\colon (G, \mathbb{M})\text{-}\mathsf{Bord} \longrightarrow \mathsf{TV}$$

of categories internal to $\mathsf{Sym}(\mathsf{Cat}^{\mathrm{f}\ell}/\mathsf{csM})$. More generally, a $(G, \mathbb{M})$-*field theory over* X is a functor $E\colon (G, \mathbb{M})\text{-}\mathsf{Bord}(X) \longrightarrow \mathsf{TV}$. If $(G, \mathbb{M})$ is the Euclidean geometry $(\mathrm{Iso}(\mathbb{E}^{d|\delta}), \mathbb{E}^{d|\delta})$, we refer to E as a *(supersymmetric) Euclidean field theory of dimension* $d|\delta$.

Let E be a $(G, \mathbb{M})$-field theory and $\Sigma \in (G, \mathbb{M})\text{-}\mathsf{Bord}_1(\emptyset, \emptyset)$, i.e., a family of closed supermanifolds with $(G, \mathbb{M})$-structure parametrized by some supermanifold S. We can evaluate E on Σ to obtain an element $E(\Sigma) \in C^\infty(S)^{ev}$. Since E is a functor between categories internal to $\mathsf{Sym}(\mathsf{Cat}^{f\ell}/\mathsf{csM})$, no information is lost by restricting Σ to be a family of *connected* closed $(G, \mathbb{M})$-manifolds. From an abstract point of view, the assignment $\Sigma \mapsto E(\Sigma)$ is a function on the moduli stack of closed connected $(G, \mathbb{M})$-manifolds.

DEFINITION 4.13. The function Z_E described above is the *partition function* of the field theory E.

If $(G, \mathbb{M})$ is the Euclidean geometry $(\mathrm{Iso}(\mathbb{E}^{2|0}), \mathbb{E}^{2|0})$ then the corresponding moduli stack has two components labeled by the (isomorphism classes of) spin structures on a torus. The function on the component corresponding to the non-bounding spin structure $++$ can be interpreted as an $SL_2(\mathbb{Z})$-invariant function on $\mathfrak{h} \times \mathbb{R}_+$ by mapping this quotient into the moduli space. Moreover, we can further restrict it to $\mathfrak{h}$, giving up the $SL_2(\mathbb{Z})$-invariance. This restriction $Z_E^{++} : \mathfrak{h} \to \mathbb{C}$ then agrees with our previous Definition 1.5.

We observe that a Euclidean field theory E of dimension $d|\delta$ determines an associated *reduced Euclidean field theory* $\bar{E}$ of dimension $d|0$ which is given as the composition

$$d|0\text{-}\mathsf{EBord} \xrightarrow{\mathcal{S}} d|\delta\text{-}\mathsf{EBord} \xrightarrow{E} \mathsf{TV} \tag{4.14}$$

$\mathcal{S}$ is the *superfication functor*, given by associating to a Euclidean manifold Y of dimension $d|0$ functorially a supermanifold $\mathcal{S}(Y)$ of dimension $d|\delta$ with Euclidean structure. We recall that a $d|0$-dimensional Euclidean structure on Y determines in particular a spin-structure on Y and hence a principal $Spin(d)$-bundle $Spin(Y) \to Y$. Then we define

$$\mathcal{S}(Y) := \Pi\left(Spin(Y) \times_{Spin(d)} \Delta\right)$$

where Δ is our choice of the complex spinor representation used in the definition of Euclidean structure of dimension $d|\delta$, $\delta = \dim_{\mathbb{C}} \Delta$. Moreover Π is the construction that turns complex vector bundles over ordinary manifolds into supermanifolds (see Example 4.2). It is not hard to see that each $\mathbb{E}^d$-chart for Y determines a $\mathbb{E}^{d|\delta}$-chart for $\mathcal{S}(Y)$. The transition function for two charts for $\mathcal{S}(Y)$ is the image of the transition function of the corresponding charts for Y under the inclusion homomorphism

$$\mathrm{Iso}(\mathbb{E}^d) = \mathbb{E}^d \rtimes Spin(d) \hookrightarrow \mathbb{E}^{d|\delta} \rtimes Spin(d) = \mathrm{Iso}(\mathbb{E}^{d|\delta})$$

REMARK 4.15. Let E be a Euclidean field theory of dimension $2|1$. Then its partition function Z_E is determined by the partition function $Z_{\bar{E}}$ of the associated reduced Euclidean spin field theory $\bar{E} = E \circ \mathcal{S}$ because the moduli stack of supertori only has one odd direction. As mentioned before, we usually only study the restriction $Z_{\bar{E}}^{++}$ to the component corresponding to the non-bounding spin structure $++$. By Propositions 3.18 and 4.16, it is in fact independent of the scale ℓ and hence determined by its restriction to $\mathfrak{h}$. This restriction was called the 'partition function' in Definition 1.5.

4.4. Modularity of the partition function of $2|1$-EFT's, part II. The goal of this subsection is to prove the following result.

PROPOSITION 4.16. *Let E be a 2-EFT which has a supersymmetric extension $\widehat{E} \in 2|1$-EFT, i.e., $E = \widehat{E} \circ \mathcal{S}$. Then E satisfies the assumption of Proposition 3.18.*

In conjunction with Proposition 3.18, this implies the first part of Theorem 1.15 for EFT's. The second part will be proved after the proof of Proposition 4.16.

PROOF. The proof of this result is based on looking at the semi-group of operators obtained by applying the field theory $\widehat{E}$ to a family of 'supercylinders' $C^s \to \mathfrak{h}^{2|1}$ parametrized by $\mathfrak{h}^{2|1} \subset \mathbb{R}^{2|1}$ (here $\mathfrak{h}^{2|1}$ is the supermanifold obtained from $\mathbb{R}^{2|1}$ by restricting the structure sheaf to the upper half plane). This family requires the choice of a scaling parameter $\ell \in \mathbb{R}_+$, and it is an extension of the family of cylinders $C^s_{\ell,\tau}$ parametrized by $\tau \in \mathfrak{h}$, since the pullback of C^s via $\mathrm{pt} \xrightarrow{\tau} \mathfrak{h} \subset \mathfrak{h}^{2|1}$ is $C^s_{\ell,\tau}$. Composition of supercylinders corresponds to the multiplication map $\mu\colon \mathbb{E}^{2|1} \times \mathbb{E}^{2|1} \to \mathbb{E}^{2|1}$ of the super Lie group $\mathbb{E}^{2|1}$ (see (4.10)) in the following sense. Composing supercylinders of our family parametrized by $\mathfrak{h}^{2|1}$ leads to a family of supercylinders parametrized by $\mathfrak{h}^{2|1} \times \mathfrak{h}^{2|1}$. This family is isomorphic to the family given by pulling back C^s via the map $\mu\colon \mathfrak{h}^{2|1} \times \mathfrak{h}^{2|1} \to \mathfrak{h}^{2|1}$, the restriction of μ to the semigroup $\mathfrak{h}^{2|1} \subset \mathbb{E}^{2|1}$.

Applying the functor $\widehat{E}$ to the family C^s we obtain a smooth map

$$f^s\colon \mathfrak{h}^{2|1} \longrightarrow \mathcal{N}(V^s)$$

to the space of nuclear operators on $V^s = E(K^s_\ell)$. Since composition of cylinders corresponds to multiplication in $\mathfrak{h}^{2|1}$, this is a semigroup homomorphism. We can write the function f^s in the form $f^s = A^s + \theta B^s$, where $A^s, B^s\colon \mathfrak{h} \to \mathcal{N}(V)$ are smooth maps and $A^s(\tau)$ (respectively $B^s(\tau)$) is an even (respectively odd) operator on the $\mathbb{Z}/2$-graded vector space V^s. The operators $A^s(\tau)$ are determined by the field theory E since $A^s(\tau) = E(C^s_{\ell,\tau})$. The homomorphism property of f^s is equivalent to the following relations (for simplicity we suppress the superscripts s throughout)

$$\begin{aligned} A(\tau_1)A(\tau_2) &= A(\tau_1+\tau_2) \\ A(\tau_1)B(\tau_2) &= B(\tau_1)A(\tau_2) = B(\tau_1+\tau_2) \\ B(\tau_1)B(\tau_2) &= -\frac{\partial A}{\partial \bar{z}}(\tau_1+\tau_2) \end{aligned} \tag{4.17}$$

In particular, $C(\tau) := B^+(\tau/2)$ is an odd operator whose square is $-\frac{\partial A^+}{\partial \bar{\tau}}(\tau)$, which proves Proposition 4.16. □

PROOF OF THEOREM 1.15, SECOND PART. We recall that according to Proposition 3.17 every integral modular function can be realized as the partition function of a $2|0$-CEFT E. It remains to show that we can construct E in such a way that it has an extension to a $2|1$-CEFT $\widehat{E}$.

We note that there is a necessary condition for the existence of $\widehat{E}$: The proof of Proposition 4.16 shows that if E has a supersymmetric extension $\widehat{E}$, then we can find smooth functions $B^s \to \mathcal{N}(V^s)^{odd}$ for $s = \pm$ such that the relations (4.17) are satisfied for $A^s(\tau) = E(C^s_\tau)$. It turns out that this condition is sufficient as well. The proof involves a study of the moduli stack of supertori. A priori, the requirement that the partition function Z_E on the stack of Euclidean tori should extend to a function on the stack of supertori could impose new restrictions on Z_E besides $SL_2(\mathbb{Z})$-equivariance. In fact, it does: Z_E extends to the moduli stack of supertori if and only if Z_E is *holomorphic*. However, the existence of B^s satisfying relations (4.17) for $A^s(\tau) = E(C^s_\tau)$ ensures that for $\tau \in \mathfrak{h}^{+s}$ the function

$$Z_E(\tau) = E(T^{+s}_\tau) = \operatorname{str} E(C^s_\tau) = \operatorname{str} A^s(\tau)$$

is holomorphic. Hence the existence of B^s satisfying the relations implies that Z_E is holomorphic restricted to $\mathfrak{h}^{++}$ and $\mathfrak{h}^{+-}$. This implies that Z_E is holomorphic on the other two components $\mathfrak{h}^{-+}$, $\mathfrak{h}^{--}$ as well, since Z_E is $SL_2(\mathbb{Z})$-equivariant and the $SL_2(\mathbb{Z})$-action permutes $\mathfrak{h}^{+-}$, $\mathfrak{h}^{-+}$ and $\mathfrak{h}^{--}$.

For the EFT E we've constructed in the proof of Proposition 3.4 respectively 3.17 it is easy to show that there are smooth families $B^s \colon \mathfrak{h} \to \mathcal{N}(V^s)$ satisfying relations (4.17): in terms of the algebraic data $(V^s, \lambda^s, \rho^s(\tau))$ that determine the 2|0-CEFT E, the operator $A^s(\tau) = E(C^s_\tau)$ is given by the composition (3.12) of $\rho^s(\tau)$ and λ^s. Since $\rho^s(\tau)$ depends *holomorphically* on τ, the function $A^s(\tau)$ is holomorphic. Hence setting $B^s \equiv 0$ the relations (4.17) are satisfied and we conclude that E has a supersymmetric extension $\widehat{E}$. □

5. Twisted field theories

In this section we will define field theories of non-zero degree, or – in physics lingo – non-zero central charge. More generally, we will define *twisted* field theories over a manifold X. As explained in the introduction we would like to think of field theories over X as representing cohomology classes for certain generalized cohomology theories. Sometimes it is *twisted* cohomology classes that play an important role, e.g., the Thom class of a vector bundle that is not orientable for the cohomology theory in question. We believe that the twisted field theories defined below (see Definition 5.2) represent twisted cohomology classes, which motivates our terminology. We will outline a proof of this for $d|\delta = 0|1$ and $1|1$.

We will describe *twisted field theories* as natural transformations between functors (see Definition 5.2). More precisely, these are functors between internal categories; their domain is our internal bordism category $d|\delta$-EBord. So our first task is to describe what is meant by a natural transformation between such functors. Then we will construct the range category and outline the construction of the relevant functors which will allow us to define Euclidean field theories of degree n. We end the section by relating Euclidean field theories of degree zero to field theories as defined in section 4.3 and by comparing our definition with Segal's definition of conformal field theories with non-trivial central charge.

5.1. Twisted Euclidean field theories. We first introduce the internal category that will serve as range category for a field theory.

DEFINITION 5.1. The internal category TA of *topological algebras* has the following object and morphism categories:

- TA_0 Is the groupoid whose objects are *topological algebras*. A topological algebra is a monoid in the symmetric monoidal groupoid TV of topological vector spaces (equipped with the projective tensor product); i.e., an object $A \in \mathsf{TV}$ together with an associative multiplication $A \otimes A \to A$. Morphisms are continuous algebra isomorphisms.

TA_1 is the category of bimodules over topological algebras. A bimodule is a triple (A_1, B, A_0), where A_0, A_1 are topological algebras, and B is an A_1-A_0-bimodule (i.e., an object $B \in \mathsf{TV}$ with a morphism $A_1 \otimes B \otimes A_0 \to B$ satisfying the usual conditions required for a module over an algebra). A morphism from (A_1, B, A_0) to (A_1', B', A_0') is a triple (f_1, g, f_0) consisting of isomorphisms $f_0 \colon A_0 \to A_0'$, $f_1 \colon A_1 \to A_1'$ of topological algebras and a morphism $f \colon B \to B'$ of topological vector spaces which is compatible with the left action of A_1 and the right action of A_0 (A_1 acts on B' via the algebra homomorphism $A_0 \to A_0'$, and similarly for A_0).

There are obvious source and target functors

$$s, t \colon \mathsf{TA}_1 \to \mathsf{TA}_0 \quad \text{given by} \quad s(A_1, B, A_0) = A_0 \quad \text{and} \quad t(A_1, B, A_0) = A_1,$$

and a composition functor

$$c \colon \mathsf{TA}_1 \times_{\mathsf{TA}_0} \mathsf{TA}_1 \longrightarrow \mathsf{TA}_1 \qquad (A_2, B, A_1), (A_1, B', A_0) \mapsto (A_2, B \otimes_{A_1} B', A_0)$$

The two categories TA_0, TA_1, the functors s, t, c and the usual associator for tensor products define a category TA of *topological algebras* internal to the strict 2-category Cat. We can do better by noting that the tensor product (in the category TV) makes TA_0, TA_1 symmetric monoidal categories and that the functors s, t, c are symmetric monoidal functors. This gives TA the structure of a category internal to the strict 2-category SymCat of symmetric monoidal categories.

Analogously to TV, there is a family version of the internal category TA, obtained by replacing algebras (respectively bimodules) by families of algebras (respectively bimodules) parametrized by some supermanifold. To come up with the correct definition of 'family' here, it is useful to recall that an algebra is a monoid in the category of vector spaces, and that a $A_1 - A_2$-bimodule is an object in the category of vector spaces which comes equipped with a left action of the monoid A_1 and a commuting right action of the monoid A_2. Since we've decided that a 'family of topological vector spaces parametrized by a supermanifold S' is a sheaf over S_{red} of topological $\mathcal{O}_S$-modules, we decree that an S-family of topological algebras (respectively bimodules) is a monoidal object in the category of S-families of topological vector spaces (respectively an object with commuting left- and right actions of monoids). The obvious forgetful functors $\mathsf{TA}_i \to \mathsf{csM}$ for $i = 0, 1$ make these categories Grothendieck fibered over the category csM of supermanifolds; the (projective) tensor product makes them symmetric monoids in $\mathsf{Cat}^{\mathsf{f}\ell}/\mathsf{csM}$.

DEFINITION 5.2. Let X be a smooth manifold and let

$$T \colon (G, \mathbb{M})\text{-}\mathsf{Bord}(X) \to \mathsf{TA}$$

be a functor between categories internal to $\mathsf{Sym}(\mathsf{Cat}^{\mathsf{f}\ell}/\mathsf{csM})$; we will refer to such a functor as *twist*. A *T-twisted $(G, \mathbb{M})$-field theory* is a natural transformation, again internal to $\mathsf{Sym}(\mathsf{Cat}^{\mathsf{f}\ell}/\mathsf{csM})$,

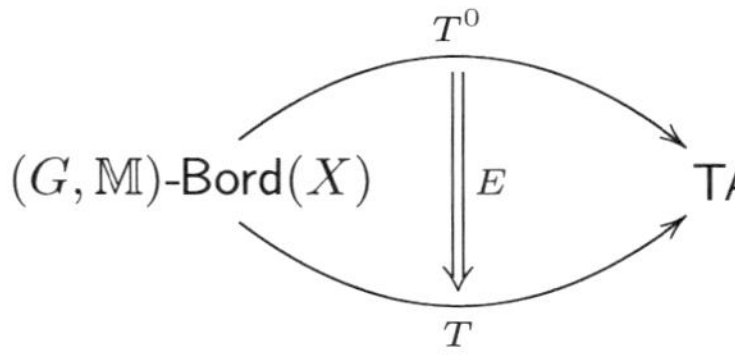

where T^0 is the *constant twist* that maps every object Y of $(G,\mathbb{M})\text{-}\mathsf{Bord}(X)_0$ to the algebra $\mathbb{C} \in \mathsf{TA}_0$ (which is the unit in the symmetric monoidal groupoid TA_0). It maps every object Σ of $(G,\mathbb{M})\text{-}\mathsf{Bord}(X)_1$ to $(\mathbb{C}, B, \mathbb{C}) \in \mathsf{TA}_1$, where B is $\mathbb{C}$ regarded as a $\mathbb{C}$-$\mathbb{C}$-bimodule (this is the monoidal unit in $\mathsf{TA}/\mathsf{csM}_1$). The functors $T_i^0\colon (G,\mathbb{M})\text{-}\mathsf{Bord}_i(X) \to \mathsf{TA}_i$ send every morphism to the identity morphism of the monoidal unit of TA_i.

We note that the symmetric monoidal structure on TA_i, $i = 0, 1$, allows us to form the tensor product $T_1 \otimes T_2$ of two twists. The tensor product $T \otimes T^0$ of any twist T and the constant twist T^0 is naturally isomorphic to T. If E_1 is a T_1-twisted field theory and E_2 is a T_2-twisted field theory, we can form the tensor product $E_1 \otimes E_2$, which is a $T_1 \otimes T_2$-twisted field theory.

Let us unravel this definition. For simplicity, we take the target X to be a point (hence no additional information) and ignore the family aspect by restricting to the fiber category $(G,\mathbb{M})\text{-}\mathsf{Bord}_{\mathrm{pt}}$ over the point pt. A twist

$$T = (T_0, T_1)\colon (G,\mathbb{M})\text{-}\mathsf{Bord}_{\mathrm{pt}} \longrightarrow \mathsf{TA}_{\mathrm{pt}}$$

associates to an object $(Y, Y^c, Y^{\pm}) \in (G,\mathbb{M})\text{-}\mathsf{Bord}_{\mathrm{pt}}$ (consisting of a $(G,\mathbb{M})$-manifold Y and a compact codimension one submanifold Y^c dividing Y into Y^+ and Y^-) a topological algebra $T_0(Y)$. To a $(G,\mathbb{M})$-bordism Σ from Y_0 to Y_1 the functor T_1 associates the $T_0(Y_1)$-$T_0(Y_0)$-bimodule $T_1(\Sigma)$.

To understand the mathematical content of the natural transformation $E = (E_0, E_1)$, we use Definition 2.19 and Diagram (2.20) in the case

$$\mathsf{C} = (G,\mathbb{M})\text{-}\mathsf{Bord}_{\mathrm{pt}} \qquad \mathsf{D} = \mathsf{TA}_{\mathrm{pt}} \qquad f = T^0 \qquad g = T \qquad n = E$$

We see that a natural transformation E is a pair (E_0, E_1), consisting of the following data (where we suppress the subscript pt):

- E_0 $(G,\mathbb{M})\text{-}\mathsf{Bord}_0 \to \mathsf{TA}_1$ is a functor; in particular, E_0 associates to each $(Y, Y^c, Y^{\pm})$ an object $E_0(Y)$ of TA_1, i.e., $E_0(Y)$ is a bimodule. The commutative triangle (2.11) implies that $E_0(Y)$ is a left module over $tE_0(Y) = g_0(Y) = T_0(Y)$ and a right module over $sE_0(Y) = f_0(Y) = T_0^0(Y) = \mathbb{C}$; in other words, $E_0(Y)$ is just a left $T_0(Y)$-module.
- According to diagram (2.20) E_1 is a natural transformation, i.e., for every bordism Σ from Y_0 to Y_1 (this is an object of C_1) we have a morphism $E_1(\Sigma)$ in D_1 whose domain (respectively range) is the image of $\Sigma \in \mathsf{C}_1$ under the functors

$$\mathsf{C}_1 \xrightarrow{g_1 \times n_0 s} \mathsf{D}_1 \times_{\mathsf{D}_0} \mathsf{D}_1 \xrightarrow{c_{\mathsf{D}}} \mathsf{D}_1 \quad \text{respectively} \quad \mathsf{C}_1 \xrightarrow{n_1 t \times f_1} \mathsf{D}_1 \times_{\mathsf{D}_0} \mathsf{D}_1 \xrightarrow{c_{\mathsf{D}}} \mathsf{D}_1.$$

More explicitly, $E_1(\Sigma)$ is a map of left $T_0(Y_1)$-modules

$$E_1(\Sigma)\colon T_1(\Sigma) \otimes_{T_0(Y_0)} E_0(Y_0) \longrightarrow E_0(Y_1) \otimes_{T_0^0(Y_1)} T_1^0(\Sigma) \cong E_0(Y_1),$$

or, equivalently, a $T_0(Y_1) - T_0(Y_0)$-bimodule map

$$E_1(\Sigma)\colon T_1(\Sigma) \longrightarrow \mathrm{Hom}(E_0(Y_0), E_0(Y_1)) \tag{5.3}$$

If $\Phi : \Sigma \to \Sigma'$ is a $(G, \mathbb{M})$-isometry, i.e., a morphism in C_1 then we obtain a commutative diagram

$$\begin{array}{ccc} T_1(\Sigma) & \xrightarrow{E_1(\Sigma)} & \mathrm{Hom}(E_0(Y_0), E_0(Y_1)) \\ {\scriptstyle T_1(\Phi)}\downarrow\cong & & {\scriptstyle E_0(\partial\Phi)}\downarrow\cong \\ T_1(\Sigma') & \xrightarrow{E_1(\Sigma')} & \mathrm{Hom}(E_0(Y_0'), E_0(Y_1')) \end{array} \tag{5.4}$$

We note that the subscripts for the functors T_i, E_i are redundant (whether we mean T_0 or T_1 is clear from the object we apply these functors to), and hence we will suppress the subscripts from now on. Summarizing for future reference, a T-twisted field theory E amounts to the following data:

(5.5) left $T(Y)$-modules $E(Y)$ and 'isometry invariant'
$T(Y_1) - T(Y_0)$-bimodule maps $T(\Sigma) \to \mathrm{Hom}(E(Y_0), E(Y_1))$

for objects $(Y, Y^c, Y^{\pm})$ and bordisms Σ from Y_0 to Y_1.

The commutativity of the octagon required in the definition of a natural transformation (Definition 2.19) amounts to the commutativity of the following diagram for any bordism Σ from Y_0 to Y_1 and bordism Σ' from Y_1 to Y_2.

$$\begin{array}{ccc} T(\Sigma') \otimes_{T(Y_1)} T(\Sigma) & \xrightarrow{E(\Sigma')\otimes E(\Sigma)} & \mathrm{Hom}(E(Y_1), E(Y_2)) \otimes_{T(Y_1)} \mathrm{Hom}(E(Y_0), E(Y_1)) \\ \downarrow\cong & & \downarrow\circ \\ T(\Sigma' \circ \Sigma) & \xrightarrow{E(\Sigma'\circ\Sigma)} & \mathrm{Hom}(E(Y_0), E(Y_2)) \end{array} \tag{5.6}$$

We note that if the twist T is the constant twist T^0, then $E(Y)$ is just a topological vector space, and $E(\Sigma)$ is a continuous linear map $E(Y_0) \to E(Y_1)$. The commutativity of the diagram above is the requirement that composition of bordisms corresponds to composition of the corresponding linear maps. In other words:

LEMMA 5.7. *The groupoid of T^0-twisted $(G, \mathbb{M})$-field theories is isomorphic to the groupoid of $(G, \mathbb{M})$-field theories as in Definition 4.12.*

REMARK 5.8. If $S, T \colon (G, \mathbb{M})\text{-}\mathsf{Bord} \to \mathsf{TA}$ are two twists, and $E \colon S \Rightarrow T$ is an invertible twist, then for each $Y \in (G, \mathbb{M})\text{-}\mathsf{Bord}$ the $T(Y) - S(Y)$-bimodule $E(Y)$ provides a Morita equivalence between the algebras $T(Y)$ and $S(Y)$; in particular, $E(Y)$ is an irreducible bimodule. If $S = T = T^0$, this implies that $E(Y)$ is a complex vector space of dimension one. This shows that only very special field theories correspond to *invertible natural transformations*.

Let E be a T-twisted $(G, \mathbb{M})$-field theory. Given $\Sigma \in (G, \mathbb{M})\text{-}\mathsf{Bord}_1(\emptyset, \emptyset)$, i.e., a family of closed supermanifolds with $(G, \mathbb{M})$-structure parametrized by some supermanifold S, we can evaluate E on Σ to obtain an element $E(\Sigma) \in \mathrm{Hom}(T(\Sigma), \mathbb{C}) = T(\Sigma)^{\vee}$ (we note that since Σ is a family of *closed* supermanifolds, $T(\Sigma)$ is just a sheaf of topological $\mathcal{O}_S$-modules (rather than bimodules over algebras associated to the incoming/outgoing boundary of Σ). Since E is a functor between categories internal to $\mathsf{Sym}(\mathsf{Cat}^{f\ell}/\mathsf{csM})$, no information is lost by restricting Σ to be a family of *connected* closed $(G, \mathbb{M})$-manifolds. From an abstract point of view, the assignment

$\Sigma \mapsto T(\Sigma)^{\vee}$ defines a sheaf of topological vector spaces over the moduli stack of closed connected $(G, \mathbb{M})$-manifolds, and $\Sigma \mapsto E(\Sigma)$ is a section of this sheaf.

DEFINITION 5.9. The section Z_E described above is the *partition function* of the twisted field theory E.

5.2. Segal's weakly conformal field theories as twisted field theories. The goal of this subsection is to show that Segal's modular functors and weakly conformal field theory can be interpreted as twists respectively twisted field theories. More precisely, special types of twists $T\colon$ 2-CBord $\to$ TA, whose domain is the 2-dimensional conformal bordism category, can be identified with *modular functors* in the sense of Segal [**Se2**, Definition 5.1]. Moreover, T-twisted conformal field theories then become *weakly conformal field theories* as in [**Se2**, Definition 5.2]. The translation works as follows.

Let $T\colon$ 2-CBord $\to$ TA be functor such that the algebra $T(S^1)$ associated to the circle is a finite direct sum of copies of $\mathbb{C}$:

$$T(S^1) = \bigoplus_{\phi\in\Phi} \mathbb{C}_\phi$$

Here the subscript ϕ is just a book keeping tool to distinguish the various copies. Let us further assume that for every 2-dimensional conformal bordism Σ the bi-module $T(\Sigma)$ is finite dimensional. If Σ is a Riemannian surface with parametrized boundary circles, we can think of Σ as a bordism from $\emptyset$ to the disjoint union of k copies of S^1, and hence $T(\Sigma)$ is a left-module over $T(\partial\Sigma) \cong T(S^1) \otimes \cdots \otimes T(S^1)$. We note that a $T(S^1)$-module can be thought of as Φ-graded vector space. Hence $T(\Sigma)$ is a $\Phi \times \cdots \times \Phi$-graded vector space. In particular, $T(\Sigma)$ can be decomposed in a direct sum of finite dimensional vector spaces; the summands are parametrized by the various ways of assigning a label in Φ to each boundary circle. These are the data of a modular functor in the sense of Segal [**Se2**, Definition 5.1]. Properties (i) and (ii) in Segal's definition follow from the fact that $\Sigma \mapsto T(\Sigma)$ is a monoidal functor. This was worked out in Hessel Posthuma's thesis. Segal requires further that $\dim T(S^2) = 1$ and that $T(\Sigma)$ depends holomorphically on Σ. The second condition can be implemented within our framework by working over the site of complex manifolds (instead of smooth manifolds).

Let E be a T-twisted conformal field theory. From our discussion above we see that E assigns to $Y = S^1 \amalg \cdots \amalg S^1$ a left $T(Y)$-module $E(Y)$. This vector space decomposes as direct sum of vector spaces, parametrized by ways to label each boundary circle by an element of Φ. To a Riemann surface Σ with boundary $\partial\Sigma$, viewed as a bordism from $\emptyset$ to $\partial\Sigma$, it assigns according to (5.5) a homomorphism

$$T(\Sigma) \longrightarrow \mathrm{Hom}(E(\emptyset), E(\partial\Sigma)) = \mathrm{Hom}(\mathbb{C}, E(\partial\Sigma)) = E(\partial\Sigma)$$

which is $T(\partial\Sigma)$-equivariant, i.e., compatible with the decomposition of these vector spaces according to ways of labeling the boundary circles. These are the data in Segal's definition of a weakly conformal field theory [**Se2**, Definition 5.2], and again it isn't hard to show that Segal's conditions on these data follow from the properties of a twisted field theory.

5.3. Field theories of degree n.

DEFINITION 5.10. **(Preliminary!)** A Euclidean field theory of dimension $d|0$ and degree $n \in \mathbb{Z}$ is a natural transformation, internal to $\mathsf{Sym}(\mathsf{Cat}^{f\ell})$,

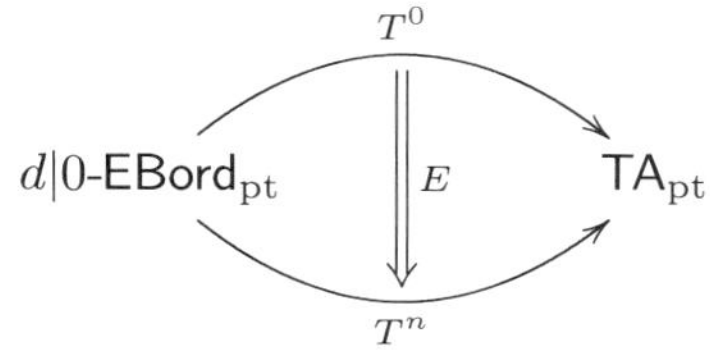

Here T^0 is the constant twist (see 5.2), and T^n is a twist we will construct below.

We recall from (5.5) that the natural transformation E in particular provides us with the following data:

- a left $T^n(Y)$-module $E(Y)$ for every object $Y \in (d|0\text{-}\mathsf{EBord}_{\mathrm{pt}})_0$;
- a $T^n(Y_1) - T^n(Y_0)$-bimodule map
$$E(\Sigma)\colon T(\Sigma) \longrightarrow \mathrm{Hom}(E(Y_0), E(Y_1))$$
for every object $\Sigma \in (d|0\text{-}\mathsf{EBord}_{\mathrm{pt}})_1(Y_0, Y_1)$.

REMARK 5.11. The definition above is preliminary in the same sense that Definition 2.25 was preliminary: we should be *working in families*, i.e., replace the internal categories $d|0\text{-}\mathsf{EBord}_{\mathrm{pt}}$ and $\mathsf{TA}_{\mathrm{pt}}$ by their (super) family versions $d|0\text{-}\mathsf{EBord}$, TA which are categories internal to $\mathsf{Sym}(\mathsf{Cat}^{f\ell}/\mathsf{csM})$ (see Def. 5.1 and Def. 4.12). We expect to construct functors $\widehat{T}^n\colon d|\delta\text{-}\mathsf{EBord} \to \mathsf{TA}$ of categories internal to $\mathsf{Sym}(\mathsf{Cat}^{f\ell}/\mathsf{csM})$ for $\delta = 0, 1$ such that the composition with the inclusion functors $d|\delta\text{-}\mathsf{EBord}_{\mathrm{pt}} \to d|\delta\text{-}\mathsf{EBord}$ is the functor T^n constructed in this section. Then $d|\delta$-EFT's of degree n for $\delta = 0, 1$ should be defined as a $\widehat{T}^n$-twisted Euclidean field theories of dimension $d|\delta$. For $d = 0$, $\delta = 1$, we gave an ad hoc construction of the twist [**HKST**, Def. 3]. The details of the construction of $\widehat{T}^n$ for $d > 0$, $\delta = 0, 1$ still have to be worked out.

We construct the twist T^n in terms of the 'basic' twists T^{-1} and T^1 by defining
$$T^n := \begin{cases} (T^1)^{\otimes |n|} & n \geq 0 \\ (T^{-1})^{\otimes |n|} & n \leq 0 \end{cases}$$
Here the tensor product $S \otimes T$ of two functors $S, T\colon d|0\text{-}\mathsf{EBord}_{\mathrm{pt}} \to \mathsf{TA}_{\mathrm{pt}}$ is defined using the symmetric monoidal structure of the categories $(\mathsf{TA}_{\mathrm{pt}})_i$ for $i = 0, 1$. This in turn is given by the (projective) tensor product (over $\mathbb{C}$) of algebras respectively bimodules. Below we will describe the construction of T^{-1}. The functor T^1 is dual to T^{-1} in the sense that $T^1 \otimes T^{-1}$ is (lax) equivalent to the 'trivial' twist T^0. In particular, if Σ is a closed Euclidean spin d-manifold, the complex line $T^1(\Sigma)$ is dual to $T^{-1}(\Sigma)$.

For Y, Σ as above, the $\mathbb{E}^{d|0}$-structure determines in particular a spin-structure and a Riemann metric on these manifolds (see Example 2.30), and hence a *Dirac operator*. We recall that the construction of the $\mathbb{Z}/2$-graded spinor bundle $S = S^+ \oplus S^-$ on a spin d-manifold Σ involves the choice of a $\mathbb{Z}/2$-graded module $N = N^{ev} \oplus N^{odd}$ over the Clifford algebra $C\ell(\mathbb{R}^d)$. Given N, the spinor bundle S is defined as the associated bundle $S = Spin(\Sigma) \times_{Spin(d)} N$, where $Spin(\Sigma) \to \Sigma$ is the principal $Spin(d)$-bundle determined by the spin structure on Σ. In [**ST1**] we chose $N = C\ell(\mathbb{R}^d)$ as a left-module over itself (leading to the 'Clifford linear Dirac operator') while here we choose $N = \mathbb{C}\ell(\mathbb{R}^d) \otimes_{\mathbb{C}\ell(\mathbb{R}^d)^{ev}} \Delta^\vee$, where Δ is the module needed as a datum in the construction of the super Euclidean group in section 4.2.

(1) $T^{-1}(Y)$ is the Clifford algebra generated by $C^\infty(Y^c, S^+_{|Y^c})$, the space of sections of the spinor bundle restricted to Y^c, equipped with the symmetric complex bilinear form that gives the boundary term of Green's formula;

(2) If Σ is a closed manifold, then $T^{-1}(\Sigma)$ is the complex line $\Lambda^{top}(\mathcal{H}^+(\Sigma))^\vee$. Here $\mathcal{H}^+(\Sigma)$ is the finite dimensional space of section of S^+ which are *harmonic* (i.e., in the kernel of the Dirac operator), and $\Lambda^{top}(V) := \Lambda^{\dim V}(V)$.

(3) If Σ is a bordism from Y_0 to Y_1 without closed components, then $T^{-1}(Y)$ is the Fockspace, a $T^{-1}(Y_1)-T^{-1}(Y_0)$-bimodule over these Clifford algebras determined by the Lagrangian subspace of $C^\infty(\partial\Sigma, S^+_{|\partial\Sigma})$ given by the boundary values of harmonic spinors.

REMARK 5.12. For $d = 1, 2$, the algebra $T^{-1}(Y)$ associated to an object $Y \in d|0\text{-}\mathsf{EBord}_{\mathrm{pt}}$ is the Clifford algebra denoted $C(Y)$ in our earlier paper [**ST1**, Def. 2.20]. Similarly, the bimodule $T^{-1}(\Sigma)$ associated to a Euclidean spin bordism $\Sigma \in (d|0\text{-}\mathsf{EBord}_{\mathrm{pt}})_1$ is the Fock space module $F(\Sigma)$ of [**ST1**, Def. 2.23]. In [**ST1**] we only considered these for $d = 1, 2$ as a side-effect of considering the Clifford-linear Dirac operator which makes the spaces $C^\infty(Y^c, S^+_{|Y^c})$ and $\mathcal{H}^+(\Sigma)$ right modules over $C\ell(\mathbb{R}^{d-1})$. The Clifford algebra $C\ell(\mathbb{R}^{d-1})$ is $\mathbb{R}$ for $d = 1$, $\mathbb{C}$ for $d = 2$, but it is a non-commutative algebra for $d > 2$. This confusing additional structure kept us from us using these data as input to the Clifford-algebra/Fockspace machine. With our current choice of N, the spinor bundle is just a complex vector bundle and we obtain a functor $T^{-1}\colon d|0\text{-}\mathsf{EBord} \to \mathsf{TA}$ for any d.

Note also that the field theories of degree in [**ST1**] are related to field theories of degree $-n$ in our current definition.

This seems an appropriate place to point out an error in our formula (2.3) derived from Green's formula for the Dirac operator D. It should read

$$\langle D\psi e_1, \phi\rangle + \overline{\langle \psi, D\phi e_1\rangle} = \langle c(\nu)\psi_| e_1, \phi_|\rangle$$

i.e., the second term should involve a complex conjugation which is missing in the formula as stated in [**ST1**]. This conjugation is necessary to make all terms $\mathbb{C}$-linear in ϕ, and $\mathbb{C}$-anti-linear in ψ (our hermitian pairings are anti-linear in the first, and linear in the second slot).

REMARK 5.13. The reader might wonder about the various duals occurring in the above construction. Concerning the dual $\Delta^\vee$ in the construction of the spinor bundle, we note that the functions on our 'model space' $\mathbb{R}^d \times \Pi\Delta$ for Euclidean structures are $C^\infty(\mathbb{R}^d) \otimes \Lambda^*(\Delta^\vee)$. In particular, using $\Delta^\vee$ rather than Δ for the construction of the spinor bundle allows us to interpret spinors on a Euclidean $d|0$-manifold Σ as *odd functions* on the associated Euclidean supermanifold $\mathcal{S}(\Sigma)$ of dimension $d|\delta$. This is a first step towards constructing the functor $\widehat{T}^{-1}$ out of the Euclidean bordism category $d|\delta\text{-}\mathsf{EBord}$.

Specializing Definition 5.9 of the partition function of a T-twisted field theory, we see that the partition function of a Euclidean field theories of dimension $2|0$ and degree n is a section Z_E of a line bundle over the moduli stack of pointed Euclidean spin tori of dimension 2. Explicitly, this stack is the quotient

$$\mathfrak{M} = SL_2(\mathbb{Z})\backslash\!\backslash \left(\mathbb{R}_+ \times (\mathfrak{h}^{++} \amalg \mathfrak{h}^{-+} \amalg \mathfrak{h}^{+-} \amalg \mathfrak{h}^{--})\right),$$

with the Euclidean torus $T^{s_1 s_2}_{\ell,\tau}$ corresponding to $(\ell,\tau) \in \mathbb{R}_+ \times \mathfrak{h}^{s_1 s_2}$; here the superscripts $s_1, s_2 \in \{+,-\}$ specify the spin structure on the torus as explained in the proof of Proposition 3.17. The group $SL_2(\mathbb{Z})$ acts on $\mathbb{R}_+ \times \mathfrak{h}$ by formula 3.1; in addition, it permutes the four copies of the upper half plane by $(A, \left(\begin{smallmatrix} s_1 \\ s_2 \end{smallmatrix}\right)) \mapsto A\left(\begin{smallmatrix} s_1 \\ s_2 \end{smallmatrix}\right)$ for $A \in SL_2(\mathbb{Z})$ and using the group isomorphism $\{+,-\} = \{+1,-1\} = \mathbb{Z}/2$. In particular, $\mathfrak{h}^{++}$ is fixed and the other three copies are permuted cyclically. The line bundle over this stack is the $SL_2(\mathbb{Z})$-equivariant line bundle over $\mathbb{R}_+ \times \coprod_{s_1 s_2} \mathfrak{h}^{s_1 s_2}$ whose fiber over $(\ell,\tau) \in \mathbb{R}_+ \times \mathfrak{h}^{s_1 s_2}$ is the line $T^{-1}(T^{s_1 s_2}_{\ell,\tau})^{\vee}$. The partition function Z_E is the equivariant section of this line bundle given by

$$Z_E(\ell,\tau) = \Big(E(T^{s_1 s_2}_{\ell,\tau})\colon T^n(T^{s_1 s_2}_{\ell,\tau}) \longrightarrow \operatorname{Hom}(T^n(\emptyset), T^n(\emptyset)) = \operatorname{Hom}(\mathbb{C},\mathbb{C}) = \mathbb{C}\Big)$$

for $\tau \in \mathfrak{h}^{s_1 s_2}$.

We note that the Weitzenböck formula [**LM**, Ch. II, Thm. 8.8] implies that harmonic spinors on any compact Euclidean manifold are *parallel* (i.e., their covariant derivative vanishes). In particular, the space of harmonic spinors on the torus $T^{s_1 s_2}_{\ell,\tau}$ can be identified with the subspace of the parallel spinors on the universal covering $\mathbb{E}^2$ which is invariant under the action of the covering group $\mathbb{Z}^2$. Spinors on $\mathbb{E}^2$ are by construction functions on $\mathbb{E}^2$ with values in $\Delta^{\vee}$ (see Remark 5.13). Hence the space of parallel spinors is the space of constant functions which can be identified with $\Delta^{\vee}$. For $(s_1,s_2) \neq (+,+)$ the $\mathbb{Z}^2$-action on the parallel spinors is non-trivial, and hence $T^{-1}(T^{s_1 s_2}_{\ell,\tau}) = \Lambda^{top}(\mathcal{H}^+(T^{s_1 s_2}_{\ell,\tau}))^{\vee}$ can be canonically identified with $\mathbb{C}$. For $(s_1,s_2) = (+,+)$, the group $\mathbb{Z}^2$ acts trivially on parallel spinors and hence $T^{-1}(T^{++}_{\ell,\tau})$ can be identified with Δ.

If E is a *conformal* Euclidean field theory of dimension $2|0$, then its partition function Z_E becomes independent of ℓ in the sense that it is invariant under a natural $\mathbb{R}_+$-action on the line bundle over the moduli stack $\mathfrak{M}$ (compare Remark 3.3 for field theories of degree zero). Then the section Z_E amounts to

(1) a function on $\mathfrak{h}^{-+}$ which is invariant under the index three subgroup $\Gamma_0(2)$ of $SL_2(\mathbb{Z})$. This subgroup fixes the spin structure $(-,+)$ and consists of matrices $A = \left(\begin{smallmatrix} a & b \\ c & d \end{smallmatrix}\right)$ with $c \equiv 0 \mod 2$.
(2) a function on $\mathfrak{h}^{++}$ with the transformation properties of a modular form of weight $-n/2$ (using a suitable choice of trivialization of the line bundle over $\mathfrak{h}^{++}$).

REMARK 5.14. A 2-dimensional conformal field theory of central charge c in the sense of Segal [**Se2**] (more precisely, a *spin* CFT, see [**Kr**]) gives a conformal Euclidean field theory of dimension $2|0$ and degree $n = c$. Its partition function as defined by Segal is a function on $\mathfrak{h}$ [**Se2**, §6] (see also [**Kr**]). Based on Segal's discussion at the end of §6, it is not hard to show that his partition function is obtained from the function in (2) above by multiplying by $\eta(\tau)^{n/2}$, where $\eta(\tau)$ is Dedekind's η-function.

6. A periodicity theorem

In this section we prove Theorem 1.16. A field theory $P \in d|0\text{-EFT}^k(X)$ is a natural transformation

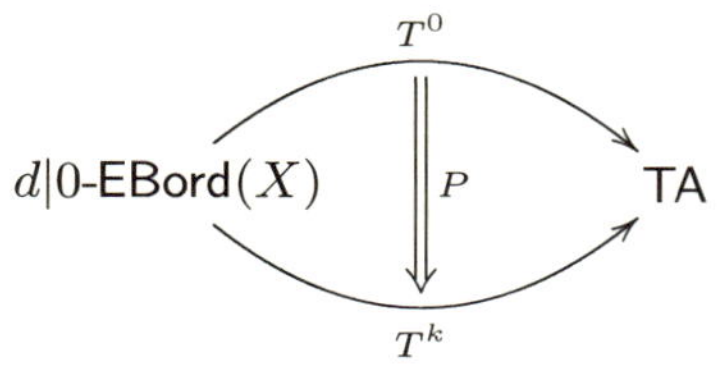

Then multiplication with P gives a functor

$$P\otimes\colon d|0\text{-EFT}^n(X) \longrightarrow d|0\text{-EFT}^{n+k}(X)$$

We note that if P is invertible in the sense that there is a natural transformation $P^{-1}\colon T^k \Rightarrow T$ such that $P\circ P^{-1}$ and $P^{-1}\circ P$ are equivalent to the identity natural transformation, then $P\otimes$ is an isomorphism, with inverse given by multiplication by the Euclidean field theory

$$T^0 \cong T^{-k} \otimes T^k \xrightarrow{\mathrm{id}_{T^{-k}} \otimes P^{-1}} T^{-k} \otimes T^0 \cong T^{-k}$$

of degree $-k$.

We recall from Equation (5.5) that P gives a left $T^k(Y)$-module $P(Y)$ for every object $Y \in d|0\text{-EBord}_0$ and a bimodule map $P(\Sigma)\colon T^k(\Sigma) \to \mathrm{Hom}(P(Y_0), P(Y_1))$ for every Euclidean bordism Σ of dimension $d|0$ from Y_0 to Y_1.

To prove the periodicity statement, we will construct an invertible conformal Euclidean field theory P of degree -48. The first step is an explicit description of the algebra $T^{-1}(K^s)$ and the bimodule $T^{-1}(C^s_\tau)$. We recall that $K^s \in 2|0\text{-CEBord}_0$ is the circle and $C^s_\tau \in 2|0\text{-CEBord}_1(K^s, K^s)$ is the cylinder with parameter $\tau \in \mathfrak{h}$. Here the superscript s specifies the spin structure: $s = +$ is the periodic spin structure, and $s = -$ is the anti-periodic spin structure. We recall that the group $\mathbb{R}$ acts on the spin manifolds K^s and C^s_τ. Ignoring the spin-structures, $\tau \in \mathbb{R}$ acts by rotation by $q := e^{2\pi i \tau}$ on the circle S^1 respectively the cylinder C_τ. We note that $1 \in \mathbb{R}$ acts trivially on the spin manifolds K^+ and C^+_τ, but by multiplication by -1 on the spinor bundle for K^- and C^-_τ, thus leading to an effective action of $\mathbb{R}/\mathbb{Z}$ for K^+, C^+_τ and an effective action of $\mathbb{R}/2\mathbb{Z}$ for K^-, C^-_τ. The $\mathbb{R}$-action on K^s as an object of $2|0\text{-CEBord}_0$ induces by functoriality an $\mathbb{R}$-action on the algebra $T^{-1}(K^s)$. We have the following $\mathbb{R}$-equivariant algebra isomorphisms:

$$T^{-1}(K^+) \cong C\ell(\mathbb{C}_0) \otimes \bigotimes_{m\in\mathbb{N}} C\ell(H(\mathbb{C}_m)) \qquad T^{-1}(K^-) \cong \bigotimes_{m\in\mathbb{N}_0+\frac{1}{2}} C\ell(H(\mathbb{C}_m)) \tag{6.1}$$

Here

(1) For a complex vector space W equipped with a $\mathbb{C}$-bilinear symmetric form $\omega\colon W \times W \to \mathbb{C}$, $C\ell(W)$ is the complex Clifford algebra generated by W;
(2) $H(V)$ for a complex vector space V is the hyperbolic form on $H(V) := V^\vee \oplus V$ defined by $\omega((f,v),(f',v')) = f(v') + f'(v)$;
(3) $\mathbb{C}_m$ for $m \in \mathbb{R}$ is a copy of the complex numbers equipped with an $\mathbb{R}$-action given by scalar multiplication by $q^m := e^{2\pi i m\tau}$ for $\tau \in \mathbb{R}$. The Clifford algebra $C\ell(\mathbb{C}_0)$ is with respect to the standard form $\omega(z,z') = zz'$ on

$\mathbb{C}_0$. The $\mathbb{R}$-action on $H(\mathbb{C}_m)$ respectively $\mathbb{C}_0$ is by isometries and hence induces an action on the Clifford algebra it generates.

(4) the tensor products are the *restricted* tensor product (i.e., the closure of finite sums of tensor products $\bigotimes_m a_m$ where $a_m = 1$ for all but finitely many m's).

The structure of the bimodule $T^{-1}(C^s_\tau)$ is determined by the following isomorphism of $T^{-1}(K^s) - T^{-1}(K^s)$-bimodules with $\mathbb{R}$-action:

$$T^{-1}(C^s_\tau) \cong T^{-1}(K^s)_{\phi(\tau)}$$

Here $T^{-1}(K^s)$ is considered as a bimodule over itself, but with the right-action twisted by the algebra homomorphism $\phi(\tau)\colon T^{-1}(C^s_\tau) \to T^{-1}(C^s_\tau)$ (i.e., the right action of $a \in T^{-1}(K^s)$ on $m \in T^{-1}(K^s)$ is given by the product $m \cdot (\phi(\tau)(a))$ in the algebra $T^{-1}(K^s)$).

Moreover, $\phi(\tau) = \bigotimes_m \phi_m(\tau)$, where $\phi_m(\tau)$ is an algebra automorphism of $C\ell(H(\mathbb{C}_m))$ (respectively $C\ell(\mathbb{C}_0)$ for $m = 0$). This automorphism is induced by the action of $\tau \in \mathfrak{h} \subset \mathbb{C}$ on $\mathbb{C}_m$ given by scalar multiplication by $e^{2\pi i m\tau}$ (so it extends the $\mathbb{R}$-action defined above).

Our goal now is to construct an invertible field theory of P of degree $-n$, i.e., an invertible natural transformation $P\colon T^0 \Rightarrow (T^{-1})^{\otimes n}$ for n as small as possible. We note that such a P gives in particular a Morita-equivalence between the algebras $T^0(K^s) = \mathbb{C}$ and $(T^{-1})^n(K^s)$. Since the algebras $C\ell(H(\mathbb{C}_m))$ and $C\ell(\mathbb{C}_0)^{\otimes 2}$ are Morita-equivalent to $\mathbb{C}$, but $C\ell(\mathbb{C}_0)$ is not, we need to assume that n is even.

To construct P, according to (5.5), we need to construct in particular a left $(T^{-1}(K^s))^{\otimes n}$-module $P(K^s)$ and a $T^{-1}(K^s)^{\otimes n} - T^{-1}(K^s)^{\otimes n}$-bimodule map

$$P(C^s_\tau)\colon (T^{-1}(C^s_\tau))^{\otimes n} \longrightarrow \mathrm{End}(P(K^s))$$

We define $P(K^s)$ to be the following left module over $T^{-n}(K^s) = (T^{-1}(K^s))^{\otimes n}$:

$$P(K^+) := M_0^{\otimes \frac{n}{2}} \otimes \bigotimes_{m\in\mathbb{N}} M_m^{\otimes n} \qquad P(K^-) := \bigotimes_{m\in\frac{1}{2}+\mathbb{N}_0} M_m^{\otimes n} \tag{6.2}$$

Here M_m (respectively M_0) is an irreducible graded module over $C\ell(H(\mathbb{C}_m))$ (respectively $C\ell(\mathbb{C}_0)^{\otimes 2}$), and we let each factor from the tensor product decomposition (6.1) act on the corresponding factor in the tensor product above. The modules M_m provide a Morita equivalence between $\mathbb{C}$ and $C\ell(H(\mathbb{C}_m))$ for $m \neq 0$ and between $\mathbb{C}$ and $C\ell(\mathbb{C}_0)^{\otimes 2}$ for $m = 0$. It follows that $P(K^s)$ provides a Morita equivalence between $\mathbb{C}$ and $T^n(K^s)$ which is a necessary feature to insure *invertibility* of P as a natural transformation from T^0 to T^n. We note that there are *two* non-isomorphic irreducible graded modules over $C\ell(H(\mathbb{C}_m))$ and $C\ell(\mathbb{C}_0)$. We will specify our choice for M_m below.

We note that any bimodule map

$$T^{-n}(C^s_\tau) = (T^{-1}(K^s)_{\phi(\tau)})^{\otimes n} \to \mathrm{End}(P(K^s))$$

is determined by the image of the unit $1 \in (T^{-1}(K^s)_{\phi(\tau)})^{\otimes n}$. Our map $P(C^s_\tau)$ is determined by specifying

$$P(C^+_\tau)(1) = q^2 \,\mathrm{id}_{M_0^{\otimes \frac{n}{2}}} \otimes \bigotimes_{m\in\mathbb{N}} b_m^{\otimes n}(\tau) \in \mathrm{End}(P(K^+)) = \mathrm{End}\left(M_0^{\otimes \frac{n}{2}} \otimes \bigotimes_{m\in\mathbb{N}} M_m^{\otimes n}\right)$$

$$P(C^-_\tau)(1) = q \bigotimes_{m\in\frac{1}{2}+\mathbb{N}_0} b_m^{\otimes n}(\tau) \qquad \in \mathrm{End}(P(K^-)) = \mathrm{End}\left(\bigotimes_{m\in\frac{1}{2}+\mathbb{N}_0} M_m^{\otimes n}\right)$$

Here $b_m(\tau)$ is an element of the Clifford algebra $C\ell(H(\mathbb{C}_m))$ which acts by left-multiplication on the module M_m. It is defined by

$$b_m(\tau) := 1 + (1-q^m)\frac{ef}{2} \in C\ell(H(\mathbb{C}_m))$$

where $\{e, f\}$ is a basis of $H(\mathbb{C}_m)$ given by $e := 1 \in \mathbb{C}_m$, and letting $f \in \mathbb{C}^\vee_{-m}$ be the dual element (in particular $\omega(e,e) = \omega(f,f) = 0$, $\omega(e,f) = 1$). It is a straightforward calculation to show that our prescription makes $P(C^s_\tau)$ a well-defined map of bimodules (we need to check that the annihilator of $1 \in T^{-n}(K^s)$ agrees with the annihilator of its desired image element $P(C^s_\tau)(1)$). It is also not hard to show that $\mathfrak{h} \to \mathrm{End}(P(K^s))$, $\tau \mapsto P(C^s_\tau)(1)$ is a homomorphism; this shows that our definition of $P(C^s_\tau)$ is compatible with composing two cylinders (in more technical terms, diagram (5.6) commutes if Σ, Σ' are two cylinders).

So far we have not constructed the bimodule maps $P(\Sigma)$ for the connected bordisms R^s_τ, L^s_τ, and $T^{s_1 s_2}_\tau$. It turns out that these are determined by $P(C^s_\tau)$. For example, an element $\gamma \in T^{-n}(C^{s_2}_\tau)$ determines an associated element $\delta \in T^{-n}(T^{s_1 s_2}_\tau)$ such that

$$P(T^{s_1 s_2}_\tau)(\delta) = \begin{cases} \mathrm{str}\, P(C^{s_2}_\tau)(\gamma) & s_1 = + \\ \mathrm{tr}\, P(C^{s_2}_\tau)(\gamma) & s_1 = - \end{cases}$$

This allows us to calculate the partition function Z_P of this putative conformal Euclidean field theory. It turns out that to ensure existence of a conformal field theory P with $P(C^s_\tau)$ as above, the *only* consistency condition that needs to be checked (in addition to additivity for gluing cylinders) is that Z_P is an $SL_2(\mathbb{Z})$-equivariant section. We recall that $T^{s_1 s_2}(T_\tau) = \mathbb{C}$ for $(s_1, s_2) \neq (+,+)$, and hence Z_P is just a function on $\mathfrak{h}^{-+} \amalg \mathfrak{h}^{+-} \amalg \mathfrak{h}^{--}$. There is a trivialization of the line bundle restricted to $\mathfrak{h}^{++}$ such that an $SL_2(\mathbb{Z})$-equivariant section corresponds to a function on the upper half plane with the equivariance properties of a modular form of weight $n/2$. Using this trivialization, the function Z_P is as follows:

$$Z_P(\tau) = \begin{cases} \left(q^{1/24}\prod_{m\in\mathbb{N}}(1-q^m)\right)^n \\ \left(2^{1/2}q^{1/24}\prod_{m\in\mathbb{N}}(1+q^m)\right)^n \\ \left(q^{-1/48}\prod_{m\in\mathbb{N}_0+\frac{1}{2}}(1-q^m)\right)^n \\ \left(q^{-1/48}\prod_{m\in\mathbb{N}_0+\frac{1}{2}}(1+q^m)\right)^n \end{cases} = \begin{cases} \eta(\tau)^n & \tau \in \mathfrak{h}^{++} \\ \left(\frac{2^{1/2}\eta(2\tau)}{\eta(\tau)}\right)^n & \tau \in \mathfrak{h}^{-+} \\ \left(\frac{\eta(\tau/2)}{\eta(\tau)}\right)^n & \tau \in \mathfrak{h}^{+-} \\ \text{something} & \tau \in \mathfrak{h}^{--} \end{cases} \tag{6.3}$$

It is well-known that $\eta(\tau)^n$ is a modular form if and only if n is a multiple of 24. However, the function Z_P is *not* $SL_2(\mathbb{Z})$-equivariant for $n = 24$: the matrix $T = \left(\begin{smallmatrix}1&1\\0&1\end{smallmatrix}\right)$ maps $\tau \in \mathfrak{h}^{+-}$ to $\tau + 1 \in \mathfrak{h}^{--}$, but $Z_P(T\tau) \neq Z_P(\tau)$ due to the factor $(q^{-1/48})^{24} = q^{-1/2}$ which changes sign under T. This forces us to take the power $n = 48$.

References

[DF] P. Deligne and D.Freed, *Supersolutions*, Quantum fields and strings: a course for mathematicians, Vol. 1 (Princeton, NJ, 1996/1997), AMS (1999) 227 – 363.

[DM] P. Deligne and J. Morgan, *Classical fields and supersymmetry*, Quantum fields and strings: a course for mathematicians, Vol. 1 (Princeton, NJ, 1996/1997), AMS (1999) 41 – 98.

[DH] C. Douglas and A. Henriques, *Topological modular forms and conformal nets*, in these proceedings.

[Du] F. Dumitrescu, *Superconnections and super-parallel transport*, To appear in Pac. J. of Math.; preprint available at arXiv:0711.2766.

[DST] F. Dumitrescu, S. Stolz and P. Teichner, *Connections as 1-dimenisonal field theories*, in preparation

[Fr] D. Freed, *Five lectures on supersymmetry*, AMS 1999.

[GS] V. Guillemin and S. Sternberg, *Supersymmetry and equivariant de Rham theory*, Springer-Verlag, Berlin, 1999. xxiv+228 pp.

[Ha] F. Han, *Supersymmetric QFT, Super Loop Spaces and Bismut-Chern Character*, preprint available at arXiv:0711.3862.

[HSST] F. Han, C. Schommer-Pries, S. Stolz and P. Teichner, *Equivariant cohomology from gauged field theories*, in preparation

[HKST] H. Hohnhold, M. Kreck, S. Stolz and P. Teichner, *Differential forms and 0-dimensional supersymmetric field theories*, Quantum Topol. 2 (2011), no. 1, 141

[HaST] F. Han, S. Stolz and P. Teichner, *The Bismut-Chern character form as dimensional reduction*, in preparation

[HoST] H. Hohnhold, S. Stolz and P. Teichner, *From minimal geodesics to supersymmetric field theories*, A celebration of the mathematical legacy of Raoul Bott, 207274, CRM Proc. Lecture Notes, 50, Amer. Math. Soc., Providence, RI, 2010

[Ho] M. Hopkins, *Algebraic Topology and Modular Forms*, Plenary Lecture, ICM Beijing 2002.

[JS] A. Joyal and R. Street, *Braided tensor categories*, Adv. Math. 102 (1993), no. 1, 20–78.

[Kr] I. Kriz, *On spin and modularity in conformal field theory*, Ann. Sci. cole Norm. Sup. (4) 36 (2003), no. 1, 57112.

[LM] H.-B. Lawson and M.L. Michelsohn, *Spin Geometry*, Princeton University Press, 1989.

[Lu] J. Lurie, *Expository article on topological field theories*, Preprint 2009.

[McL] S. Mac Lane, *Categories for the working mathematician*, Second edition. Graduate Texts in Mathematics, 5. Springer-Verlag, New York, 1998. xii+314 pp

[M] N. Martins-Ferreira, *Pseudo-categories*, J. Homotopy Relat. Struct. 1 (2006), no. 1, 47–78

[MV] R. Meise and D. Vogt, *Introduction to functional analysis*, Translated from the German by M. S. Ramanujan and revised by the authors. Oxford Graduate Texts in Mathematics, 2. The Clarendon Press, Oxford University Press, New York, 1997. x+437 pp.

[Se1] G. Segal, *Elliptic Cohomology*, Séminaire Bourbaki 695 (1988) 187–201.

[Se2] G. Segal, *The definition of conformal field theory*, Topology, geometry and quantum field theory, 423 – 577, London Math. Soc. Lecture Note Ser., 308, Cambridge Univ. Press, Cambridge, 2004.

[SP] C. Schommer-Pries, *The classification of two-dimensional extended topological field theories*, Berkeley thesis, 2009

[SST] C. Schommer-Pries, S. Stolz and P. Teichner, *Twisted cohomology via twisted field theories*, in preparation

[ST1] S. Stolz and P. Teichner, *What is an elliptic object?* Topology, geometry and quantum field theory, 247–343, London Math. Soc. Lecture Note Ser., 308, Cambridge Univ. Press, Cambridge, 2004.

[ST2] S. Stolz and P. Teichner, *Traces in monoidal categories*, To appear in Transactions of the AMS, 2011, available on the arxive.

[ST3] S. Stolz and P. Teichner, *Modularity of supersymmetric partition functions*, in preparation

[ST4] S. Stolz and P. Teichner, *Modular forms from twisted supersymmetric Euclidean field theories*, in preparation

[ST5] S. Stolz and P. Teichner, *Presentations of super Euclidean bordism categories*, in preparation

[ST6] S. Stolz and P. Teichner, *Constructing Euclidean field theories with prescribed partition function*, in preparation

[Th] W. Thurston, *Three-Dimensional Geometry and Topology*, Volume 1, Princeton University Press, 1997.

[Vi] A. Vistoli, *Grothendieck topologies, fibered categories and descent theory*, Fundamental algebraic geometry, 1–104, Math. Surveys Monogr., 123, Amer. Math. Soc., Providence, RI, 2005.

[Wi] E. Witten, *The index of the Dirac operator on loop space*, Elliptic curves and modular forms in Alg. Top., Princeton Proc. 1986, LNM 1326, Springer, 161–181.

[Za] D. Zagier, *Note on the Landweber-Stong elliptic genus*, Elliptic curves and modular forms in Alg. Top., Princeton Proc. 1986, LNM 1326, Springer, 216–224.

Department of Mathemetics, University of Notre Dame, South Bend, Indiana, 46556

Department of Mathematics, University of California, Berkeley, CA, 94720 and Max Planck Institute for Mathematics, Bonn, Germany

Proceedings of Symposia in Pure Mathematics
Volume **83**, 2011

TOPOLOGICAL MODULAR FORMS AND CONFORMAL NETS

CHRISTOPHER L. DOUGLAS AND ANDRÉ G. HENRIQUES

1. Introduction. The generalized cohomology theory K-theory has had an especially rich history in large part because it admits a geometric definition in terms of vector bundles. This definition has provided a framework for a spectacular array of generalizations of K-theory, including equivariant K-theory, algebraic K-theory, twisted K-theory, L-theory, and KK-theory, among others, with applications to such diverse topics as indices of elliptic operators, classification of manifolds, topology of diffeomorphism groups, geometry of algebraic cycles, and classification of C^*-algebras.

Since its introduction by Hopkins and Miller, the cohomology theory TMF, topological modular forms, has been seen as a higher form of K-theory—here 'higher' has been variously interpreted to refer to categorification or taking loop spaces, though at root it refers to the increase of the exponent controlling the formal group laws associated to the cohomology theory. As the present definitions of TMF are predominantly homotopy-theoretic in character [HMi, HMa, Lu], it has been a persistent dream of algebraic topologists to develop a geometric definition for TMF analogous to the one for K-theory. Various potential definitions have been investigated [BDR, HK, Se, ST1], particularly ones utilizing bundles of linear categories, or considering vector bundles on loop spaces, or involving models for quantum and conformal field theory. A particularly concise, if as yet ill-defined, proposal for such a geometric model, is the $\mathit{Diff}(S^1)$-equivariant K-theory of the free loop space:

$$TMF^*(M) \overset{?}{\simeq} K^*_{\mathit{Diff}(S^1)}(LM).$$

A $\mathit{Diff}(S^1)$-action on a vector bundle, over a free loop space, can be interpreted as a shadow of an action by a category of geometric surfaces, which itself is closely related to the structures appearing in a two-dimensional conformal field theory. Building primarily on work of Segal and Stolz-Teichner, we propose that conformal nets, a mathematical model for certain conformal field theories, could play a prominent role in a field-theoretically-inspired geometric definition of TMF.

Though the 0th group $K^0(X)$ of the generalized cohomology theory K-theory can be described in terms of equivalence classes of vector bundles on the space X, to geometrically describe all the groups $K^n(X)$ requires considering bundles of modules for the Clifford algebras $\mathit{Cliff}(n)$ [Kar]. There is a family of fundamental conformal

2010 *Mathematics Subject Classification.* 55N34, 81T40, 81T05.

The first author was partially supported by a Miller Research Fellowship.

nets, the free fermions *Fer*(n), that are conformal-field-theoretic analogues of the ordinary Clifford algebras. Moreover, there is a notion of boundary condition for a conformal net that is analogous to the notion of module for an algebra. We therefore speculate that bundles of fermionic boundary conditions will be the basic underlying objects of a geometric model for *TMF*.

As orientations of manifolds and vector bundles are central to the theory of ordinary homology, providing fundamental classes, Thom classes, Euler classes, and Poincaré duality, among other structures, so spin structures on manifolds and vector bundles are the key notions of orientation for K-theory. Furthermore, string structures are the basic orientation relevant for the theory of topological modular forms; in particular, manifolds with string structures have *TMF*-fundamental classes and *TMF*-Euler classes.

A spin structure on $\mathbb{R}^n$ can be conveniently encoded in the context of Clifford algebras as an invertible bimodule from *Cliff*(n) to itself; analogously, a string structure on $\mathbb{R}^n$ has a natural description, using conformal nets, as an invertible bimodule (called, in this context, an invertible defect) from *Fer*(n) to itself. Leveraging this notion of a string structure, we can construct the basic data of the *TMF*-Euler class of a string vector bundle.

The 8-fold periodicity of the real K-theory groups is intimately related to the 8-fold Morita periodicity of the real Clifford algebras, and we imagine there is a similar relationship between the 576-fold periodicity of *TMF* and an appropriate notion of periodicity for the free fermion nets. Specifically, we conjecture that the net of n fermions is equivalent, in the 3-category of conformal nets, to the net of $n + 576$ fermions. Using an action of the stable homotopy groups of spheres on the homotopy of the 3-category of conformal nets, we establish a lower bound on the periodicity of the fermions, proving that if n fermions is equivalent to $n + k$ fermions, then 24 must divide k.

2. Topological modular forms. We discuss a few facts about the ring of topological modular forms, particularly concerning its periodicity and relationship to the ring of ordinary modular forms. The reader should refer to Goerss' article [Go] for a thorough overview of the field, and to Behrens' paper [Be] and Bauer's paper [Ba] for details about, respectively, the construction of *TMF* and the computation of its homotopy.

Associated to each elliptic curve C there is an elliptic cohomology theory E_C, and in an appropriate sense, the cohomology theory *TMF* functions as a universal elliptic cohomology theory. This universal property provides a map from the ring of topological modular forms, that is from the *TMF*-cohomology of a point, to the ring of modular forms, that is to the ring of sections of certain line bundles over the moduli stack of elliptic curves. The ring of weakly holomorphic integral modular forms MF_* is $\mathbb{Z}[c_4, c_6, \Delta^{\pm 1}]/(c_4^3 - c_6^2 - 1728\,\Delta)$—see for instance Deligne's computation [De]—and the map

$$\phi : TMF_*(pt) \rightarrow MF_*$$

is a rational isomorphism. In fact, the kernel and cokernel of the map ϕ are both 2- and 3-torsion groups. Note that ϕ is grading-preserving, provided we take the degrees of c_4, c_6, and Δ to be 8, 12, and 24 respectively; these are twice the classical gradings of those elements.

Because the discriminant $\Delta \in MF_*$ is invertible, the ring of modular forms is periodic of period 24; multiplication by Δ provides the periodicity isomorphism from MF_n to MF_{n+24}, for any n. There is no topological modular form lifting the modular form Δ, and so the periodicity of modular forms is not reproduced directly in *TMF*. However, the power Δ^{24} does lift to a topological modular form, and the ring TMF_* is therefore periodic of period $24 \times 24 = 576$.

This situation comparing modular forms to topological modular forms, in which a classical periodicity is dilated, is entirely analogous to the situation in K-theory. One construction of real K-theory is as follows: to each curve carrying a formal group structure isomorphic to the multiplicative formal group, there is an associated cohomology theory equivalent to complex K-theory; real K-theory is, in an appropriate sense, the universal cohomology theory mapping to all the resulting versions of complex K-theory. This universal property provides a map from the coefficient ring $KO_*(pt)$ of real K-theory to the ring of sections of certain line bundles over the moduli stack of multiplicative curves. That ring of sections is $\mathbb{Z}[a^{\pm 1}]$, where the periodicity generator a has degree 4. The map

$$\psi : KO_*(pt) \ \rightarrow \ \mathbb{Z}[a^{\pm 1}]$$

is a rational isomorphism, and in fact has kernel and cokernel being 2-torsion groups. Though the generator a is not in the image of the map ψ, the power a^2 does lift to a class in real K-theory; thus, real K-theory is periodic of period $4 \times 2 = 8$.

As listing all 576 coefficient groups of *TMF* might tax the reader's attention, we write out these groups through degree 28, to give a sense of their structure in this range and of their relation to the entire first period of the ring of classical modular forms.

The homotopy groups of *TMF* in low degrees:

π_0	π_1	π_2	π_3	π_4	π_5	π_6	π_7	π_8	π_9
$\mathbb{Z}[x]$	$\mathbb{Z}/2[x]$	$\mathbb{Z}/2[x]$	$\mathbb{Z}/24$	$\mathbb{Z}[x]$	0	$\mathbb{Z}/2$	0	$\mathbb{Z}[x] \oplus \mathbb{Z}/2$	$\mathbb{Z}/2[x] \oplus \mathbb{Z}/2$

π_{10}	π_{11}	π_{12}	π_{13}	π_{14}	π_{15}	π_{16}	π_{17}	π_{18}	π_{19}
$\mathbb{Z}/2[x] \oplus \mathbb{Z}/3$	0	$\mathbb{Z}[x]$	$\mathbb{Z}/3$	$\mathbb{Z}/2$	$\mathbb{Z}/2$	$\mathbb{Z}[x]$	$\mathbb{Z}/2[x] \oplus \mathbb{Z}/2$	$\mathbb{Z}/2[x]$	0

π_{20}	π_{21}	π_{22}	π_{23}	π_{24}	π_{25}	π_{26}	π_{27}	π_{28}
$\mathbb{Z}[x] \oplus \mathbb{Z}/24$	$\mathbb{Z}/2$	$\mathbb{Z}/2$	0	$24\mathbb{Z} + x\mathbb{Z}[x]$	$\mathbb{Z}/2[x]$	$\mathbb{Z}/2[x]$	$\mathbb{Z}/12$	$\mathbb{Z}[x] \oplus \mathbb{Z}/2$

This table indicates that $\pi_0(TMF)$ is the ring $\mathbb{Z}[x]$, and the remaining homotopy groups are described as $\pi_0(TMF)$-modules. The element $x \in \pi_0(TMF)$ maps, under ϕ, to the modular form $c_4^3\,\Delta^{-1}$. The 24th homotopy group is isomorphic to the ideal of $\mathbb{Z}[x]$ generated by 24 and x; as a $\mathbb{Z}[x]$-module, it has two generators, which we might suggestively call $[24\Delta]$ and $[c_4^3]$ subject to the relation $x[24\Delta] = 24[c_4^3]$. Besides $[24\Delta], [c_4^3] \in \pi_{24}(TMF)$, the only elements in this range that map nontrivially into MF_* are the $\mathbb{Z}[x]$-module generators of the $\mathbb{Z}[x]$ factor in degrees $0, 4, 8, 12, 16, 20, 28$, which hit respectively $1, 2c_6c_4^2\Delta^{-1}, c_4, 2c_6, c_4^2, 2c_6c_4, 2c_6c_4^2 \in MF_*$.

3. Conformal nets. The geometric model for K-theory depends not only on the existence of the Clifford algebras *Cliff*(n), but also on having the Clifford algebras situated in a 2-category of algebras, bimodules, and intertwiners—this 2-category provides the crucial notions of Clifford modules and Morita equivalence. As part of their program to develop a geometric model for *TMF* as a space of two-dimensional quantum field theories, Stolz and Teichner [ST1] realized they would need a 3-category encoding some kind of 'higher Clifford algebras'. In order that this 3-category interact well with existing notions of one-dimensional quantum field theories, they hoped the 3-category would have a prescribed relationship to the 2-category of von Neumann algebras over the complex numbers. Specifically, they asked the following:

Question. *(Stolz-Teichner, 2004) Is there an interesting symmetric monoidal 3-category $\mathcal{C}$ delooping the 2-category vN of von Neumann algebras, in the sense that* $\mathrm{Hom}_{\mathcal{C}}(1,1) = vN$*?*

As it is certainly possible to take a connected deloop BvN of vN, in this question 'interesting' amounted to wanting there to be natural objects of $\mathcal{C}$ besides the unit object, ideally a family of objects with periodicity properties analogous to those of Clifford algebras.

In the forthcoming papers [BDH1,2,3], we introduce a composition operation on defects between conformal nets and show that conformal nets themselves form a 3-category delooping von Neumann algebras. Conformal nets are one of the existing mathematical models for conformal field theory; traditionally, a conformal net is defined (with inspiration from algebraic quantum field theory) as a collection of algebras of operators associated to the subintervals of the standard circle. There are many other formalisms for conformal field theory, for instance vertex algebras [FBZ], chiral algebras [BD], and algebras over (partial) operads [Se, Hu, FRS]. Among these frameworks, conformal nets provide a particularly appropriate context for investigating the higher categorical structure of conformal field theory. As such, we borrow terminology from conformal field theory in naming the 1-, 2-, and 3-morphisms of our 3-category CN of conformal nets; the following table lists a few of the key pieces of categorical structure, and the associated names:

The 3-category of conformal nets	
Objects	Conformal nets
Arrows, $\mathcal{A} \to \mathcal{B}$	Defects between the nets $\mathcal{A}$ and $\mathcal{B}$
Arrows from $\mathcal{A}$ to the unit object 1	Boundary conditions for $\mathcal{A}$
2-morphisms between arrows, $\mathcal{A} \Downarrow \mathcal{B}$	Sectors between $\mathcal{A}$-$\mathcal{B}$-defects
2-morphisms from $\mathrm{id}_{\mathcal{A}} : \mathcal{A} \to \mathcal{A}$ to itself	Representations of $\mathcal{A}$
3-morphisms between 2-morphisms	Intertwiners of sectors

Before describing nets and defects in more detail, we mention the kind of 3-categorical structure we consider. On the one hand, the higher categorical structure of conformal nets is somewhat stricter than a maximally weak symmetric monoidal 3-category; on the other hand, conformal nets naturally have a bit more structure than would be recorded in an ordinary 3-category. We address both these features by building the 3-categorical structure of conformal nets as an internal bicategory in the 2-category of symmetric monoidal categories. Such internal bicategories, described in [BDH1], distinguish two kinds of 1-morphisms and two kinds of 2-morphisms. This is analogous to the notion of bicategory given by internal categories in the 2-category of categories, investigated by Shulman [Sh], which records two distinct kinds of 1-morphisms; for instance, in the bicategory of algebras, the ring homomorphisms and the bimodules between algebras are two structurally distinct types of morphisms. In the internal bicategory of conformal nets, the 1-morphisms are split into the homomorphisms of nets and the defects between nets, and the 2-morphisms are split into the homomorphisms of defects and the sectors between defects.

As mentioned above, the classical notion of conformal net [Kaw, GF, Lo] focuses on a collection of algebras associated to subintervals of the standard circle. In order to facilitate the construction of the higher categorical structure of nets, we introduce instead a notion of coordinate-free conformal net [BDH2], in which the collection of algebras is indexed by abstract intervals, without reference to the circle. This more flexible notion is reminiscent of structures that have been considered in algebraic quantum field theory on curved 4-dimensional space-times [BFV].

Definition. *A coordinate-free conformal net is a continuous covariant functor*

$$\mathcal{A}\colon \left\{ \begin{array}{c} \text{oriented intervals,} \\ \text{embeddings} \end{array} \right\} \to \left\{ \begin{array}{c} \text{von Neumann algebras,} \\ \text{injective linear maps} \end{array} \right\}$$

taking orientation-preserving embeddings to homomorphisms and orientation-reversing embeddings to antihomomorphisms; here an 'interval' is a 1-manifold diffeomorphic to $[0,1]$*. The functor* $\mathcal{A}$ *is subject to the following conditions, for* I *and* J *subintervals of the interval* K*:*

- Locality: *If* $I, J \subset K$ *have disjoint interiors, then the images of* $\mathcal{A}(I)$ *and* $\mathcal{A}(J)$ *are commuting subalgebras of* $\mathcal{A}(K)$*.*
- Strong additivity: *If* $K = I \cup J$*, then the algebra* $\mathcal{A}(K)$ *is topologically generated by* $\mathcal{A}(I)$ *and* $\mathcal{A}(J)$*.*
- Split property: *If* $I, J \subset K$ *are disjoint, then the map from the algebraic tensor product* $\mathcal{A}(I) \otimes_{alg} \mathcal{A}(J) \to \mathcal{A}(K)$ *extends to the spatial tensor product.*
- Inner covariance: *If* $\varphi\colon I \to I$ *is a diffeomorphism that restricts to the identity in a neighborhood of* ∂I*, then* $\mathcal{A}(\varphi)$ *is an inner automorphism of* $\mathcal{A}(I)$*.*
- Vacuum sector: *Suppose* $J \subsetneq I$ *contains the boundary point* $p \in \partial I$*. Let* $\bar{J}$ *denote* J *with the reversed orientation.* $\mathcal{A}(J)$ *acts on* $L^2(\mathcal{A}(I))$ *via the left action of* $\mathcal{A}(I)$*, and* $\mathcal{A}(\bar{J}) \cong \mathcal{A}(J)^{op}$ *acts on* $L^2(\mathcal{A}(I))$ *via the right action of* $\mathcal{A}(I)$*. We require that the action of* $\mathcal{A}(J) \otimes_{alg} \mathcal{A}(\bar{J})$ *on* $L^2(\mathcal{A}(I))$ *extends to an action of* $\mathcal{A}(J \cup_p \bar{J})$*.*

In the vacuum sector property in this definition, $L^2(A)$ refers to the Haagerup space, that is the standard form, of the von Neumann algebra A [Ha]—this is a Hilbert space, equipped with the structure of an A-A-bimodule, that functions as an identity bimodule for Connes fusion over A. We do not explicitly demand the existence of a vacuum vector, nor, more dramatically, do we require that the net be positive energy. Note that classical conformal nets can be promoted to give examples of coordinate-free conformal nets. Another source of examples is Minkowskian conformal field theories: nets of algebras on 2-dimensional Minkowski space-time provide (non-positive-energy) conformal nets when restricted to a spatial slice [KLM, LR].

Morphisms between coordinate-free conformal nets are called defects, and their definition is inspired by the notions of topological and conformal defects in quantum and conformal field theory.

Definition. *Let $\mathcal{A}$ and $\mathcal{B}$ be coordinate-free conformal nets. An $\mathcal{A}$-$\mathcal{B}$-defect is a functor*

$$D\colon \left\{ \begin{array}{c} \text{intervals in } \mathbb{R} \text{ whose boundary} \\ \text{does not contain } 0, \text{ inclusions} \end{array} \right\} \to \left\{ \begin{array}{c} \text{von Neumann algebras,} \\ \text{homomorphisms} \end{array} \right\}$$

such that $D|_{\mathbb{R}_{>0}}$ is given by $\mathcal{A}$, and $D|_{\mathbb{R}_{<0}}$ is given by $\mathcal{B}$. It is subject to the following conditions:

- Isotony: *If $I \subset J$ and $0 \in I$, then the corresponding map $D(I) \to D(J)$ is injective.*
- Locality: *If $I \cap J$ is a point, then the images of $D(I)$ and $D(J)$ commute in $D(I \cup J)$.*
- Strong additivity: *$D(I \cup J)$ is topologically generated by $D(I)$ and $D(J)$.*

The defect functor D is also subject to a version of the vacuum sector axiom, which we do not reproduce here.

Defects are the analogs for conformal nets of bimodules between algebras. The basic underlying objects in a geometric model for K-theory are bundles, not of bimodules but of modules for Clifford algebras. When one of the two nets $\mathcal{A}$ or $\mathcal{B}$ is trivial (that is, is the constant functor $I \mapsto \mathbb{C}$), the notion of a defect specializes to the conformal-net analog of a module. This notion is called a boundary condition for the net, and is given as follows.

Definition. *A boundary condition for the coordinate-free conformal net $\mathcal{A}$ is a functor*

$$D\colon \left\{ \begin{array}{c} \text{intervals in } \mathbb{R}_{\geq 0}, \\ \text{inclusions} \end{array} \right\} \to \left\{ \begin{array}{c} \text{von Neumann algebras,} \\ \text{homomorphisms} \end{array} \right\}$$

whose restriction to $\mathbb{R}_{>0}$ is given by $\mathcal{A}$. It must satisfy isotony, locality, strong additivity, and vacuum sector conditions; the last of these is:

- Vacuum sector: *Suppose $0 \in I \subset \mathbb{R}_{\geq 0}$. Let J be a subinterval of I that contains the boundary point $p \in \partial I$, $p \neq 0$. The algebras $\mathcal{A}(J)$ and $\mathcal{A}(\bar{J})$ act on $L^2(D(I))$ via the left and right action of $D(I)$, respectively. We require that the action of $\mathcal{A}(J) \otimes_{alg} \mathcal{A}(\bar{J})$ on $L^2(D(I))$ extends to an action of $\mathcal{A}(J \cup_p \bar{J})$.*

Note that defects and boundary conditions are well established in the conformal field theory literature [FRS, LR]. Previous notions do not, however, permit boundary conditions for chiral conformal field theories, as the above notion does; the basic conformal nets we will use in our investigation of *TMF* will necessarily be chiral CFTs.

The notion of boundary condition for a net allows us to consider bundles of boundary conditions as potential geometric representatives of *TMF*-cohomology classes. To investigate this idea further, we need a family of nets that functions as an analog of the Clifford algebras. Indeed we can construct such a family, namely the free fermions.

4. KO-theory and Clifford algebras, *TMF* and fermions. Both in the homotopy-theoretic construction [HMi] and in the field-theoretic approach to cohomology theories [ST1], there are vivid parallels between real K-theory, that is KO, and *TMF*. As mentioned above, a crucial ingredient in a geometric understanding of KO is the family *Cliff*(n) of Clifford algebras. We propose that the free fermion coordinate-free conformal nets $Fer(n)$ play a role for TMF analogous to the role of Clifford algebras for KO.

The family of Clifford algebras has three essential properties: 1) it extends to a functor from real vector spaces to algebras, so in particular the nth algebra $Cliff(n)$ carries an action of the nth orthogonal group; 2) this functor is exponential, in that $Cliff(V \oplus W) = Cliff(V) \otimes Cliff(W)$; 3) these algebras encode the fundamental transformation group for KO, in the sense that the automorphism group of a vector space V equipped with a Morita equivalence between $Cliff(V)$ and $Cliff(n)$ is the spin group $Spin(V)$. We would like our fermion family of nets to satisfy analogous properties.

Note that the relevant Clifford algebras are naturally $\mathbb{Z}/2$-graded algebras, and the fermion nets must similarly be considered as $\mathbb{Z}/2$-graded conformal nets. As described in the last section, a conformal net is a functor from a category of oriented intervals to the category of von Neumann algebras; a $\mathbb{Z}/2$-graded conformal net is a functor from a category of spin intervals to the category of $\mathbb{Z}/2$-graded von Neumann algebras. A spin interval is an oriented interval equipped with a square root S of its complexified cotangent bundle. In addition to satisfying the $\mathbb{Z}/2$-graded analogs of the previously described axioms, the functor of a $\mathbb{Z}/2$-graded net must send the spin involution of an interval to the grading involution of the von Neumann algebra associated to that interval. See [DH] for a precise definition of the $\mathbb{Z}/2$-graded version of conformal nets.

As a $\mathbb{Z}/2$-graded net, the fermion $Fer(1)$ provides an algebra $Fer(1)(I)$, defined as follows, for each spin interval I. The orientation of I provides a canonical real subbundle S_+ of the complex spinor bundle S of the interval I, and therefore a real Hilbert space $L^2(I, S_+)$ of spinor-valued L^2 functions on the interval; the algebra $Fer(1)(I)$ is the completion, in an appropriate Fock representation, of the complexified Clifford C^*-algebra $Cliff_{\mathbb{C}}(L^2(I; S_+))$ of this space of sections [DH, To, Was]. The family $Fer(n)$ of nets does indeed satisfy our three desired properties: 1) by tensoring the spinors with a vector space, the construction produces a functor from vector spaces to nets, $V \mapsto Fer(V)$; 2) because the Clifford algebra construction on L^2 functions is exponential, the fermion construction itself is exponential; 3) the fermions encode the fundamental transformation group for TMF, namely the string group, in the sense that the automorphism group of a vector space V equipped with an invertible defect between $Fer(V)$ and $Fer(n)$ is the string group $String(V)$. This last property is described in more detail in the next section.

Clifford algebra $Cliff(n)$	Free fermion $Fer(n)$
$Cliff(n)$ has an action of $O(n)$	$Fer(n)$ has an action of $O(n)$
$Cliff$ is an exponential functor: $Cliff(V \oplus W) = Cliff(V) \otimes Cliff(W)$	Fer is an exponential functor: $Fer(V \oplus W) = Fer(V) \otimes Fer(W)$
$Cliff(n)$ can be used to define $Spin(n)$	$Fer(n)$ can be used to define $String(n)$

Roughly speaking, a class in the nth real K-theory of a space X, that is $KO^n(X)$, is represented by a bundle of $Cliff(n)$-modules over X. A $Cliff(n)$-module can be viewed as a homomorphism, in the 2-category Alg of algebras, from the Clifford algebra to the trivial algebra; that is, the module is an element of $\mathrm{Hom}_{Alg}(Cliff(n), 1)$.

As the fermions are the *TMF* analogs of Clifford algebras, we now consider homomorphisms in the 3-category *CN* of nets from the fermions to the trivial net; these elements of $\mathrm{Hom}_{CN}(\mathit{Fer}(n), 1)$ are boundary conditions for the fermions. Bundles of such boundary conditions provide, we believe, the core data for a representative of a *TMF*-cohomology class.

Cohomology theory	KO^*	TMF^*
The cohomological degree is controlled by	The Clifford algebras $\mathit{Cliff}(n)$	The free fermion conformal nets $\mathit{Fer}(n)$
Cohomology classes of degree n are represented by	Bundles of $\mathit{Cliff}(n)$-modules	Bundles of $\mathit{Fer}(n)$-boundary conditions

It is not, in fact, the case that any class in $KO^*(X)$ can be represented by a bundle of $\mathit{Cliff}(n)$-modules. Describing a complete geometric definition of *KO* involves elaborating and modifying the idea of bundles of modules, in one of a few possible directions: for instance, one can consider bundles of Hilbert spaces with Clifford algebra actions and compatible fiberwise Fredholm operators, or one can use quasibundles (that is finite-dimensional but not necessarily locally trivial bundles) of Clifford modules. A complete geometric description of *TMF* classes will certainly involve analogous refinements to the notion of bundles of fermionic boundary conditions. Though what refinements are needed is at present unresolved, one can use the geometric field-theoretic viewpoint of Stolz and Teichner [ST2] as a source of clear inspiration concerning potential directions for investigation. Indeed, when we consider the Stolz-Teichner perspective (that *TMF* classes should be bundles of twisted two-dimensional geometric field theories) in the context of conformal nets and fermions, the bundle of field theories when evaluated on points would produce a bundle of fermionic boundary conditions. The kind of information encoded in the 1- and 2-dimensional parts of such field theories provides a guide to potential elaborations of the notion of bundles of boundary conditions.

5. String structures and the *TMF*-Euler class. String manifolds have *TMF*-fundamental classes [AHR], and in this sense string structures play a role for *TMF*-cohomology analogous to that played by spin structures for *KO* or by orientations for ordinary cohomology. In the paper [DH] we provide the first direct connection between the conformal net of free fermions and *TMF* by proving that the fermions elegantly encode the notion of a string structure on a vector bundle.

Recall that the string group $\mathit{String}(n)$ is a topological group whose homotopy type is the 3-connected cover of $O(n)$. It fits into a tower of connective covers of the orthogonal group, obtained by successively killing the lowest remaining homotopy group:

$$O(n) \underset{\substack{\uparrow \\ \text{kill } \pi_0}}{\longleftarrow} SO(n) \underset{\substack{\uparrow \\ \text{kill } \pi_1}}{\longleftarrow} \mathit{Spin}(n) \underset{\substack{\uparrow \\ \text{kill } \pi_3}}{\longleftarrow} \mathit{String}(n).$$

There are a number of existing models for the string group, for instance those in [ST1, BCSS, Wal, SP].

From a homotopy theoretic perspective, an orientation on a vector bundle is a lift of the classifying map of the bundle from $BO(n)$ to $BSO(n)$; similarly a spin structure or string structure is a lift of the classifying map from $BO(n)$ to respectively $B\mathit{Spin}(n)$ or $B\mathit{String}(n)$. A more concrete geometric notion of orientation is the following: an orientation on a vector bundle is a trivialization of the top exterior power of the bundle. Not surprisingly, for geometric applications, having this geometric characterization of the notion of orientation is essential. As we will describe presently, such a characterization exists for spin structures, and one of the contributions of our fermionic approach in [DH] is the construction of an analogously geometric characterization of string structures.

Let V be a real n-dimensional vector space equipped with an inner product. A typical description of a spin structure on V is as a nontrivial double cover of the oriented orthonormal frame bundle of V. This description is reasonably concrete, but not entirely geometric because double covers are themselves a rather topological construction. We can refine this double-cover notion of spin structure by specifying a geometric structure S associated to the vector space V such that the natural projection map $\pi : \mathrm{Aut}(V,S) \to \mathrm{Aut}(V) = SO(V)$ is a nontrivial double cover. Here, an automorphism of the pair (V,S) is by definition an automorphism of V, together with an isomorphism between S and its pullback along that automorphism of V. An example of such a structure S is an invertible *Cliff*(V)-*Cliff*(n) bimodule, that is a Morita equivalence between the Clifford algebras of V and of $\mathbb{R}^n$; we also insist that this invertible bimodule S is equipped with an inner product and that automorphisms of S be unitary. To see that the automorphism group of this structure is as desired, observe that

$$\ker\big(\pi\colon \mathrm{Aut}(V,S) \to \mathrm{Aut}(V)\big) = \mathrm{Aut}_{\mathit{Cliff}(V)\text{-}\mathit{Cliff}(n)\text{-bimod}}\big(S\big) = \{\pm 1\}.$$

A computation of the boundary homomorphism of the sequence $\{\pm 1\} \to \mathrm{Aut}(V,S) \to \mathrm{Aut}(V)$ shows that the cover is nontrivial.

The main theorem of [DH] shows that replacing the Clifford algebras by the free fermions in the above description does indeed produce a characterization of a string structure. Specifically, a string structure on a vector space V is an invertible *Fer*(V)-*Fer*(n) defect D. Here 'invertible' must be interpreted as meaning weakly invertible in the 3-category *CN* of conformal nets; said differently, the structure is an equivalence in *CN* between V-fermions *Fer*(V) and n-fermions *Fer*(n). As in the spin case, to verify that the structure is as desired, one considers the kernel of the projection to the special orthogonal group:

$$\ker\big(\pi\colon \mathrm{Aut}(V,D) \to \mathrm{Aut}(V)\big) = \mathrm{Aut}_{\mathit{Fer}(V)\text{-}\mathit{Fer}(n)\text{-defect}}\big(D\big) = \{\mathbb{Z}/2\text{-gr complex lines}\}.$$

This kernel K is not a group but a groupoid, and its homotopy type is $\mathbb{Z}/2 \times BS^1$, as we would expect for the kernel of the projection from $\mathit{String}(V)$ to $SO(V)$. An elaborate computation shows that the boundary homomorphism $\pi_3(SO(n)) \to \pi_2(K)$ is an isomorphism, for n at least 5, and this implies that the homotopy type of $\mathrm{Aut}(V,D)$ is indeed that of $\mathit{String}(V)$.

These geometric models for spin and string structures allow a particularly direct description of, respectively, the KO-Euler class of a spin vector bundle and the TMF-Euler class of a string vector bundle. Let V be a vector bundle over X equipped with a spin structure S; that is, S is a bundle over X whose fiber at $x \in X$ is an invertible *Cliff*(V_x)-*Cliff*(n) bimodule S_x. Recall that certain $KO^n(X)$ classes can be represented by bundles of *Cliff*(n)-modules. The KO-Euler class $e_V \in KO^n(X)$ of the bundle V is represented by the bundle S, viewed simply as a right *Cliff*(n)-module bundle:

$$e_V = \big[S \circlearrowleft \mathit{Cliff}(n)\big] \in KO^n(X).$$

That is, to obtain the Euler class of V from the spin structure S, we only need to forget the left *Cliff*(V) action on S.

To adapt this description to the *TMF*-Euler class $e_V \in TMF^n(X)$ of a string vector bundle V over X, we need to know how to take a $Fer(V)$-$Fer(n)$-defect and 'forget the action of $Fer(V)$'. This can be accomplished as follows:

Definition. *Given an $\mathcal{A}$-$\mathcal{B}$-defect E, we let $\vec{E}$: {intervals in $\mathbb{R}_{\geq 0}$} $\to$ {vN algebras} be the boundary condition for $\mathcal{B}$ given by*

$$\vec{E}(I) := \begin{cases} \mathcal{B}(I) & \text{if } 0 \notin I, \\ E([-1,0] \cup I) & \text{if } 0 \in I. \end{cases}$$

Now let V be a vector bundle equipped with a string structure D; that is, D is a bundle over X whose fiber at $x \in X$ is an invertible $Fer(V_x)$-$Fer(n)$ defect D_x. Forgetting the $Fer(V_x)$ action on each fiber produces a bundle $\vec{D}$ of $Fer(n)$-boundary conditions, with fibers $\vec{D}_x$. We expect that this bundle $\vec{D}$ of boundary conditions, perhaps with additional data, will represent the *TMF*-Euler class $e_V \in TMF^n(X)$ of the string vector bundle V.

6. Periodicity of the fermions. There are two distinct notions of equivalence between algebras: algebra isomorphism and Morita equivalence. The first notion arises by viewing algebras as the objects of a 1-category whose morphisms are algebra homomorphisms. The second notion arises by viewing algebras as the objects of a 2-category whose morphisms are bimodules and whose 2-morphisms are maps of bimodules.

There are, analogously, two distinct notions of equivalence between conformal nets. The first notion, isomorphism of nets, arises by considering nets as objects of a 1-category, with morphisms the natural transformations of the functors defining the nets. The existence of a 3-category of nets provides a completely new, second notion of equivalence of nets, which we refer to as *CN*-equivalence: two nets $\mathcal{A}$ and $\mathcal{B}$ are *CN*-equivalent if there exist defects $D\colon \mathcal{A} \to \mathcal{B}$ and $E\colon \mathcal{B} \to \mathcal{A}$ such that both composites DE and ED are Morita equivalent to the identity defects.

Though the essential nature and properties of *CN*-equivalence are as yet enigmatic, we can establish certain relationships between the notion of *CN*-equivalence and the representation categories of nets. Note that a representation of a net $\mathcal{A}$ is a Hilbert space equipped with compatible actions of the algebras $\mathcal{A}(I)$ for I a subinterval of the standard circle. The representation category $\mathrm{Rep}(\mathcal{A})$ of a net $\mathcal{A}$ can be expressed in purely categorical terms within the 3-category *CN* as

$$\mathrm{Rep}(\mathcal{A}) = \mathrm{Hom}_{\mathrm{Hom}_{CN}(\mathcal{A},\mathcal{A})}(\mathrm{id}_\mathcal{A}, \mathrm{id}_\mathcal{A}).$$

As a result, the *CN*-equivalence class of a net encodes the representation category, in the sense that if two nets have inequivalent representation categories, then they cannot be *CN*-equivalent. *CN*-equivalence also provides a specific characterization of the kernel of the functor taking a net to its representation category. Kawahigashi, Longo, and Müger [KLM] showed that a conformal net has trivial representation category if and only if the μ-index of the net is equal to one. We have proven [DH] that a conformal net $\mathcal{A}$ has μ-index equal to one if and only if it is *CN*-invertible, that is if there is another net $\mathcal{B}$ such that $\mathcal{A} \otimes \mathcal{B}$ is *CN*-equivalent to the trivial net

1. The fermion net has trivial representation category, so in particular $Fer(n)$ is an invertible net.

The real Clifford algebra $\mathit{Cliff}(1)$ is (Morita) invertible, in the 2-category of algebras, and it generates a $\mathbb{Z}/8$ subgroup of the group of Morita equivalence classes of invertible algebras; that is, for any n, there is a Morita equivalence

$$\mathit{Cliff}(n) \;\simeq\; \mathit{Cliff}(n+8).$$

This 8-fold periodicity of the Clifford algebras is the direct algebraic correspondent of the Bott periodicity of KO-theory. In a curious reversal, the periodicity of TMF is known [HMa, Ba], namely $TMF^n(X) \simeq TMF^{n+576}(X)$, but the corresponding algebraic periodicity has not yet been established. On the view that the fermions are the TMF analogs of the Clifford algebras, we make the following conjecture.

Conjecture. *For every n, there exists a CN-equivalence between the conformal nets $Fer(n)$ and $Fer(n+576)$.*

In their original question asking for a 3-category delooping von Neumann algebras, as described in section 3, Stolz and Teichner indicated that one desideratum for the 3-category would be that it contain a $\mathbb{Z}/576$ subgroup in its group of equivalence classes of invertible objects. If the fermion has minimal period 576, then the fermion conformal nets would provide such a subgroup.

This fermion periodicity conjecture remains quite mysterious, even from the point of view of physics. As described in the next section, though, we can establish a lower bound of 24 on the periodicity of fermions.

7. Fermions and stable homotopy. There is an invariant of invertible conformal nets taking values in the 24th roots of unity in S^1—this invariant arises from an action of the third homotopy group of the sphere spectrum on the geometric realization of the 3-groupoid of invertible conformal nets. Our lower bound on the periodicity of the fermions is established by computing that this invariant on $Fer(1)$ is a primitive 24th root of unity.

Given a symmetric monoidal 3-category C, let $C^\times$ denote the symmetric monoidal 3-groupoid of invertible objects, invertible morphisms, invertible 2-morphisms, and invertible 3-morphisms of C. (If the 3-category C has a natural notion of adjoint on its 3-morphisms, as CN does, we let $C^\times$ refer to the 3-groupoid whose 3-morphisms are not only invertible but in fact unitary.) The geometric realization $|C^\times|$ of $C^\times$ will have the structure of a spectrum, and the homotopy groups of this spectrum can be described as

$$\begin{aligned}
\pi_0\big(|C^\times|\big) &= \text{equivalence classes of invertible objects of } C,\\
\pi_1\big(|C^\times|\big) &= \text{equivalence classes of invertible morphisms from the unit 1 to itself},\\
\pi_2\big(|C^\times|\big) &= \text{equiv. classes of invertible 2-morphisms from the identity } \mathrm{id}_1 \text{ to itself},\\
\pi_3\big(|C^\times|\big) &= \text{invertible 3-morphisms from the double identity } \mathrm{id}_{\mathrm{id}_1} \text{ to itself},\\
\pi_n\big(|C^\times|\big) &= 0 \quad \text{for} \quad n \geq 4.
\end{aligned}$$

As any spectrum is a module over the sphere spectrum $\mathbb{S}$, the homotopy groups $\pi_*\big(|C^\times|\big)$ are a module over $\pi_*(\mathbb{S})$, the ring of stable homotopy groups of spheres. The most interesting piece of this module structure is the homomorphism

$$\nu : \pi_0\big(|C^\times|\big) \longrightarrow \pi_3\big(|C^\times|\big),$$

given by the action of the generator ν of $\pi_3(\mathbb{S}) = \mathbb{Z}/24$.

As the identity sector on the identity defect on the unit net is simply the Hilbert space of complex numbers, the third homotopy group $\pi_3\big(|CN^\times|\big)$ of the 3-groupoid of invertible conformal nets is S^1. The above action of the generator ν of the third stable homotopy group of spheres provides an invariant

$$\nu : \big\{\text{Invertible conformal nets}\big\} \to S^1,$$

which is multiplicative in the sense that $\nu(\mathcal{A} \otimes \mathcal{B}) = \nu(\mathcal{A})\nu(\mathcal{B})$. Because the class $\nu \in \pi_3(\mathbb{S}) = \mathbb{Z}/24$ has order 24, this invariant necessarily takes values in the 24th roots of unity.

Theorem. *The image of Fer*(1) *under the map* ν *is a primitive* 24th *root of unity.*

Proof. The homotopy groups of $|CN^\times|$ are as follows:

$\pi_0(\lvert CN^\times\rvert)$	$\pi_1(\lvert CN^\times\rvert)$	$\pi_2(\lvert CN^\times\rvert)$	$\pi_3(\lvert CN^\times\rvert)$	$\pi_{\geq 4}(\lvert CN^\times\rvert)$
?	$\mathbb{Z}/2$	$\mathbb{Z}/2$	S^1	0

The first $\mathbb{Z}/2$ corresponds to the two Morita equivalence classes of $\mathbb{Z}/2$-graded central simple algebras over $\mathbb{C}$: *Cliff*(0) and *Cliff*(1). The second $\mathbb{Z}/2$ corresponds to the two isomorphism classes of $\mathbb{Z}/2$-graded lines: the even line and the odd line. The last non-zero homotopy group S^1 corresponds to the linear isometries of $\mathbb{C}$.

Let $CN^\times_{\text{top}}$ be the topological 3-category whose underlying 3-category is $CN^\times$, and whose spaces of 3-morphisms have been topologized as subspaces of mapping spaces between Hilbert spaces. The homotopy groups of the geometric realization of $CN^\times_{\text{top}}$ are as follows:

$\pi_0(\lvert CN^\times_{\text{top}}\rvert)$	$\pi_1(\lvert CN^\times_{\text{top}}\rvert)$	$\pi_2(\lvert CN^\times_{\text{top}}\rvert)$	$\pi_3(\lvert CN^\times_{\text{top}}\rvert)$	$\pi_4(\lvert CN^\times_{\text{top}}\rvert)$	$\pi_{\geq 5}(\lvert CN^\times_{\text{top}}\rvert)$
$\cong \pi_0(\lvert CN^\times\rvert)$	$\mathbb{Z}/2$	$\mathbb{Z}/2$	0	$\mathbb{Z}$	0

Indeed we have a fibration sequence

$$|CN^\times| \to |CN^\times_{\text{top}}| \to K(\mathbb{R}, 4),$$

where $\mathbb{R}$ is given the discrete topology. Let F denote the image of *Fer*(1) in $\pi_0(|CN^\times|)$. It follows from the above fiber sequence that the equation $\nu F = e^{2k\pi i/24}$ in $\pi_3\big(|CN^\times|\big)$ is equivalent to the Toda bracket relation $k \in \langle 24, \nu, F\rangle$ in $\pi_4(|CN^\times_{\text{top}}|)$. (Note that because $\pi_4(\mathbb{S}) = 0$, this Toda bracket is a $24\mathbb{Z}$-torsor inside $\pi_4(|CN^\times_{\text{top}}|) \cong \mathbb{Z}$.)

We will compute the Toda bracket $\langle 24, \nu, F\rangle$ in $|CN^\times_{\text{top}}|$ by relating it to a Toda bracket in $\mathbb{Z} \times BO$. The space $\coprod BO(n)$ is a classifying space for real vector spaces, so the free fermion functor $V \mapsto Fer(V)$, from real vector spaces to conformal nets, induces a map

$$\coprod_{n \geq 0} BO(n) \to |CN^\times_{\text{top}}|.$$

Because the target $|CN^\times_{\text{top}}|$ is group complete, this map extends to a map of spectra

$$\rho : \mathbb{Z} \times BO \to |CN^\times_{\text{top}}|$$

from the group completion $\Omega B(\coprod BO(n)) \simeq \mathbb{Z} \times BO$. This map sends the generator ι of $\pi_0(\mathbb{Z} \times BO)$ to the class of the free fermion $F \in \pi_0(|CN^\times_{\text{top}}|)$; that is $\rho(\iota) = F$.

The spectrum $\mathbb{Z} \times BO$ has the Toda bracket relation $\omega \in \langle 24, \nu, \iota \rangle$ where $\omega \in \pi_4(\mathbb{Z} \times BO) \cong \mathbb{Z}$ is a generator. As Toda brackets are natural, we have the relation

$$\rho(\omega) \in \rho(\langle 24, \nu, \iota \rangle) \subset \langle 24, \nu, F \rangle \subset \pi_4(|CN^{\times}_{\mathrm{top}}|).$$

It therefore suffices to check that $\rho(\omega)$ is a generator of $\pi_4(|CN^{\times}_{\mathrm{top}}|)$. An involved computation carried out in [DH] shows that ρ induces an isomorphism on π_4. Altogether, we conclude that $\nu F = e^{\pm 2\pi i/24} \in \pi_3(|CN^{\times}|) \cong S^1$, as desired. □

As the invariant $\nu F \in \pi_3(|CN^{\times}|)$ only depends on the CN-equivalence class of the fermion, the above result provides a lower bound on the periodicity of $Fer(1)$ in the group $\pi_0(|CN^{\times}|)$ of invertible conformal nets.

Corollary. *If $Fer(n)$ is CN-equivalent to the trivial conformal net, then n is a multiple of 24.*

In the notes [Dr], Drinfel'd says that there is a braided monoidal category (the Ising category) associated to the free fermion, that is 16-periodic with respect to an operation of 'reduced tensor product'. Combining this mod-16 invariant with the above mod-24 invariant should show that the period of the free fermion, if finite, is in fact a multiple of 48.

References

[ABS] Atiyah, M. F.; Bott, R.; Shapiro, A. *Clifford modules*, Topology 3 (1964), 3–38.

[AHR] Ando, M.; Hopkins, M.; Rezk, C. *Multiplicative orientations of KO-theory and of the spectrum of topological modular forms*, preprint, 2010.

[BD] Beilinson, A.; Drinfel'd, V. *Chiral algebras*, American Mathematical Society Colloquium Publications, 51.

[BDR] Baas, N.; Dundas, B.; Rognes, J. *Two-vector bundles and forms of elliptic cohomology*, Topology, geometry and quantum field theory, 18–45, London Math. Soc. Lecture Note Ser., 308, 2004.

[BCSS] Baez, J.; Crans, A. S.; Stevenson, D.; Schreiber, U. *From loop groups to 2-groups*, Homotopy, Homology and Applications 9 (2007), 101–135.

[BDH1,2,3] Bartels, A.; Douglas, C.; Henriques A. *Internal bicategories*; *Conformal nets I: Fusion of defects*; *Conformal nets II: The 3-category*, in preparation.

[Ba] Bauer, T. *Computation of the homotopy of the spectrum tmf*, in *Groups, homotopy and configuration spaces*, Geometry and Topology Monographs 13 (2008), 11–40.

[Be] Behrens, M. *Notes on the construction of tmf.* Proceedings of the Talbot workshop 2007, available at http://math.mit.edu/conferences/talbot/2007/tmfproc/

[BFV] Brunetti, R.; Fredenhagen, K.; Verch, R. *The generally covariant locality principle, a new paradigm for local quantum field theory*, Comm. Math. Phys. 237 (2003), 31–68.

[De] Deligne, P. *Courbes elliptiques: formulaire d'après J. Tate.* In Modular functions of one variable, IV (Proc. Internat. Summer School, Univ. Antwerp, 1972), 53–73. LNM, Vol. 476. Springer, Berlin, 1975.

[DH] Douglas, C.; Henriques A. *Geometric string structures*, in preparation, available at http://www.math.uu.nl/people/henrique/PDF/TringWP.pdf.

[Dr] Drinfel'd, V. *Reduced tensor product*, private communication, 2009.

[FBZ] Frenkel, E.; Ben-Zvi, D. *Vertex algebras and algebraic curves*, Mathematical Surveys and Monographs, 88. American Mathematical Society, 2001.

[FRS] Fuchs, J.; Runkel, I.; Schweigert, C. *TFT construction of RCFT correlators. I, II, III, IV, and V*, Nuclear Phys. B 646, 2002, 353–497; NPB 678, 2004, 511–637; NPB 694, 2004, 277–353; NPB 715, 2005, 539–638; and Theo. Appl. Cat., 16, 2006, 342–433.

[GF] Gabbiani, F.; Fröhlich, J. *Operator algebras and conformal field theory*, Comm. Math. Phys. 155 (1993), 569–640.

[Go] Goerss, P. G. *Topological modular forms [after Hopkins, Miller and Lurie].* Séminaire Bourbaki. Vol. 2008/2009. Exposés 997–1011. Astérisque 332 (2010), Exp. No. 1005, viii, 221–255

[Ha] Haagerup, U. *The standard form of von Neumann algebras.* Math. Scand. 37, 1975, 271283.

[HMa] Hopkins, M.; Mahowald, M. *From elliptic curves to homotopy theory*, 1998.

[HMi] Hopkins, M; Miller. H. *Elliptic curves and stable homotopy theory*, 1999.

[HK] Hu, P.; Kriz, I. *Conformal field theory and elliptic cohomology*, Adv. Math. 189 (2004), 325–412.

[Hu] Huang, Y.-Z. *Two-dimensional conformal geometry and vertex operator algebras*, Progress in Mathematics, 148. Birkhäuser, 1997.

[Kar] Karoubi, M.; *K-theory. An introduction.* Grund. Math. Wiss., Band 226. Springer, 1978.

[Kaw] Kawahigashi, Y. *Conformal field theory and operator algebras*, preprint, 2007, to appear in the Proceedings of ICMP (Rio de Janeiro, 2006).

[KLM] Kawahigashi, Y.; Longo, R.; Müger, M. *Multi-interval subfactors and modularity of representations in conformal field theory*, Comm. Math. Phys. 219 (2001), 631–669.

[Lo] Longo, R. *Lectures on Conformal Nets*, book preprint, available at http://www.mat.uniroma2.it/~longo/Lecture_Notes.html.

[LR] Longo, R.; Rehren, K.-H. *Local fields in boundary conformal QFT*, Rev. Math. Phys. 16 (2004), 909–960.

[Lu] Lurie, J. *A Survey of Elliptic Cohomology*, Algebraic Topology, 219–277, Abel Symp. 4, Springer, Berlin, 2009.

[SP] Schommer-Pries, C. *Central Extensions of Smooth 2-Groups and a Finite-Dimensional String 2-Group*, preprint, 2009, arXiv:0911.2483.

[Se] Segal, G. *The definition of conformal field theory*, Topology, geometry and quantum field theory, 421–577, London Math. Soc. Lecture Note Ser., 308, Cambridge Univ. Press, Cambridge, 2004.

[Sh] Shulman, M. *Framed bicategories and monoidal fibrations*, Theory Appl. Categ. 20 (2008), 650–738.

[ST1] Stolz, S.; Teichner, P. *What is an elliptic object?*, Topology, geometry and quantum field theory, London Math. Soc. LNS 308, Cambridge Univ. Press, 2004, 247–343.

[ST2] Stolz, S.; Teichner, P. *Super symmetric Euclidean field theories and generalized cohomology, a survey*, preprint, 2008.

[To] Toledano Laredo, V. *Fusion of Positive Energy Representations of* $LSpin_{2n}$, thesis, University of Cambridge, 1997, 157 pages.

[Wal] Waldorf, K. *String connections and Chern-Simons theory*, preprint, 2009, arXiv:0906.0117.

[Was] Wassermann, A. *Operator algebras and conformal field theory. III. Fusion of positive energy representations of* $LSU(N)$ *using bounded operators*, Invent. Math. 133 (1998), 467–538.

Mathematical Institute, 24–29 St Giles', Oxford, OX1 3LB, United Kingdom
E-mail address: `cdouglas@maths.ox.ac.uk`
URL: `http://people.maths.ox.ac.uk/cdouglas`

Mathematisch Instituut, Universiteit Utrecht, 3508 TA Utrecht, The Netherlands
E-mail address: `a.g.henriques@uu.nl`
URL: `http://www.staff.science.uu.nl/~henri105`

Titles in This Series

83 **Hisham Sati and Urs Schreiber, Editors,** Mathematical Foundations of Quantum Field Theory and Perturbative String Theory

82 **Michael Usher, Editor,** Low-Dimensional and Symplectic Topology (University of Georgia, Athens, Georgia, 2009)

81 **Robert S. Doran, Greg Friedman, and Jonathan M. Rosenberg, Editors,** Superstrings, geometry, topology, and C*-algebras

80.2 **D. Abramovich, A. Bertram, L. Katzarkov, R. Pandharipande, and M. Thaddeus, Editors,** Algebraic Geometry (Seattle, 2005)

80.1 **D. Abramovich, A. Bertram, L. Katzarkov, R. Pandharipande, and M. Thaddeus, Editors,** Algebraic Geometry

79 **Dorina Mitrea and Marius Mitrea, Editors,** Perspectives in Partial Differential Equations, Harmonic Analysis and Applications: A Volume in Honor of Vladimir G. Maz'ya's 70th Birthday

78 **Ron Y. Donagi and Katrin Wendland, Editors,** From Hodge Theory to Integrability and TQFT

77 **Pavel Exner, Jonathan P. Keating, Peter Kuchment, Toshikazu Sunada, and Alexander Teplyaev, Editors,** Analysis on graphs and its applications

76 **Fritz Gesztesy (Managing editor), Percy Deift, Cherie Galvez, Peter Perry, and Wilhelm Schlag, Editors,** Spectral theory and mathematical physics: A Festschrift in honor of Barry Simon's 60th birthday, Parts 1 and 2 (California Insitute of Technology, Pasadena, CA, March 27–31, 2006)

75 **Solomon Friedberg (Managing editor), Daniel Bump, Dorian Goldfeld, and Jeffrey Hoffstein, Editors,** Multiple Dirichlet series, automorphic forms, and analytic number theory (Bretton Woods, New Hampshire, July 11–14, 2005)

74 **Benson Farb, Editor,** Problems on mapping class groups and related topics, 2006

73 **Mikhail Lyubich and Leon Takhtajan, Editors,** Graphs and patterns in mathematics and theoretical physics (Stony Brook University, Stony Brook, NY, June 14–21, 2001)

72 **Michel L. Lapidus and Machiel van Frankenhuijsen, Editors,** Fractal geometry and applications: A jubilee of Benoît Mandelbrot, Parts 1 and 2 (San Diego, California, 2002 and École Normale Supérieure de Lyon, 2001)

71 **Gordana Matić and Clint McCrory, Editors,** Topology and Geometry of Manifolds (University of Georgia, Athens, Georgia, 2001)

70 **Michael D. Fried and Yasutaka Ihara, Editors,** Arithmetic fundamental groups and noncommutative algebra (Mathematical Sciences Research Institute, Berkeley, California, 1999)

69 **Anatole Katok, Rafael de la Llave, Yakov Pesin, and Howard Weiss, Editors,** Smooth ergodic theory and its applications (University of Washington, Seattle, 1999)

68 **Robert S. Doran and V. S. Varadarajan, Editors,** The mathematical legacy of Harish-Chandra: A celebration of representation theory and harmonic analysis (Baltimore, Maryland, 1998)

67 **Wayne Raskind and Charles Weibel, Editors,** Algebraic K-theory (University of Washington, Seattle, 1997)

66 **Robert S. Doran, Ze-Li Dou, and George T. Gilbert, Editors,** Automorphic forms, automorphic representations, and arithmetic (Texas Christian University, Fort Worth, 1996)

65 **M. Giaquinta, J. Shatah, and S. R. S. Varadhan, Editors,** Differential equations: La Pietra 1996 (Villa La Pietra, Florence, Italy, 1996)

64 **G. Ferreyra, R. Gardner, H. Hermes, and H. Sussmann, Editors,** Differential geometry and control (University of Colorado, Boulder, 1997)

63 **Alejandro Adem, Jon Carlson, Stewart Priddy, and Peter Webb, Editors,** Group representations: Cohomology, group actions and topology (University of Washington, Seattle, 1996)

62 **János Kollár, Robert Lazarsfeld, and David R. Morrison, Editors,** Algebraic geometry—Santa Cruz 1995 (University of California, Santa Cruz, July 1995)

ISBN 978-0-8218-5195-1

9 780821 851951

PSPUM/83